LONG TERM EVOLUTION

3GPP LTE Radio and Cellular Technology

INTERNET and COMMUNICATIONS

This new book series presents the latest research and technological developments in the field of Internet and multimedia systems and applications. We remain committed to publishing high-quality reference and technical books written by experts in the field.

If you are interested in writing, editing, or contributing to a volume in this series, or if you have suggestions for needed books, please contact Dr. Borko Furht at the following address:

Borko Furht, Ph.D.
Department Chairman and Professor
Computer Science and Engineering
Florida Atlantic University
777 Glades Road
Boca Raton, FL 33431 U.S.A.

E-mail: borko@cse.fau.edu

Volume 1 Handbook of Internet Computing
Volume 2 UNIX Administration: A Comprehensive Sourcebook for Effective Systems & Network Management
Volume 3 Wireless Sensor Networks: Architectures and Protocols
Volume 4 Multimedia Security Handbook
Volume 5 Handbook of Multimedia Computing
Volume 6 Handbook of Internet and Multimedia Systems and Applications
Volume 7 Wireless Internet Handbook: Technologies, Standards, and Applications
Volume 8 Handbook of Video Databases: Design and Applications
Volume 9 Handbook of Wireless Local Area Networks: Applications, Technology, Security, and Standards
Volume 10 Handbook of Mobile Broadcasting: DVB-H, DMB, ISDB-T, and MediaFLO
Volume 11 Long Term Evolution: 3GPP LTE Radio and Cellular Technology

LONG TERM EVOLUTION

3GPP LTE Radio and Cellular Technology

Edited by Borko Furht • Syed A. Ahson

CRC Press is an imprint of the
Taylor & Francis Group, an **informa** business
AN AUERBACH BOOK

CRC Press
Taylor & Francis Group
6000 Broken Sound Parkway NW, Suite 300
Boca Raton, FL 33487-2742

First issued in paperback 2019

ISBN-13: 978-1-4200-7210-5 (hbk)
ISBN-13: 978-0-367-38571-2 (pbk)

Library of Congress Cataloging-in-Publication Data

Long Term Evolution : 3GPP LTE radio and cellular technology / editors, Borko Furht, Syed A. Ahson.
p. cm. -- (Internet and communications)
Includes bibliographical references and index.
ISBN 978-1-4200-7210-5 (hardcover : alk. paper)
1. Universal Mobile Telecommunications System. 2. Mobile communication systems. 3. Cellular telephone systems. I. Furht, Borko. II. Ahson, Syed.

TK5103.4883.L66 2009
621.3845--dc22 2009004034

Visit the Taylor & Francis Web site at
http://www.taylorandfrancis.com

and the CRC Press Web site at
http://www.crcpress.com

Contents

Preface

This book provides technical information about all aspects of 3GPP LTE. The areas covered range from basic concepts to research-grade material, including future directions. The book captures the current state of 3GPP LTE technology and serves as a source of comprehensive reference material on this subject. It has a total of 12 chapters authored by 50 experts from around the world. The targeted audience includes professionals who are designers or planners for 3GPP LTE systems, researchers (faculty members and graduate students), and those who would like to learn about this field.

The book has the following objectives:

- to serve as a single comprehensive source of information and as reference material on 3GPP LTE technology;
- to deal with an important and timely topic of emerging technology of today, tomorrow, and beyond;
- to present accurate, up-to-date information on a broad range of topics related to 3GPP LTE technology;
- to present material authored by experts in the field; and
- to present the information in an organized and well-structured manner.

Although the book is not precisely a textbook, it can certainly be used as a textbook for graduate courses and research-oriented courses that deal with 3GPP LTE. Any comments from readers will be highly appreciated.

Many people have contributed to this handbook in their unique ways. The first and foremost group that deserves immense gratitude is the highly talented and skilled researchers who have contributed the 12 chapters. All of them have been extremely cooperative and professional. It has also been a pleasure to work with Rich

O'Hanley, Jessica Vakili, and Judith Simon of CRC Press, and we are extremely gratified for their support and professionalism. Our families have extended their unconditional love and strong support throughout this project, and they deserve very special thanks.

Borko Furht
Boca Raton, Florida

Syed Ahson
Seattle, Washington

The Editors

Borko Furht is chairman and professor of computer science and engineering at Florida Atlantic University (FAU) in Boca Raton, Florida. He is the founder and director of the Multimedia Laboratory at FAU, funded by the National Science Foundation. Before joining FAU, he was a vice president of research and a senior director of development at Modcomp, a computer company in Fort Lauderdale, Florida, and a professor at the University of Miami in Coral Gables, Florida. He received his PhD degree in electrical and computer engineering from the University of Belgrade, Yugoslavia. His research interests include multimedia systems and applications, video processing, wireless multimedia, multimedia security, video databases, and Internet engineering. He is currently principal investigator or coprincipal investigator and leader of several large multiyear projects, including "One Pass to Production," funded by Motorola, and "Center for Coastline Security Technologies," funded by the U.S. government as a federal earmark project.

Dr. Furht has received research grants from various government agencies, such as NSF and NASA, and from private corporations, including IBM, Hewlett Packard, Racal Datacom, Xerox, Apple, and others. He has published more than 25 books and about 250 scientific and technical papers, and he holds two patents. He is a founder and editor-in-chief of the *Journal of Multimedia Tools and Applications* (Kluwer Academic Publishers, now Springer). He is also consulting editor for two book series on multimedia systems and applications (Kluwer/Springer) and Internet and communications (CRC Press). He has received several technical and publishing awards and has consulted for IBM, Hewlett Packard, Xerox, General Electric, JPL, NASA, Honeywell, and RCA. He has also served as a consultant to various colleges and universities. He has given many invited talks, keynote lectures, seminars, and tutorials. He has been program chair as well as a member of program committees at many national and international conferences.

Syed Ahson is a senior software design engineer with Microsoft. As part of the Mobile Voice and Partner Services Group, he is busy creating new and exciting end-to-end mobile services and applications. Prior to joining Microsoft, he was a senior staff software engineer with Motorola, where he played a significant role in the creation of several iDEN, CDMA, and GSM cellular phones. Ahson has extensive experience with wireless data protocols, wireless data applications, and cellular telephony protocols. Before he joined Motorola, he was a senior software design engineer with NetSpeak Corporation (now part of Net2Phone), a pioneer in VoIP telephony software.

Ahson has published more than 10 books on emerging technologies such as WiMAX, RFID, mobile broadcasting, and IP multimedia subsystems. His recent books include *IP Multimedia Subsystem Handbook* and *Handbook of Mobile Broadcasting: DVB-H, DMB, ISDB-T, and MediaFLO.* He has authored several research articles and teaches computer engineering courses as adjunct faculty at Florida Atlantic University, Boca Raton, Florida, where he introduced a course on Smartphone technology and applications. Ahson received his MS degree in computer engineering in 1998 from Florida Atlantic University and received his BSc degree in electrical engineering from Aligarh University, India, in 1995.

Contributors

M. Andersson
Bluetest AB
Gothenburg, Sweden

J. Carlsson
SP Technical Research Institute of Sweden
Boras, Sweden

Tao Chen
Devices R&D
Nokia
Oulu, Finland

José Tomás Entrambasaguas
University of Málaga
Málaga, Spain

Xun Fan
Philips Research Asia
Shanghai, China

Gábor Fodor
Ericsson Research
Stockholm, Sweden

A. Forck
Fraunhofer Institute for Telecommunications
Heinrich-Hertz-Institut
Berlin, Germany

Gerardo Gómez
University of Málaga
Málaga, Spain

Thomas Haustein
Nokia Siemens Networks
Munich, Germany

Tero Henttonen
Devices R&D
Nokia
Helsinki, Finland

Volker Jungnickel
Fraunhofer Institute for Telecommunications
Heinrich-Hertz-Institut
Berlin, Germany

Markku Kuusela
Devices R&D
Nokia
Helsinki, Finland

Lihua Li
Beijing University of Posts and Telecommunications
Beijing, China

F. Javier López-Martinez
University of Málaga
Málaga, Spain

Petteri Lundén
Devices R&D
Nokia
Helsinki, Finland

Jijun Luo
Nokia Siemens Networks
Munich, Germany

Esa Malkamäki
Devices R&D
Nokia
Helsinki, Finland

Wolfgang Mennerich
Nokia Siemens Networks
Munich, Germany

David Morales-Jiménez
University of Málaga
Málaga, Spain

Jussi Ojala
Devices R&D
Nokia
Helsinki, Finland

C. Orlenius
Bluetest AB
Gothenburg, Sweden

Basuki E. Priyanto
Aalborg University
Aalborg, Denmark

András Rácz
Ericsson Research
Budapest, Hungary

Norbert Reider
Budapest University of Technology
and Economics
Budapest, Hungary

Juan J. Sánchez
University of Málaga
Málaga, Spain

Rainer Schoenen
Aachen University
North Rhine-Westphalia, Germany

Martin Schubert
Fraunhofer Institute for
Telecommunications
Heinrich-Hertz-Institut
Berlin, Germany

Egon Schulz
Nokia Siemens Networks GmbH
Munich, Germany

Troels B. Sorensen
Aalborg University
Aalborg, Denmark

András Temesváry
Budapest University of Technology
and Economics
Budapest, Hungary

Lars Thiele
Fraunhofer Institute for
Telecommunications
Heinrich-Hertz-Institut
Berlin, Germany

V. Venkatkumar
Nokia Siemens Networks GmbH
Munich, Germany

S. Wahls
Fraunhofer Institute for
Telecommunications
Heinrich-Hertz-Institut
Berlin, Germany

Bernhard Walke
Aachen University
North Rhine-Westphalia, Germany

Haiming Wang
Devices R&D
Nokia
Beijing, China

T. Wirth
Fraunhofer Institute for Telecommunications
Heinrich-Hertz-Institut
Berlin, Germany

A. Wolfgang
Chalmers University of Technology
Gothenburg, Sweden

Gang Wu
Philips Research Asia
Shanghai, China

Xiaodong Xu
Beijing University of Posts and Telecommunications
Beijing, China

Ping Zhang
Beijing University of Posts and Telecommunications
Beijing, China

Wolfgang Zirwas
Nokia Siemens Networks
Munich, Germany

Chapter 1

Introduction

Borko Furht and Syed Ahson

Mobile users are demanding higher data rate and higher quality mobile communication services. The 3rd Generation Mobile Communication System (3G) is an outstanding success. The conflict of rapidly growing users and limited bandwidth resources requires that the spectrum efficiency of mobile communication systems be improved by adopting some advanced technologies. It has been proved, in both theory and in practice, that some novel key technologies such as MIMO (multiple input, multiple output) and OFDM (orthogonal frequency division multiplexing) improve the performance of current mobile communication systems. Many countries and organizations are researching next-generation mobile communication systems, such as the ITU (International Telecommunication Union), European Commission FP (Framework Program), WWRF (Wireless World Research Forum), Korean NGMC (Next-Generation Mobile Committee), Japanese MITF (Mobile IT Forum), and China Communication Standardization Association (CCSA). International standards organizations are working for standardization of the E3G (Enhanced 3G) and 4G (the 4th Generation Mobile Communication System), such as the LTE (long term evolution) plan of the 3rd Generation Partnership Project (3GPP) and the AIE (air interface of evolution)/UMB (ultramobile broadband) plan of 3GPP2.

The 3GPP LTE release 8 specification defines the basic functionality of a new, high-performance air interface providing high user data rates in combination with low latency based on MIMO, OFDMA (orthogonal frequency division multiple access), and an optimized system architecture evolution (SAE) as main enablers. At the same time, in the near future increasing numbers of users will request mobile broadband data access everywhere—for example, for synchronization of e-mails, Internet access, specific applications, and file downloads to mobile devices like personal digital assistants (PDAs) or notebooks. In the future, a 100-fold increase in mobile data traffic is expected, making further improvements beyond LTE release 8 necessary and possibly ending in new LTE releases or in a so-called IMT (international mobile telecommunications) advanced system. The LTE successor of the third-generation mobile radio network will delight customers more than ever. The need for radio coverage will be a primary goal in the rollout phase, whereas a high capacity all over the radio cell will be the long-term goal. High spectral efficiency is crucial for supporting the high demand of data traffic rates that future mobile user terminals will generate, while the available spectrum is still a scarce and limiting resource at each geographic location.

This book provides technical information about all aspects of 3GPP LTE. The areas covered range from basic concepts to research-grade material, including future directions. The book captures the current state of 3GPP LTE technology and serves as a source of comprehensive reference material on this subject. It has a total of 12 chapters authored by 50 experts from around the world.

Chapter 2 ("Evolution from TD-SCDMA to FuTURE") describes a project called Future Technologies for a Universal Radio Environment (FuTURE) that was launched in China. The goal of the project is to support theoretical research, applicable evaluation, and 4G trial system development of the proposed technologies for Chinese beyond 3G/4G mobile communications. The FuTURE plan aims at the trends and requirements of wireless communication in the next 10 years and expects to play an active role in the process of 4G standardization.

The FuTURE project is composed of two branches: the FuTURE FDD (frequency division duplex) system and the FuTURE TDD (time division duplex) system, both of which are investigating and demonstrating advanced techniques for systems to meet the application requirements around the year 2010. For both TDD and FDD, the trial platform contains six access points (APs) and 12 distributed antenna arrays; each AP is equipped with two spaced antenna arrays. High-definition reactive video services have been performed and demonstrated on FuTURE trial systems.

In the research and development of the FuTURE TDD system, a series of innovations in basic theory has been proposed, such as the flat wireless access network structure that has been taken into practice; the group cell, which is a novel generalized cellular network architecture, and the slide handover strategy based on it; the soft fractional frequency reuse to improve the spectrum efficiency; the reliable end-to-end QoS (quality of service) mechanism; the antenna array based on the generalized cellular network architecture to increase the system capacity with the spectrum efficiency of 7 b/Hz; the modular MIMO to support various environments; and the scalable modular structure of system, which can support a data rate up to 330 Mb/s. The FuTURE TDD system provides an overall solution that can fulfill the requirements of E3G/4G and adopts lots of key technologies in many aspects, including cellular network architecture, physical layer, medium access control (MAC) layer, radio resource management, and hardware design. As a result, some positive effects have been brought out in the areas of system design, pivotal algorithm, and trial system development.

The LTE physical layer is targeted to provide improved radio interface capabilities between the base station and user equipment (UE) compared to previous cellular technologies like universal mobile telecommunications system (UMTS) or high-speed downlink packet access (HSDPA). According to the initial requirements defined by the 3GPP (3GPP 25.913), the LTE physical layer should support peak data rates of more than 100 Mb/s over the downlink and 50 Mb/s over the uplink. A flexible transmission bandwidth ranging from 1.25 to 20 MHz will provide support for users with different capabilities. These requirements will be fulfilled by employing new technologies for cellular environments, such as OFDM or multiantenna schemes (3GPP 36.201). Additionally, channel variations in the time/frequency domain are exploited through link adaptation and frequency-domain scheduling, giving a substantial increase in spectral efficiency. In order to support transmission in paired and unpaired spectra, the LTE air interface supports both FDD and TDD modes.

Chapter 3 ("Radio-Interface Physical Layer") presents a detailed description of the LTE radio-interface physical layer. Section 3.1 provides an introduction to the physical layer, focusing on the physical resources' structure and the set of procedures defined within this layer. This section provides a general overview of the different multiantenna technologies considered in LTE and its related physical layer procedures. Link adaptation techniques, including adaptive modulation, channel coding, and channel-aware scheduling, are described in Section 3.2. The topic of multiple antenna schemes for LTE is tackled in Section 3.3, which also provides an overview of the different multiantenna configurations and multiantenna-related procedures defined in LTE. Section 3.4 addresses other physical layer procedures like channel estimation, synchronization, and random access. This section is focused on those procedures that allow the reception and decoding of the frame under a certain bit error rate (BER) level: synchronization and channel estimation. Furthermore, this section also presents the random access (RA) structure and procedure necessary for initial synchronization in the uplink prior to UE data transmission.

Chapter 4 ("Architecture and Protocol Support for Radio Resource Management") discusses the radio resource management (RRM) functions in LTE. This chapter gives an overview of the LTE RAN (radio access network) architecture, including an overview of the OFDM-based radio interface. The notion of radio resource in LTE is defined, and the requirements that the 3GPP has set on the spectral efficient use of radio resources are presented. Section 4.2 gives an overview of the multiantenna solutions in general and then discusses the different MIMO variants in the context of LTE, also addressing the required resource management functions used for the antenna port configuration control. Section 4.3 discusses the most important measurements in LTE, grouping them into UE and eNode B measurements and, when appropriate, draws an analogy with well-known wideband code division multiple access (WCDMA) measurements.

Chapter 4 shows that a number of advanced RRM functions are needed in today's wireless systems; these are being developed and standardized in particular for the 3GPP LTE networks, in order to fulfill the ever-increasing capacity demands by utilizing the radio interface more efficiently. Considering the facts that the available radio spectrum is a limited resource and the capacity of a single radio channel between the UE and the network is also limited by the well-known theoretical bounds of Shannon, the remaining possibility to increase the capacity is to increase the number of such "independent" radio channels in addition to trying to approach the theoretical channel capacity limits on each of these individual channels. Advanced radio resource management techniques play a key role in achieving these goals. This chapter presents methods and examples of how advanced RRM functions are realized in LTE and how these functions together make LTE a high-performance, competitive system for many years to come.

Multiple input, multiple output techniques have been integrated as one of the key approaches to provide the peak data rate, average throughput, and system performance in 3GPP LTE. Based on the function of the multiple transmission symbol streams in MIMO, the operation modes of multiple transmit antennas at the cell site (denoted as MIMO mode) are spatial division multiplexing (SDM), precoding, and transmit diversity (TD).

Based on the allocation of the multiple transmission streams in MIMO, the MIMO mode is denoted as single user (SU)-MIMO if the multiple transmission symbol streams are solely assigned to a single UE and multiuser (MU)-MIMO if the spatial division multiplexing of the modulation symbol streams for different UEs uses the same time-frequency resource. Because the LTE downlink is an OFDM system, the MIMO modes proposed in LTE are MIMO-OFDM schemes. Chapter 5 ("MIMO OFDM Schemes for 3GPP LTE") introduces the two main categories of MIMO-OFDM schemes in LTE—SU-MIMO and MU-MIMO—and the related physical channel procedures. The performance of selected codebooks is provided to show their advantages.

Uplink transmission in the LTE of the UMTS terrestrial radio access system (UTRA LTE) has numerous physical layer advantages in comparison to UTRA WCDMA, mainly to achieve two to three times better spectral efficiency. These include flexible channel bandwidth up to 20 MHz, flexible user resource allocation in both time and frequency domains, and a shorter time transmission interval (TTI) of 1 ms. Specifically challenging for the uplink is that these enhancements are to be achieved, preferably, with reduced power consumption to extend the battery life and cell coverage. The radio access technique is one of the key issues in the LTE uplink air interface. In LTE, OFDMA has been selected as the multiple-access scheme for downlink and single-carrier frequency division multiple access (SC-FDMA) for uplink. OFDM is an attractive modulation technique in a cellular environment to combat frequency selective fading channels with a relatively low-complexity receiver. However, OFDM requires an expensive and inherently inefficient power amplifier in the transmitter due to the high peak-to-average power ratio (PAPR) of the multicarrier signal.

Chapter 6 ("Single-Carrier Transmission for UTRA LTE Uplink") presents the key techniques for LTE uplink as well as the baseline performance. Radio access technology is the key aspect in LTE uplink, and two radio access schemes, SC-FDMA and OFDMA, are studied. The performance results are obtained from a detailed UTRA LTE uplink link-level simulator. The simulation results show that both SC-FDMA and OFDMA can achieve a high spectral efficiency; however, SC-FDMA benefits in obtaining lower PAPR than OFDMA, especially for low-order modulation schemes. A 1×2 SIMO (single input, multiple output) antenna configuration highly increases the spectral efficiency of SC-FDMA, making the performance of SC-FDMA with a minimum mean square error (MMSE) receiver comparable to OFDMA, especially at high coding rates. The peak spectral efficiency results for SC-FDMA confirm that it meets the requirement to achieve a spectral efficiency improvement of two or three times that of 3GPP release 6.

LTE uplink numerology is introduced in Section 6.2. The description of both radio access techniques is given in Section 6.3. The PAPR evaluation for each radio access scheme is discussed in Section 6.4. Section 6.5 describes the link-level model used for the evaluations in this work, including the specific parameter settings. Afterward, the LTE uplink performance with various key techniques is presented, including link adaptation (Section 6.6), fast H-ARQ (Section 6.7), antenna configuration (Section 6.8), flexible frequency allocation (Section 6.9), and typical channel estimation (Section 6.10). Finally, turbo equalization is presented in Section 6.11 as a performance enhancement technique for SC-FDMA transmission. The chapter concludes by showing the impact of the nonlinear power amplifier to both in-band and out-of-band performance.

In the future, mobile network operators (MNOs) will have to provide broadband data rates to an increasing number of users with lowest cost per bit and probably as flat rates as known from fixed network providers. Intercell interference is the most limiting factor in current cellular radio networks, which means that any type of practical, feasible interference mitigation will be of highest importance to tackle the expected 100-fold traffic challenge. As the only means actually known to overcome interference, cooperation is therefore a very likely candidate for integration into an enhanced LTE system. In early 2008, a study item was launched at 3GPP searching for useful enhancements of LTE release 8, which might end in a new LTE release R9. In parallel, there are strong global activities for the definition of the successor of IMT2000, so-called IMT Advanced systems, where it is likely that some form of cooperation will be included. Accordingly, research on cooperative antenna systems has become one of the hottest topics, and it will form the input of ongoing standardization activities. In IEEE 802.16j, the relaying task group defines mobile multihop relay (MMR) solutions for WiMAX systems; meanwhile, cooperative relaying has been adopted as one type of relaying.

Chapter 7 ("Cooperative Transmission Schemes") introduces different cooperative antenna (COOPA) concepts like intra- and inter-NB cooperation and distributed antenna systems (DASs) based on remote radio heads (RRHs) or as self-organizing networks, addressing different aspects of the previously mentioned issues. This chapter is organized into five sections: The second section provides a basic analysis of cooperative antenna systems, the third classifies the basic types of cooperation systems, the fourth is concerned with implementation issues, and the fifth concludes the chapter. The motivation for cooperation, theoretical analysis, and simulation results is presented in this chapter, giving a clear view of the high potential of cooperative antenna systems and promising performance gains on the order of several hundred percent or even more. Different implementation strategies have been presented and some important issues for the core element of any cooperation area have been addressed for the example of intersector cooperation. The newly proposed COOPA HARQ (hybrid automatic repeat request) concept allows reducing feedback overhead and, at the same time, solves the critical issue of feedback delay.

In recent years, the interest in multihop-augmented, infrastructure-based networks has grown because relaying techniques help improve coverage and capacity without too much cost for infrastructure. Recently, radio technologies from industry and academia, such as IEEE 802.16 (WiMAX—worldwide interoperability for microwave access), HiperLAN/2, and the Winner-II system, have included multihop support right from the start.

Chapter 8 ("Multihop Extensions to Cellular Networks—the Benefit of Relaying for LTE") describes how relaying can be introduced into the LTE protocol stack, the basic blocks, and the extensions. Different approaches are covered for exploiting the benefits of multihop communications, such as solutions for radio range extension (trading coverage range for capacity) and solutions to combat shadowing at high radio frequencies. It is shown that relaying can also enhance capacity (e.g., through the exploitation of antenna gains, spatial diversity, or by suitable placement against shadowing due to obstructions). Further, multihop is presented as a means to reduce infrastructure deployment costs. The performance is presented and studied in several scenarios here.

In the 3GPP WCDMA/HSPA (high-speed packet access) standards, the radio link control (RLC) sublayer is defined on top of the MAC sublayer and under other, higher sublayers such as RRC (radio resource control) and PDCP (packet data convergence protocol). Its main function is to guarantee reliable data transmission by means of segmentation, ARQ (automatic retransmission request), in-sequence delivery, etc. The RLC layer has three functional modes: TM (transparent mode), UM (unacknowledged mode), and AM (acknowledged mode), which enable services with different speed and reliability for the upper layer.

Chapter 9 ("User Plane Protocol Design for LTE System with Decode-Forward Type of Relay") examines the legacy UMTS RLC protocols and discusses its evolution for the future broadband network. Different operating schemes of RLC for the multihop relay-enhanced cell (REC) are investigated. The AM mode of RLC for the unicast traffic on the downlink user plane is studied in detail. Because the AM mode incorporates the ARQ functionality, it is by far the most important operation mode for reliable data transfer. The conclusion is that the two-hop RLC scheme performs poorly compared to the per-hop RLC scheme due to the asymmetric capacity on the two hops. This chapter concludes that even with the more efficient per-hop RLC scheme, a flow control mechanism on the BS (base station)–relay node (RN) hop has to be developed. However, the topic of flow control, as well as the impact of the HARQ/ARQ cross-layer interaction, would be left for our future study.

Recently, there has been an increasing interest in using cellular networks for real-time (RT) packet-switched (PS) services such as voice over Internet protocol (VoIP). The reason behind the increased interest in VoIP is to use VoIP in all-IP networks instead of using circuit-switched (CS) speech. This would result in cost savings for operators because the CS-related part of the core network would not be needed anymore. In conventional cellular networks, RT services (e.g., voice) are carried over dedicated channels because of their delay sensitivity, while non-real-time (NRT) services (e.g., Web browsing) are typically transported over time-shared channels because of their burstiness and lower sensitivity to the delay. However, with careful design of the system, RT services can be efficiently transported over time-shared channels as well. A potential advantage of transmission speech on a channel previously designed for data traffic is the improved efficiency in terms of resource sharing, spectrum usage, provision of multimedia services, and network architecture. The challenge is to port VoIP services on wireless networks while retaining the QoS of circuit-switched networks and the inherent flexibility of IP-based services.

Long term evolution of 3GPP work targeting to release 8 is defining a new packet-only wideband radio-access technology with flat architecture aiming to develop a framework toward a high-data-rate, low-latency, and packet-optimized radio access technology called E-UTRAN (evolved universal terrestrial radio access network). Its air interface is based on OFDMA for downlink (DL) and SC-FDMA for uplink (UL). Because E-UTRAN is purely packet-switched radio access technology, it does not support CS voice at all, stressing the importance of efficient VoIP traffic support in E-UTRAN. However, supporting VoIP in E-UTRAN faces the very same challenges as those for any other radio access technology: (1) the tight delay requirement and intolerable scheduling overhead combined with the frequent arrival of small VoIP packets and (2) the scarcity of radio resources along with control channel restriction. Thus, designing effective solutions to meet the stringent QoS requirements such as packet delay and packet loss rate (PLR) of VoIP in E-UTRAN is crucial.

Chapter 10 ("Radio Access Network VoIP Optimization and Performance on 3GPP HSPA/LTE") aims to give an overview of the challenges faced in implementing VoIP service over PS cellular networks, in particular 3GPP HSPA and LTE.

Generally accepted solutions leading to an efficient overall VoIP concept are outlined, and the performance impact of various aspects of the concept is addressed. In Section 10.3, characteristics of VoIP traffic are given by describing the properties and functionality of the used voice codec in HSPA/LTE and by presenting the requirements and the used quality criteria for VoIP traffic. The most important mechanisms to optimize air interface for VoIP traffic in (E-) UTRAN are presented in Section 10.4. Section 10.5 provides a description for the most important practical limitations in radio interface optimization, and existing solutions (if any) to avoid these limitations are covered in Section 10.6. The chapter concludes with a system simulation-based performance analysis of VoIP service, including a comparison between HSPA and LTE. Thus, some optimized solutions addressing the VoIP bottlenecks in the leading cellular systems such as HSPA and LTE were proposed and studied. The control channel overhead reduction solutions adopted by 3GPP are especially attractive because they provide very good performance.

Chapter 11 ("Early Real-Time Experiments and Field Trial Measurements with 3GPP-LTE Air Interface Implemented on Reconfigurable Hardware Platform") verifies that the 3GPP LTE air interface can be operated indoors as well as outdoors, achieving very high data rates. Robustness to channel conditions has also been shown for coverage areas of up to 800 m with mobility. Furthermore, the chapter demonstrates the benefits of the new 3GPP-LTE air interface with a new cross-layer MAC architecture of link adaptation, transmit MIMO mode selection, and modulation and coding scheme (MCS) level selection, all of which are frequency dependent. The main focus of this chapter is on discussion of multiple antenna gains in the downlink for a single-user case and single-cell scenario. Measurement results with the authors' MIMO configuration test-bed have shown throughput exceeding 100 Mb/s, a significant increase over the existing single-antenna system.

The reverberation chamber has for about three decades been used for electromagnetic compatibility (EMC) testing of radiated emissions and immunity. It is basically a metal cavity that is sufficiently large to support many resonant modes, and it is provided with means to stir the modes so that a statistical field variation appears. It has been shown that the reverberation chamber represents a multipath environment similar to what we find in urban and indoor environments. Therefore, during recent years it has been developed as an accurate instrument for measuring desired radiation properties for small antennas as well as active mobile terminals designed for use in environments with multipath propagation.

Chapter 12 ("Measuring Performance of 3GPP LTE Terminals and Small Base Stations in Reverberation Chambers") uses a hands-on approach, starting by reviewing basic properties of reverberation chambers and giving an overview of ongoing research and benchmarking activities. It then describes how to calibrate reverberation chambers and how they are used to measure antenna efficiency, total radiated power (TRP), and total isotropic sensitivity (TIS)—important parameters for all wireless devices with small antennas, including 3GPP LTE terminals and small base stations. It also describes how diversity gain and MIMO capacity can be measured directly in the Rayleigh environment (in about 1 minute, compared to hours or more using other technologies), using a multiport network analyzer. The first-ever repeatable system throughput measurements in a reverberation chamber are described.

In order to evaluate the performance of multiantenna wireless communication systems, classical measurement techniques known from antenna measurements, such as anechoic chambers, are no longer sufficient. The measurement setups for (multi-) antenna measurements presented in the first part of this chapter show how reverberation chambers can be used as an alternative to current techniques. The accuracy of the reverberation chamber measurements has been analyzed and compared to classic antenna measurements when this is possible.

The methods developed for single-antenna characterization can be taken a step further to evaluate the performance of different multiantenna setups and algorithms—for example, diversity combining. These diversity measurements require a fading environment and are therefore perfectly suited for reverberation chambers. If base-station simulators are used for measurements, it is even possible to investigate the performance of different antenna setups—not only in terms of power gains, but also in terms of BER. Current state-of-the-art measurements in reverberation chambers even include throughput measurements of wireless systems.

Although link measurements of wireless systems in reverberation chambers are nowadays well established and even made their way into standardization, there are still many ongoing research activities that relate, for example, to improving the accuracy of the chamber further. With the evolution of wireless standards, the attractiveness of reverberation chambers for link/system evaluation increases. Possible extensions of existing measurement techniques with focus on LTE and beyond systems are presented at the end of this chapter. A lot of potential for further development lies particularly in manipulating the channel in the reverberation chamber to a larger extent and to using the chamber for multiple terminal and multiple base-station setups.

The chapter ends by describing several new ways of measuring 3GPP LTE parameters and technologies that mainly are simulated in software today as real-life measurements (e.g., drives tests) are too complicated and time consuming. However, many of these parameters (e.g., multiuser MIMO, opportunistic scheduling, and multiple base-station handovers) should be easy to measure in the reverberation chamber.

Chapter 2

Evolution from TD-SCDMA to FuTURE

Ping Zhang, Xiaodong Xu, and Lihua Li

Contents

2.1 FuTURE Project in China

2.1.1 Background of FuTURE Project

In recent years, along with the continuous development of network and communication technologies and the global commercialization of the 3G (3rd Generation Mobile Communication System), the support for the data service and unsymmetrical service has encountered a greater improvement compared with the 2G (2nd Generation Mobile Communication System), and also the diversity of services is better. At the same time, the requirements for higher data rate and higher quality wireless communication service are brought forward by users. On the one hand, the conflict of rapidly growing numbers of users and limited bandwidth resources becomes outstanding, so the spectrum efficiency of system should be improved by adopting some advanced technologies. On the other hand, it has been proved in both theory and practice that some novel key technologies, such as MIMO (multiple input, multiple output) and OFDM (orthogonal frequency division multiplexing), can improve the performance of current mobile communications systems. Also, many problems need to be improved in the current system while providing service with high data rates and high performance. Therefore, the research of next-generation mobile communication systems has been put into the calendar.

Many countries and organizations have already carried out the research of next-generation mobile communication systems, such as ITU (International Telecommunication Union), European Commission FP (Framework Program), WWRF (Wireless World Research Forum), Korean NGMC (Next Generation

Mobile Committee), Japanese MITF (Mobile IT Forum), and China Communication Standardization Association (CCSA). Some international standards organizations are working for the standards of the E3G (enhanced 3G) and 4G (4th Generation Mobile Communication System), such as the LTE (long term evolution) plan of the 3GPP (3rd Generation Partnership Project) and the AIE (air interface of evolution)/UMB (ultramobile broadband) plan of 3GPP2. The ITU also made the calendar of the 4G communication system, IMT (International Mobile Telecommunications) Advanced. WRC'07 (World Radio-Communication Conference 2007), which was held at the end of 2007, has already allocated multiband spectrum resources for future 4G systems. The 4G is coming.

When research on next-generation mobile communications had just stepped into its startup period, a project called Future Technologies for a Universal Radio Environment (FuTURE) was launched in China. The goal of the project is to support theoretical research, applicable evaluation, and 4G trial system development of the proposed technologies for the Chinese Beyond 3G (B3G)/4G mobile communications. The FuTURE plan aims at the trends and requirements of wireless communication in the next 10 years, and it expects to play an active role in the process of 4G standardization. FuTURE is listed in the 863 Program of China's Science and Technology Development Plan in the Tenth Five Years and formally started in 2002.

The 863 Program is a state-level scientific program. It was proposed in March 1986 and launched in 1987. In April 2001, the China State Council agreed on further carrying out 863 programs on the tenth-five year period. Information technology is an important field of 863 programs. The FuTURE project was initially launched as a part of China's 863 Program in the wireless communications area.

At the end of October 2006, the FuTURE project was successfully tested and certified by the Ministry of Science and Technology (MOST) in Shanghai, representing China's first Beyond 3G mobile communication trial networks of TDD (time division duplex) and FDD (frequency division duplex) systems.

2.1.2 Features of FuTURE System

The next-generation communication system faces requirements beyond year 2010, which will be mainly mobile data service. It should have wider coverage and supports high mobility. The FuTURE system differs radically from the 3G system in the key technologies, where the core network may have a continuous evolution while the air interface should have a revolutionary development. The reason will be detailed as follows.

The packet data service will be in a dominant position, whose "percent occupied" in the overall traffic will increase from 10–20% currently to 80% in the future, while the market of voice service will shrink gradually. The traditional cellular mobile communication system is designed for the requirement of voice service, and it is difficult to apply the code division multiple access (CDMA) technology directly to the 4G system because of its constraint in the aspects of acquisition and synchronization. This means that a fully new wireless transmission method and its corresponding network structure should be designed for the 4G system to meet the requirements of future mobile communication services. With respect to the evolution of E3G standards, this trend has emerged in the standardization process of 3GPP LTE, and OFDM technology has been accepted by E3G standards to accommodate the burst transmission of packet data. The function of CDMA is being weakened, and it is applied more in the allocation of radio resources. It can be predicted that the OFDM will be a competitive air interface technology in the 4G system because of its high spectrum efficiency and flexible spectrum usage.

The peak data rate of the 4G system should be tens or hundreds of times higher than that of the 3G system, and it will be 100 Mb/s–1 Gb/s or even more. The frequency band of 2 GHz for traditional cellular mobile communication is now not enough, so the spectrum resource with higher frequency will be used. But the problem is that the attenuation of electromagnetic waves will increase, and the transmitted power should be increased accordingly tens or hundreds of times, while the cell coverage would be decreased if the traditional cellular mobile communication technologies are still adopted. The novel cellular network infrastructure needs to be researched to solve this problem.

The frequency resource used by mobile communication is limited, and in order to provide higher data rates for users within the limited frequency bandwidth, the overall spectrum efficiency of the system should become 10 times higher by adopting some new technologies. It has been proved by information theory that multiantenna technology such as MIMO can greatly improve the capacity of mobile communication systems, and the research result guides the development direction of the future mobile communication technologies. But in the practical system—especially in the mobile terminal with limited volume— there are still many challenging problems in theory and technology to be solved for implementing the MIMO.

The peak data rate of the 4G system will be 100 Mb/s or more, but the actual data rate the user requires may vary from 10 kb/s to 100 Mb/s dynamically. In order to meet this requirement, the radio resource allocation strategy of the future mobile communication system must be flexible enough to accommodate the dynamical variety efficiently.

For the convergence of network and service, many kinds of wireless access systems will be included in the 4G system, such as cellular mobile communication systems, wideband wireless access systems, satellite systems, and rover systems. To satisfy the mobility requirements of the user, the 4G system should support an open network interface, provide interconnectivity between all kinds of wireless access systems, support seamless roaming between different networks, and ensure continuity of service for users.

Based on the preceding considerations, the FuTURE system will have the following features in radio transmission technology:

- all-IP-based architecture on the interconnectivity of an IP v.6 core network;
- separated control plane and user plane and wireless access networks transparent to the core network;
- flat network structure and cellular network topology suites for high-frequency electromagnetic wave transmission and multiantenna technology;
- air interface that suits the packet burst service, with the peak data rate of 100 Mb/s–1 Gb/s and service with dynamic range by allocating the radio resource flexibly;
- for high-speed movement, frequency spectrum efficiency reaching 2–5 b/s/Hz; for low-speed movement, reaching 5–10 b/s/Hz or higher;
- maximum frequency bandwidth of 20 MHz and frequency carrier of 3.5 GHz;
- for voice services, a bit error rate (BER) of no more than 1E-03; for data services, of no more than 1E-06;
- support of vehicular speed up to 250 km/h;
- peak-to-average ratio (PAR) of radio transmission signals of less than 10 dB;
- end-to-end QoS guarantee better than that of the current telecommunication real-time service;
- support of the characters of self-organization, relay, multihop, and reconfiguration; and
- easy-to-develop and -apply new services and support of service convergence.

The FuTURE mobile communication system will meet the users' requirements of communication with anyone, anywhere, in any way, and at any time, and the user will obtain versatile services, such as virtual reality, videoconferencing, accurate positioning, and high-definition real-time pictures. A lot of advanced technologies, including OFDM, MIMO, and distributed network structure, will be adopted in the 4G system, and a new air interface will be provided to bring better experiences for terminal users.

2.1.3 FDD and TDD Branches of FuTURE Project

The FuTURE project is composed of two branches: the FuTURE FDD system and FuTURE TDD system, both of which are investigating and demonstrating advanced techniques for systems to meet the application requirements around the year 2010. The two branches will be evaluated on a uniform trial platform. For both TDD and FDD, the trial platform contains six access points (APs) and 12 distributed antenna arrays; each AP is equipped with two spaced antenna arrays. With excellent radio transmission performance, high-definition reactive video services have been performed and demonstrated on FuTURE trial systems. Although both TDD and FDD have their own advantages and disadvantages, TDD has some superiority in supporting asymmetry services and multihop functions.

2.2 FuTURE TDD System

Beijing University of Posts and Telecommunications (BUPT) is responsible for the uplink design and trial system integration of the FuTURE TDD system. The university has finished the work of researching and developing China's first 4G TDD mobile communication system with the cooperation of the UESTC (University of Electronic Science and Technology of China), the HUST (Huazhong University of Science and Technology), and the SJTU (Shanghai Jiaotong University). The FuTURE 4G TDD system adopts advanced link transmission technologies such as MIMO and OFDM, which provide the peak data rate of 122 Mb/s on the 3.45-GHz carrier frequency, with the bandwidth of 17.27 MHz and the spectrum efficiency of about 7 b/Hz. It supports IPv6, simultaneous multichannel VOD (video on demand), HDTV (high-definition television), high-speed data download (file transfer protocol, FTP), Internet browsing, instant messaging, and VoIP (voice over Internet protocol) with data rates up to 1,000 times higher and lower power consumption than before.

2.2.1 *Main Features of FuTURE TDD System*

The FuTURE TDD system employs time division duplex mode, where the uplink and downlink work on the same frequency but at different time slots, and then realizes bidirectional communication. Compared with the FDD system, the TDD system will bring advantages in many aspects:

- More flexible spectrum usage. The TDD system does not require paired frequency bands, so it has better flexibility without the constraint of spectrum usage that FDD systems have.
- Better support for unsymmetrical services. The TDD system is more suitable for unsymmetrical service by allocating the time slots for uplink and downlink adaptively according to the traffic.
- Reciprocity between the uplink and downlink. In the TDD system, the uplink and downlink work on the same frequency; hence, the electromagnetic wave propagation characteristics of uplink and downlink are almost the same, so as the channel parameters. This reciprocity makes it possible that the channel estimation of uplink can be directly used by the downlink and vice versa, and it brings many advantages to TDD systems in the aspects of power control, application of transmission preprocessing, smart antennas, and transmission diversity.
- The system offers better support for the technologies of multihop, relay, and self-organization.

There are also some challenges to the TDD system. For example, it requires more accurate timing and synchronization. But it can be predicted that the TDD technology will be used widely in the 4G and that these problems will be resolved. At the same time, the FuTURE TDD system is backward compatible with the TD-SCDMA (time division–synchronous code division multiple access) standard proposed by China, which is one of the international 3G standards in some aspects such as frame structure. Thus, the 3G system based on the TD-SCDMA standard can easily evolve to the FuTURE 4G system.

2.2.2 Promising Technologies Implemented in FuTURE TDD System

In the research and development of the FuTURE TDD system, a series of innovations in basic theory has been proposed, such as

- flat wireless access network structure, which has been put into practice;
- group cell, which is a novel generalized cellular network architecture, and the slide handover strategy based on it;
- soft fractional frequency reuse to improve the spectrum efficiency;
- reliable end-to-end QoS (quality of service) mechanism;
- antenna array based on the generalized cellular network architecture to increase the system capacity, with a spectrum efficiency of 7 b/Hz;
- modular MIMO to support various environments; and
- scalable modular structure of the system, which can support data rates up to 330 Mb/s.

The FuTURE TDD system provides an overall solution that can fulfill the requirements of E3G/4G and adopts lots of key technologies in many aspects, including cellular network architecture, physical layer, medium access control (MAC) layer, radio resource management, and hardware design. As a result, some positive effects have been brought out in the areas of system design, pivotal algorithm, and trial system development.

2.2.2.1 Network Architecture and Cellular Topology

In the research of the FuTURE TDD system, the generalized distributed cellular architecture–group cell is brought out as a novel cellular network architecture; the slide handover strategy and the generalized distributed access network structure based on it are proposed. The cellular network architecture of the group cell can take full advantage of the multiantenna technology, and it fits for the advanced technologies in the physical layer to solve the problem of frequent handover results from higher carrier frequency and, consequently, smaller cell coverage. All the available resources in the system based on the group cell are scheduled and allocated uniformly by the system, so it is easier to optimize the overall resource allocation and avoid the interferences to increase the system capacity and more favorable to use advanced digital signal processing (DSP) technologies to improve the system performance efficiently. In addition, the slide handover can be accomplished by the physical layer adaptively to avoid complex interactive high-level signaling and increase the handover speed. In this way, the user will always be in the center of cell, so the cell edge effect will be eliminated.

The generalized distributed radio access network is a flat and simplified one with high efficiency, where the control plane is apart from the user plane and the radio resources are managed mainly by the control domain. Compared with the network structure currently used, it can reduce the transmission latency of signaling and data to improve system performance, meet the requirement of all-IP, and save the cost of network construction.

The separation of control plane and user plane simplifies network architecture and the signaling load. The generalized distributed cellular structure enlarges the coverage and fully uses the advantages brought by multiantenna techniques. The slide handover strategy makes users always feel that they are at the center of the cell to avoid cell edge effect and resist intercell interference.

2.2.2.1.1 Generalized Distributed Radio Access Network

In a generalized distributed radio access network (RAN), the radio network controller (RNC) in the 3G UTRAN (universal terrestrial radio access network) is separated into two elements. The RAN-specific part migrates into the RAN server, whereas the cell-specific control functions become a part of the AP, which connects antenna elements by optical fiber. Cell-specific processing of user traffic (packet data convergence protocol, PDCP; radio link control, RLC; MAC; etc.) is exclusively performed by the AP. The antenna elements are only responsible for signal transmission and reception. Each AP selects some of them to serve a certain mobile terminal (MT) to construct a generalized distributed cell (group cell). This generalized distributed radio access network architecture is presented in Figures 2.1 and 2.2.

2.2.2.1.2 Definition and Construction Method of Generalized Distributed Cell–Group Cell

Based on multiantenna transmission techniques, the generalized distributed cell is characterized by several adjacent cells that use the same resources (such as frequency, code, or time slot) to communicate with a specific MT and use different resources to communicate with different MTs. The generalized distributed cell is different from the cluster system in current cellular communication systems in that the cells in the cluster use different resources to communicate with MTs.

Figure 2.2 describes a typical generalized distributed cell-based wireless communication system in which each AP has several separated antenna elements (AEs).

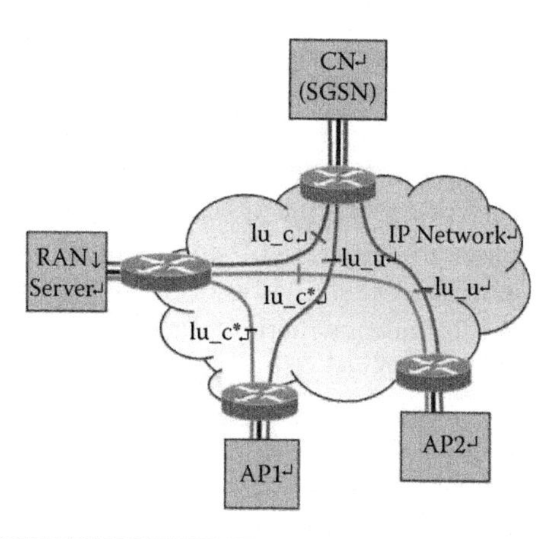

Figure 2.1 Generalized distributed radio access network.

AP1 has 9 AEs and AP2 connects with 10 AEs. If the AEs in the area are indexed by 1–9 and the size of the generalized distributed cell is 3, three generalized distributed cells are connected with AP1 in this area: generalized distributed cell 1 of AEs 1, 2, and 3 (area), generalized distributed cell 2 of AEs 4, 5, and 6 (area), and generalized distributed cell 3 of AEs 7, 8, and 9 (area). This is a fixed

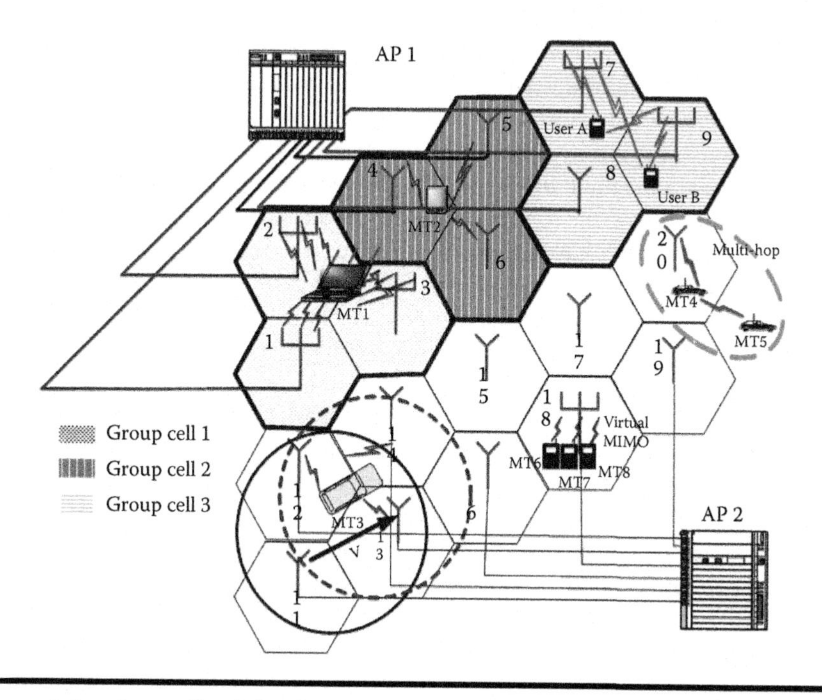

Figure 2.2 Generalized distributed cell structure.

generalized distributed cell structure and the AE of each generalized distributed cell is fixed. With the movement of the MT, different fixed generalized distributed cells will be selected.

Considering the situation in AP2, with the movement of MT4, the AE of the generalized distributed cell that serves the MT4 can also move, corresponding with it. As shown in Figure 2.2, antennas 11, 12, and 13 of AP2 are used for MT4 in time slot 1. With the movement of the MT4, in time slot 2, antennas 12, 13, and 14 will be selected by AP2. The construction of the generalized distributed cell is dynamically changed instead of fixed. This is the slide generalized distributed cell structure.

The construction of slide generalized distributed cells may be viewed as the process of sliding windows. The several cells in one generalized distributed cell could be regarded as in one window and the window could change dynamically in size, shape, and slide speed due to the moving speed and direction of the MT. When the MT moves at relatively rapid speed, the size of the slide window will become larger so as to keep up with the movement of the mobile and decrease the number of handover times. When the speed of the MT is relatively slow, the size of the slide window will become smaller to reduce the waste of resources. If the MT changes its moving direction, the direction of slide window would change at the same time. As a new handover strategy, slide handover changes adaptively in correspondence with the generalized distributed cell structure. Different MTs may correspond to different generalized distributed cells.

Based on the generalized distributed cell architecture, the MIMO technique has many more applications, which indicate the unification of the physical layer and the network architecture. In the case of one antenna in each cell, the generalized distributed cell can be regarded as the spaced MIMO structure (generalized distributed cells 4, 5, and 6 in Figure 2.2). When there are AEs distributed in each cell connected to the same AP, it changes to another structure called a distributed MIMO (AEs 1, 2, and 3). Additionally, UWB technology helps to realize the communication among several MTs nearby. Based on this, an AE in a cell and MTs within the corresponding coverage compose virtual MIMO structure (AE 18 and MTs 6, 7, and 8).

The signals could be transmitted and received by all the antennas of the group by techniques such as MIMO, STC, and OFDMA (orthogonal frequency division multiple access). Therefore, the system's ability to resist interference could be improved, the handover times could be greatly decreased, and the system capacity could be increased. Furthermore, because the signal sources of the same generalized distributed cell are identical, the MT does not need handover in intrageneralized distributed cells. Only if the MT moves out of coverage of the current generalized distributed cell does the handover between the generalized distributed cells–generalized distributed cell handover occur. The generalized distributed cell handover can effectively avoid the frequent handover between cells. Generalized distributed cell handover is also classified into fixed handover and slide handover, corresponding to the construction method of generalized distributed cells. Based on the slide handover strategy, the generalized distributed cell after handover could have some common areas with former generalized distributed cells. As shown in Figure 2.2, with the movement of MT4, the current generalized distributed cell (AEs 11, 12, and 13) that serves MT4 changes into other generalized distributed cells (AEs 12, 13, and 14).

2.2.2.1.3 Separation of Control Plane and User Plane

Because the control plane and user plane scale differently in future data communication, they need to be processed and carried separately. In this contribution, the control traffic from CN will address a (centralized) RAN server, whereas the user traffic is directly routed to an extended AP. This AP terminates the Iu interface for the user traffic and performs the necessary radio-specific processing. The control part of Iu is terminated in the RAN server.

In order to optimize the access system and to reduce network cost, a strict separation of control and user plane is necessary. The evolved architecture should take this into account and offer the flexibility for future adaptation. The proposed generalized distributed cell architecture uses a centralized node (RAN server) only in the control plane and decentralizes the user plane handling in AP.

Iu traffic should be split into a control part (Iu_c) and a user part (Iu_u). Iu_c terminates at the RAN server and Iu_u at the AP. Control information between the RAN server and AP should be transferred via the new Iu_c* (new interface to be further specified) interface.

2.2.2.1.4 Entity of Generalized Distributed RAN

RAN server. Except for the user traffic handling, the RAN server behaves similarly to the former RNC. It manages mobility inside the RAN and the necessary Iu bearers for control and user traffic (Iu_c and Iu_u). For the control part of the Iu interface, the RAN server behaves like a regular RNC from the CN point of view. For the user part of the Iu interface, the AP acts as the former RNC. Furthermore, the RAN server manages micromobility (i.e., mobility inside RAN such as paging and AP relocation) and radio mobility (i.e., mobility between adjacent APs) via Iu_c*.

AP. When antenna arrays are distributed in each cell connected to the same AP, they change to another structure called distributed MIMO (antennas 1, 2, and 3). Each AP connects several antenna elements and, by selecting some of them to serve a certain MT, constructs a generalized distributed cell. Antenna elements are responsible only for signal transmission and reception. The signal process is done in the signal processing unit. AP also contains the cell-specific radio resource management. This enables the AP to manage its radio resources autonomously. On demand, they are requested from the RAN server via the Iu_c* interface.

2.2.2.2 Physical Layer Techniques

To obtain the peak data rate of more than 100 Mb/s, the FuTURE TDD system adopts many key technologies in the physical layer, such as OFDM, MIMO, PAPR (peak-to-average power ratio) reduction, and link adaptation. To realize the transmission data rate up to 1 Gb/s, deep studies have been carried out on the physical layer technologies by introducing some basic theories, such as information theory, norm theory, and matrix theory.

In the FuTURE TDD system, the technologies of OFDM combined with MIMO are used to utilize the characters of OFDM fully under the MIMO structure. The OFDM technology has better capability of anti-multipath interference and higher spectrum efficiency than nonorthogonal multicarrier solutions. Moreover, it can be realized by simple discrete Fourier transform (DFT), so it will be an important technology in the next mobile communication system. Meanwhile, the system adopts the turbo code with bit rate of 100 Mb/s coupled with puncturing and soft demodulation technologies to enhance the system performance and decrease the transmission power; it realizes wideband wireless access with low power consumption and high peak data rate. Additionally, in order to utilize fully the spatial resource to transmit data, the MIMO detection technology, which has good performance, is used at the receiver. Four antennas in the MT and eight in the AP form a 4×8 MIMO system. At the same time, the technology of V-BLAST (vertical Bell Laboratory layered space–time) codes, combined with interference cancellation, is adopted. The technologies mentioned earlier increase spectrum efficiency and improve system performance because of the spatial multiplex and spatial diversity. The overall system performance has a further improvement by adopting other key technologies.

Considering the application of OFDM and MIMO in B3G mobile communication systems, the FuTURE TDD Special Working Group pays much attention to the following topics:

- frame design;
- multiple access schemes; and
- key physical layer techniques, such as MIMO and LA.

2.2.2.2.1 Frame Structure

The frame structure has a drastic influence on system performance, and the wireless frame structure designed here is depicted in Figure 2.3. The frame is composed of eight burst time slots (TSs), where TS0 is designed for the downlink dedicated signaling, including the system information broadcast, paging, etc. The dedicated time slot (TS1) is used for both uplink and downlink frame and frequency synchronization. The remainders are designed for data transmission. Moreover, advanced techniques such as LA (link adaptation) and JT could take advantage of the fact that

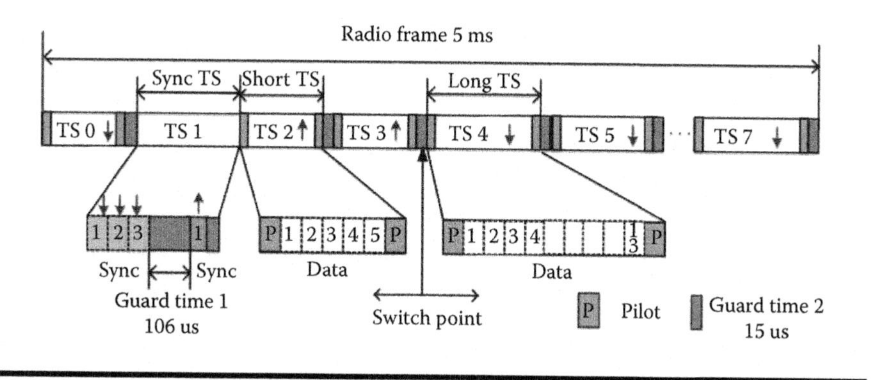

Figure 2.3 Frame structure for FuTURE TDD system.

CSI (channel state information) is reciprocal for the TDD system. The backward compatibility with one of 3G standards—TD-SCDMA—is considered in system design, especially in frame structure design. The reason is that the large-scale field test of TD-SCDMA with TDD will be finished in 2004. It will be a promising standard supporting 3G services. Obviously, smooth evolution from 3G to B3G system is also an important consideration.

The parameters and characteristics of wireless frames are listed as follows:

- The duration of a radio frame is 5 ms, and it could reduce the complexity of adaptive modulation.
- The guard time between uplink and downlink is 106 μs, and it is possible to support a cellular radius as large as 15.9 km. If a multiantenna technique is adopted, it is possible to enlarge the cellular radius.
- There are two types of time slots: short and long. The unequal length for downlink and uplink not only can decrease the cost for guard time, but also can guarantee the flexibility of resource allocation.
- The alterable switch point can flexibly support the service requirements. The change of data ratio after changing switching point can be seen clearly in Figure 2.3. Regarding the asymmetric tendency of future services, the time slot ratio between the uplink and the downlink is about 1:4. This is a default mode.

2.2.2.2.2 Block Structure for Link Layer

In Figure 2.4, the block structure for uplink design is plotted. Here, MIMO, OFDM, and LA are adopted for FuTURE TDD radio transmission, in which P/S is the parallel-to-serial transform module and S/P is the converse operation. CP represents the adding of cyclic prefix (CP).

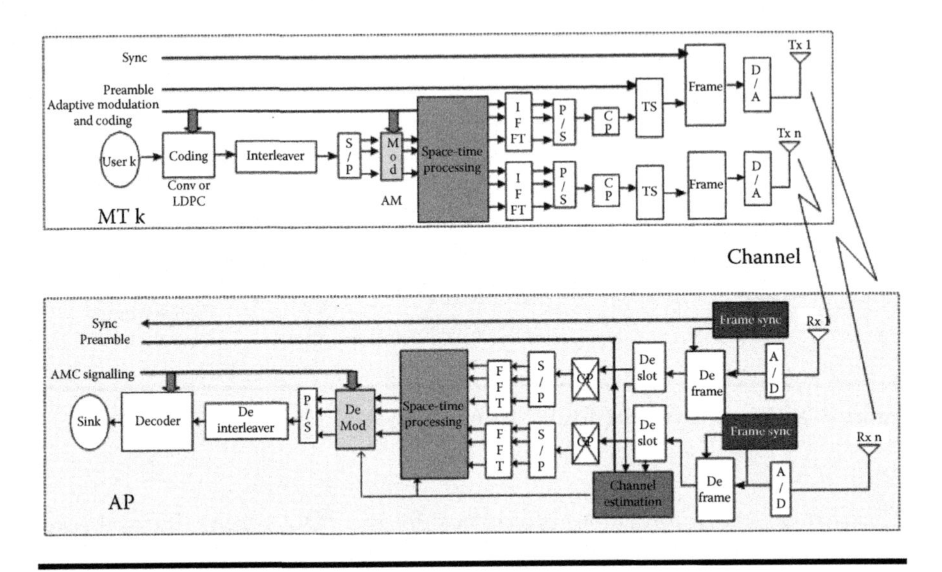

Figure 2.4 The block structure of FuTURE TDD uplink.

2.2.2.2.3 Multiple Access Techniques

Multicarrier transmission, such as OFDM, should be exploited in the FuTURE TDD system to support high data rate transmission. A group of subcarriers is employed as the basic resource unit in an OFDMA system. A subcarrier group is not necessarily composed of adjacent subcarriers; the interleaved subcarriers in frequency domain are also supported, as shown in Figure 2.5. In addition, OFDMA can be combined with time division multiple access (TDMA) and spatial division medium access (SDMA) in order to provide flexible multiple access and resource allocation, as shown in Figure 2.6.

2.2.2.2.4 OFDM Modulation

OFDM is an attractive method of transmitting high-rate information over highly dispersive mobile radio channels by dividing the serial input data stream into a number of parallel streams and transmitting these low-rate parallel streams simultaneously. As unquestionable proof of its maturity, OFDM was standardized as the European digital audio broadcast (DAB) as well as the digital video broadcast (DVB) scheme. It was also selected as the high-performance local area network (HIPERLAN) transmission technique as well as part of the IEEE802.11 wireless local area network (WLAN) standard. Furthermore, it has become the E3G mobile radio standard.

OFDM converts a frequency-selective channel into a parallel collection of frequency-flat subchannels. The subcarriers have the minimum frequency separation

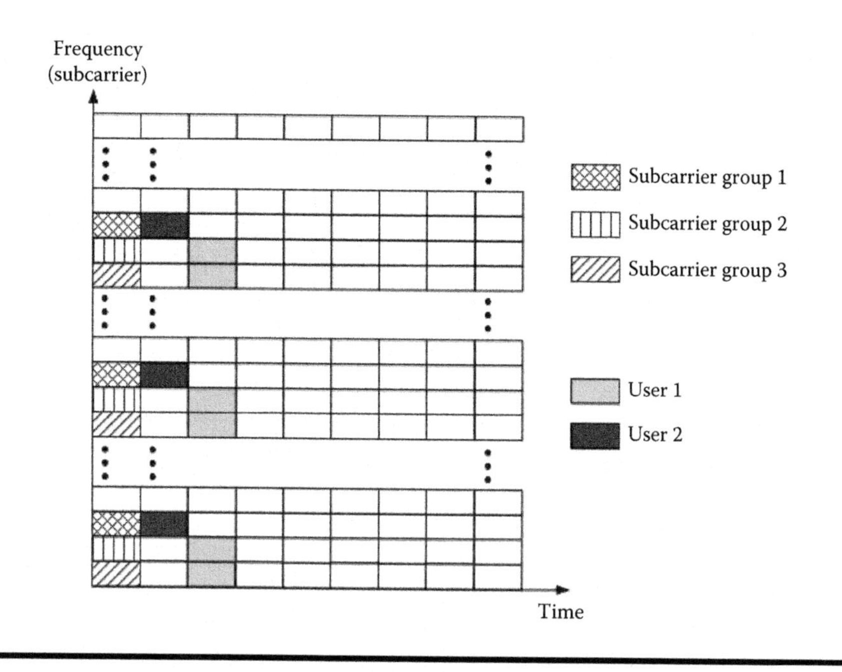

Figure 2.5 Two-dimensional resource allocation.

required to maintain orthogonality of their corresponding time domain waveforms, yet the signal spectra corresponding to the different subcarriers overlap in frequency. Hence, the available bandwidth is used very efficiently. OFDM is robust against multipath fading because the intersymbol interference (ISI) is completely eliminated by introducing a CP or a guard interval of each OFDM symbol. Meanwhile, the intercarrier interference (ICI) is also avoided, which is so crucial in high data rate transmission. Moreover, the CP enables the receiver to use fast signal processing transforms such as a fast Fourier transform (FFT) for OFDM implementation, thus dramatically reducing the complexity.

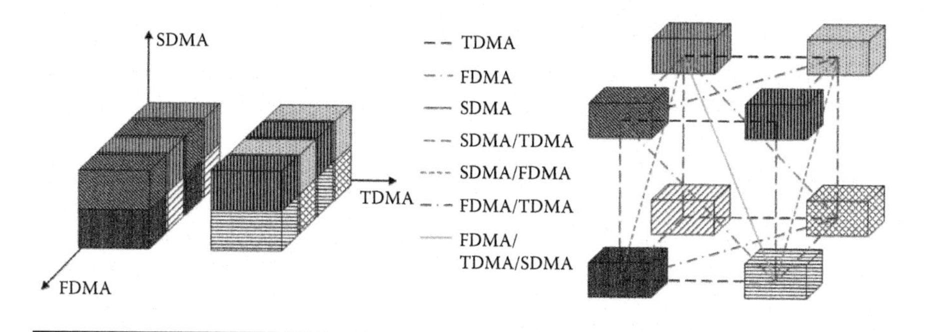

Figure 2.6 Flexible combinations of multiple access schemes.

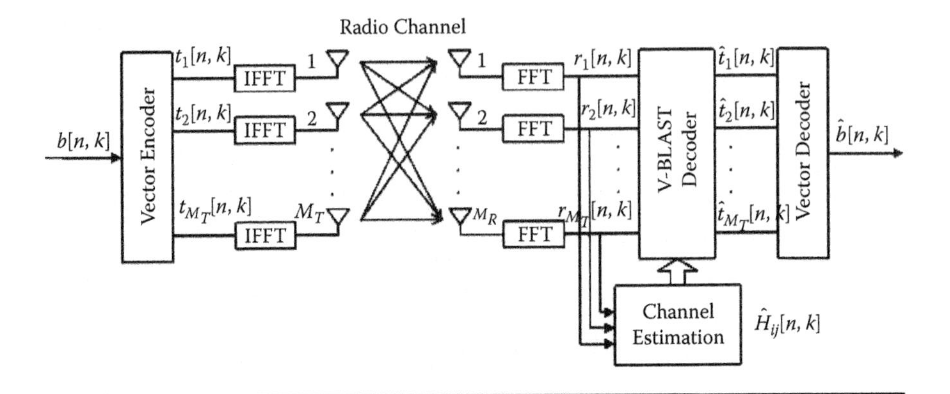

Figure 2.7 The structure of the MIMO-OFDM system.

OFDM offers fully scalable bandwidth to suit varying spectrum allocation and frequency-domain scheduling. Meanwhile, OFDM could be easily combined with other multiple schemes, including FDMA (frequency division multiple access), OFDMA, TDMA, CDMA, and SDMA, in order to support multirate services and to achieve a frequency reuse factor of one. OFDM-CDMA assigns a subset of orthogonal codes to each user; thus, information symbols are spread in either frequency domain or time domain. In OFDMA systems, the signal of each user is transmitted via a set of subcarriers within the duration of several OFDM symbols. In our proposal, the interleaved OFDMA/TDMA is employed in forward link and the localized OFDMA/TDMA in reverse link.

OFDM is also very suitable for MIMO transmission. As each subcarrier is flat fading, the MIMO signal processing for the frequency selective fading channel can be performed for each subcarrier, which is substantially simplified. The structure of the V-BLAST-based MIMO-OFDM system can be expressed as in Figure 2.7.

2.2.2.2.5 Modulation Scheme

The FuTURE TDD system supports QPSK (quadrature phase shift keying), 16 QAM (quadrature amplitude modulation), and 64 QAM. Even higher-order modulation, such as 128 QAM, is also considered for the downlink.

2.2.2.2.6 MIMO and Transmit Diversity

MIMO structure should be adopted in the FuTURE TDD system in order to achieve high data transmission rates. As a research hotspot and potential technique for future communication, MIMO is able to enhance the spectrum efficiency and improve transmission performance by taking advantages of spatial resources. Analysis and simulations prove that MIMO can provide high spectrum efficiency of 20–40 bits/s/Hz. As an important branch, MIMO techniques based on transmit diversity (such as STTC; space–time block code, STBC) can obtain performance gain by repetition. MIMO techniques based on spatial multiplexing (such as D-BLAST and V-BLAST) increase the system capacity significantly by serial-to-parallel transmission.

For correlated MIMO environments, channel information should be fed back and precoding is performed at the transmitter. As another kind of MIMO structure application, beam-forming should also be considered and analyzed according to practical requirements. Furthermore, advanced MIMO techniques should be paid more attention because the number of receiving antennas may be fewer than those of transmitting antennas in the forward link.

Based on some special network architectures, such as generalized distributed antenna architecture, the MIMO technique could embody its superiority in more application forms (e.g., spaced MIMO, distributed MIMO, and virtual MIMO). Additionally, adaptive MIMO schemes should be considered in the FuTURE TDD system.

2.2.2.2.7 Link Adaptation

The FuTURE TDD system should allow for link adaptation, which involves not only AMC but also adaptive MIMO and other advanced adaptive transmission techniques. In addition, the signaling overhead in both forward and reverse link should be considered.

Adaptive modulation and coding (AMC). Adaptive modulation and coding as one of the effective link adaptation schemes should be considered and adopted in the FuTURE TDD system to improve the system capacity with high spectral efficiency. Systems in TDD mode have channel reciprocity because both forward and reverse links are accommodated in the same frequency band. Consequently, the system can estimate the CSI from reception. Hence, TDD systems are more suitable to adopt the AMC technique.

By employing the AMC technique, the power of the transmitted signal is held constant over a frame interval, and the modulation and coding schemes (MCSs) can be chosen adaptively according to channel conditions. The bearer service (BS) may compute the best combination of MCSs either based on postdetection SNR measurement reported by AT or based on signal-to-noise (SNR) measurement of the (reciprocal) reverse channel. Some MCSs and their MCRs (modulation coding rates) are listed in Table 2.1. In a system with AMC, users in favorable positions (close to the cell site) are typically assigned higher-order modulation with higher code rates (e.g., 64 QAM with R = 3/4 turbo codes), while users in unfavorable positions (close to the cell boundary) are assigned lower-order modulation with lower code rates (e.g., QPSK with R = 1/2 turbo codes).

AMC can benefit the system from several aspects. A higher data rate is available for communication under good channel conditions, which consequently

Table 2.1 AMC Schemes and Modulation Coding Rates

MCS	*Modulation*	*Coding*	*MCR (bits/symbol)*
MCS1	QPSK	1/3	0.667
MCS2	QPSK	1/2	1
MCS3	QPSK	3/4	1.5
MCS4	16 QAM	1/2	2
MCS5	16 QAM	3/4	3
MCS6	64 QAM	3/4	4.5

increases the average throughput of the system. Communication quality is guaranteed in cases in which selection of MCSs is in accordance with channel condition. Interference variation is reduced due to link adaptation based on variations in the modulation/coding scheme instead of variations in transmission power. In systems with TDD mode, AMC is more convenient for application and the scheduling of various MCSs is flexible with low latency.

Hybrid automatic repeat request (HARQ). Hybrid ARQ is an important link adaptation technique. In AMC schemes, C/I measurements or similar measurements are used to set the modulation and coding format; in HARQ, link layer acknowledgments are used for retransmission decisions. AMC by itself does provide some flexibility to choose an appropriate MCS for the channel conditions according to measurements either based on AT measurement reports or network determined. However, an accurate measurement is required and there is a delay. Therefore, the ARQ mechanism is still required. HARQ automatically adapts to the instantaneous channel conditions and is insensitive to the measurement error and delay. Combining AMC with HARQ leads to the best of both fields: AMC provides the coarse data rate selection, and HARQ provides for fine data rate adjustment based on channel conditions.

Adaptive MIMO. In realistic wireless transmission environments, channel characteristics of MIMO systems behave in time, frequency, and space dimensions. In order to guarantee system performance in various channel conditions, adaptive MIMO, which is able to choose appropriate MIMO schemes adaptively based on different channel environments, should be supported in the FuTURE TDD system. Technology of transmit beam-forming (TxBF) is an advisable choice in high spatial correlated channels. In low correlation cases, technology of spatial division multiplexing (SDM) is more suitable.

Therefore, adaptive MIMO should be supported to maintain high performance under various channel environments. The switching criterion according to different channel conditions for choosing appropriate MIMO schemes such as TxBF or SDM should be further discussed. For simple demonstration, the primary diagram of adaptive MIMO is shown in Figure 2.8.

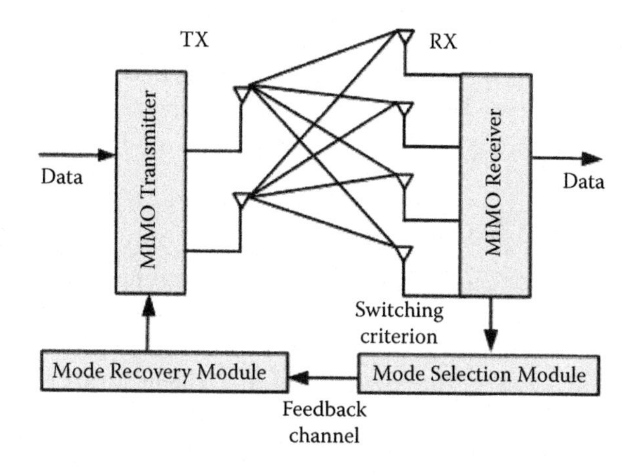

Figure 2.8 Primary diagram of adaptive MIMO.

Additionally, the adaptive power allocation can be combined with the V-BLAST technique to bring good performance. Because the overall bit error rate (BER) performance of spatial multiplexing systems is limited by early detection stages, allocating a low data rate for early detection stages may reduce the probability of error propagation and thereby improve the averaged BER performance for all detection stages. Thus, different transmit antennas of the V-BLAST system should be allocated different transmit data rates if better BER performance is expected. One important advantage of the method is that the optimal ordering does not need to be determined because the allocation of the data rate assumes the receiver detects the streams in a predefined order. When the data rate on different transmit antennas increases by degrees in some predetermined order, detection order at receivers would execute at the same order. Consequently, the complexity of the conventional successive cancellation receiver of V-BLAST systems is greatly reduced. A simple way for the modified V-BLAST is to select a different modulation constellation for each transmit data stream. Then the first decoded stream will typically have the smallest constellation size because it is decoded first. Similarly, employing channel coding with different data rates would also be effective. In this case, the first stream to be decoded would adopt the channel-coding scheme of the lowest rate.

Adaptive MIMO can guarantee system performance and throughput under various MIMO channel conditions and make full use of the MIMO channel. Adaptive MIMO is more convenient and needs less time delay in TDD mode.

Adaptive antenna selection. As an effective technique to fulfill different service requirements of users, adaptive antenna selection should be considered in B3G-TDD systems. According to various kinds of services, the transmit antennas could be selected for transmission adaptively based on channel conditions and data rates required.

Adaptive power allocation. The FuTURE TDD system should support adaptive power allocation. Different proportions of power could be allocated to different subcarrier groups to exploit system performance optimization. In forward link, the scheme for power allocation can be adjusted according to channel information fed back from AT. In TDD mode, taking advantage of reciprocal radio channels, adaptive power allocation can be carried out more flexibly with little time delay.

2.2.2.2.8 Random Access Procedure

Scheduled access and contention-based access should be supported. The contention-based random access scheme is based on the contention windows (CWs) strategy. The length of the CW is decided by the number of retransmission requests and the QoS levels of MTs' services. The advantage of this scheme is that we can get lower frame delay with a small number of users' access requests, even if the user number increases and collisions arise; the frame delay is also controlled in an acceptable range.

The dynamic allocation scheme is that BS increases the number of random access channel (RACH) subchannels of the next MAC frame if collided RACH subchannels exist in the current frame and decreases it with the proper weighting factor in the case of successful attempts. Thus, it can dynamically increase the subchannels' number of random access channels in the case of a large number of access attempts and decrease the subchannels when there are fewer access requests. The proposed scheme, based on results of access attempts, can get higher throughput compared to fixed allocation schemes. The scheme is also adoptable into this system.

2.2.2.2.9 Link Performance of FuTURE TDD System

In the simulation, 3.5-GHz carrier frequency and 20-MHz system bandwidth are considered. The bandwidth of each subcarrier is 19.5 kHz and 832 subcarriers are employed to transmit information. In order to provide a guard band for D/A conversion, nulls are placed at the end of the spectrum. To implement OFDM modulation/demodulation, 1,024-point inverse fast Fourier transform (IFFT)/FFT is used. Some of the service-specific parameters are presented in Table 2.2. The simulation channel is a six-path Rayleigh fading channel with an exponential power-delay profile defined by FuTURE. The maximum delay spread is 10 μs.

Table 2.2 Parameters for Different Services

Data Rate	30 Mb/s	50 Mb/s	100 Mb/s
Modulation	QPSK	16 QAM	16 QAM/64 QAM
Channel coding	Turbo	Turbo	Turbo
Coding rate	1/2	2/5	3/5
Antennas at AP	8	8	8
Antennas at MT	4	4	4

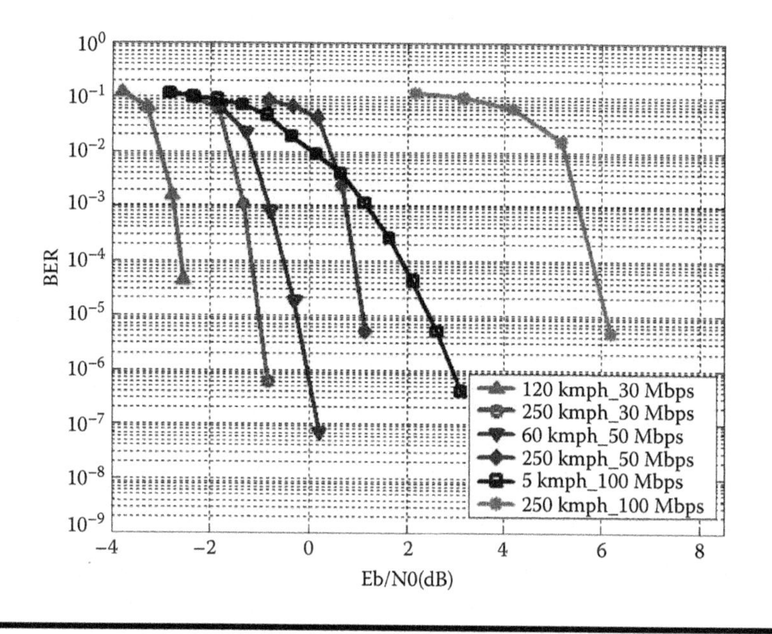

Figure 2.9 Performance of various data services (outdoors, 250 km/h).

Simulation results of various data services at different mobility are shown in Figure 2.9. It is proved by simulation that the FuTURE TDD radio transmission link can support more than 100-Mb/s data rate transmission. Meanwhile, supporting high vehicle speed (250 km/h) with large delay spread (10 μs) is required. At the same time, the B3G link can provide reliable transmission for a large scale of high data rate transmission.

2.2.2.3 Layer 2 and Layer 3 Techniques

The FuTURE TDD system has proposed a series of innovations in the field of radio resource management and has obtained many achievements in the technology of resource allocation and management for next-generation wireless communication systems based on the basic theories of extension set, fuzzy set, game theory, and joint optimization. Such achievements are mainly concentrated in the aspects of high-efficiency frequency reuse strategy, access control strategy for multiantenna systems, and cell edge user performance improvement.

Physical layer (PHY) techniques establish a high-speed radio transmission platform, while radio resource management (RRM) techniques ensure high reliability and high efficiency for radio transmission. The target for RRM design is to provide the highest system capacity and data throughput through optimizing limited radio resources. The main issue is to investigate efficient QoS-oriented resource allocation strategies to optimize jointly the usage of radio resources, such as time, frequency, space, and power. Radio resource management algorithms should support various classes of traffic while guaranteeing their required QoS.

The basic requirements to MAC layer include:

- support of TDD evolution and compatible TDD/FDD hybrid systems;
- support of a much higher data rate;
- support of much higher spectrum efficiency;
- support of a wide range of QoS and mobility;
- reasonable complexity and cost; and
- trade-off between backward compatibility and performance improvement.

The services and functions of MAC include:

- mapping between logical channels and physical channels;
- selection of appropriate transport format for each transport channel, depending on source rate;
- priority handling between data flows of MT;
- priority handling between MTs by means of dynamic scheduling;
- identification of MTs on common transport channels;
- multiplexing/demultiplexing of upper layer protocol data units (PDUs) into/from transport blocks delivered to/from the physical layer on common transport channels;
- multiplexing/demultiplexing of upper layer PDUs into/from transport block sets delivered to/from the physical layer on dedicated transport channels;
- traffic volume measurement;
- transport channel type of switching;
- ciphering for transparent mode RLC; and
- access service class selection for RACH and CPCH (common packet channel) transmission.

2.2.2.4 Intercell Interference Mitigation

According to the features of next-generation mobile communication systems, the theory and the corresponding implementation method of soft fractional frequency reuse (SFFR) were first proposed based on the theories of extension set. FuTURE TDD employed SFFR to ensure the QoS for those users at the cell edge under; to resolve the problem of intercell interference exits in the next-generation mobile communication system, which mainly adopted FDMA technology; and to greatly improve the utilization of limited frequency resources.

As an important strategy to avoid intercell interference, frequency reuse has already been studied for tens of years. Although CDMA-based systems, with a frequency reuse factor of one, have been developed in 3G systems, it appears that OFDMA-based systems will still play a dominant role in the next-generation mobile systems. Therefore, an efficient frequency reuse strategy, as well as enhancing performance for the cell-edge users, is still left as an open research issue for those FDMA-based systems.

To use the spectrum resource more sufficiently and enable the cell-edge users to obtain better performance, SFFR is proposed for OFDMA-based systems. Based on SFFR, the basis of mitigating intercell interference through resource planning/coordination is to classify the users into cell-edge users and inner-cell users according to their geometry factor. The threshold of the geometry factor can be determined by actual traffic distribution in one certain cell.

Considering the following rules, users in each cell are divided into two major groups. One group is for cell-edge users, and the other group is for inner-cell users. Assign the whole frequency band for all users in each cell as follows: Split the whole frequency band into two parts, G and F. According to SFFR, frequency is assigned as follows: Frequency set G and a subset of F are available for inner-cell users; for cell-edge users, SFFR strategy is illustrated by Figure 2.10.

Frequency reuse is a basic technology in wireless systems. In this proposal, SFFR strategy is proposed for intercell interference mitigation based on extension set theory. SFFR can improve spectral efficiency and improve performance of cell-edge users. It can also be used to accomplish the frequency plan for the whole network.

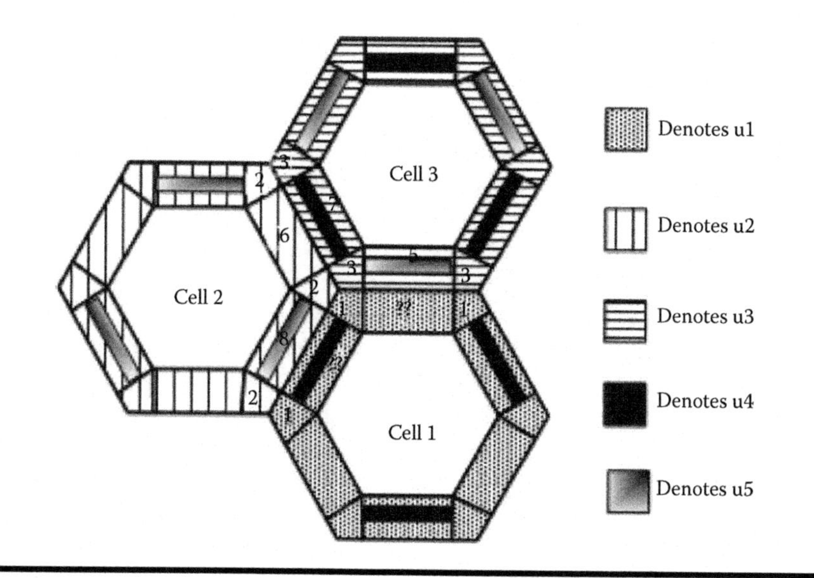

Figure 2.10 SFFR scheme for three-cell model.

2.2.3 FuTURE TDD Trial System and Trial Network

2.2.3.1 Hardware Implementation for FuTURE TDD Trial System

Based on the research of basic theories for the FuTURE TDD system, the hardware demonstration system has been developed and field tested in multicell networks. This hardware platform supports the peak data rate of more than 100 Mb/s and can meet users' requirements for various future services with different QoS, including HDTV, VoIP, data, video, FTP, and Internet.

The FuTURE TDD system adopts an overall infrastructure of the general hardware platform, standard ATCA (advanced telecommunications computing architecture) case, and backplane and flexible modular board structure. This kind of hardware platform structure is suited not only for the TDD mode but also for the FDD mode—not only the AP, but also the MT—and it can be used to research wireless transmission technology and test new wireless network structures. Based on the concept of modular design, the system is divided into various modules according to their functions, so it is easy to improve a module's design or update its functions without affecting other boards. The main modules are connected by using the peer-to-peer distributed interconnection structure based on the exchange; therefore, it is convenient to add or remove a module and adjust the system scale.

The system scale and system performance are scalable. Although extreme reduction of the direct coupling between various factors in the overall infrastructure design has been considered, the system can provide the required abilities of expanded connectivity and processing (for example, two four-element antenna arrays can be united into an eight-element one). The concept of modular MIMO is also proposed to adjust or expand flexibly the number of antenna arrays, the element number of every antenna array, the number of users, and the user data rate.

The technology of fully distributed parallel processing is adopted to reduce the dependence on CPUs and to avoid the bottleneck of network throughput and processing performance introduced by centralized processing technology. As a result, it increases the system's effective throughput capacity and improves the system's flexibility, expandability, and adaptability.

The whole trial system is based on SDR (software-defined radio) technology, where the baseband signal is processed in FPGA and DSP while the processing of control, wireless, and network protocols is implemented in programmable communication processors. This infrastructure based on SDR can sufficiently meet the system's requirement for further evolution. The system supports reconfiguration/dynamic configuration, dynamic update of the system function and signal processing algorithm (reloadable), remote update through network, and the OTA (over-the-air) and dynamic update technologies for the signal processing and control software of the MT.

2.2.3.2 Multicell and Multiuser Trial Network

The 4th Generation Mobile Communication System of the FuTURE TDD plan and the corresponding field trial system were checked and accepted in Shanghai City in October 2006 by a group of experts from the Ministry of Science and Technology of China, the Ministry of Information Industry of China, the Chinese Academy of Sciences, and the Chinese Academy of Engineering. They verified that the first B3G TDD system of China was implemented successfully. The field test results of the corresponding trial system had already validated its advantages in the aspects of multicellular network, support for high mobility, and high-speed data service.

In the research and development of the FuTURE TDD system, a series of advanced technologies was adopted, such as distributed network architecture based on RoF, slide handover, soft fractional frequency reuse, modular MIMO, and future generation computer services (FCGS). Also, the first 4G field trial system based on a distributed radio network was established in Shanghai, where a scheme of a combined network of TDD and FDD was introduced and then some key technologies in the system were analyzed, including the unification of software and hardware trial platform, wide coverage for 40- to 100-Mb/s data rate, a high mobility test, spectrum efficiency of 2–10 b/s/Hz, low power consumption, and good electromagnetic compatibility (EMC) performance.

Figure 2.11 shows the overall structure of FuTURE TDD trial systems, with three APs and three MTs included, where the distributed network structure was

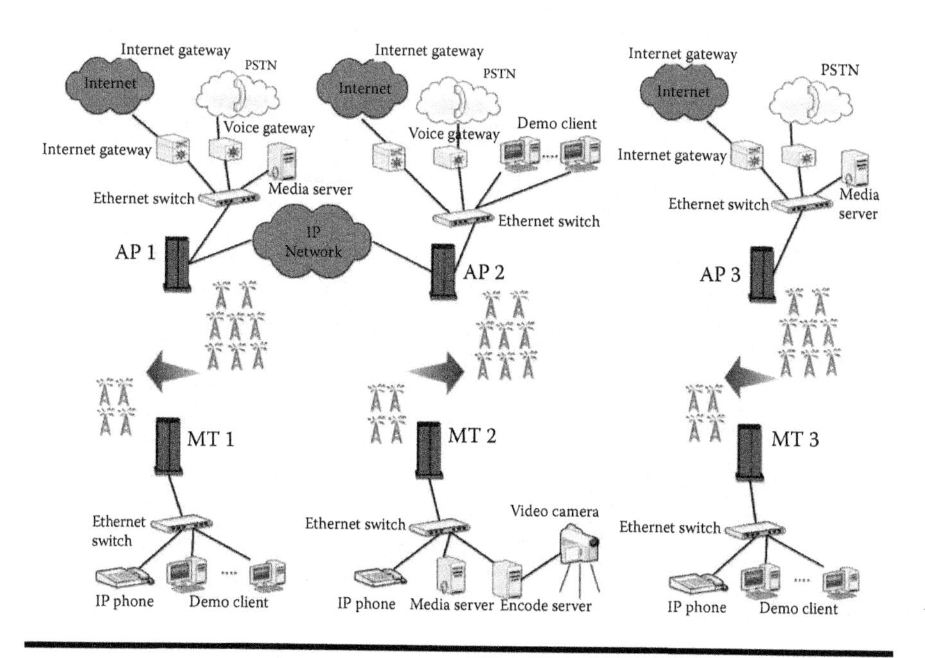

Figure 2.11 The overall structure of the FuTURE TDD trial system.

Figure 2.12 The hardware and antenna unit in the FuTURE TDD trial system.

introduced. The mobility test was performed between AP1 and AP2, including a handover test and a high-speed mobile performance test for the terminal. Figure 2.12 shows the antenna deployment in the FuTURE TDD trial system, including the AP, MT, antenna unit, and its field allocation. Figure 2.13 shows the

Figure 2.13 The test vehicle of the FuTURE TDD trial system.

test vehicle of the FuTURE TDD trial system, which supported vehicle speed up to 50–100 km/h.

2.2.3.3 Field Test Results of FuTURE TDD Trial System

For the FuTURE TDD trial network, the campus of Shanghai University of Engineering Science was selected as a typical urban environment with low vehicle speed. The campus map is shown in Figure 2.14. In this map, two-dimensional labels are used for location identifiers in order to facilitate the test process.

2.2.3.3.1 BLER Performance Test

In the FuTURE TDD trial network, just as Figure 2.14 shows, four distributed antenna array sites are located at each corner of the campus. For each site, four antennas are set up and eight antennas from two sites are connected to one AP. The test vehicle is equipped with four antennas and moves from point (1, 1) around the campus to point (5, 1). According to instant measurement results, the BLER performance of uplink is demonstrated in Figure 2.15. It can be seen that the FuTURE TDD trial system could support a peak data rate of 100 Mb/s along most paths around the campus. However, the data transmission rate should be reduced on the road between the rest house and the library, which also requires the application of seamless adaptive MIMO transmission.

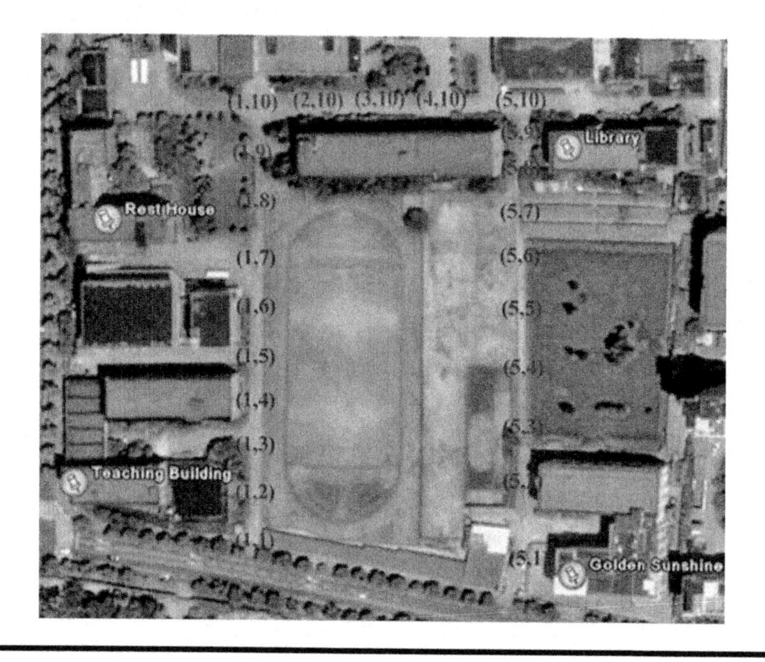

Figure 2.14 Test map of campus for test of FuTURE TDD trial system.

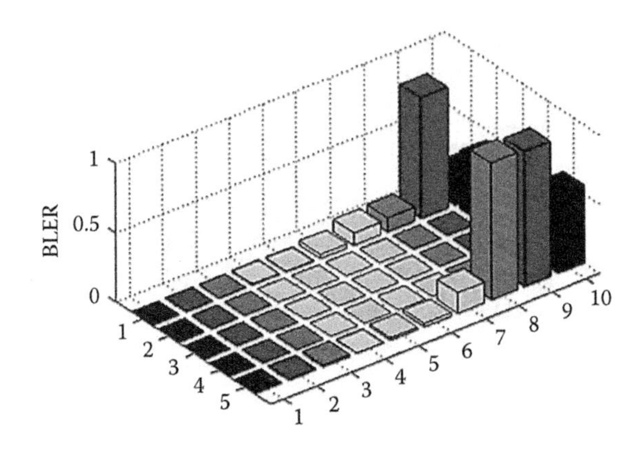

Figure 2.15 BLER performance around the campus.

2.2.3.3.2 Single-Frequency Networking Test

In the FuTURE TDD trial network, a simplified form of the SFFR scheme was performed and tested based on the distributed cellular network. Figure 2.16 shows the scenario of networking and the test procedure with a stationary terminal and a mobile terminal. According to the figure, when MT2 moves around the campus, it is first located in the same cell with MT1, but then in different cells. Taking advantage of SFFR, the two terminals first occupy half of the frequency resources

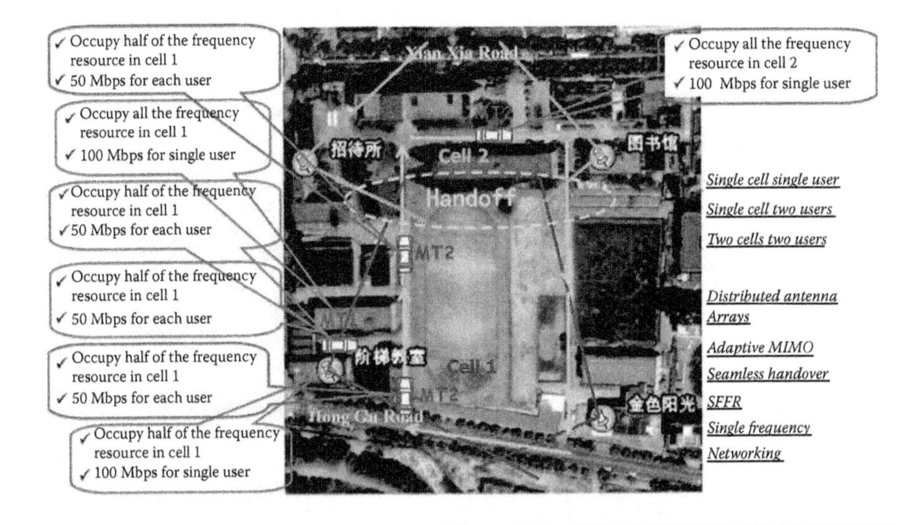

Figure 2.16 Test of SFFR with single-frequency networking.

Figure 2.17 HDTV service demo and videoconference service demo.

in one cell and then occupy all the frequency resources in each cell, which makes the scheme of single-frequency networking feasible.

2.2.3.3.3 Service Demonstration Test Results

In order to test the capability of the service load of the FuTURE TDD trial system, several kinds of different services were performed simultaneously on the platform in which HDTV service, videoconference, and high-speed FTP download services were all supported by wireless communication.

2.2.4 Achievements and Significance of FuTURE TDD System

The ITU has named the 4G system "IMT Advanced" and will start its standardization around 2008. The research in basic theory and development of a corresponding trial system for the FuTURE TDD system has been accomplished. The first field trial system based on a distributed radio network has been established successfully in Shanghai, and it has the basic characteristics of the 4G TDD mobile communication system. This system can provide radio transmission with a peak data rate of 100 Mb/s for high-mobility terminals and the demonstration of HDTV (Figure 2.17); the hardware platform can support wireless transmission with a data rate up to gigabits per second. It has the highlights of high spectrum efficiency and low transmission power, which represent the evolutionary trends of the next-generation mobile communication technologies. There are many innovations in radio network architecture, transmission, and radio resource management technologies for the next-generation wireless communication.

In the development of the trial system, the FuTURE TDD special working groups proposed 129 advanced technologies with independent intellectual properties, and most of them have been taken into practice. At the same time, some accumulated intellectual properties are devoted to 3GPP LTE and 3GPP2

AIE/UMB, and several standard proposals have been submitted to CCSA, 3GPP, and 3GPP2. Among them, BUPT has submitted more than 10 proposals to CCSA with 6 accepted, 3 proposals to ITU, 10 proposals to 3GPP2, and 30 proposals to 3GPP. Among them, three proposals have been accepted by ITU, seven by 3GPP, and one by 3GPP2 UMB release v2.0. Such work has pushed the evolution of the 3G technology, including TD-SCDMA, and laid the foundation of developing next-generation wideband wireless mobile communication technologies and participating in international competition on behalf of China. Additionally, the FuTURE TDD special working group plays an active role in the FuTURE Forum and maintains several documents about wireless transmission technology, spectrum management, radio resource management, and network structure.

The successful implementation of the FuTURE TDD system indicates that China is now in an advantageous position in the research and development process of the 4G Mobile Communication System. Based on this platform, commercial products can be developed to enter the R&D services market of wideband mobile communication technology and realize new types of multimedia services impossible before, such as the HDTV and super-high-speed download services. It will provide strong support for the successful development of the Chinese mobile communications industry and national socioeconomic level. The key technologies in the FuTURE TDD system can greatly improve the performance of high-speed wideband wireless access for users and are valuable to improving people's cultural lives, with the result of bringing positive social benefits.

2.3 TD-SCDMA, LTE, and FuTURE TDD

TD-SCDMA was proposed by the Chinese Academy of Telecommunication Technology (CATT) in 1998, accepted by the ITU as a 3G standard in November 1999, and accepted by the 3GPP as the low chip rate (LCR) TDD mode of universal terrestrial radio access (UTRA) in March 2001. TD-SCDMA has adopted many advanced technologies, such as synchronous CDMA, smart antenna, joint detection, SDR, baton handover, and dynamic channel allocation (DCA). TD-SCDMA originated in China and is strongly supported by the Chinese government. It is one member in the 3G family.

LTE is a project launched by 3GPP. It began in 2004 and aims to accompany a brand new standard to satisfy the requirements for the next 5–10 years. After the study item and work item, one version for such a system has been worked out. TDD and FDD modes are supported. It takes MIMO, OFDM, etc. to bring to the system the ability to raise data rates as high as 100 Mb/s. It can be seen as a system midway between 3G and 4G systems. The two standards support TDD mode; hence, it is necessary to make a comparison between them and FuTURE TDD.

Table 2.3 Comparisons of TD-SCDMA, LTE TDD, and FuTURE TDD Systems

Aspects	*TD-SCDMA*	*LTE TDD*	*FuTURE TDD*
Duplex	TDD	TDD	TDD
Multiple access	TDMA, CDMA	TDMA, OFDMA	TDMA, OFDMA
Bandwidth	5 MHz	1.4–20 MHz	20 MHz
Peak data rate	2 Mb/s	100 Mb/s	100 Mb/s
Frequency efficiency	0.4 b/s/Hz	5 b/s/Hz	5 b/s/Hz
Antenna configuration	Smart antenna	MIMO	MIMO
Supported services	VoIP, basic multimedia services, etc.	VoIP, high data rate multimedia services	VoIP, high data rate multimedia services
Channel coding	Turbo coding	Turbo coding	Turbo coding
Modulation	QPSK, 8 PSK, 16 QAM	QPSK, 16 QAM, 64 QAM	16 QAM
MIMO technique	Beam forming	CDD, STBC, precoding	V-BLAST
Retransmission	HARQ	HARQ	HARQ

2.3.1 Similarities among TD-SCDMA, LTE TDD, and FuTURE TDD

The FuTURE TDD system can be seen as an evolution system for TD-SCDMA. Because the same advanced techniques are adopted as in LTE, they exhibit the same transmission ability. The three systems have much in common and are shown in Table 2.3, in which the main technical configurations can be found.

From Table 2.3, we can find the similarities among all the systems. From the basic aspects, they can all support TDD mode; hence, the advantages of TDD (such as flexible frame structure, high efficiency, and flexible frequency band utilization) can be obtained when one of the systems is employed. The three systems can all support TDMA when a multiple access scheme is concerned. On the other hand, we investigate some specific techniques. The three systems all adopt turbo coding as the channel coding technique, support high-order modulation such as 16 QAM, and employ HARQ as the retransmission strategy. As far as other specific techniques are concerned, FuTURE TDD is more like LTE. The same 20-MHz bandwidth is adopted as the maximum bandwidth. Because of the same adoption of MIMO and OFDM techniques, high-frequency efficiency can be achieved—as high as 5 b/s/Hz—for both LTE and FuTURE TDD. They can both support 100-Mb/s transmission and high data rate multimedia services.

2.3.2 Enhancement of FuTURE TDD to TD-SCDMA

Many similarities can be found between FuTURE TDD and TD-SCDMA, but FuTURE TDD can outperform TD-SCDMA because it is designed as a Beyond-3G system. The conclusion can also be proved from Table 2.3.

- The frequency efficiency for FuTURE TDD is 5 b/s/Hz—much higher than for TD-SCDMA. This gives FuTURE TDD the ability to support high data rate multimedia services. The data rate can achieve 100 Mb/s, which can surely satisfy the requirement for all kinds of services in the future.
- A new multiple access scheme is employed, which is named OFDMA. With the scheme, more users may access the wireless network because different users can be distinguished in both time and frequency domains. Hence, FuTURE TDD can solve the problem of wireless access for a very large number of users.
- FuTURE TDD was developed after TD-SCDMA; hence, it can obtain more advanced techniques. The employment of MIMO and OFDM techniques brings FuTURE TDD better performance. This helps FuTURE TDD support wireless transmission even in environments with high moving speed and large distances from the base station. It can save much power, improve transmission quality, and bring better service experience to users.

2.4 Evolution Map from TD-SCDMA to FuTURE

Due to the difference between TD-SCDMA and FuTURE, the course of evolution will be long. In this section, we will introduce the evolution, where key technology and system evolutions will be introduced.

2.4.1 Key Technology Evolution Map

TD-SCDMA is based on CDMA, utilizes a single antenna on the user equipment (UE) side, adopts low chip rate transmission, and employs joint detection to bring performance improvement. In comparison, the MIMO-OFDM technique is adopted in FuTURE, the sampling rate is relatively much higher, and FuTURE adopts advanced detection techniques to benefit the system. Hence, the evolved systems should be gradually equipped with some techniques.

2.4.1.1 Generalized Distributed Antenna Arrays Combined with Dynamical Channel Allocation

The capacity of TD-SCDMA is mainly limited by intercell interference, rather than by the intracell interference for the smart antenna with joint detection adopted; thus, the effect of cell breadth is not serious. If the intercell interference can be decreased, then the system capacity will be further improved. A distributed antenna can be used to improve the coverage in indoor scenarios and mitigate fast fading. For the generalized distributed antenna arrays (GDAAs), the whole cell is divided into several subcells and every subcell is centrally covered by an independent antenna array. The radio resource of one cell is shared by three of its subcells. Then the transmit power of every subcell is reduced for a small coverage radius and the intercell interference is reduced as well; thus, the capacity can be improved. Furthermore, DCA can be combined with a smart antenna and GDAA so as to improve the system performance using SDMA.

2.4.1.2 Improved Joint Detection

Joint detection is one of the key techniques of TD-SCDMA; it can eliminate MAI and ISI. The enhancement and improvement of joint detection can provide the potential for further performance improvement. When SDMA is implemented with a smart antenna, the same codes can be reused in the same cell, and interference may exist between different users who use the same codes because the smart antenna cannot eliminate the side lobe perfectly. If joint detection can take into account this kind of interference and eliminate it, then SDMA can obtain higher efficiency. Furthermore, if joint detection can eliminate the intercell interference from the neighboring cells, then the system capacity and spectrum efficiency will be improved greatly. Thus, the enhancement of joint detection has potential for improving the system performance.

2.4.1.3 Higher Spreading Factor and Chip Rate

The potential benefits of higher chip rate include system-level gains from trunking efficiency, link-level gains from the ability to better resolve channel paths, the ability to support more accurate location services, higher peak bit rate and cell throughput, and an improved ability to reject narrowband interferers. Thus, with a higher spreading factor and chip rate—for example, SF = 32 and 2.56 Mc/s—more gain from spreading processing, multipath resolution, and interference elimination can be obtained for TD-SCDMA, and better system performance can be expected. The benefit of 7.68 Mc/s for UTRA TDD has been investigated in detail. Therefore, a higher spreading factor and higher chip rate TD-SCDMA can provide a possible solution for further improving the system performance.

2.4.1.4 Virtual Antenna Array

Usually, due to limited size and cost, it is difficult to mount multiple antennas at the UE. By using a virtual antenna array (VAA), several idle UEs near the master UE can cooperate to work for the master UE and construct a VAA to improve the system coverage and data-transmission capability. When the distance among UEs is quite large compared to the wavelength, fading on the antennas of VAA is independent and thus the spatial multiplexing and diversity gain can be exploited to improve the system performance. The connection among UEs can be built through Bluetooth or ultrawideband (UWB) technology in a free-frequency band.

To obtain the gain of the VAA, the received baseband signal of the VAA must be forwarded to the master UE and processed centrally there. However, the transmission of the baseband signals from the VAA to the master UE requires quite a powerful UWB air interface. Here, two examples of two supplementary UEs with two antennas each are presented. It is assumed that the quantification of the baseband signal needs 12 b for one floating number of the I or Q signal. The sampling rates of the baseband signal are 1.28 and 5 MHz, respectively, for these two examples. To forward the baseband signals from the VAAs to the master UE, the data rate needed is calculated as follows.

2.4.1.5 MIMO OFDMA

MIMO is a very promising technology to improve the spectrum efficiency. However, because the much broader frequency band will be used, more multipaths will be separated, and thus the conventional single-carrier system will have much higher complexity in MIMO detection. OFDM is a good candidate to solve such problems. Due to the excellent capability of mitigating frequency-selective fading and ISI, OFDM is very suitable for high data rate transmission in wideband wireless channels. Meanwhile, the implementation of OFDM is very simple using IFFT/FFT, and AMC on every subcarrier can theoretically make full use of the spectrum. (Factually, AMC on every subcarrier will consume too much signaling; AMC usually is based on a group of subcarriers, called a chunk in 3GPP, in order to reduce the signaling consumption. The optimal size of a chunk is a trade-off between the AMC efficiency and the signaling consumption.) All these features make OFDM very competitive for LTE and B3G systems. Combining OFDM with MIMO, the frequency selective fading MIMO channel can be separated into many flat fading MIMO subchannels, and thus MIMO detection can be simplified. On the other hand, the dynamic range of the service data rate is quite large (e.g., from several kilobits per second to 100 Mb/s) and flexible multiple access (MA) methods are necessary to provide a flexible data rate (e.g., OFDMA/TDMA).

2.4.2 System Evolution Map

When we consider the system evolution, we have to take technology, requirements of users, and other factors such as the consideration from the manufacturers into

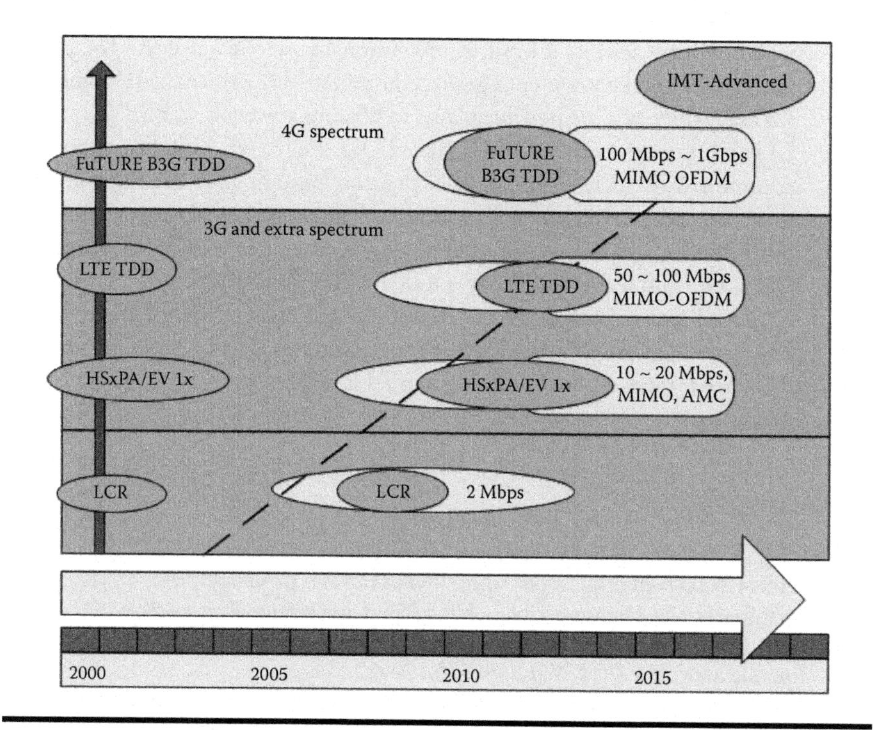

Figure 2.18 Evolution map from TD-SCDMA to FuTURE TDD system.

consideration. We can illustrate the evolution map from Figure 2.18. There should be three steps for evolution from TD-SCDMA to FuTURE TDD:

1. TD-SCDMA will evolve to HSxPA/EV 1x. Nowadays, the deployment of TD-SCDMA is under way in China. People in the 2G age will experience the multimedia services provided by TD-SCDMA very soon. It can support services with data rates up to 2 Mb/s, and the chip rate is 1.28 Mc/s. This course will last for several years. Then the requirement for wireless access will be higher. In order to satisfy the requirement without completely updating the services, it is best for operators to evolve the system to HSxPA/EV 1x, which can adopt more advanced techniques such as MIMO and AMC to achieve higher transmission capability. The chip rate will be higher accordingly.
2. HSxPA/EV 1x will evolve to LTE TDD. As multimedia, Internet, etc. develop, further higher data rate services will be required (as high as 50–100 Mb/s). CDMA cannot perform better due to its technical disadvantages. Hence, MIMO-OFDM will act as the main technique, and LTE-TDD will gradually replace HSxPA/EV 1x. This step is of great importance,

which can be seen as a kind of revolution for wireless access. The operators have to make great changes to equipment. Due to the employment of MIMO, there will be new problems in placing the evolved BS.

3. LTE TDD will evolve to FuTURE TDD. FuTURE TDD has much in common with LTE TDD. Hence, the two systems may exist at the same time. With the development of software-defined radio and the commonality of the systems, it is easy for operators to change between FuTURE TDD and LTE TDD. On the other hand, FuTURE TDD adopts many more advanced techniques, such as GDAA and virtual MIMO. Hence, it can bring many good points, which are the characteristics of 4G, named IMT Advanced. Hence, the evolution from LTE TDD to FuTURE TDD is also necessary.

Bibliography

Adel, A. M., A. J. Saleh, J. R. Rustakl, and R. S. Roman. 1987. Distributed antennas for indoor radio communications, *IEEE Transactions on Communications* 35 (12).

Bell Laboratories. 1971. High-capacity mobile telephone system technical report, December.

Erdal, C. and C. Ersoy. 2002. Application of 3G PCS technologies to rapidly deployable mobile networks. *IEEE Network* 16 (5): 20–27.

Foschini, G. J. 1996. Layered space–time architecture for wireless communication in a fading environment when using multielement antennas. *Bell Labs Technical Journal* 41–59.

Foschini, G. J. and M. J. Gans. 1998. On limits of wireless communications in a fading environment when using multiple antennas. *Wireless Personal Communication* 311–335.

Frattasi, S., H. Fathi, F. H. P. Fitzek, R. Prasad, and M. D. Katz. Defining 4G technology from the user's perspective. *IEEE Network* 20 (1).

Gao, Xiqi, Shidong Zhou, and Wuyang Zhou. 2007. FuTURE FDD system: Key technologies and experimental validation. *FuTURE Forum* 2 (1): 50–53.

Ghosh, A., D. R. Wolter, J. G. Andrews, and R. Chen. 2005. Broadband wireless access with WiMax/802.16: Current performance benchmarks and future potential. *IEEE Communications Magazine* Feb.

Guangyi, Liu, Jianhua Zhang, Ping Zhang, Ying Wang, Xiantao Liu, and Shuang Li. 2006. Evolution map from TD-SCDMA to FuTURE B3G TDD. *IEEE Communications Magazine* March: 54–61.

IEEE 802.16. http://groups.ieee.org/group/802/16/

ITU. 2007. Circular letter for IMT Advanced.

Klein, S. G. 1991. On the capacity of cellular CDMA system. *IEEE Transactions on Vehicular Technology* 40 (2).

Lee, W. C. Y. 1991. Smaller cells for greater performance. *IEEE Communications Magazine* 29 (11): 19–23.

Mihailescu, C., X. Lagrange, and Ph. Godlewski. 1998. Locally centralized dynamic resource allocation algorithm for the UMTS in Manhattan environment. *Ninth IEEE International Symposium on Personal, Indoor and Mobile Radio Communications* 1: 420–423.

Tao, Xiaofeng. 2002. Wireless communications beyond 3G. 2002. *International Forum on Future Mobile Telecommunications & China-EU Postconference on Beyond 3G,* Beijing, China, Nov. 20–22, 2002.

———. 2003. Novel cell infrastructure and handover mode. Ninth WWRF Conference, Zurich, Switzerland, July 1–2, 2003.

Tao, Xiaofeng, Zuojun Dai, and Chao Tang. 2004. Generalized cellular network infrastructure and handover mode-group cell and group handover. *Acta Electronic Sinica* 32 (12A): 114–117.

Tao, Xiaofeng, Li Ni, Zuojun Dai, Baoling Liu, and Ping Zhang. 2003. Intelligent group handover mode in multicell infrastructure. *IEEE PIMRC2003,* Beijing, China, Sept. 7–10, 2003.

Tao, Xiaofeng, Dan Shang, Li Ni, and Ping Zhang. 2003. Group cells and slide handover mode. *ICCT 2003,* Beijing, China, April 9–11, 2003.

Ware, H. 1983. The competitive potential of cellular mobile telecommunications. *IEEE Communications Magazine* November: 19.

Wolniasky, P. W., G. J. Foschini, G. D. Golden, and R. Valenzuela. 1998. V-BLAST: An architecture for realizing very high data rates over the rich scattering wireless channel. *Proceedings of ISSE'98* 295–300.

Xu, X., X. Tao, C. Wu, Y. Wang, and P. Zhang. 2006. Capacity analysis of multiuser diversity in generalized distributed cellular architecture. *IEEE International Conference on Intelligent Transportation Systems Telecommunication,* June: 533–536.

Xu, X., C. Wu, X. Tao, Y. Wang, and P. Zhang. 2007. Maximum utility principle access control for Beyond 3G mobile system. *Wireless Communications and Mobile Computing,* to appear June 2007.

Xu, X., X. Tao, C. Wu, and P. Zhang. 2006. Group cell architecture for future mobile communication system. *ICN/ICONS/MCL 2006,* April, 2006: 199–203.

———. 2006. Capacity and coverage analyses for the generalized distributed cellular architecture–group cell. *IEEE International Conference on Communication, Circuits and Systems (ICCCAS),* June 2006: 847–851.

———. 2006. Fast cell group selection scheme for improving downlink cell edge performance. *IEEE International Conference on Communication, Circuits and Systems (ICCCAS),* June 2006: 1382–1386.

Yanikomeroglu, H., and E. S. Sousa. 1997. Antenna interconnection strategies for personal communication systems. *IEEE Journal on Selected Areas in Communications* 15 (7): 1327–1336.

Yu, X., G. Chen, M. Chen, and X. Gao. 2005. The FuTURE Project in China. *IEEE Communications Magazine* 43 (1): 70–75.

Zahariadis, T. B. 2004. Migration toward 4G wireless communications. *IEEE Wireless Communications* 11 (3).

Zhang, P., X. Tao, D. Shang, and L. Ni. 2002. Group cell and group handover. *Proceedings of the 9th Mainland–Taiwan Workshop on Wireless Communication,* Beijing, China, Sept. 2002.

Zhang, P., X. Tao, J. Zhang, Y. Wang, L. Li, and Y. Wang. 2005. The visions from FuTURE beyond 3G TDD. *IEEE Communication Magazine* 43 (1): 38–44.

Zhao, X., J. Kivinen, P. Vainikainen, and K. Skog. 2002. Propagation characteristics for wideband outdoor mobile communications at 5.3 GHz. *IEEE Journal on Selected Areas in Communications* 20 (3): 507–514.

Zhou, S., M. Zhao, X. Xu, J. Wang, and Y. Yao. 2003. Distributed wireless communication system: A new architecture for future public wireless access. *IEEE Communications Magazine* 41 (3): 108–113.

3GPP, REV-WS005. 2005. DoCoMo's view on 3G evolution and requiements-3G LTE scenario—Super 3G-x. LTE workshop, Tokyo, March 7.

3GPP, TR23.882. 2006. 3GPP system architecture evolution: Report on technical options and conclusions.

3GPP, TR25.814. 2006. Physical layer aspects for evolved universal terrestrial radio access (UTRA).

3GPP, TR25.892. 2004. Feasibility study for orthogonal frequency division multiplexing (OFDM) for UTRAN enhancement.

3GPP, TR 25.912. 2007. Feasibility study for evolved UTRA and UTRAN.

3GPP. TR25.913. 2005. Requirements for evolved UTRA (E-UTRA) and evolved UTRAN (E-UTRAN).

3GPP2, ATE. 2006. Evolution of cdma2000 networks.

3GPP2, cdma2000. 2005. Enhanced packet data air interface system—System requirements document.

3GPP2, C30-20060731-040R4. 2006. Joint proposal for 3GPP2 physical layer for AIE FDD spectra.

Chapter 3

Radio-Interface Physical Layer

Gerardo Gómez, David Morales-Jiménez, F. Javier López-Martínez, Juan J. Sánchez, and José Tomás Entrambasaguas

Contents

3.1 Physical Layer Overview

The long term evolution (LTE) physical layer is targeted to provide improved radio interface capabilities between the base station* and user equipment (UE) compared to previous cellular technologies like the universal mobile telecommunications system (UMTS) or high-speed downlink packet access (HSDPA).

According to the initial requirements defined by the 3rd Generation Partnership Project (3GPP; 3GPP 25.913), the LTE physical layer should support peak data rates of more than 100 Mb/s over the downlink and 50 Mb/s over the uplink. A flexible transmission bandwidth ranging from 1.25 to 20 MHz will provide support for users with different capabilities. These requirements will be fulfilled by employing

* In LTE, the base station is called enhanced NodeB (eNodeB).

new technologies for cellular environments, such as orthogonal frequency division multiplexing (OFDM) or multiantenna schemes (3GPP 36.201).

Additionally, channel variations in the time/frequency domain are exploited through link adaptation and frequency-domain scheduling, giving a substantial increase in spectral efficiency. In order to support transmission in paired and unpaired spectra, the LTE air interface supports both frequency division duplex (FDD) and time division duplex (TDD) modes.

This chapter presents a detailed description of the LTE radio-interface physical layer. For that purpose, this section provides an introduction to the physical layer, focusing on the physical resource structure and the set of procedures defined within this layer. Link adaptation techniques, including adaptive modulation, channel coding, and channel aware scheduling, are described in Section 3.2. The topic of multiple antenna schemes for LTE is tackled in Section 3.3. Finally, Section 3.4 addresses other physical layer procedures like channel estimation, synchronization, and random access.

3.1.1 Physical Resource Structure

Physical resources in the radio interface are organized into radio frames. Two radio frame structures are supported: type 1, applicable to FDD, and type 2, applicable to TDD. Radio frame structure type 1 is 10 ms long and consists of 10 subframes of length $T_{subframe}$ = 1 ms. A subframe consists of two consecutive time slots, each of 0.5 ms duration, as depicted in Figure 3.1.

One slot can be seen as a time-frequency resource grid, composed of a set of OFDM subcarriers along several OFDM symbol intervals. The number of OFDM subcarriers (ranging from 128 to 2,048) is determined by the transmission bandwidth, whereas the number of OFDM symbols per slot (seven or six) depends on the cyclic prefix length (normal or extended). Hereinafter, we will assume a radio frame type 1 with normal cyclic prefix (i.e., seven OFDM symbols per slot).

In cases of multiple antenna schemes, a different resource grid is used for each transmit antenna. The minimum resource unit allocable by the scheduler to a user is delimited by a physical resource block (PRB). Each PRB corresponds to 12

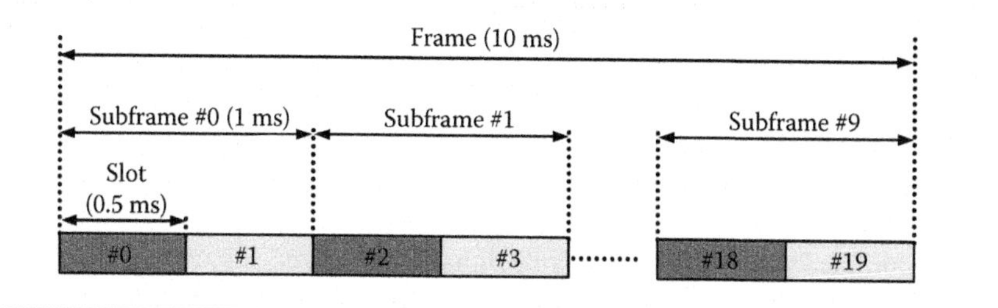

Figure 3.1 Radio frame structure type 1.

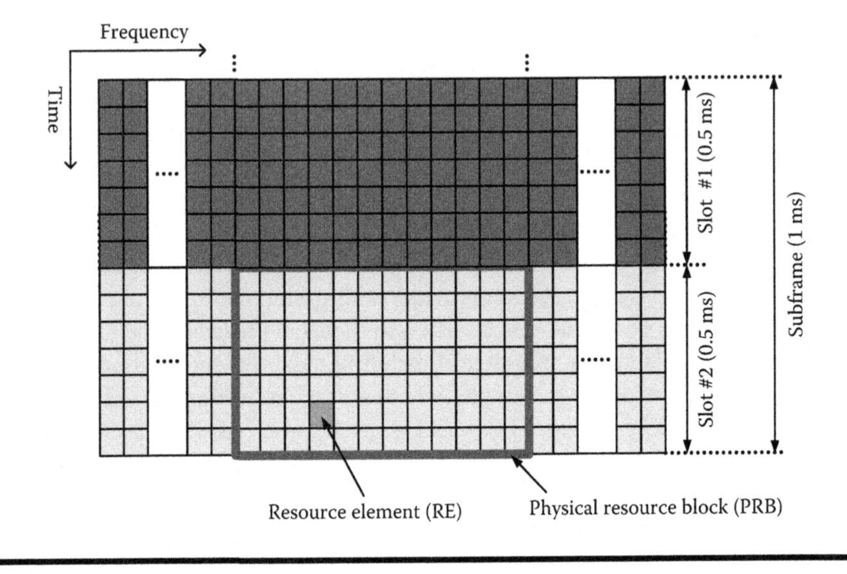

Figure 3.2 Subframe structure.

consecutive subcarriers for one slot (i.e., a PRB contains a total of 7 × 12 resource elements (REs)* in the time-frequency domain, as shown in Figure 3.2).

3.1.2 Reference Signals

In LTE downlink, special reference signals are used to facilitate downlink channel estimation procedures. In the time domain, reference signals are transmitted during the first and third last OFDM symbols of each slot. In the frequency domain, reference signals are spread over every six subcarriers. Therefore, an efficient channel estimation procedure may apply a two-dimensional time-frequency interpolation to provide an accurate estimation of the channel frequency response within the slot time interval.

When a downlink multiantenna scheme is applied, one reference signal is transmitted from each antenna in such a way that the mobile terminal is able to estimate the channel quality corresponding to each path. In this case, reference signals corresponding to each antenna are transmitted on different subcarriers so that they do not interfere with each other. In addition, resource elements used for transmitting reference signals on a specific antenna are not reused on other antennas for data transmission. An example of reference signal allocation for two-antenna downlink transmission is illustrated in Figure 3.3.

The complex values of the reference signals in the time-frequency resource grid are generated as the symbol-by-symbol product of a two-dimensional orthogonal

* A resource element corresponds to an OFDM subcarrier in the time-frequency grid.

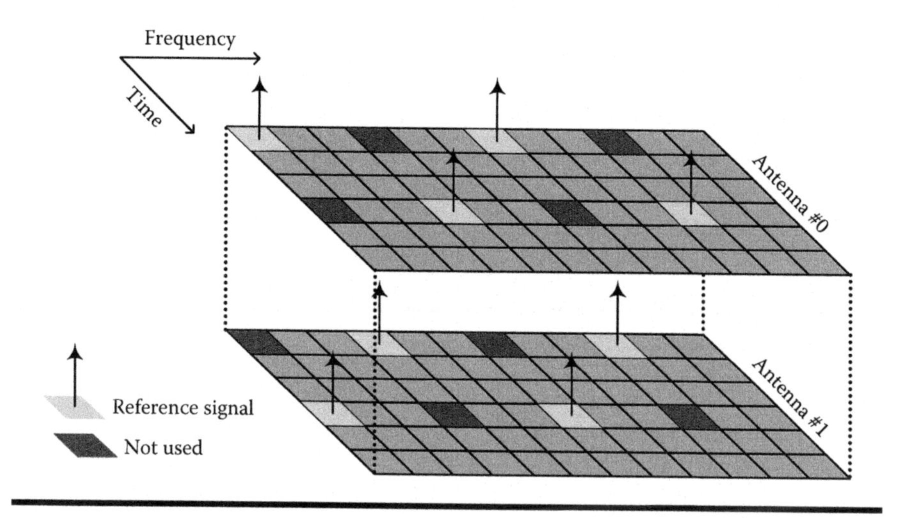

Figure 3.3 Example of reference symbol allocation for two-antenna transmission in downlink.

sequence and a two-dimensional pseudorandom sequence. This two-dimensional reference signal sequence also determines the cell identity to which the terminal is connected. There are 504 reference signal sequences defined in the LTE specification, corresponding to 504 different cell identities. Additionally, frequency shifting is applied to the reference signals in order to provide frequency diversity.

Reference signals are also used in the uplink to facilitate coherent demodulation (then called *demodulation signals*) as well as to provide channel quality information for frequency dependent scheduling (referred to as *sounding signals*). Demodulation signals are transmitted on the fourth OFDM symbol of each UL slot along the transmission bandwidth allocated to a particular user, whereas sounding signals utilize a larger bandwidth to provide channel quality information on other frequency subcarriers. The uplink reference signals are based on constant amplitude zero autocorrelation (CAZAC) sequences. Further details on CAZAC sequences can be found in Section 3.4.1.1.

3.1.3 Synchronization Signals

The base station periodically sends synchronization signals in the downlink so that the mobile terminals may be always synchronized. These signals also help the terminal during the cell search and handover procedures. Synchronization signals consist of two portions:

- *Primary synchronization signals* are used for timing and frequency acquisition during cell search. The sequence used for the primary synchronization signal is generated from a frequency-domain Zadoff–Chu sequence (see

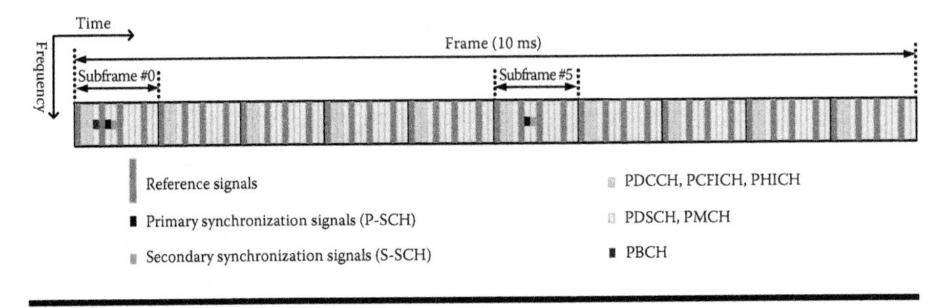

Figure 3.4 Downlink radio frame structure.

Section 3.4 for further details) and transported over the primary synchronization channel (P-SCH).

- *Secondary synchronization signals* are used to acquire the full cell identity. They are generated from the concatenation of two 31-length binary sequences. Secondary synchronization signals are allocated on the secondary synchronization channel (S-SCH).

Synchronization signals are transmitted on the 72 center subcarriers (around the DC subcarrier) within the same predefined slots, in the last two OFDM symbols in the first slot of subframes 0 and 5 (twice per 10 ms). Figure 3.4 shows the time allocation of both the synchronization and reference signals within the downlink radio frame.

3.1.4 Physical Channels

LTE supports a wide set of physical channels that are responsible for carrying information from higher layers (both user data and control information). The complete set of downlink/uplink physical channels, together with a brief explanation of their purpose, is listed in Table 3.1. A simplified diagram showing the location of the physical channels and signals in the radio frame is provided in Figure 3.4 (downlink) and Figure 3.5 (uplink).

For each physical channel, specific procedures are defined for channel coding, scrambling, modulation mapping, antenna mapping, and resource element mapping. A general structure of the whole downlink processing sequence is illustrated in Figure 3.6. This sequence is determined by the following processes:

1. *Cyclic redundancy check (CRC).* The first step in the processing sequence is the CRC attachment. A CRC code is calculated and appended to each transport block (TB),* thus allowing for receiver-side detection of residual

* A transport block is defined as the data accepted by the physical layer to be jointly encoded.

Table 3.1 LTE Physical Channels

Direction	*Channel Type*	*Description*	
Downlink	PDSCH	Physical downlink shared channel	Carries downlink user data from upper layers as well as paging signaling
	PMCH	Physical multicast channel	Used to support point-to-multipoint multimedia broadcast multicast service (MBMS) traffic
	PBCH	Physical broadcast channel	Used to broadcast a certain set of cell or system-specific information
	PCFICH	Physical control format indicator channel	Determines the number of OFDM symbols used for the allocation of control channels (PDCCH) in a subframe
	PDCCH	Physical downlink control channel	Carries scheduling assignments, uplink grants, and other control information; the PDCCH is mapped onto resource elements in up to the first three OFDM symbols in the first slot of a subframe
	PHICH	Physical HARQ indicator channel	Carries the hybrid automatic repeat request (HARQ) ACK/NAK
Uplink	PUSCH	Physical uplink shared channel	Carries uplink user data from upper layers; resources for the PUSCH are allocated on a subframe basis by the scheduler
	PUCCH	Physical uplink control channel	PUCCH carries uplink control information, including channel quality indication (CQI), HARQ ACK/NACK, and uplink scheduling requests
	PRACH	Physical random access channel	Used to request a connection setup in the uplink

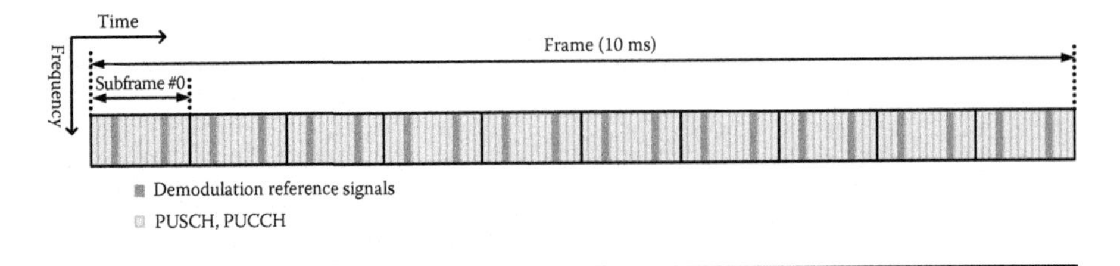

Figure 3.5 Uplink radio frame structure.

errors in the decoded TB. The corresponding error indication, reported via uplink, can be used by the downlink hybrid automatic repeat request (HARQ) protocol to perform a retransmission.

2. *Channel coding.* The goal of this process is to increase reliability in the transmission by adding redundancy to the information vector, resulting in a longer vector of coded symbols. This functionality includes code block segmentation, turbo or convolutional coding (depending on the channel type), rate matching, and code block concatenation (see Section 3.2 for further details).
3. *Scrambling.* Scrambling of the coded data helps to ensure that the receiver-side decoding can fully utilize the processing gain provided by the channel coding. Scrambling in LTE downlink consists of multiplying (exclusive or operation) the sequence of coded bits (taken as input) by a bit-level scrambling sequence.
4. *Modulation mapping.* The block of scrambled bits is modulated into a block of complex-valued modulation symbols. Because LTE uses adaptive modulation and coding (AMC) to improve data throughput, the selected modulation scheme is based on the instantaneous channel conditions for each user. The allowed modulation schemes for downlink and uplink are shown in Table 3.2.
5. *Antenna mapping.* Signal processing related to multiantenna transmission is performed at this stage. This procedure is responsible for mapping and

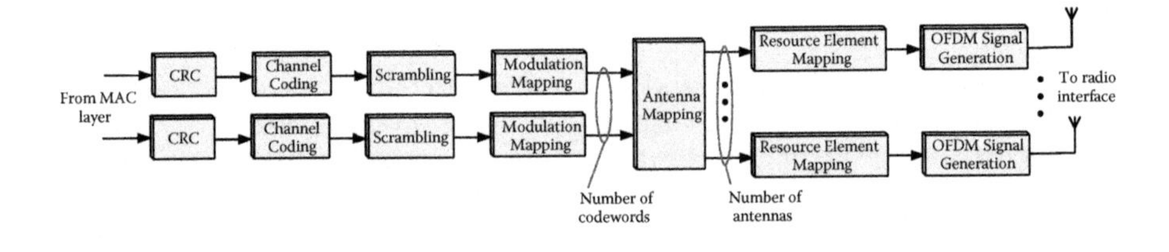

Figure 3.6 Physical layer processing sequence in the base station (downlink).

Table 3.2 Modulation Schemes

Direction	*Channel*	*Modulation Schemes*
Downlink	PDSCH	QPSK, 16 QAM, 64 QAM
	PMCH	QPSK, 16 QAM, 64 QAM
	PBCH	QPSK
	PCFICH	QPSK
	PDCCH	QPSK
	PHICH	BPSK
Uplink	PUSCH	QPSK, 16 QAM, 64 QAM
	PUCCH	BPSK, QPSK
	PRACH	*u*th Root Zadoff–Chu

precoding the modulation symbols to be transmitted onto the different antennas.* The antenna mapping can be configured in different ways to provide different multiantenna schemes, including transmit diversity, beam forming, and spatial multiplexing. More details—regarding the antenna mapping procedure and, in general, related to LTE multiantenna schemes—are provided in Section 3.3.

6. *Resource element mapping.* Modulation symbols for each antenna will be mapped to specific resource elements in the time-frequency resource grid.
7. *OFDM signal generation.* The last step in the physical processing chain is the generation of time-domain signals for each antenna, which is addressed in detail in the next section.

In addition to the functionalities previously described, HARQ with soft combining is jointly used with CRC and channel coding to allow the terminal to request retransmissions of erroneously received transport blocks. When a (re)transmission fails, incremental redundancy is used to enable the combination of successively received radio blocks until the full block is correctly decoded.

Previous functionalities may be dynamically configured, depending on the type of physical channel being processed. As an example, the downlink control channels (physical downlink control channel—PDCCH) are convolutionally encoded and use a transmit diversity configuration in the antenna mapping functionality. However, for downlink data channels (i.e., physical downlink shared channel—PDSCH—or physical multicast channel—PMCH), turbo coding is applied and other multiantenna schemes such as spatial multiplexing may be employed.

* LTE supports up to four transmit antennas.

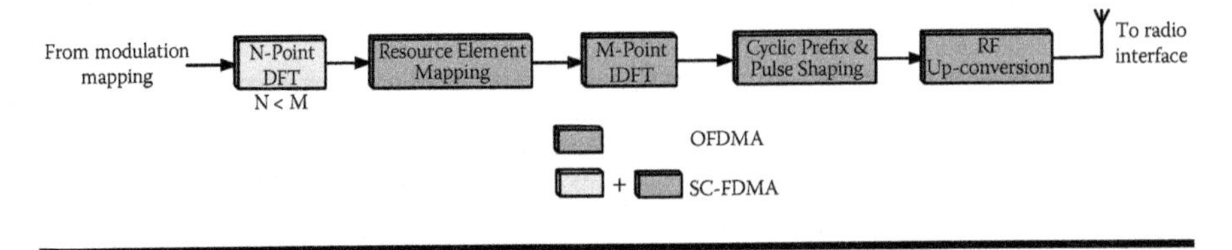

Figure 3.7 Processing sequence in the transmission modem (uplink and downlink).

Also, some of the physical layer functionalities are adapted to the higher layer (e.g., scheduling) decisions in order to perform link adaptation or channel-aware scheduling. The modulation and coding scheme, resource allocation, and multiantenna-related information are employed to configure the corresponding functionalities. Therefore, a cross-layer design is required to support efficient and adaptive physical layer functionalities.

3.1.5 OFDM/SC-FDMA Signal Generation

The LTE downlink transmission scheme is based on orthogonal frequency division multiple access (OFDMA), which is a multiuser version of the OFDM modulation scheme. In the uplink, single carrier frequency division multiple access (SC-FDMA) is used, which can be also viewed as a linearly precoded OFDM scheme known as discrete Fourier transform (DFT)-spread OFDM. However, SC-FDMA has been selected for the uplink due to the lower peak-to-average power ratio (PAPR) of the transmitted signal compared to OFDM. Low PAPR values benefit the terminal in terms of transmit power efficiency, which also translates into increased coverage.

The processing sequence in the signal generation process is quite similar in downlink and uplink, as shown in Figure 3.7. The main difference comes from the elimination of the antenna mapping process and the addition of a DFT-spread block, which is the key process for the PAPR reduction.

The following processing blocks are involved in the signal generation sequence:

- *N-point DFT.* This block performs the DFT-spread OFDM operation, which provides a high degree of similarity with the downlink OFDM scheme and the possibility to use the same system parameters.
- *Resource element mapping.* Modulation symbols for each antenna will be mapped to specific resource elements in the time-frequency resource grid.
- *M-point inverse discrete Fourier transform (IDFT).* OFDM symbols are converted into the time domain to be transmitted through the air interface.

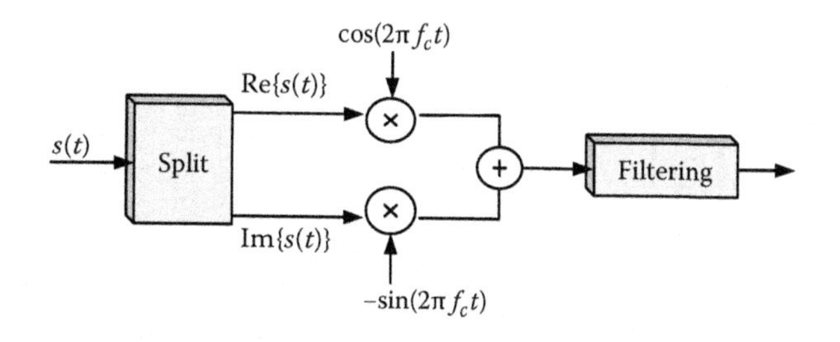

Figure 3.8 Modulation and up-conversion.

- *Cyclic prefix (CP).* For each OFDM symbol to be transmitted, the last N_{cp} samples of the symbol are prepended to the symbol as a cyclic prefix with the aim of improving the robustness to multipath effect.
- *RF (radio frequency) up-conversion.* Complex-valued OFDM base-band signal for each antenna is mapped to the carrier frequency (f_c), as depicted in Figure 3.8.

In general, the use of OFDM in both downlink and uplink (in a DFT-spread OFDM form) is an adequate transmission scheme for several reasons:

- OFDM improves the spectral efficiency of the system.
- OFDM provides a high degree of robustness against channel frequency selectivity.
- Frequency-domain link adaptation and scheduling are allowed.
- LTE can easily support different spectrum allocations by varying the number of OFDM subcarriers.

Table 3.3 summarizes the main OFDM modulation parameters* for different spectrum allocations. As mentioned before, LTE supports a scalable transmission bandwidth that leads to a different number of occupied subcarriers (assuming the same subcarrier spacing). Two different CP lengths are allowed. A longer CP is more suitable for environments with very extensive delay spread (e.g., very large cells). Equivalently, for multimedia/multicast single-frequency network (MBSFN)-based transmission, all enhanced NodeBs are transmitting the same signal to the terminals, which increase the delay spread of the equivalent channel. In this particular case, the extended cyclic prefix is therefore typically needed. However, a

* In FDD configuration, the uplink uses the same frame structure and modulation parameters as the downlink.

Table 3.3 OFDM Modulation Parameters

Transmission BW	*1.4 MHz*	*3 MHz*	*5 MHz*	*10 MHz*	*15 MHz*	*20 MHz*
Slot duration	0.5 ms (Two-slot compound; 1-ms subframe)					
Subcarrier spacing	15 kHz					
Sampling frequency	1.92 MHz (1/2 × 3.84 MHz)	3.84 MHz	7.68 MHz (2 × 3.84 MHz)	15.36 MHz (4 × 3.84 MHz)	23.04 MHz (6 × 3.84 MHz)	30.72 MHz (8 × 3.84 MHz)
DFT size	128	256	512	1024	1536	2048
Number of occupied subcarriers[a]	85	181	301	601	901	1201
Number of OFDM symbols per subframe (short/long CP)	Seven symbols (short CP) or six symbols (long CP) per slot					
CP length (μs/sample) short	(4.69/9) × 6, (5.21/10) × 1	(4.69/18) × 6, (5.21/20) × 1	(4.69/36) × 6, (5.21/40) × 1	(4.69/72) × 6, (5.21/80) × 1	(4.69/108) × 6, (5.21/120) × 1	(4.69/144) × 6, (5.21/160) × 1
CP length (μs/sample) Long	(16.67/32)	(16.67/64)	(16.67/128)	(16.67/256)	(16.67/384)	(16.67/512)

[a] Including DC subcarrier.

longer CP is less efficient from an overhead point of view, so we have to consider these two factors simultaneously.

3.2 Link Adaptation Techniques

In LTE, as in many other cellular communication systems, the quality of the received signal depends, among other factors, on the path loss, the presence of interfering signals, the sensitivity of the receiver, and the multipath propagation phenomenon. All these phenomena (that are inherent to wireless environments) may cause strong variations in the quality of the received signal. The main purpose of link adaptation techniques is to compensate such variations in order to guarantee the required quality of service (QoS) of each UE (e.g., user data rate, packet error rate, latency) as well as to maximize the system throughput. Several techniques used in LTE to cope with channel variations will be introduced in the next subsections.

In the uplink, the different UEs transmit to a common receiver, the eNodeB, which is able to measure the quality of the receiver signals. Based on these measurements (among other factors), the eNodeB makes decisions regarding scheduling, selection of modulation and coding scheme, or transmission power control.

In the downlink, adaptation techniques require certain feedback information from the receiver in order to adapt adequately the transmitted signal to the channels' conditions. In that sense, the UE is responsible for measuring and reporting periodically the instantaneous channel quality of one (or a group of) resource block(s) to the transmitter by means of channel quality indicators (CQIs).* These indicators are designed to achieve an efficient trade-off between link-adaptation performance and uplink signaling overhead. The CQI reporting frequency is adjustable in terms of subframe units and may be defined on a per-UE or per-UE-group basis.

The reported CQIs from UEs indicating the downlink channel quality can be used at the eNodeB for the following purposes:

- selection of modulation and coding scheme;
- time/frequency selective scheduling;
- interference management; and
- transmission power control for physical channels.

The eNodeB establishes the time and frequency resources that can be used by each UE to report CQI values. In LTE, a UE is configured by higher layers to feed back a wideband CQI as well as one CQI for each sub-band in a predefined set of sub-bands (see Figure 3.9). The specifications also allow the UE to select only the M

* Information referring to multiple antenna transmission is also reported and used to adapt transmitted signal to channel conditions. These procedures are described in detail in Section 3.2.5.

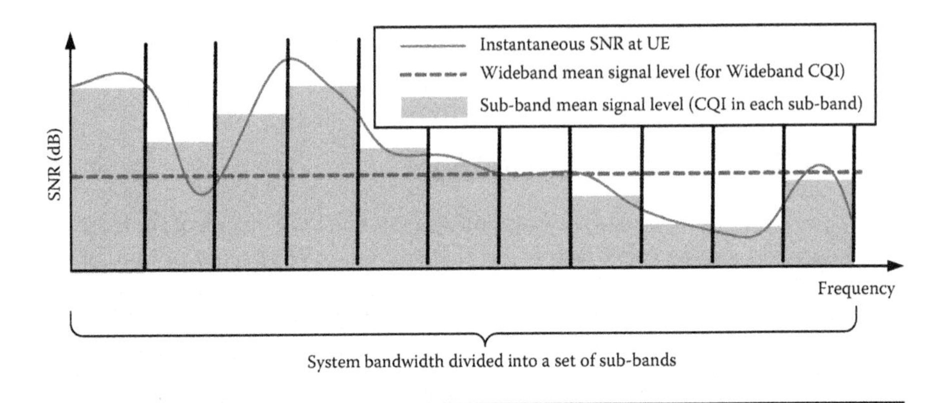

Figure 3.9 Wideband CQI versus sub-band CQI.

best sub-bands within the predefined set to be reported. In both cases, the CQI of each sub-band is encoded differentially with respect to the wideband CQI.

The CQI reporting may be periodic or aperiodic, with a minimum reporting interval of one subframe for the latter. The aperiodic report is performed by each UE after receiving an indication sent by the eNodeB in a scheduling grant. In both cases, the time and frequency resources that can be used by the UE to report are controlled by the eNodeB.

Although beyond the scope of this section, an important practical issue related to CQI reporting is worthy of mention. Due to several factors, a received CQI may be outdated, inexact, or even corrupted, which potentially leads to a degradation of system performance (Morales-Jimenez et al. 2008). For instance, potential delays introduced by the feedback channel imply that reported CQIs from the receivers do not match the current channel state at the time of transmission. In addition, an inexact channel estimation, quantification errors, or potential errors introduced in the feedback channel may lead to an erroneous CQI at the transmitter end.

3.2.1 Adaptive Modulation and Coding

Different-order quadrature amplitude modulation (QAM) modulations have been chosen for LTE, ranging from binary phase key shifting (BPSK) to 64 QAM (see Table 3.2). Higher-order modulations allow for achieving higher data rates at the expense of decreasing robustness (i.e., they are suitable under good channel conditions).

The adaptive modulation technique is aimed to improve data throughput while keeping the bit error rate (BER) below a predefined target value (BER_T). In order to achieve that, the transmitted signal is modified according to the instantaneous channel quality.

Additionally, the coding scheme may be dynamically modified to match the instantaneous channel conditions for each user, then being denoted as AMC. In this case, both modulation and coding schemes are jointly changed by the transmitter to adapt the transmitted signal to the varying channel conditions in both time and frequency domains. The coding scheme establishes the amount of redundancy that is added to the transmitted data in order to increase transmission reliability. The ratio between original data length and coded data length establishes the effective coding rate R. Hence, good link quality conditions lead to the use of a high-order modulation scheme (more bits per symbol) as well as a low coding rate (less redundancy on transmitted data). In contrast, lower modulation schemes together with higher coding rates are needed under poor link conditions in order to maintain the requested QoS level. In LTE specifications (3GPP 36.213), a predefined set of modulation and coding schemes (MCSs) corresponding to each CQI level is defined.

3.2.2 Channel Coding

Since Claude Shannon published his landmark paper (1948, 379), channel coding has been widely used in digital communications to increase reliability in the transmission through a nonideal channel. Channel coding techniques make it possible to detect and correct some of the errors introduced by the channel at the expense of adding redundant information to the transmitted data.

The complete channel coding scheme may be seen as a complex process in which error detection/correction techniques, rate matching, and channel interleaving are applied to transmitted information before being mapped onto the physical media (see Figure 3.10). The channel coding process chain in LTE (3GPP 36.212) is described in the next subsections.

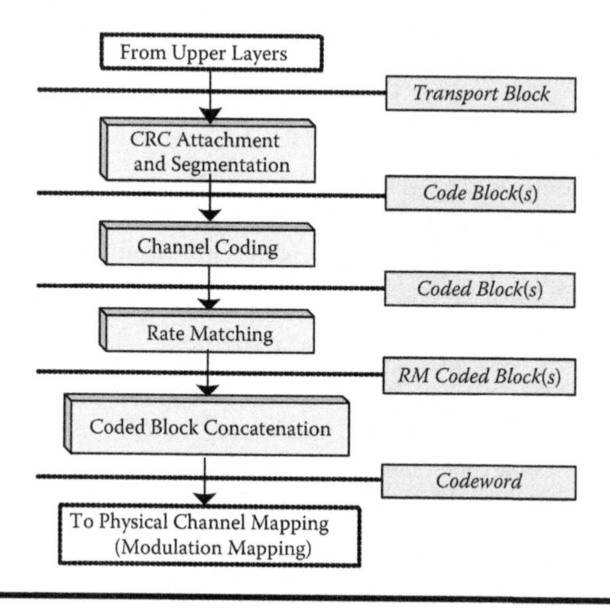

Figure 3.10 Simplified channel coding scheme in LTE.

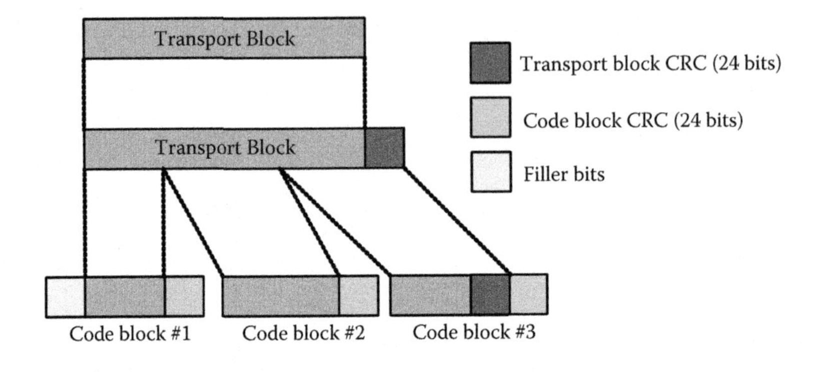

Figure 3.11 Detail of the segmentation of a transport block after CRC bits are appended.

3.2.3 *CRC and Segmentation*

A fixed 24-b CRC is calculated for each transport block coming from layer 2. The calculated CRC is appended to the transport block in order to check the data integrity at the receiver end. If the input sequence length is shorter than 40 bits, filler bits are added to the beginning of the code block. Otherwise, if the resulting bit sequence is longer than the maximum allowed code block size (6144 bits), a segmentation process must be carried out. After the segmentation, filler bits are added to the last segment, if needed. Finally, an additional CRC sequence of 24 b is then attached to each resulting segment. Otherwise, if the input sequence length is shorter than 40 b, filler bits are added to the beginning of the code block. The complete process is depicted in Figure 3.11.

3.2.4 *Coding Techniques in LTE*

Two different coding techniques have been chosen for channel coding in LTE: tail biting convolutional coding for control channels and turbo coding for data channels. The final bit sequence obtained as a result of the coding processes is referred to as a *coded block.*

3.2.4.1 *Tail Biting Convolutional Coding*

Convolutional codes, first introduced in Elias (1954, 29), allow for achieving a coding rate of $1/m$ because a set of m coded bits $(d_k^{(0)} \ldots d_k^{(m-1)})$ is generated for each input bit, c_k. Resulting coded bits depend on both the current bit and a number of previous bits determined for the constraint length, k, of the encoder. In LTE, control channels are coded with a tail biting convolutional code with a coding rate 1/3 and a constraint length of 7, as depicted in Figure 3.12.

At the receiver end, a decoding process is requested to recover the original sequence of data bits. This process is usually based on the Viterbi algorithm (Viterbi

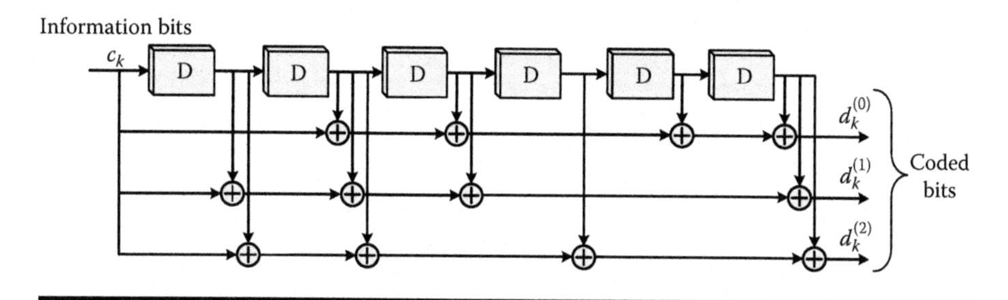

Figure 3.12 Tail biting convolutional coding scheme.

1967, 260), which has the advantage of performing the maximum likelihood decoding. Additionally, the Viterbi algorithm is highly parallelizable and therefore ideal for hardware decoder implementation.

3.2.4.2 *Turbo Coding*

Turbo codes, first proposed by Berrou, Glavieux, and Thitimajshima (1993, 1064), enable reliable communications with efficiencies close to the theoretical limit predicted by Shannon (1948, 379). Since their introduction, turbo codes have been proposed for low-power systems (e.g., satellite communications or interference limited scenarios like third-generation cellular technology).

LTE turbo encoding is based on two eight-state constituent encoders and one turbo code internal interleaver. The sequence of original input bits, c_k, is encoded by the first encoder; the input to the second encoder is an interleaved version of the original sequence, c'_k. Therefore, the resulting coded sequence is the result of combining the information sequence of bit (systematic bits, S) with the two sequences of parity bits (P1 and P2). Hence, the overall code rate is 1/3 even though the rate of each encoder is 1/2, because their outputs are combined with input data bits (see Figure 3.13).

Finally, a trellis termination operation is performed to force the encoders back to a zero initial state after coding a transport block. Once all the information bits are encoded, the tail bits from the shift register feedback in each encoder are padded after the encoded information bits. The tail bits are used to terminate each constituent encoder while the other constituent encoder is disabled (its correspondent switch in position 2).

In order to reduce the complexity in both turbo encoder and decoder, a quadratic polynomial permutation (QPP) interleaver is used instead of release-6 turbo interleaver (3GPP 36.212). This kind of interleaver (Takeshita 2006, 1249) holds the maximum contention-free property. That is a great advantage from a hardware implementation perspective because contention in interleaver memory access is avoided and, therefore, it is possible to parallelize decoding with a single extrinsic memory. The reason for this interleaving process is that the performances of

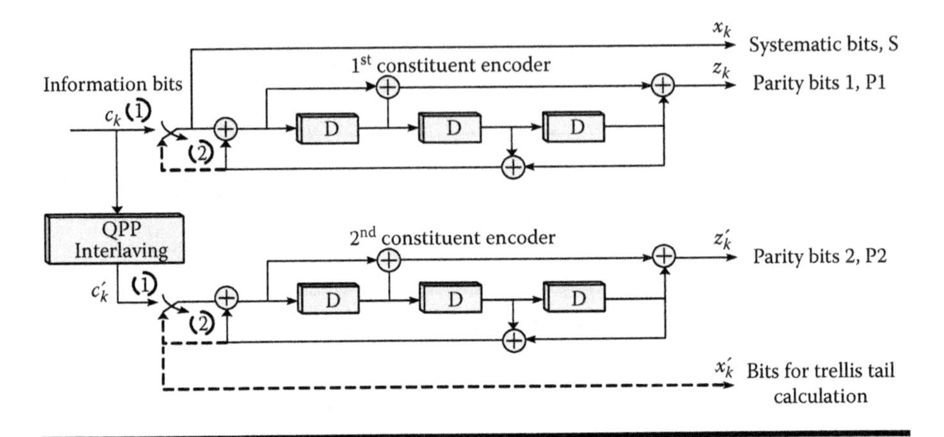

Figure 3.13 Turbo code coding scheme proposed for LTE.

turbo and convolutional codes are improved when the errors introduced by the radio channel are statistically independent because they are random error correcting codes.

3.2.4.3 Rate Matching

The purpose of the rate matching process is to adjust the coded block size to the allocated physical resources. The first step in this process consists of an interleaving of the systematic and parity bits in the coded block provided as an output of the turbo encoder process. This interleaving process is aimed to mitigate the effects of error bursts likely to appear in radio mobile channels due to the time-variant fading effects. Consequently, a dual de-interleaving operation is needed at the receiver side to recover original coded blocks. The result of the interleaving process is stored into a circular buffer where a puncturing process is applied to decrease the effective coding rate according to the MCS corresponding to the considered CQI level.

The complete rate matching process is depicted in Figure 3.14. This process allows different coding rates to be achieved (i.e., a higher coding rate to improve robustness against the effects of the channel at the expense of a lower spectral efficiency, or vice versa). Additionally, the rate matching process is used in LTE to give support to the HARQ protocol, as described in Section 3.2.5.

3.2.4.4 Coded Block Concatenation

This process consists of the sequential concatenation of the rate matching outputs for the different code blocks. The result—namely, a code word—is the coded version of the original transport block to be transmitted.

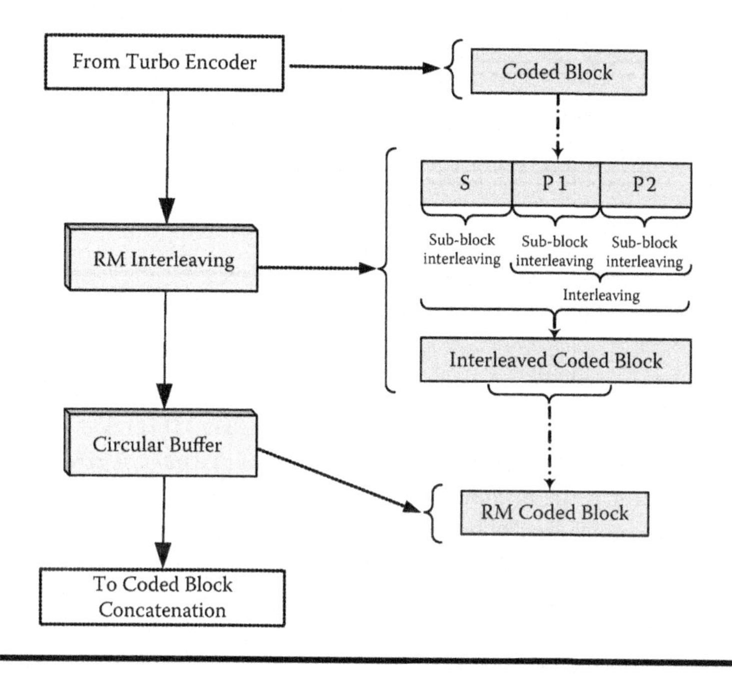

Figure 3.14 Detail of interleaving process for a code word generated with turbo codes in LTE.

3.2.5 *Hybrid Automatic Repeat Request (HARQ)*

Hybrid automatic repeat request, a combination of traditional ARQ and error correction codes (such as the turbo code introduced in a previous subsection), is a common element in the physical layer of current mobile communication systems such as HSPA (high-speed packet access) or LTE. At the expense of a higher complexity in its implementation than required for a traditional ARQ strategy, HARQ achieves a better performance than ordinary ARQ techniques. In addition to the error-detection information added to the data, forward error correction (FEC) bits are appended before transmission. Whenever an error is detected, hybrid ARQ is requested to perform a retransmission.

Typically, two types of HARQ retransmission strategies are implemented. In the first one, known as *chase combining* or *type I HARQ*, each retransmission is identical to the original transmission. Once a transport block is received, its integrity is checked by calculating its CRC and comparing it with the CRC sequence. A decoding error raises a retransmission request that is answered by the transmitter by sending again the same transport block. An incorrectly received transport block is stored at the receiver in order to be combined with the retransmitted transport block to increase the probability of successful decoding (see Figure 3.15). Different criteria, such as the link quality associated with each transport block, may be applied to decide how the transport blocks are combined.

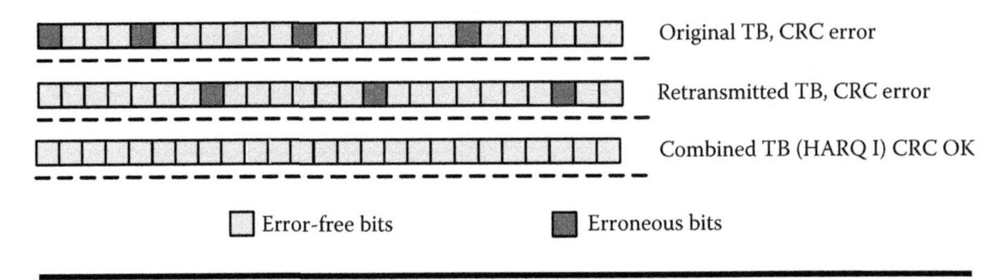

Figure 3.15 Example of chase combining HARQ.

In the second one, known as *incremental redundancy* or *type II HARQ*, a coded transport block (i.e., a code word) is sent to the receiver. This code word is compounded by most of the systematic bits and a few parity bits for error correction. If an error occurs in the decoding process, a retransmission request is sent to the transmitter. Additional parity bits are transmitted and combined upon arrival with the previous received version of the coded transport block, resulting in a lower coding rate. The process is repeated until a successful decoding is performed or until the transport block is discarded when the retransmission limit is reached (see Figure 3.16).

Incremental redundancy has the potential of achieving a better performance than chase combining (Frenger, Parkvall, and Dahlman 2001, 1829). However, the former implies a greater complexity in its implementation than the latter and requires a larger receiver buffer size.

In LTE, HARQ retransmission strategy is only supported for share channels in uplink (physical uplink shared channel—PUSCH) and downlink (PDSCH), and it is based on multiple parallel stop-and-wait processes. Transport blocks are decoded upon arrival. An acknowledge (ACK) command is reported to the transmitter when a transport block is successfully decoded. On the other hand, a negative acknowledge (NACK) is reported in cases of unsuccessful decoding, implying a retransmission request for the particular transport block. Once an ACK/NACK message is received at the transmitter, its timing is used to associate it with its correspondent HARQ process.

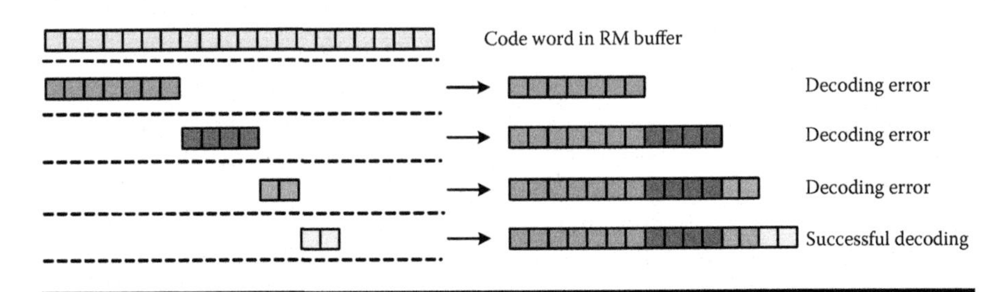

Figure 3.16 Example of incremental redundancy HARQ at code word level.

LTE downlink and uplink HARQ operations are based on different protocols. On the one hand, downlink HARQ is based on an asynchronous protocol, in which retransmissions may occur at any time. Additionally, each process is identified with an explicit HARQ process number. On the other hand, uplink HARQ is based on a synchronous protocol, in which the retransmission instant is fixed by the initial transmission time and therefore the process number can be implicitly derived from it.

3.2.6 Channel-Aware Scheduling

A fundamental characteristic of wireless channels is the fading of the received signal due to constructive and destructive interference between multipaths. Consequently, the short-term capacity of a channel changes randomly over time. In an OFDM-based technology like LTE, the signal level associated with each OFDM subcarrier also varies along the frequency domain. Additionally, in a multiuser environment, channel quality varies asynchronously for different users. Such variations in (time-frequency) channel conditions can be exploited to increase system throughput. An opportunistic scheduler assigns the channel to the user with the best channel conditions at a given time. By exploiting this transmission multiuser diversity, one can achieve higher total system capacity as the number of users increases (Entrambasaguas et al. 2007, 24).

If we consider the QoS requirements of the different data flows, they cannot be guaranteed by assigning a percentage of time to each user (i.e., round-robin policy) because each user may experience a different average link capacity. Hence, advanced scheduling algorithms have been defined to consider different aspects—such as the instantaneous channel conditions, QoS requirements, pending retransmissions, or UE capabilities—in order to allocate transmission turns to users adequately. In LTE, the eNodeB is responsible for implementing the scheduling policies in both downlink and uplink scenarios. The UEs are informed through control channels about the resource allocation decisions on a subframe rate.

An example of how channel information is used to manage the resource allocation among users in the downlink is shown in Figure 3.17. In this figure, the upper three-dimensional graph depicts the time-frequency evolution of the instantaneous signal-to-noise ratio (SNR) experienced by two different users. These SNR values will determine the CQI levels to be reported to the eNodeB, as described in previous sections. The reported information is used at the eNodeB scheduler to decide the time-frequency resources used to transmit to each UE (depicted in the lower graph). As it was noted at the beginning of this section, the information about quality of the link, in terms of CQI, may not match exactly the current state of the channel. In order to mitigate this source of error in the scheduling process, a predictive algorithm might be applied.

In order to achieve a higher system performance in LTE, scheduling is tightly integrated with link adaptation and HARQ processes.

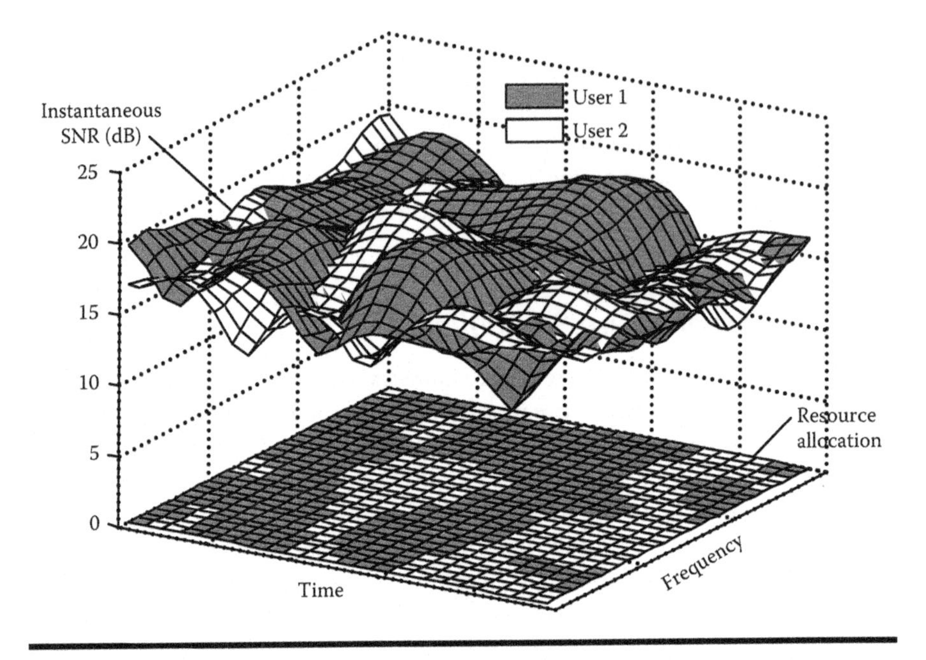

Figure 3.17 Example of channel-aware scheduling for two users.

3.3 Multiple Antennas

3.3.1 Introduction

The use of multiple antennas at the transmitter or the receiver side in combination with signal processing may significantly improve the system performance. Multiple antenna technologies are necessary to achieve the initial LTE requirements in terms of coverage, QoS, and targeted data rates (3GPP 25.913). Some of the LTE key targets, such as the number of supported users per cell or user peak data rates, cannot be fulfilled without considering multiple antenna schemes. Thus, rather than being a later addition to the standard, multiple antenna techniques are part of the LTE standardization.

The radio link is typically affected by the well-known multipath phenomenon (Goldsmith 2005), which produces constructive and destructive interferences (or signal fading) at the receiver. By having multiple antennas at the transmitter or the receiver side, multiple radio "channels" or paths are established (i.e., one radio path between each transmit and receive antenna). Different paths will ideally experience uncorrelated fading whether the distance between antenna elements is relative* or by applying different polarization directions to the transmit antennas. These different

* Distance between antenna elements to achieve uncorrelated fading depends on both the carrier frequency of the radio communication and the geometry of the deployment scenario.

paths may be exploited in different ways in order to obtain spatial diversity from the ideally uncorrelated fading or to transmit multiple streams simultaneously.

The basic principle of multiantenna diversity is to provide the receiver with multiple copies of the transmitted signal in order to combat instantaneous fading on the radio channel. Spatial diversity may be achieved by having multiple antennas at either the transmitter or the receiver side. A large group of so-called *receive diversity* schemes (Hutter et al. 2000, 707) takes advantage of the multiple receive antennas in order to improve the effective SNR. However, receive diversity is out of the scope of this chapter and we will focus on *transmit diversity* schemes, where multiple antennas are applied at the transmitter (Kaiser 2000, 1824; Goldsmith 2005). In general, multiantenna diversity plays a significant role in extending coverage and enhancing link robustness.

Additionally, multiple antennas may be employed to shape the overall antenna radiation pattern in a way that maximizes the transmitter or receiver beam in a desired direction (e.g., target receiver or transmitter). These schemes are referred to as *classical beam forming* or smart antenna systems and may be used to increase the received signal strength or to suppress dominant interfering signals. Therefore, the use of beam-forming techniques improves the SINR (signal-to-interference-plus-noise ratio), which results in coverage and, eventually, a capacity enhancement.

In both transmit diversity and beam forming, one data stream is transmitted through multiple antennas in order to improve the SNR or SINR conditions at the receiver. However, data rates are mainly bandwidth limited when SNR reaches a certain level. In these cases, spatial multiplexing (SM) techniques allow transmitting multiple data streams simultaneously by having multiple antennas at both the transmitter and the receiver sides. Spatial multiplexing schemes, sometimes referred to as MIMO* (multiple input, multiple output), result in a significant increase of spectral efficiency and data rates.

The different types of multiantenna schemes and their corresponding achievable gains are summarized in Table 3.4. Also, it is possible to distinguish between those configurations that do not need knowledge of the channel at the transmitter—namely, *open-loop* schemes—and those needing information about the channel, also called *closed-loop* schemes. Open-loop-based configurations have lower complexity and signaling overhead than closed-loop solutions, as well as lower performance. On the other hand, closed-loop-based schemes take advantage of the channel information at the transmitter by applying some signal processing prior to transmission in order to optimize data reception (e.g., weighting the transmitted signals by considering the channel conditions experienced at each antenna).

As already mentioned in Section 3.1, LTE radio resources are organized into the two-dimensional time-frequency grid provided by OFDM. When transmitting through multiple antennas, the spatial dimension is added to the grid, as shown

* The term MIMO is also used in general to refer to any transmission scheme with multiple antennas at both the transmitter and the receiver sides.

Table 3.4 Different Multiantenna Schemes and Corresponding Benefits

Tx Data Streams	*Multiantenna Scheme*	*Gain*	*Benefits*
One	Transmit diversity	Diversity gain	Link robustness Coverage
	Classical beam forming	Power or antenna gain	Coverage Capacity
Multiple	Spatial multiplexing	Capacity gain	Spectral efficiency Data rates

in Figure 3.18. The aforementioned multiantenna schemes can be applied independently to each OFDM subcarrier, which individually experiences a flat fading or non-time-dispersive channel. In LTE, multiantenna signal processing is applied independently to each frequency sub-band or group of consecutive OFDM subcarriers, thus making it possible to have different transmission schemes for different frequency scheduled users.

A multiantenna configuration is determined by the multiantenna scheme to be applied (e.g., spatial multiplexing) and other parameters such as the number of transmit and receive antennas. Different configurations may be applied for different

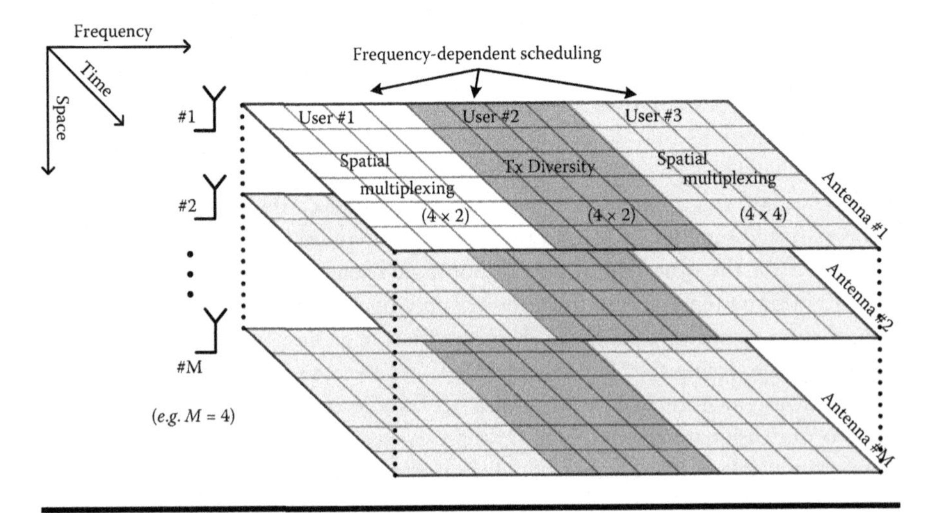

Figure 3.18 Example of LTE multiantenna transmission with M = 4 transmit antennas.

scheduled users within the same time slot, as illustrated in Figure 3.18. The baseline antenna configuration in LTE consists of two antennas at the eNodeB and two antennas at the UE, whereas other configurations, up to a maximum of four transmit and four receive antennas, are supported as well.

The LTE standard defines a set of possible multiantenna configurations to be applied in the downlink transmission. However, uplink configurations are not detailed at this stage of the specifications (3GPP 36.212). Moreover, the number and complexity of multiantenna schemes for the uplink will be considerably reduced with respect to downlink, as expected from LTE initial requirements (e.g., lower data rates for the uplink). Therefore, downlink transmission will be assumed throughout this section unless otherwise stated.

This section provides a general overview of the different multiantenna technologies considered in LTE and its related physical layer procedures. The LTE-supported schemes that will be presented in this section are classified into two groups: transmit diversity schemes, where only one data stream is transmitted with the aim of enhancing link robustness or coverage, and spatial multiplexing schemes, where multiple data streams are transmitted simultaneously in order to increase data rates. Finally, closed-loop- or precoding-based configurations will be introduced.

3.3.2 Transmit Diversity

Transmit diversity schemes rely on the use of multiple antennas at the transmitter side in order to obtain spatial diversity when transmitting one data stream. The basic principle is to provide the receiver with multiple copies of the transmitted signal as a means to combat instantaneous fading, resulting in a significant gain in instantaneous SNR. Therefore, transmit diversity is commonly applied when users experience bad channel conditions, with the objective of extending coverage (support for larger cells) or enhancing QoS issues.

In order to provide spatial diversity, it is necessary to have low mutual correlation between the channels associated with the different transmit antennas (i.e., relatively large antenna spacing). Under this assumption, both open-loop and closed-loop schemes are used to achieve spatial diversity in LTE. Open-loop schemes aim to exploit spatial diversity blindly by means of transmitting different versions of the signal through different transmit antennas. On the other hand, closed-loop schemes employ the information about the downlink channel at the transmitter to apply adequate antenna weights (i.e., phase and amplitude of the signals transmitted through different antennas are properly adjusted). In this case, multiple transmit antennas provide beam forming in addition to diversity. Hereinafter, the term *beam forming** will be used to refer to closed-loop transmit diversity schemes.

* The difference should be noted between *beam forming*, with low mutual antenna correlation, and *classical beam forming* or *smart antennas*, where diversity cannot be provided due to the high mutual antenna correlation.

In LTE, open-loop transmit diversity is achieved by means of cyclic delay diversity and space frequency block coding. The next two subsections provide more details about these diversity techniques. Additionally, closed-loop transmit diversity schemes, such as beam forming or transmit antenna selection, are considered in LTE.

3.3.2.1 Cyclic Delay Diversity (CDD)

In general, delay diversity consists of transmitting the same signal with different relative delays through the different transmit antennas (Bauch and Malik 2004, 17). This is completely transparent to the mobile terminal, which only sees a single radio link with additional time dispersion and, equivalently, additional frequency selectivity. Thus, delay diversity can be seen as a means to transform antenna or space diversity into frequency diversity, which is especially desirable in scenarios with reduced multipath propagation. Cyclic delay diversity is a special type of delay diversity, where cyclic shifts are applied (instead of linear delays) to the signal transmitted over the different antennas. An example of CDD for two-antenna configuration is illustrated in Figure 3.19a, whereas extended configurations with more than two transmit antennas can be achieved by applying a different shift to each antenna. CDD operates blockwise and thus it is applicable to block-based transmission schemes such as OFDM.

A cyclic shift of the time domain signal is equivalent to a frequency-dependent phase shift of the frequency domain OFDM symbol, as illustrated in Figure 3.19b. That is, complex modulation symbols, denoted as S_j, are mapped onto the resources of the first antenna, whereas a phase-shifted version of the same modulation symbols is mapped onto the second antenna. The fact that CDD is transparent to the UE makes this diversity technique very suitable for being combined with other multiantenna schemes. Thus far in LTE, CDD is applied in conjunction with spatial multiplexing.

3.3.2.2 Space Frequency Block Coding (SFBC)

Space frequency block coding was first proposed in Dehghani et al. (2004, 543) as an extension of traditional space-time block coding for OFDM systems. In SFBC,

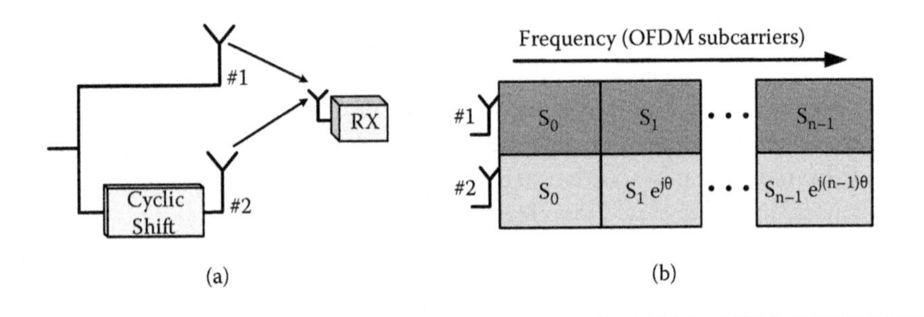

Figure 3.19 Cyclic delay diversity for two-antenna configuration.

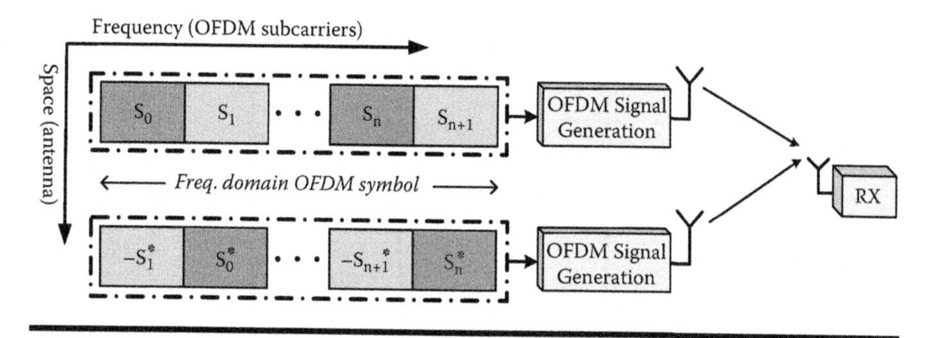

Figure 3.20 Space frequency block coding (SFBC) for two-antenna configuration.

the basic idea of the Alamouti scheme (1998, 1451) is applied in frequency domain instead of time domain. Figure 3.20 illustrates the SFBC operation for the particular two-antenna configuration. SFBC operation is performed on pairs of complex valued modulation symbols. Hence, each pair of symbols has an associated pair of frequency resources (i.e., a pair of OFDM subcarriers). Modulation symbols are mapped directly onto the available resources of the first antenna. However, mapping of each pair on the second antenna is reversely ordered, sign reversed, and complex conjugated, as illustrated in Figure 3.20.

For a proper reception of the transmitted modulation symbols, the mobile terminal must be informed about SFBC transmission and a simple linear operation (Dehghani et al. 2004, 543) has to be applied to the received signal. Contrarily to cyclic delay diversity, SFBC provides diversity on the level of modulation symbols.

3.3.3 Spatial Multiplexing

The objective of spatial multiplexing is to increase data rates by transmitting different data streams over the different parallel channels provided by the multiple transmit and receive antennas. Let us assume a system with M transmit and N receive antennas, as depicted in Figure 3.21. Then the MIMO radio channel consists of $M \times N$ ideally uncorrelated paths. This configuration theoretically provides $L = \min(M, N)$ parallel channels that allow simultaneously transmission of L data streams. Signals from transmit antennas cause interferences to each other at the receiver, and hence spatially multiplexed streams are overlapped due to propagation through the MIMO channel, as seen in Figure 3.21. Therefore, a MIMO processing stage is needed at the receiver in order to suppress interfering signals. This processing stage can be seen as a detection of the originally transmitted data streams from the received signals, denoted as y_j.

Additionally, if some knowledge of the MIMO channel is available at the transmitter, the different parallel channels can be made independent of each other,

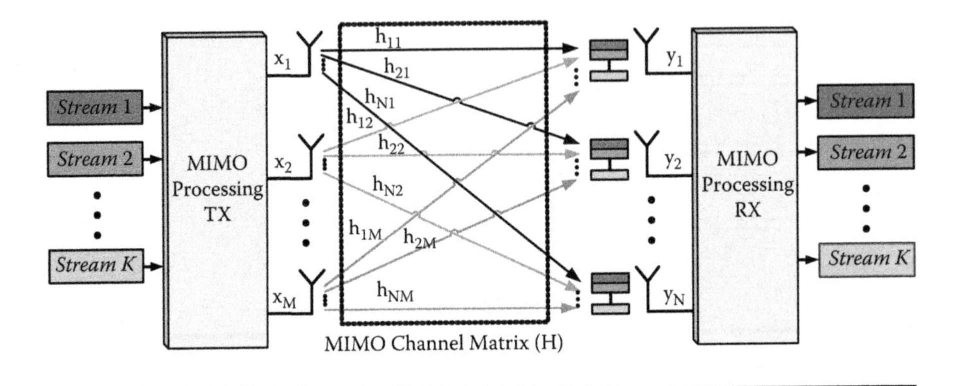

Figure 3.21 Spatial multiplexing of different data streams over different transmit antennas.

thus reducing significantly the interfering signals at the receiver. This is possible by means of closed-loop- or precoding-based spatial multiplexing, as discussed in Section 2.3.

The theoretical maximum rate increase factor is L in a rich scattering environment with high spatial diversity (i.e., uncorrelated paths). However, there is no appreciable gain with a predominant line-of-sight component or, equivalently, when a high correlation between different paths occurs. Under such conditions, the received signals have no spatial signature and the originally multiplexed signals cannot be recovered at the receiver (parallel transmission of data streams becomes unfeasible).

Therefore, spatial multiplexing is strongly conditioned by the MIMO channel matrix (**H**). From an algebraic point of view, the rank of **H** determines the maximum achievable number of spatially multiplexed signals—that is, the number of parallel channels for reliable data transmission. Assuming R to be the rank of **H,** the number of multiplexed data streams or spatial multiplexing order (K) must satisfy:

$$K \leq R \leq L \tag{3.1}$$

That is, the multiplexing gain is limited to the rank of the channel matrix. It is clear from Equation (3.1) that when **H** has full rank (i.e., $R = L$—for example, in a rich scattering scenario), the maximum theoretical gain can be achieved. However, highly correlated paths imply that **H** is ill conditioned or rank deficient, so the spatial multiplexing order (number of multiplexed data streams) may be significantly limited.

In general, spatial multiplexing schemes can be classified depending on whether the different multiplexed streams are jointly or separately coded. According to this criterion, there are single code-word* (SCW) and multiple code-word (MCW)

* Strictly speaking, a code word is defined as a coded transport block. However, the term *code word* will be used in a wide sense to refer an independently coded data stream hereinafter.

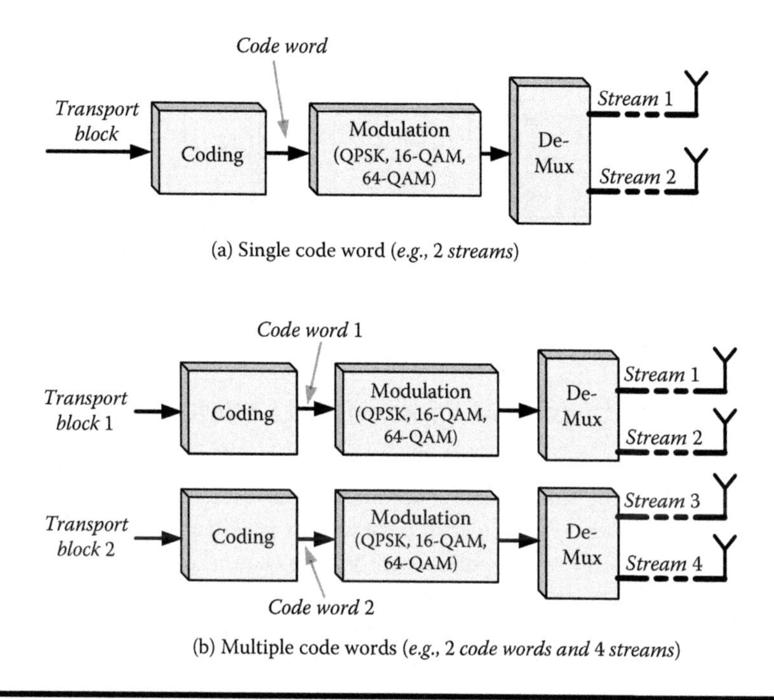

Figure 3.22 Single code word and multiple code word multiplexing schemes.

schemes. In the single code-word approach, one coded and modulated transport block is spatially multiplexed into several data streams (Figure 3.22a). In the multiple code-word case (Figure 3.22b), different coding and modulation processes are implemented for each of the transport blocks that are simultaneously transmitted. In this case, it is possible to have a different coding or modulation rate for each code word (i.e., data stream independently coded). Consequently, an MCW approach allows for applying different FEC and HARQ processes to the different code words.

In LTE, spatial multiplexing follows an MCW approach with two code words and a maximum of four transmitted data streams (Figure 3.22b). That is, two code words are simultaneously transmitted and each code word may be spatially multiplexed into one or more data streams. Therefore, the number of multiplexed streams will be limited to R, and only two code words will be allowed.

Different code words transmitted in the same time-frequency resource can be allocated either to the same user or to different users for achieving different purposes. The first case is referred to as space division multiplexing (SDM) and is applied in order to increase user data rates. On the other hand, multiple code words can also provide multiple access (i.e., different code words are allocated to different users sharing the same radio resource).

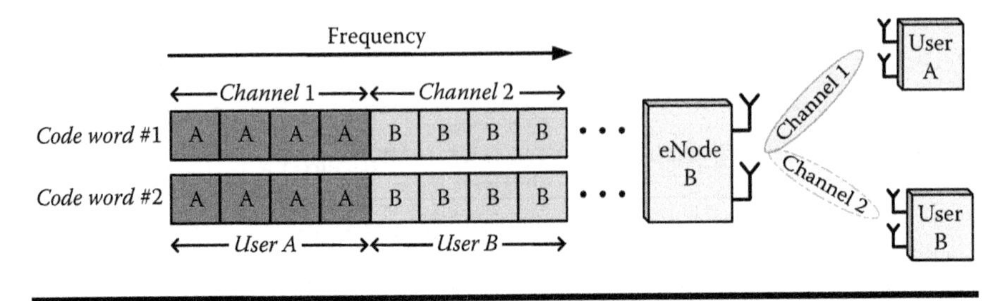

Figure 3.23 Example of single-user MIMO with two code words.

3.3.3.1 Space Division Multiplexing (SU-MIMO)

The basic idea of space division multiplexing, also referred to as single-user MIMO (SU-MIMO), is to transmit several code words to the same user in the same radio resource ("time frequency") in order to increase user peak data rates. As depicted in Figure 3.23, there is only one scheduled user on each channel or frequency sub-band.

SU-MIMO transmission allows for applying the so-called *per stream rate control* by independently adapting each of the transmitted code words. The mobile terminal reports channel quality indicators for each of the received code words; this information is used to adapt the modulation and coding rates independently per code word. Also, the fact that several streams are transmitted to the same user allows for implementing a stream-by-stream successive interference cancellation (SIC) strategy at the receiver (Wubben and Kammeyer 2006, 2183; Kyungchun, Chun, and Hanzo 2007, 2438). Successfully decoded streams are suppressed from the received signal (interference cancellation), thus avoiding error propagation in the reception of the remaining streams. Hence, the application of SIC strategies allows efficient recovery of the different data streams. SU-MIMO is very suitable for users experiencing good channel conditions and, in general, throughput is increased in the center of the cell (near the eNodeB).

3.3.3.2 Multiple Access

When multiple antennas are used to support spatially multiplexed users in the same radio resource, the term *space division multiple access* (SDMA) is adopted. This scheme, commonly referred to as multiuser MIMO (MU-MIMO) (Jiang and Hanzo 2007, 1430), is under consideration in LTE.

Figure 3.24 shows an example of MU-MIMO with two code words being transmitted by two antennas. The same time-frequency resource (or channel) is shared by users A and B (i.e., transmission on different antennas is allocated to different users). The way in which different code words are allocated to users is determined by the medium access control (MAC) layer according to different scheduling algorithms (Jiang and Hanzo 2007, 1430). MU-MIMO transmission allows increase of system capacity or, equivalently, support of a larger number of users per cell.

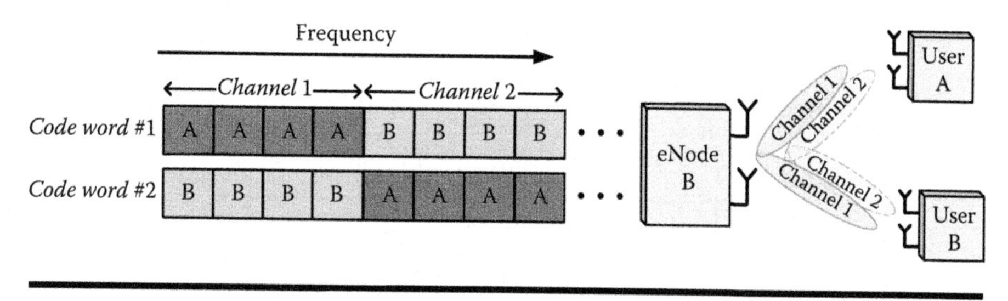

Figure 3.24 Example of multiuser MIMO with two code words.

3.3.3.3 Precoding-Based Spatial Multiplexing

In general, as discussed earlier, the number of multiplexed signals (K) will be limited to the rank of the channel matrix (R), which is less than or equal to the number of transmit antennas (M). With some information about the downlink channel at the transmitter, precoding can be applied in order to improve isolation among parallel channels. Hence, interfering signals will be suppressed and effective SINR for each of the data streams will rise, thus increasing system capacity. Precoding can be seen as a means to achieve a diagonal effective channel matrix $\mathbf{D}$ and thus to have R independent channels for transmission over M transmit antennas. Precoding is based on the singular-value decomposition (SVD) of the MIMO channel (Goldsmith 2005):

$$\mathbf{H} = \mathbf{U}\mathbf{D}\mathbf{V}^{\mathrm{H}} \tag{3.2}$$

where $\mathbf{U}$ ($N \times R$) and $\mathbf{V}$ ($M \times R$) are matrices whose columns are orthonormal* and $\mathbf{D}$ is a ($R \times R$) diagonal matrix. Additionally, the diagonal elements of $\mathbf{D}$ are the R strongest eigenvalues of $\mathbf{H}^{\mathrm{H}}\mathbf{H}$. By applying the precoding matrix $\mathbf{V}$ prior to transmission and the shaping matrix $\mathbf{U}^{\mathrm{H}}$ immediately after reception, a diagonalization of the channel matrix is attained. The vector of symbols after receiver shaping ($\tilde{\mathbf{y}}$) can be expressed by considering Equation (3.2) as follows:

$$\begin{aligned}
\tilde{\mathbf{y}} &= \mathbf{U}^{\mathrm{H}}(\mathbf{H}\mathbf{x} + \mathbf{n}) = \mathbf{U}^{\mathrm{H}}(\mathbf{U}\mathbf{D}\mathbf{V}^{\mathrm{H}}\mathbf{x} + \mathbf{n}) \\
\tilde{\mathbf{y}} &= \mathbf{U}^{\mathrm{H}}(\mathbf{U}\mathbf{D}\mathbf{V}^{\mathrm{H}}\mathbf{V}\tilde{\mathbf{x}} + \mathbf{n}) = \mathbf{U}^{\mathrm{H}}\mathbf{U}\mathbf{D}\mathbf{V}^{\mathrm{H}}\mathbf{V}\tilde{\mathbf{x}} + \mathbf{U}^{\mathrm{H}}\mathbf{n} \\
\tilde{\mathbf{y}} &= \mathbf{D}\tilde{\mathbf{x}} + \mathbf{U}^{\mathrm{H}}\mathbf{n}
\end{aligned} \tag{3.3}$$

where $\tilde{\mathbf{x}}$ is the vector of modulation symbols and $\mathbf{x}$ is the vector of precoded symbols to be transmitted on the antenna ports. It is clear from Equation (3.3) that the

* Orthonormal columns of each U and V imply $\mathbf{U}\mathbf{U}^{\mathrm{H}} = \mathrm{IN}$ and $\mathbf{V}^{\mathrm{H}}\mathbf{V} = \mathrm{IM}$.

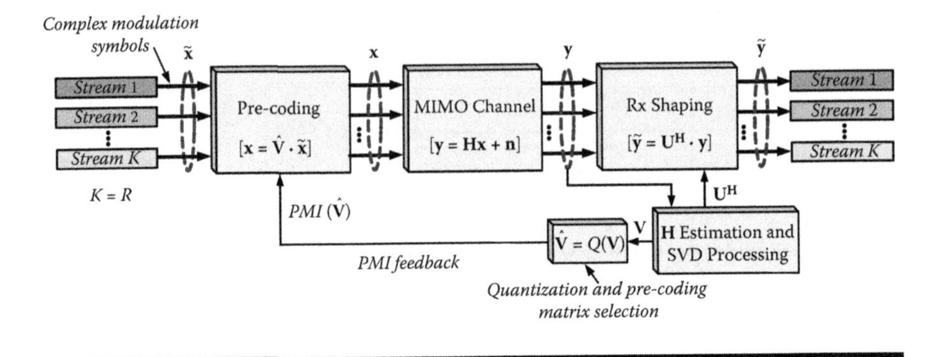

Figure 3.25 Basic idea of codebook-based unitary precoding.

equivalent channel matrix is the diagonal matrix **D**, which implies R independent parallel channels for data transmission. The eigenvalues of $\mathbf{H}^H\mathbf{H}$ will determine the gain of each parallel channel. The overall process, including transmit precoding, propagation through MIMO channel, and receiver shaping, is illustrated in Figure 3.25.

As stated before, both the transmitter and the receiver need knowledge of the downlink channel in order to perform the precoding and shaping processes. However, in FDD systems, the downlink channel cannot be estimated at the transmitter, so the receiver estimates should be reported via uplink transmissions. Furthermore, the uplink bandwidth is severely constrained in real systems and the limited feedback does not allow for reporting the true estimates of **H**. Therefore, the SVD processing is carried out at the receiver and a quantized version of the precoding matrix ($\hat{\mathbf{v}}$) is reported to the transmitter, as shown in Figure 3.25. The precoding matrix $\hat{\mathbf{v}}$ is selected from a predefined set of precoding matrices referred to as a *codebook.*

LTE employs a precoded spatial multiplexing scheme with codebook-based feedback from users. Both the UE and the eNodeB are aware of the predefined set of precoding matrices; consequently, only a matrix identifier called PMI (precoding matrix indicator) is reported (see Figure 3.25). Precoding in LTE is used for two purposes:

- In general, orthogonal transmission on the parallel channels is achieved and thus the interference among transmitted signals at the receiver is significantly reduced.
- The precoding stage at eNodeB also serves for mapping the K multiplexed data streams to the M transmit antennas. The application of precoding in combination with extra antennas ($K < M$) provides beam forming in addition to spatial multiplexing. Actually, the same precoding framework is used for beam forming and precoded spatial multiplexing in LTE. In beam

forming, only one data stream is transmitted and the generic $(M \times K)$ precoding matrix becomes an $(M \times 1)$ beam-forming vector. It should be noted that the larger the codebook size is, the higher the beam-forming gain that will be achieved.

3.3.4 LTE Multiantenna Procedures

LTE multiantenna transmission relies on the use of the aforementioned transmit diversity and spatial multiplexing schemes. Both open-loop and closed-loop solutions based on these schemes will be applied to achieve diversity or multiplexing gains. The LTE standard defines a set of possible multiantenna configurations for data transmission over the MIMO radio interface. The multiantenna configuration to be applied will be determined by the number of available transmit antennas, specific user capabilities, and current MIMO channel conditions. Therefore, multiantenna transmission must be dynamically configured by upper layers for each physical channel and corresponding user, according to the particular user information such as channel state, speed, or QoS issues. The procedure to apply a specific multiantenna configuration for transmission of a physical channel is defined by so-called *antenna mapping,* which can be considered as the LTE multiantenna framework.

This subsection provides an overview of the different multiantenna configurations and multiantenna-related procedures defined in LTE. First, antenna mapping procedures will be introduced for the sake of clarity, and then the defined multiantenna configurations and different modes of operation will be presented. Finally, switching between modes of operation and adaptation procedures will be detailed.

3.3.4.1 Antenna Mapping

Multiantenna processing is applied at the eNodeB in order to transmit data according to some specific multiantenna configuration. In LTE, the downlink physical channel processing chain includes a multiantenna processing stage referred to as *antenna mapping* (see Section 3.1.4). The complex modulation symbols of each code word are taken as inputs, and the complex values to be mapped on each antenna are the result after the antenna mapping procedure. The way antenna mapping is carried out is configured by higher layers and explicitly determines the multiantenna configuration that will be applied to each physical channel. Thus, antenna mapping represents the multiantenna transmission framework in LTE and is composed of the following procedures:

1. *Layer mapping.* The purpose of this stage is to map the different code words (independently coded streams) onto the multiplexed streams (layers*) that

---

* Conceptually, a layer corresponds to a spatially multiplexed channel. Strictly speaking, the concept of layer should not be used when no spatial multiplexing is applied (e.g., transmit diversity).

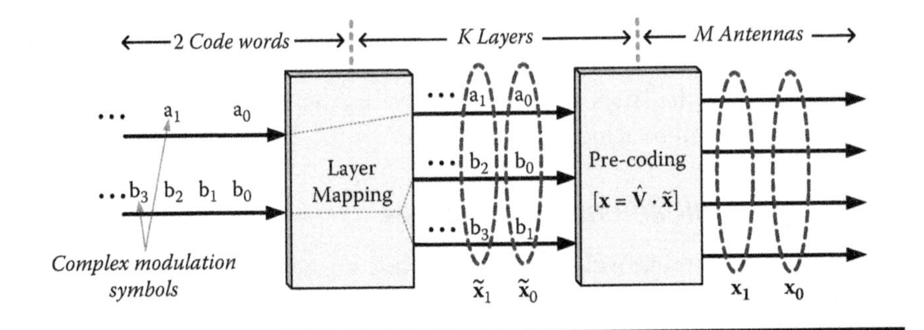

Figure 3.26 Antenna mapping for spatial multiplexing with two code words, three layers, and four antennas.

will be finally transmitted. A maximum of two code words is supported, whereas the number of layers is limited to the rank of the channel matrix. Hence, complex modulation symbols from each code word are mapped onto the different layers of the transmission. Symbols from the same code word can be mapped to one or several layers, as shown in Figure 3.26. Therefore, the number of layers is always greater than or equal to the number of code words.

2. *Precoding.* One modulation symbol from each layer is extracted and the resulting vector is processed according to the applied multiantenna scheme. This stage includes the precoding process (as presented in Section 3.3.3.3) for spatial multiplexing and the mapping procedure defined by SFBC (Section 3.3.2.2) for transmit diversity. The output of the precoding stage is an M-size vector of complex values to be mapped on each of the M antenna ports. This stage can be seen as a means to create a set of virtual antennas (or layers) from physical antennas. Hence, the full base station power can be used irrespective of the number of transmission layers.

An example of antenna mapping, including layer mapping and precoding, is illustrated in Figure 3.26 for spatial multiplexing with two code words, $K = 3$ layers and $M = 4$ transmit antennas. In this case, the first code word is directly mapped to the first layer, while the second code word is demultiplexed into the second and third layers. Finally, the precoding stage performs the conversion from three layers to the four antennas of the eNodeB.

3.3.4.2 LTE Multiantenna Configurations

The different multiantenna schemes supported in LTE have been presented in Sections 3.3.2 and 3.3.3 for achieving diversity and multiplexing gains, respectively. However, these schemes are only the ingredients to be combined in order to obtain a certain multiantenna configuration. Concretely, three configurations have been defined for LTE downlink:

- *Transmit diversity* consists of applying an SFBC scheme (as detailed in Section 3.3.2.2) according to the number of antennas of the eNodeB. There is only one code word and the number of layers is equal to the number of transmit antennas. From a conceptual point of view, layer mapping is not performed because there is no spatial multiplexing. Hence, the layer mapping procedure acts as a complementary part of precoding to perform the SFBC processing. Both layer mapping and precoding functionalities are defined in a different way when transmit diversity is applied (3GPP 36.212).
- *Beamforming* is the closed-loop transmit diversity solution adopted for LTE downlink. It is based on codebook precoding as explained in Section 3.3.2.2. Regarding the antenna mapping procedure, beam forming is considered as a particular case of spatial multiplexing with only one code word and one layer. Thus, the layer mapping stage is transparent.
- *Spatial multiplexing* employs a SU-MIMO approach with two code words and codebook-based precoding (see Section 3.3.3). In addition, CDD is applied in order to provide delay diversity. That is, once the precoding is performed, a different cyclic shift is applied to the signal transmitted over the different antennas. Therefore, spatial multiplexing configuration in LTE can be seen as a hybrid multiplexing-diversity transmission mode.

Table 3.5 summarizes the multiantenna configurations defined for LTE downlink, together with their corresponding setup of code words, layers, and transmit antennas. Spatial multiplexing allows for enhancing peak data rates of users experiencing good channel conditions (i.e., high SNR in a rich scattering scenario). However, transmit diversity is applied for users experiencing a low SNR in order to enhance coverage or link robustness. Control channels such as PDCCH are

Table 3.5 Multiantenna Configurations for LTE Downlink

Multiantenna Configuration	*Applied Schemes*	*No. Code Words*	*No. Layers*	*No. Antennas*
Transmit diversity	SFBC	1	2	2
			4	4
Beam forming	Codebook-based beam forming	1	1	2 or 4
Spatial multiplexing	SU-MIMO with codebook-based precoding, and CDD	2	2	2
			2, 3, or 4	4

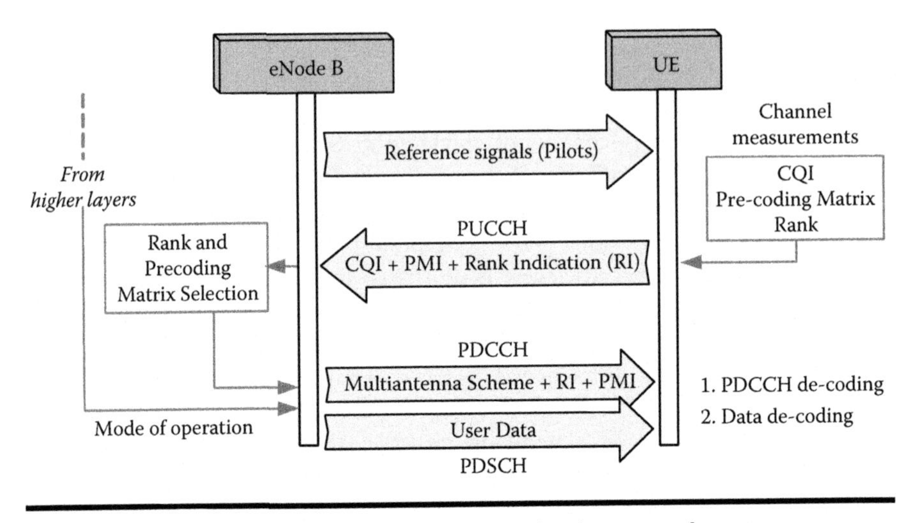

Figure 3.27 Rank adaptation and precoding selection procedures.

transmitted with transmit diversity configuration because a correct reception is the goal rather than an increased data rate for these channels.

On the other hand, both spatial multiplexing and beamforming configurations are based on precoding, and thus the UE needs to know the precoding matrix (or vector) for properly recovering transmitted data. The applied configuration, including number of layers and precoding matrix, is decided at the eNodeB, which takes into account the information reported by the UE (i.e., suitable rank and preferred precoding matrix). However, the final eNodeB decision may differ with respect to the UE preferred values; therefore, the UE must be informed about the applied configuration (see Figure 3.27). As a consequence, closed-loop configurations cannot be used to transmit PDCCH (which contains L1 signaling) and will only be used for user data transmission (PDSCH).

The aforementioned antenna mapping procedure, as well as multiantenna configurations, is not defined for uplink transmission at this stage of the LTE specifications. However, future LTE releases may include the use of uplink MU-MIMO as collaborative or distributed MIMO. That is, the same time-frequency resources are shared by two UEs, which transmit on a single antenna port to the multiple receive antennas of the eNodeB.

3.3.4.3 Multiantenna Adaptation Procedures

The time-variant channel matrix and user-specific conditions (e.g., speed) imply that several adaptation or switching procedures are needed in order to apply the appropriate multiantenna configuration for each user. The number of layers or spatial multiplexing order is dynamically adapted to the rank of the channel matrix. The UE reports the rank of the channel matrix to the eNodeB, which in turn selects the appropriate rank (number of layers) of the transmission. This decision is based on the reported RI (rank indicator), as well as on other parameters as the reported CQI. As an example, a user under poor channel conditions (i.e., low CQI values) will be transmitted on rank 1 even if the channel matrix has rank 2. This procedure, referred to as *rank adaptation,* is illustrated in Figure 3.27.

Closed-loop-based configurations, such as precoded spatial multiplexing and beam forming, exhibit good performance when the terminal speed is not very high. However, for high-mobility scenarios, the UE reports become obsolete sooner and hence adaptive precoding may be outdated. Therefore, two modes of operation can be identified when considering low- or high-mobility scenarios:

- *Closed-loop operation* implies adaptive precoding and rank adaptation (1, 2, 3, or 4) based on the UE reports. A rank 1 transmission corresponds to beam forming, whereas a rank 2, 3, or 4 is associated with spatial multiplexing configuration. The UE reports must be reliable for a proper adaptation process; therefore, this mode is only suitable for users moving at low or medium speed.
- *Open-loop operation* is employed for high-mobility scenarios and consists of applying either transmit diversity or fixed (nonadaptive) precoding. Switching between these alternatives is performed by means of rank adaptation (i.e., transmit diversity for rank 1 and spatial multiplexing with fixed precoding for rank 2). This operation mode allows for increasing throughput of high-mobility users experiencing good channel conditions.

As stated earlier, multiantenna configurations are adapted to channel variations and user-specific conditions (e.g., user speed). Adaptation or switching between different configurations or modes of operation is performed at two levels:

- *Dynamic switching* (or physical layer switching) occurs in the context of rank adaptation. This does not mean that rank adaptation is performed at the same rate as precoding adaptation. It simply means that the switching across different ranks does not require higher layer signaling. Dynamic switching can be seen as the adaptation of a certain multiantenna configuration to the current channel conditions.
- *Semistatic switching* implies higher layer signaling and may occur across different modes of operation (e.g., between open-loop and closed-loop operations).

3.4 Physical Layer Procedures

In the previous sections, a general overview of the LTE physical layer has been presented, according to the parameters and procedures defined by the 3GPP (3GPP 36.211; 3GPP 36.212; 3GPP 36.213). However, from the receiver point of view, some mechanisms are not explicitly specified in the standard because their implementation is vendor specific. This section is focused on those procedures that allow the reception and decoding of the frame under a certain BER level: synchronization and channel estimation. Furthermore, this section also presents the random access (RA) structure and procedure, which is necessary for initial synchronization in the uplink prior to UE data transmission.

3.4.1 Synchronization

Although the use of OFDM in current and future communication systems is widely extended due to its appealing properties, from a practical point of view, some drawbacks exist. One of the major issues is related to the strict requisites on time and frequency synchronization; this section is thus dedicated to analyzing these aspects on OFDM-based systems.

There are two main points related to synchronization in OFDM-based systems, as depicted in Figure 3.28. The first is *symbol synchronization* or *time synchronization,* which consists of how the receiver is able to determine the exact instant at which the OFDM symbol starts—that is, the correct position of the DFT window. The second is *frequency synchronization,* which tries to eliminate the carrier frequency offset (CFO) caused by the mismatch from the RF local oscillators (f_C) and the Doppler shift (f_D).

Most parts of the algorithms for time synchronization are based on the calculation of correlation metrics that exploit some periodicity of the OFDM signal—that is, due to the cyclic prefix (CP) (Van de Beek, Sandell, and Borjesson 1997, 1800)

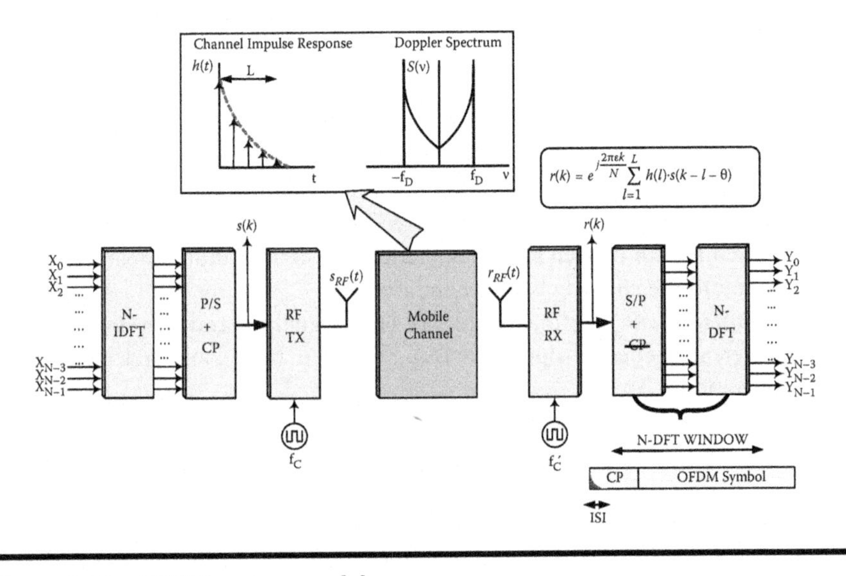

Figure 3.28 OFDM system model.

or to a known training sequence (Schmidl and Cox 1997, 1613). The obtained metrics also allow for having an estimation of the CFO in the time domain. Morelli, Kuo, and Pun (2007, 1394) present a comprehensive survey on synchronization techniques for OFDM-based systems.

The OFDM system model follows the scheme described in Figure 3.28. The received signal, $r(k)$, is affected by a CFO, ε (namely, the frequency offset relative to the subcarrier spacing, Δf); an uncertainty on the beginning of the OFDM symbols denoted as θ; and multipath propagation due to a channel impulse response, $h(t)$, with length L. The time and frequency synchronization consists of the estimation of θ and ε, respectively. Timing errors result in intersymbolic interference (ISI), whereas imperfect estimations of CFO lead to intercarrier interference (ICI), degrading the system performance.

The problem of time synchronization (Figure 3.29) in an OFDM-based system is focused on finding the instant at which the OFDM symbol starts, whose estimation is denoted as $\hat{\theta}$. The portion of the cyclic prefix corrupted by ISI depends on the length of the channel response, but, in general, a valid range of values exists for acquiring time synchronization inside the ISI free zone. If the value of $\hat{\theta}$ is located within this region, data subcarriers, X_k, will not be affected by ISI, but rather by a phase shift proportional to k index and $\Delta\theta$. This phase shift causes a rotation of the received subcarriers in the I–Q plane and can be easily compensated in the frequency domain. However, for values of $\hat{\theta}$ located outside the ISI free zone, the DFT window will be contaminated by the contiguous symbol, causing dispersion on the received subcarriers and degrading the error rate.

On the other hand, frequency synchronization (Figure 3.29) is related to the correction of the CFO that affects the received signal; the estimation of this parameter

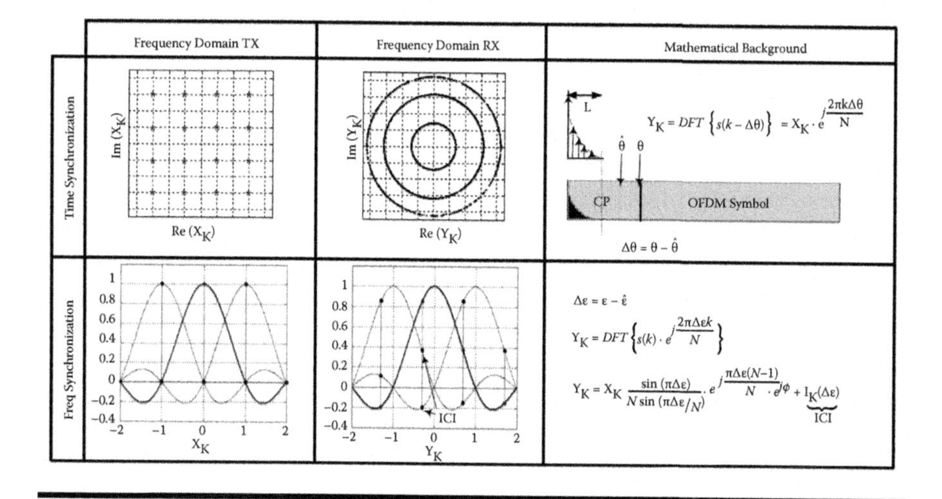

Figure 3.29 Effects of an imperfect time and frequency synchronization in OFDM.

is denoted as $\hat{\varepsilon}$. In the case of an imperfect correction of ε, the time domain received signal after CFO compensation is multiplied by a complex residual carrier; this leads to a shift in the frequency domain of the received subcarriers, Y_K, resulting in the loss of orthogonality among them. The expression for Y_K under imperfect CFO estimation, extracted from Morelli et al. (2007, 1394), is shown in Figure 3.29.

Time and frequency synchronization can be seen as dual problems: A deviation on the estimation of θ causes a time shift of the received signal in the time domain, which corresponds to a product by a complex exponential in the frequency domain. On the other hand, an inaccurate estimation of ε results in a residual modulation of the received signal by an error subcarrier, which corresponds to a shift of the received subcarriers in the frequency domain.

3.4.1.1 Synchronization in OFDMA DL

As mentioned in previous sections, dedicated channels exist for synchronization purposes in LTE downlink: *P-SCH* and *S-SCH* (see Figure 3.4). These channels may be used by the UE to acquire time and frequency synchronization from the received signal and also for frame alignment.

The P-SCH is composed of a 62-length Zadoff–Chu sequence, $d_u[n]$, generated in the frequency domain (3GPP 36.211). The expression for this sequence is given in Equation (3.4), where the Zadoff–Chu sequence index u can take the values 25, 29, or 34:

$$d_u(n)=\begin{cases} e^{-j\frac{\pi u n(n+1)}{63}} & n=0,1,\ldots,30 \\ e^{-j\frac{\pi u(n+1)(n+2)}{63}} & n=31,32,\ldots,61 \end{cases} \tag{3.4}$$

Zadoff–Chu sequences (Frank, Zadoff, and Heimiller 1962, 381; Chu 1972, 531) are specially generated pseudorandom polyphase sequences with perfect correlation properties. These properties make P-SCH particularly interesting for obtaining accurate synchronization metrics.

Prior to frequency domain processing of P-SCH, an initial coarse synchronization is needed in order to determine an initial position of the DFT window. This initial estimation can be extracted either from the CP of the downlink signal or from time-domain-matched filtering with the IDFT of the Zadoff–Chu sequence (Figure 3.30).

Once coarse synchronization has been performed in the time domain, CFO estimation can be accurately obtained from matched filtering in the frequency domain. This process is illustrated in Figure 3.31.

If necessary, it is also possible to obtain synchronization metrics from the downlink reference signal by following the method proposed in Bo, Jian-hua, and Yon (2004, 56), although such a signal is not transmitted for synchronization purposes.

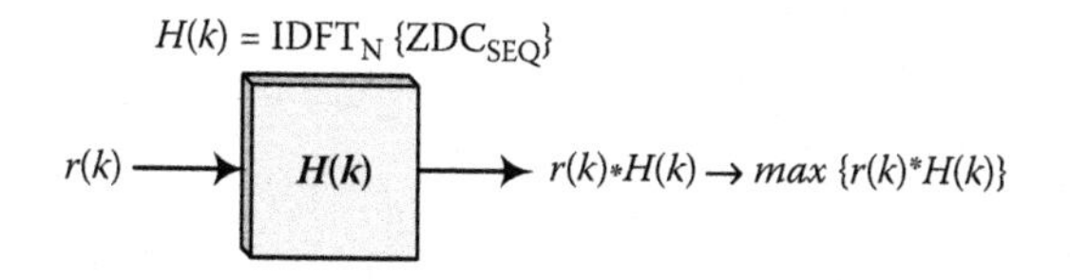

Figure 3.30 Time-domain-matched filtering for P-SCH-based time synchronization.

3.4.1.2 Synchronization in SC-FDMA UL

In the uplink, the signal received by the eNodeB is composed by the addition of the individual signals transmitted by the UEs. Thus, the uplink time and frequency synchronization becomes a complex multiparameter estimation problem. In Morelli et al. (2007, 1394), several uplink scenarios are presented that introduce some simplifications with the aim of reducing the complexity of the synchronization process.

Although LTE uses SC-FDMA for uplink transmission of data and control channels, the specific physical signals used for synchronization purposes are transmitted within an OFDMA scheme. Thus, the algorithms and mechanisms for synchronization are the same as the methods presented in the preceding section.

During the random access procedure, the random access preamble transmitted by the UE is used by the eNodeB for achieving initial time and frequency synchronization and also for timing advance correction. This initial process will be detailed more extensively in Section 3.4.3.

After this initial synchronization from the random access preamble, the eNodeB can use either the cyclic prefix or the demodulation reference signal for detection and synchronization tracking because the structure of the latter (Figure 3.32) can be seen as a training sequence (a preamble or midamble) in order to obtain reliable synchronization metrics.

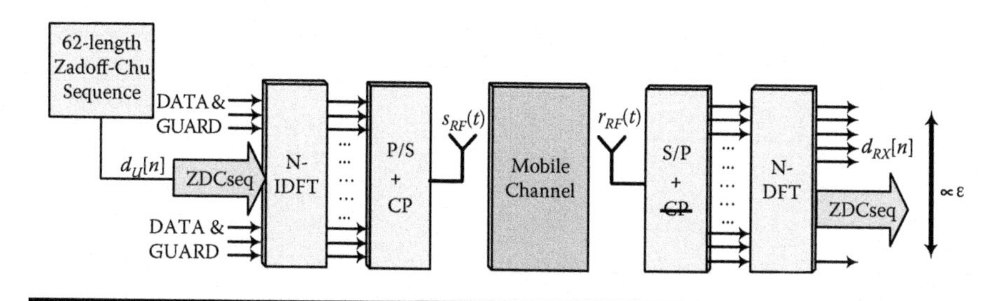

Figure 3.31 Frequency-domain-matched filtering for P-SCH-based frequency synchronization.

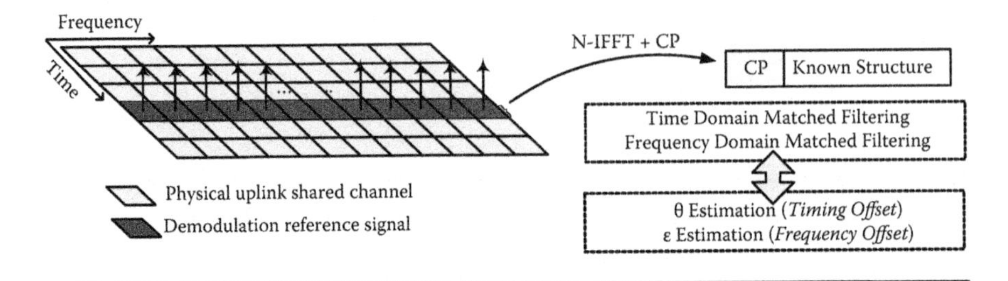

Figure 3.32 Applicability of UL demodulation reference signal for synchronization purposes.

3.4.1.3 Cell Search Procedure

Prior to an uplink initial transmission or a handover procedure, a cell search process is needed. Cell search is the procedure by which a UE synchronizes with a certain cell and detects its physical layer *cell ID* and *cell ID group*. In the LTE downlink, the signals P-SCH and S-SCH are transmitted to facilitate this process.

In Figure 3.33, a sequence of the operations performed during the cell search procedure is presented:

1. This step consists of finding the center frequency (f_C) of the RF carrier from the set of supported frequencies defined by the standard.
2. Then an initial synchronization is needed in order to locate the position of the DFT window at the receiver, thus being able to decode the physical channels.
3. The next step is the P-SCH detection, which consists of the detection of the Zadoff–Chu root, according to Equation (3.4). In this stage, the cell ID is acquired and the frame is synchronized within a 5-ms basis. Additionally, fine time and frequency synchronization can be performed, thanks to the good correlation properties of Zadoff–Chu sequences.

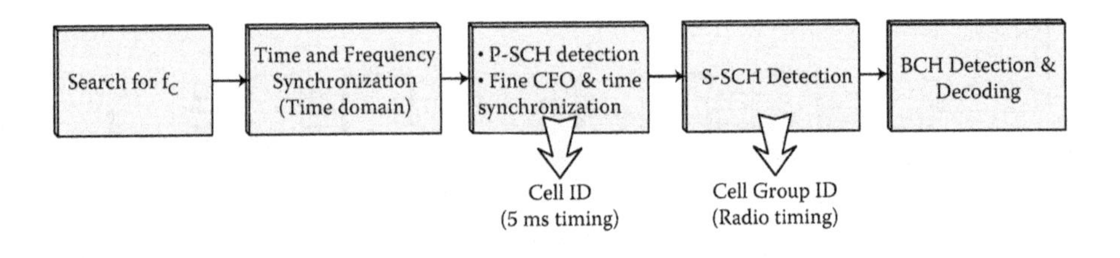

Figure 3.33 Cell search procedure.

4. S-SCH detection is performed, which allows completion of the frame synchronization process—that is, the radio timing for the downlink transmission—and also obtaining the cell group ID.
5. Finally, the cell search procedure is completed, and the UE is able to decode the broadcast channel (BCH) prior to the transmission of an uplink random access preamble.

3.4.1.4 Other Synchronization Procedures

Apart from time-frequency synchronization and cell search, additional issues related to synchronization in LTE, such as radio link monitoring, intercell synchronization, or timing advance strategies, have been specified in 3GPP 36.213. The latter is particularly critical from the point of view of the synchronization algorithms for the uplink; an efficient timing advance mechanism must force the user signals to be aligned in time at the eNodeB receiver (a quasi-synchronous scenario) (Morelli et al. 2007, 1394), thus simplifying considerably the uplink reception. When the eNodeB detects that the signal from a certain UE is misaligned, the eNodeB transmits a timing advance command in order to force the UE to adjust its uplink transmission timing. This command is defined to be a multiple of $16 \cdot T_s$ (sampling time) and is relative to the previous uplink timing.

For high-mobility users, uplink timing has to be adjusted continuously. This fact may even cause a residual misalignment because UE position when the timing advance is computed by the eNodeB differs from UE position when the timing advance adjustment reaches the UE. In this scenario, more advanced reception techniques may be required in the uplink.

3.4.2 Channel Estimation

It is well known that one of the main benefits of OFDM is its inherent immunity against multipath effects. Thanks to the decomposition of the transmitted signal on orthogonal subcarriers along the available bandwidth, it is possible to consider the frequency selective channel as approximately flat for every subcarrier. This assumption allows for easily performing the channel equalization process in the frequency domain by means of multiplications by constant values.

In the general case, mobile channels are considered to be time variant and frequency selective. This fact requires a dynamic estimation of the channel frequency response at the receiver. The most extended strategy for channel estimation in OFDMA relies on the use of pilot subcarriers, distributed along the transmission bandwidth, whose transmitted values are known at the receiver (Coleri et al. 2002, 223). This strategy allows an easy estimation of the channel frequency response for this set of frequencies.

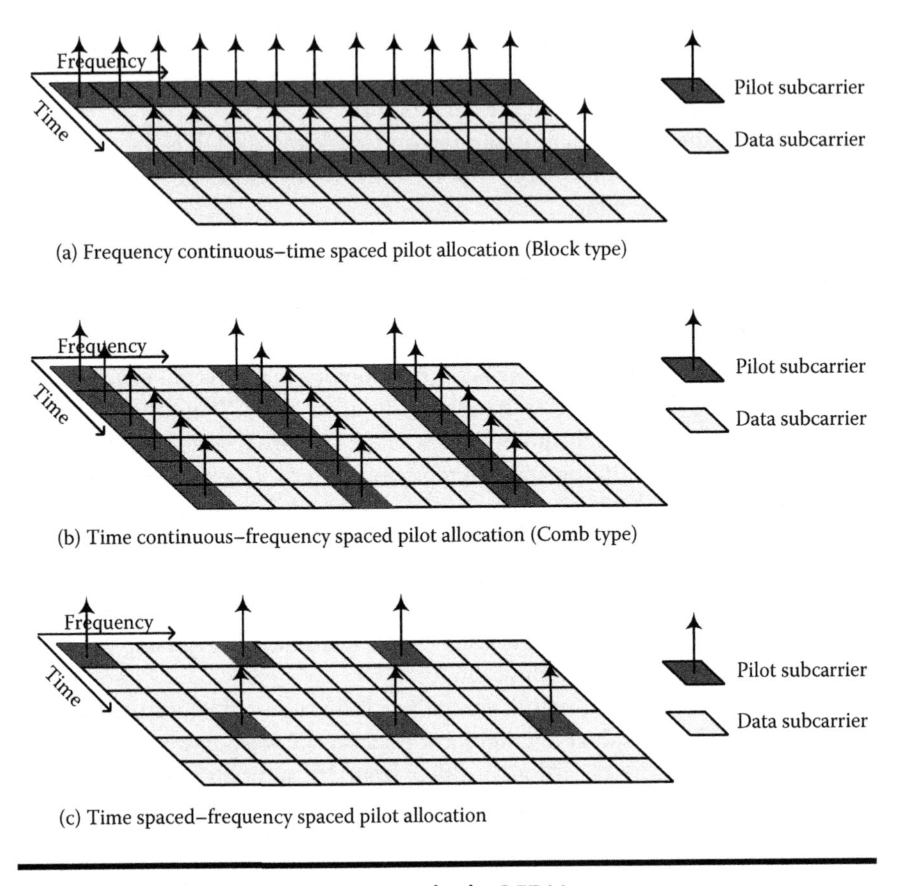

Figure 3.34 Pilot arrangement strategies in OFDM.

When pilot subcarriers are allocated in the time-frequency domain of an OFDMA system, different possibilities exist, as depicted in Figure 3.34. The first strategy (Figure 3.34a)—namely, *block type pilot arrangement*—consists of allocating a complete OFDM symbol for pilot transmission. Thus, an estimation of the channel frequency response can be obtained by least squares (LS), minimum mean square error (MMSE), or similar methods. This approach is a suitable strategy for slow time-varying channels.

An alternative is *comb type pilot arrangement,* shown in Figure 3.34b. In this technique, pilot subcarriers are transmitted only in some frequencies, but continuously in time. For the rest of the frequencies, the channel frequency response is obtained by linear (López-Martínez et al. 2007, 987), second-order (Hsieh and Wei 1998, 217), or other interpolation methods.

The last strategy (Figure 3.34c) is a combination of the former methods, where pilot subcarriers are spaced in both time and frequency. In general, the time spacing between symbols with pilots is given by channel time variability—that is, the coherence time of the channel. On the other hand, frequency spacing between pilot subcarriers is determined by the frequency selectivity of the channel, which can be quantified by its coherence bandwidth (Pätzold 2002; Goldsmith 2005).

3.4.2.1 Channel Estimation in OFDMA DL

In LTE downlink, cell-specific downlink reference signals are transmitted on one or several antenna ports of the eNodeB in order to facilitate the demodulation of the OFDMA signal at the UE. The structure of this reference signal was presented in Section 3.1.2 and corresponds to a time-frequency spaced pilot arrangement scheme (see Figure 3.34c). Thus, two-dimensional time and frequency interpolation is required at the receiver.

For MIMO transmission, the time-frequency allocation of reference signals on each antenna is targeted to avoid interfering with each other (see Figure 3.3). Therefore, MIMO channel estimation may be addressed as the estimation of $M \times N$ individual channels, where M is the number of transmit antennas and N is the number of receive antennas. This way, channel estimation in MIMO can be tackled as a combination of different single-input, single-output (SISO) estimations (Figure 3.35).

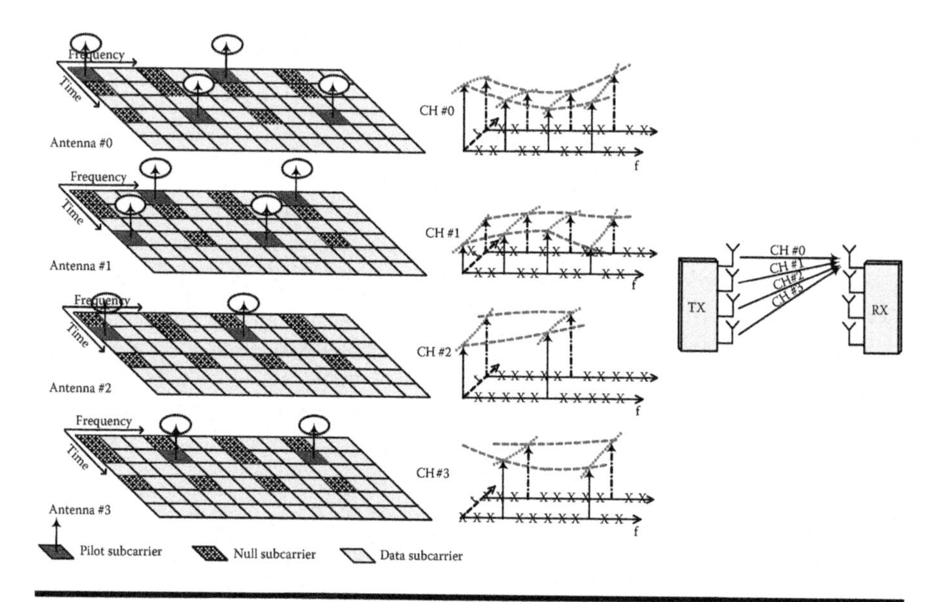

Figure 3.35 Channel estimation in DL-OFDMA from MIMO reference signal (4 × 4 case).

3.4.2.2 Channel Estimation in SC-FDMA UL

In the case of LTE uplink, reference signals are transmitted with two different objectives. The first is to facilitate channel estimation at the eNodeB prior to demodulation (*demodulation reference signal*). The second, the *sounding reference signal* (SRS), helps to provide information of the uplink channel response in a range of frequencies greater than the allocated bandwidth for a particular UE. This allows the eNodeB to implement uplink frequency-dependent scheduling strategies. Both variants of the uplink reference signal are based on Zadoff–Chu sequences (3GPP 36.211).

The demodulation reference signal is transmitted in the fourth SC-FDMA symbol of the slot (see Figure 3.5), and its size is equal to the uplink allocated bandwidth for that UE. Demodulation reference signal structure can be easily identified with a block type pilot arrangement scheme (see Figure 3.34), so classical channel estimation mechanisms for this strategy may be used. Also, time interpolation is not necessary to perform because the allocated resources for the UE are likely to be frequency hopped by the eNodeB uplink scheduler.

The SRS is transmitted within a bandwidth greater than that allocated for this UE. This way, the eNodeB can have out-of-band information of the channel frequency response for this individual user, which facilitates the implementation of channel-aware scheduling mechanisms in the uplink. The transmission of an SRS may interfere with the PUSCH of other users*; hence, a broadcast parameter is defined to indicate the possible location of SRS subframes so that the UEs can know where to puncture their PUSCH transmission. On the other hand, an SRS is not allowed to interfere with PUCCH. The parameters that define the SRS are bandwidth, duration, periodicity, location in the subframe, cyclic shift for the Zadoff–Chu sequence, and transmission comb.

3.4.3 Random Access

Once a UE is synchronized to the downlink transmission after the cell search procedure, it is able to request the eNodeB for resource allocation to begin the transmission or to recover from a loss of synchronism in the uplink. In addition, this mechanism allows the eNodeB to estimate and adjust the UE transmission timing within a fraction of the cyclic prefix.† This process—namely, the *random access procedure*—is based on the transmission of a special symbol (random access preamble) mapped on PRACH. Although PRACH can only be transmitted in certain time and frequency resources, the eNodeB is not explicitly prohibited from allocating these resources for data transmission in the uplink.

* Different UEs use different Zadoff–Chu sequences when transmitting SRSs. Because these sequences are orthogonal, they do not interfere with each other, although they may be transmitted in the same time/frequency resources.

† This is the reason that the random access procedure described in this section is referred to as "nonsynchronized."

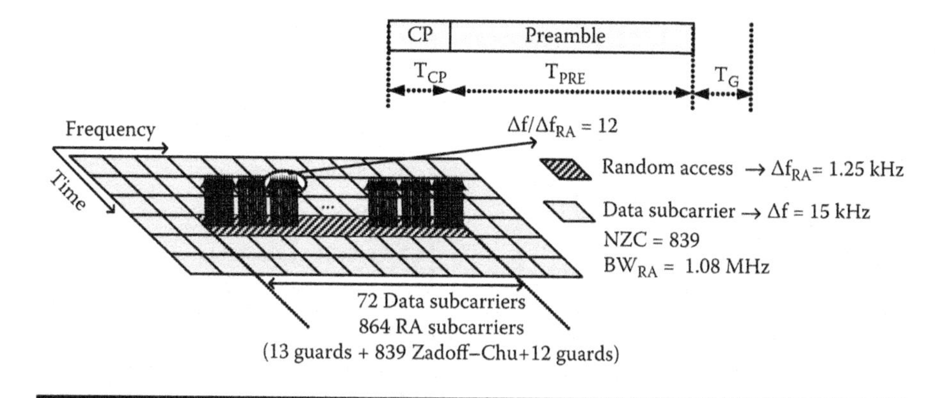

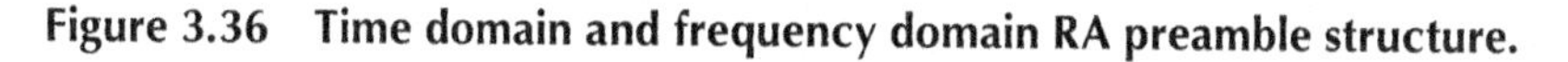

Figure 3.36 Time domain and frequency domain RA preamble structure.

3.4.3.1 Random Access Preamble

The physical layer random access preamble, whose time-frequency structure is illustrated in Figure 3.36, consists of a cyclic prefix of length T_{CP} and a preamble part of length T_{PRE}. The parameter values are listed in 3GPP 36.213 and depend on the frame structure and the random access configuration, whereas higher layers are in charge of controlling the preamble format.

The bandwidth for the random access preamble is 1.08 MHz, allocated in the central subcarriers. Thus, it can be used for any transmission bandwidth in the cell. Because uplink transmission for a particular UE has not been aligned yet with the uplink timing through timing advance correction, a guard time, T_G, is required to avoid interfering with other UEs. Guard bands in the frequency domain are inserted on the random access preamble in order to avoid data interference at preamble edges. Additionally, a smaller frequency spacing, Δf_{RA}, is used to provide a higher granularity for synchronization.

In order to achieve a good detection, a low false-alarm probability, and accurate time and frequency synchronization in the uplink, Zadoff–Chu sequences with zero correlation zone are used (3GPP 36.213). Zadoff–Chu sequences with zero correlation zone are generated from one of several root Zadoff–Chu sequences, defined in Expression (3.5), where the length of Zadoff–Chu sequence N_{ZC} is equal to 839 for a type-1 frame structure:

$$x_u[n] = e^{-j\frac{\pi un(n+1)}{N_{ZC}}}, \quad 0 \leq \mathrm{n} \leq N_{ZC} - 1 \tag{3.5}$$

An RA preamble with a zero correlation zone is defined by a cyclic shift, given in Expression (3.6), that is applied to one of the root Zadoff–Chu sequences from Expression (3.5):

$$x_{u,v}[n] = x_u[(n + C_v)\,\mathrm{mod}(N_{ZC})] \tag{3.6}$$

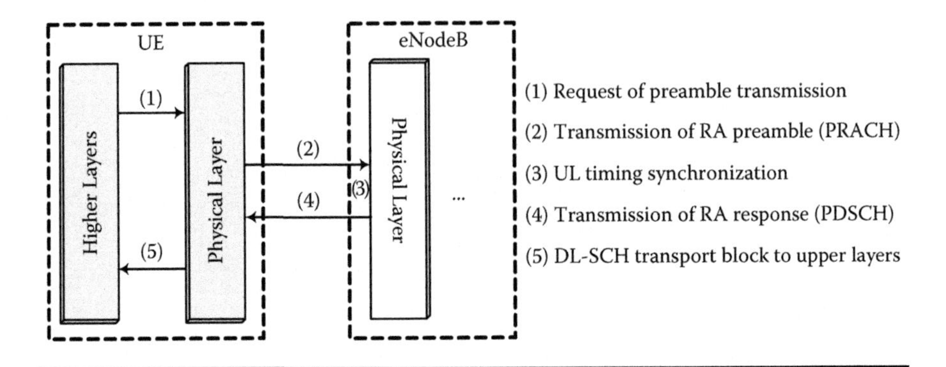

Figure 3.37 Random access procedure from the physical layer perspective.

The size of the cyclic shift, C_v, must be larger than the maximum round-trip delay, so the number of root Zadoff–Chu sequences and the number of generated Zadoff–Chu sequences with zero correlation zone are changed, depending on cell radius. That is, the required number of different root Zadoff–Chu sequences grows with the cell radius.

3.4.3.2 Random Access Procedure

After downlink synchronization is acquired by the UE, the physical layer must receive some configuration parameters from upper layers before the initiation of the nonsynchronized physical random access procedure. Particularly, the following parameters must be indicated: PRACH configuration and frequency allocation and preamble format, as well as other parameters for determining the root sequences and their cyclic shifts in the preamble sequence.

The random access procedure is depicted in Figure 3.37. It can be observed that from the physical layer point of view, the random access procedure only comprises the transmission of random access preamble and random access response. The remaining messages necessary for the establishment of terminal identity and other aspects are scheduled by upper layers for transmission on PDSCH. Thus, these messages are not considered part of the physical layer random access procedure.

References

Alamouti, S. M. 1998. A simple transmit diversity technique for wireless communications. *IEEE Journal on Selected Areas in Communications* 16 (8): 1451–1458.

Bauch, G., and J. S. Malik. 2004. Orthogonal frequency division multiple access with cyclic delay diversity. Paper presented at the ITG Workshop on Smart Antennas, March 18–19, Munich, Germany: 17–24.

Berrou, C., A. Glavieux, and P. Thitimajshima. 1993. Near Shannon limit error-correcting coding and decoding: Turbo codes. Paper presented at the IEEE International Conference on Communications, May 23–26, Geneve, Switzerland: 1064–1070.

Bo, A. I., G. E. Jian-hua, and W. Yon. 2004. Symbol synchronization technique in COFDM systems. *IEEE Transactions on Broadcasting* 50 (1): 56–62.

Chu, D. 1972. Polyphase sequences with good correlation properties. *IEEE Transactions on Information Theory* 18 (4): 531–532.

Coleri, S., M. Ergen, A. Puri, and A. Bahai. 2002. Channel estimation techniques based on pilot arrangement in OFDM systems. *IEEE Transactions on Broadcasting* 48 (3): 223–229.

Dehghani, M. J., R. Aravind, S. Jam, and K. M. M. Prabhu. 2004. Space-frequency block coding in OFDM systems. Paper presented at the IEEE Region 10 Conference TENCON-2004, November 21–24, Chiang Mai, Thailand: 543–546.

Elias, P. 1954. Error free decoding. *IRE Transactions on Information Theory* IT-4:29–37.

Entrambasaguas, J. T., M. C. Aguayo-Torres, G. Gómez, and J. F. Paris. 2007. Multiuser capacity and fairness evaluation of channel/QoS aware multiplexing algorithms. *IEEE Network Magazine* 21 (3): 24–30.

Frank, R., S. Zadoff, and R. Heimiller. 1962. Phase shift pulse codes with good periodic correlation properties. *IEEE Transactions on Information Theory* 8 (6): 381–382.

Frenger, P., S. Parkvall, and E. Dahlman. 2001. Performance comparison of HARQ with chase combining and incremental redundancy for HSDPA. Paper presented at the IEEE VTS 54th Vehicular Technology Conference, October 7–11, Atlantic City, NJ: 1829–1833.

Goldsmith, A. J. 2005. *Wireless communications*. Cambridge: Cambridge University Press.

Hsieh, M. H., and C.-H. Wei. 1998. Channel estimation for OFDM systems based on comb-type pilot arrangement in frequency selective fading channels. *IEEE Transactions on Consumer Electronics* 44 (1): 217–225.

Hutter, A. A., J. S. Hammerschmidt, E. de Carvalho, and J. M. Cioffi. 2000. Receive diversity for mobile OFDM systems. Paper presented at the IEEE Wireless Communications and Networking Conference, September 23–28, Chicago, IL: 707–712.

Jiang, M., and L. Hanzo. 2007. Multiuser MIMO-OFDM for next-generation wireless systems. *Proceedings of the IEEE* 95 (7): 1430–1469.

Kaiser, S. 2000. Spatial transmit diversity techniques for broadband OFDM systems. Paper presented at the IEEE Global Telecommunications Conference, November 27–30, San Francisco, CA: 1824–1828.

Kyungchun, L., J. Chun, and L. Hanzo. 2007. Optimal lattice-reduction aided successive interference cancellation for MIMO systems. *IEEE Transactions on Wireless Communications* 6 (7): 2438–2443.

López-Martínez, F. J., E. Martos-Naya, J. T. Entrambasaguas, and M. García-Abril. 2007. Low complexity synchronization and frequency equalization for OFDM systems. Paper presented at the 14th IEEE International Conference on Electronics, Circuits and Systems, December 11–14, Marrakesh, Morocco: 987–990.

Morales-Jiménez, D., J. J. Sánchez, G. Gómez, M. C. Aguayo-Torres, and J. T. Entrambasaguas. 2008. Imperfect adaptation in next generation OFDMA cellular systems. Accepted for publication in *Journal of Internet Engineering* 1 (3).

Morelli, M., C. J. Kuo, and M. O. Pun. 2007. Synchronization techniques for orthogonal frequency division multiple access (OFDMA): A tutorial review. *Proceedings of the IEEE* 95 (7): 1394–1427.

Pätzold, M. 2002. *Mobile fading channels.* New York: John Wiley & Sons.

Schmidl, T. M., and D. C. Cox. 1997. Robust frequency and timing synchronization for OFDM. *IEEE Transactions on Communications* 45 (12): 1613–1621.

Shannon, C. E. 1948. A mathematical theory of communication. *Bell System Technical Journal* 27: 379–423; 623–656.

Takeshita, O. Y. 2006. On maximum contention-free interleavers and permutation polynomials over integer rings. *IEEE Transactions on Information Theory* 52 (3): 1249–1253.

Van de Beek, J. J., M. Sandell, and P. O. Borjesson. 1997. ML estimation of time and frequency offset in OFDM systems. *IEEE Transactions on Signal Processing* 45 (7): 1800–1805.

Viterbi, A. J. 1967. Error bounds for convolutional codes and an asymptotically optimum decoding algorithm. *IEEE Transactions on Information Theory* 13 (2): 260–269.

Wubben, D., and K. D. Kammeyer. 2006. Low complexity successive interference cancellation for per-antenna-coded MIMO-OFDM schemes by applying parallel-SQRD. Paper presented at the IEEE VTS 63rd Vehicular Technology Conference, May 7–10, Melbourne, Australia: 2183–2187.

3GPP Specification TR 25.912. 2006. Feasibility study for evolved universal terrestrial radio access (UTRA) and universal terrestrial radio access network (UTRAN). Release 7 v7.2.0.

3GPP Specification TR 25.913. 2006. Requirements for evolved UTRA (E-UTRA) and evolved UTRAN (E-UTRAN). Release 7 v7.3.0.

3GPP Specification TS 36.201. 2007. Long term evolution (LTE) physical layer: General description. Release 8 v8.1.0.

3GPP Specification TS 36.211. 2007. Physical channels and modulation. Release 8 v8.1.0.

3GPP Specification TS 36.212. 2007. Multiplexing and channel coding. Release 8 v8.1.0.

3GPP Specification TS 36.213. 2007. Evolved universal terrestrial radio access (E-UTRA): Physical layer procedures. Release 8 v8.1.0.

Chapter 4

Architecture and Protocol Support for Radio Resource Management (RRM)

Gábor Fodor, András Rácz,
Norbert Reider, and András Temesváry

Contents

4.1 Introduction

In this chapter we discuss the radio resource management (RRM) functions in long term evolution (LTE). The term *radio resource management* is generally used in wireless systems in a broad sense to cover all functions that are related to the assignment and the sharing of radio resources among the users (e.g., mobile terminals, radio bearers, user sessions) of the wireless network. The type of the required resource control, the required resource sharing, and the assignment methods are primarily determined by the basics of the multiple access technology, such as frequency division multiple access (FDMA), time division multiple access (TDMA), or code division multiple access (CDMA) and the feasible combinations thereof. Likewise, the smallest unit in which radio resources are assigned and distributed among the entities (e.g., power, time slots, frequency bands/carriers, or codes) also varies depending on the fundamentals of the multiple access technology employed on the radio interface [23].

The placement and the distribution of the RRM functions to different network entities of the radio access network (RAN), including the functional distribution between the terminal and the network as well as the protocols and interfaces between the different entities, constitute the *RAN architecture.* Although the required RRM functions determine, to a large extent, the most suitable RAN architecture, it is often an engineering design decision how a particular RRM function should be realized. For example, whether intercell interference coordination or handover control is implemented in a distributed approach (in each base station) or in a centralized fashion both can be viable solutions. We will discuss such design issues throughout this chapter.

In LTE, the radio interface is based on the orthogonal frequency division multiplexing (OFDM) technique. In fact, OFDM serves both as a modulation technique and as a multiple access scheme. Consequently, many of the RRM functions can be derived from the specifics of the OFDM modulation. In the rest of this section we give an overview of the LTE RAN architecture, including an overview of the

OFDM-based radio interface. Subsequently, we define and introduce the notion of radio resource in LTE and present the requirements that the 3GPP (3rd Generation Partnership Project) has set on the spectral efficient use of radio resources, which entail the presence of certain RRM functions in the system.

For a comprehensive overview and detailed description of the overall LTE system, the reader is referred to Dahlman et al. [1].

4.1.1 The LTE Architecture

Before discussing the details of the LTE architecture, it is worth looking at the general trends in radio link technology development, which drive many of the architectural changes in cellular systems today. The most important challenge in any radio system is to combat the randomly changing radio link conditions by adapting the transmission and reception parameters to the actual link conditions (often referred to as the channel state.) The better the transmitter can follow the fluctuations of the radio link quality and adapt its transmission accordingly (modulation and coding, power allocation, scheduling), the better it will utilize the radio channel capacity. The radio link quality can change rapidly and with large variations, which are primarily due to the quickly fading fluctuations on the radio link, although other factors such as mobility and interference fluctuations also contribute to these. Because of this, the various radio resource management functions have to operate on a time scale matching that of the radio link fluctuations. As we will see, the LTE requirements on high (peak and average) data rates, low latency, and high spectrum efficiency are achieved partly due to the radio resource control functions being located close to the radio interface, where such instantaneous radio link quality information is readily available.

In addition to the quickly changing radio link quality, the bursty nature of typical packet data traffic imposes a challenge on the radio resource assignment and requires a dynamic and fast resource allocation, taking into account not only the instantaneous radio link quality but also the instantaneous packet arrivals. As a consequence, a general trend in the advances of cellular systems is that the radio-specific functions and protocols get terminated in the base stations, and the rest of the radio access network entities are radio access technology agnostic. Thus, the radio access network exhibits a distributed architecture without a central radio resource control functionality.

The LTE architecture is often referred to as a *two-node architecture* because, logically, only two nodes are involved—in the user and control plane paths—between the user equipment and the core network. These two nodes are (1) the base station, called eNode B, and (2) the serving gateway (S-GW) in the user plane and the mobility management entity (MME) in the control plane, respectively. The MME and the GW belong to the core network, called evolved packet core (EPC) in 3GPP terminology. The GW executes generic packet processing functions similar to router functions, including packet filtering and classification. The MME terminates the

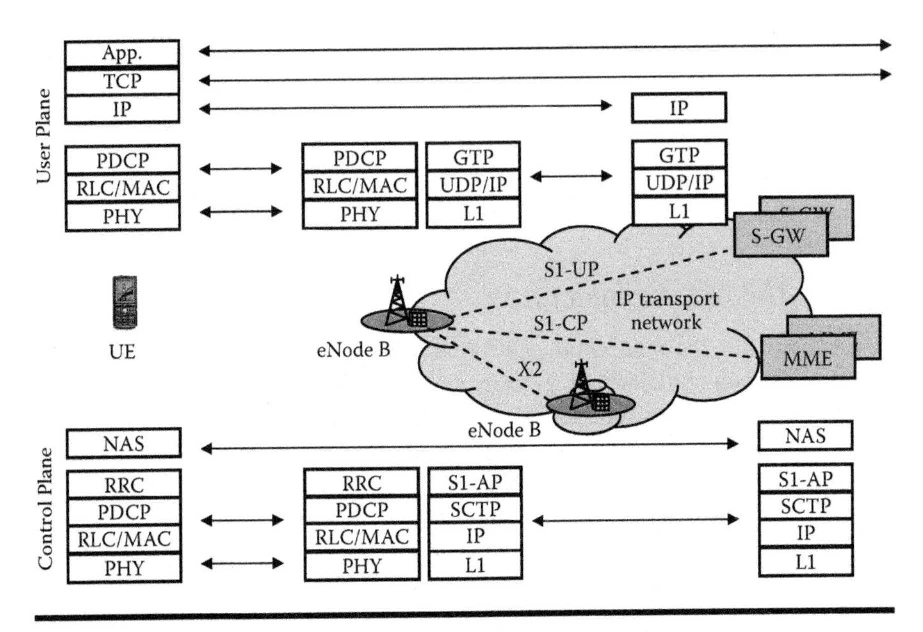

Figure 4.1 The 3GPP long term evolution (LTE) RAN architecture.

so-called nonaccess stratum signaling protocols with the user equipment (UE) and maintains the UE context, including the established bearers, the security context, and the location of the UE. In order to provide the required services to the UE, the MME talks to the eNode B to request resources for the UE. It is important to note, however, that the radio resources are owned and controlled solely by the eNode B and that the MME has no control over the eNode B radio resources. Although the MME and the GW are LTE-specific nodes, they are radio agnostic.

The LTE architecture is depicted in Figure 4.1, which shows the control plane and user plane protocol stacks between the UE and the network. As it can be seen in the figure, the radio link specific protocols, including radio link control (RLC) [2] and medium access control (MAC) [3] protocols, are terminated in the eNode B. The packet data convergence protocol (PDCP) layer [4], which is responsible for header compression and ciphering, is also located in the eNode B. In the control plane, the eNode B uses the radio resource control (RRC) protocol [21] to execute the longer time-scale radio resource control toward the UE. For example, the establishment of radio bearers with certain quality of service (QoS) characteristics, the control of UE measurements, or the control of handovers is supported by RRC. Other short time-scale radio resource controls toward the UE are implemented via the MAC layer or the physical layer control signaling (e.g., the signaling of assigned resources and transport formats via physical layer control channels).

The services are provided to the UE in terms of evolved packet system (EPS) *bearers*. The packets belonging to the same EPS bearer get the same end-to-end treatment in the network. A finite set of possible *QoS profiles*—in other words, packet treatment characteristics—is defined and are identified by so-called labels. A label identifies a certain set of packet treatment characteristics (i.e., scheduling weights, radio protocol configurations such as RLC acknowledge or unacknowledge mode, hybrid automatic repeat request [HARQ] parameters, etc.). Each EPS bearer is associated with a particular QoS class (i.e., with a particular QoS label). There are primarily two main bearer types: guaranteed bit rate (GBR) and non-GBR bearers. For GBR bearers, the network guarantees a certain bit rate to be available for the bearer at any time. The bearers, both GBR and non-GBR, are further characterized by a maximum bit rate (MBR), which limits the maximum rate that the network will provide for the given bearer.

The end-to-end EPS bearer can be further broken down into a radio bearer and an access bearer. The radio bearer is between the UE and the eNode B, and the access bearer is between the eNode B and the GW. The access bearer determines the QoS that the packets get on the transport network; the radio bearer determines the QoS treatment on the radio interface. From an RRM point of view, the radio bearer QoS is in our focus because the RRM functions should ensure that the treatment that the packets get on the corresponding radio bearer is sufficient and can meet the end-to-end EPS bearer-level QoS guarantees.

In summary, we can formulate the primary goal of RRM as to control the use of radio resources in the system such that the QoS requirements of the individual radio bearers are met and the overall used radio resources on the system level are minimized. That is, the ultimate goal of RRM is to satisfy the service requirements at the smallest possible cost for the system.

4.1.2 The Notion of Radio Resource in LTE

The radio interface of LTE is based on the OFDM technology, in which the radio resource appears as one common shared channel, shared by all users in the cell. The scheduler, which is located in the eNode B, controls the assignment of time-frequency blocks to UEs within the cell in an orthogonal manner so that no two UEs canthus intracell interference is avoided. One exception, though, is multiuser spatial multiplexing, also called multiuser MIMO (multiple input, multiple output), when multiple UEs with spatially separated channels the uplink of LTE. More details of MIMO support in LTE are discussed in Section 4.2.7. Such a scheduler function is needed for both the uplink (UL) and the downlink (DL) so that it is compatible with frequency domain duplexing (FDD) and time domain duplexing (TDD) modes.

Figure 4.2 shows the resource grid of the uplink and downlink shared channels. The smallest unit in the resource grid is the resource element (RE), which corresponds to one subcarrier during one symbol duration. These resource elements are organized into larger blocks in time and in frequency, where seven of such symbol durations constitute one *slot* of length 0.5 ms and 12 subcarriers during one slot

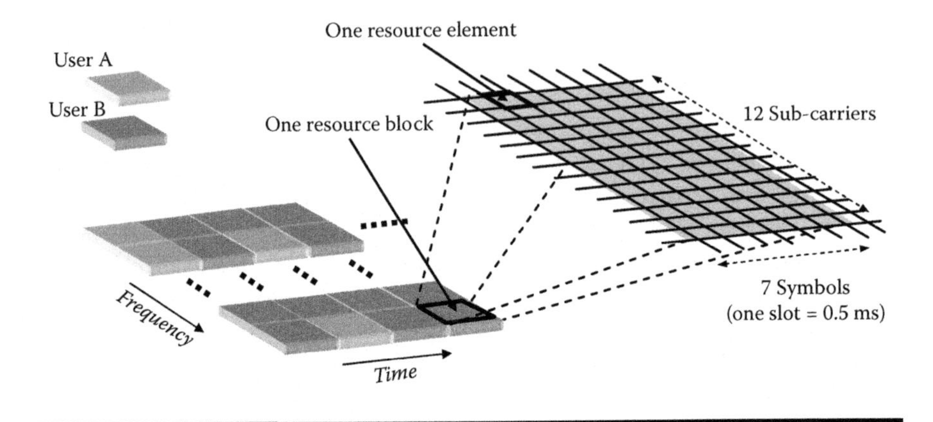

Figure 4.2 Uplink/downlink resource grid.

form the so-called resource block (RB). Two consecutive time slots are called a *subframe* and 10 of such subframes create a *frame,* which is of 10-ms length. The scheduler can assign resource blocks only in pairs of two consecutive RBs (in time); that is, the smallest unit of resource that can be assigned is two RBs.

There is, however, one important difference between the feasible assignments on the UL and DL shared channels. Because in the UL the modulation uses the single carrier FDMA (SC-FDMA) concept, the allocation of RBs per UE has to be on consecutive RBs in frequency. The SC-FDMA modulation basically corresponds to a discrete Fourier transform (DFT) precoded OFDM signal, where the modulation symbols are mapped to consecutive OFDM carriers. The primary motivation for using the SC-FDMA scheme in the UL is to achieve better peak-to-average power ratios. For more details on the layer 1 (L1) radio interface parameters, modulation, and coding schemes, see reference 5.

Because the LTE physical layer is defined such that it supports various multi-antenna MIMO schemes [7] such as transmit diversity and spatial multiplexing, the virtual space of radio resources is extended with a third dimension corresponding to the antenna port, in addition to the classical time and frequency domains. This essentially means that a time-frequency resource grid is available per antenna port. In the downlink, the system supports multistream transmission on up to four transmit antennas. In the uplink, no multistream transmission is supported from the same UE, but multiuser MIMO transmission is possible.

Based on the preceding discussion, we can define the abstract resource element in LTE as the three tuple of [time, frequency, antenna port]. Thus, the generic radio resource assignment problem in LTE can be formulated as finding an optimal allocation of the [time, frequency, antenna port] resource units to UEs so that the QoS requirements of the radio bearers are satisfied while minimizing the use of the radio resources. A closely related function to resource assignment is link adaptation (LA), which selects transport format—that is, modulation and coding scheme (MCS)—and allocates power to the assigned [time, frequency, antenna port] resource. Primarily, the scheduler in the eNode B executes the preceding resource assignment function, although the antenna configuration can be seen as a somewhat separated function from the generic scheduler operation. The scheduler selects the time-frequency resource to assign to a particular UE based on the channel conditions and the QoS needs of that UE. Then the LA function selects MCS and allocates power to the selected time-frequency resources. More information on the scheduler and on the LA function is presented in Sections 4.2.2 and 4.2.3. The antenna configuration, such as the MIMO mode and its corresponding parameters (e.g., the precoding matrix), can be controlled basically separately from the time-frequency assignments of the scheduler, although the two operations are not totally independent. More details on the antenna configuration control are discussed in Section 4.2.7.

In an ideal case, the assignment of [time, frequency, antenna port] resources and the allocation of MCS and power setting would need to be done in a network-wide manner on a global knowledge basis in order to obtain the network-wide optimum assignment. However, for obvious reasons, this is infeasible in practical conditions because such a solution would require a global "super scheduler" function operating based on global information. Therefore, in practice, the resource assignment is performed by distributed entities operating on a cell level in the individual eNode Bs. However, this does not preclude some coordination between the distributed entities in neighbor eNode Bs—an important aspect of the RRM architecture that needs to be considered in LTE. Such neighbor eNode B coordination can be useful in the case of various RRM functions such as intercell interference coordination (ICIC). These aspects will be discussed in the sections focusing on the particular RRM function later in this chapter.

We can differentiate the following main RRM functions in LTE, each of which will be discussed separately in Section 4.2:

- radio bearer control (RBC) and radio admission control (RAC);
- dynamic packet assignment–scheduling;
- link adaptation and power allocation;
- handover control;
- intercell interference coordination;
- load balancing;
- MIMO configuration control; and
- MBMS (multicast broadcast multimedia services) resource control.

4.1.3 Radio Resource Related Requirements

Prior to the development of the LTE concept, the 3GPP defined a number of requirements that this new system should fulfill. These requirements vary depending on whether they are related to the user perceived performance or to the overall system efficiency and cost. Accordingly, there are requirements on the peak user data rates, user plane and control plane latency, and spectrum efficiency. The requirements on the spectral efficiency or on the user throughput including average and cell edge throughputs are formulated as relative measures to baseline HSPA (high-speed packet access)—that is, the 3GPP release-6 standards suite—performance. For example, achieving a spectral efficiency and user throughput of at least two to three times that of the HSPA baseline system is required. The downlink and uplink peak data rates should reach at least 100 and 50 Mb/s (in a 20-MHz band), respectively. For the full set of requirements, see reference 6.

It is clear that fulfilling such requirements can be possible only with highly efficient radio resource management techniques able to squeeze out the most from the instantaneous radio link conditions by adapting to the fast fluctuations of the radio link and by exploiting various diversity techniques. With respect to adapting to radio link fluctuations, fast link adaptation and link quality dependent scheduling have high importance; in terms of diversity, the various MIMO schemes, such as transmit diversity, spatial multiplexing, and multiuser MIMO, play a key role.

4.2 Radio Resource Management Procedures

In order to meet the RRM related requirements for LTE, 3GPP TS 36.300 [14] and 3GPP TR R3.018 [16] list the RRM functions that need to be supported in LTE. In this chapter, we list and discuss these functions with the understanding that the interplay of the various RRM functions is an important aspect, although it is typically not subject to standardization. For instance, ICIC may result in limitations in the usage and power setting of certain resource blocks that affect the operation of the dynamic resource allocation (scheduler). Also, radio bearer control may have an impact on the operation of RAC by manipulating threshold values that the RAC takes into account when making an admission decision. Likewise, at the time of writing, the interaction between intercell power control relying on overload indication and intercell interference coordination relying on traffic load indication is currently under study by the 3GPP.

4.2.1 Radio Bearer Control (RBC) and Radio Admission Control (RAC)

The establishment, maintenance, and release of radio bearers (as defined in 3GPP TR 25.813 [17]) involve the configuration of radio resources associated with them. When setting up a radio bearer for a service, RBC takes into account the overall resource situation in LTE, the QoS requirements of in-progress sessions, and the QoS requirement for the new service (see Figure 4.3). RBC is also concerned with the maintenance of radio bearers of in-progress sessions at the change of the

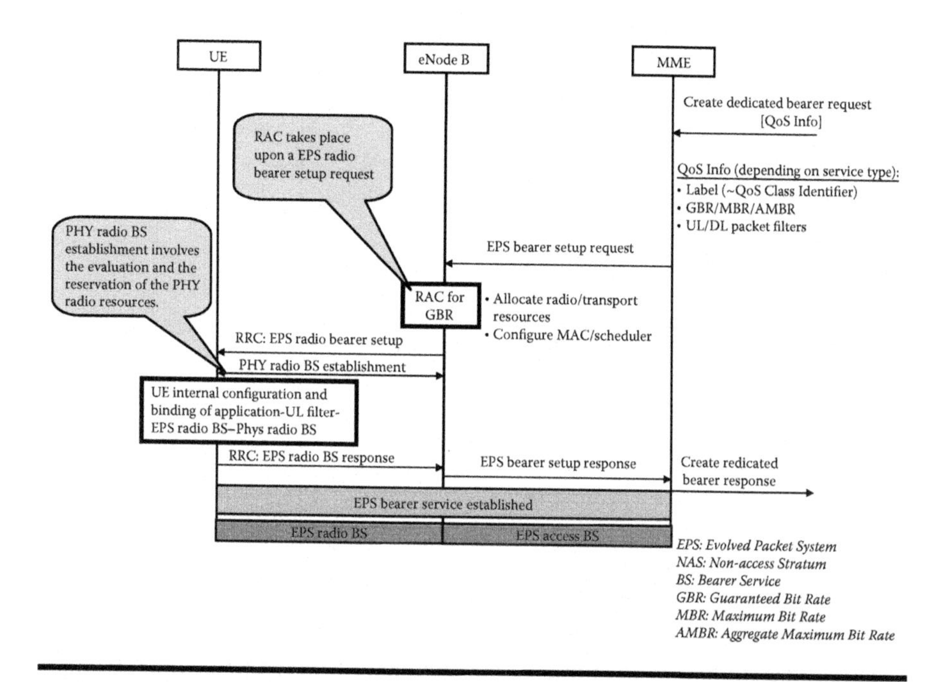

Figure 4.3 Radio admission control (RAC) in conjunction with evolved packet system (EPS) bearer service establishment. From the LTE radio access network's perspective, the mobility management entity of the core network requests an EPS bearer (characterized by a set of QoS parameters). The RAN exercises admission control for guaranteed bit rate (GBR) services and, in case of admission, establishes the underlying physical radio bearer service that will support the requested EPS bearer.

radio resource situation due to mobility or for other reasons. RBC is involved in the release of radio resources associated with radio bearers at session termination, handover, or other occasions. It is important to realize that RBC and "setting up" a radio bearer in LTE do not imply the static assignment and dedication of radio resources to users or user data flows. For example, when the RAC is executed upon a radio bearer setup, the RAN assesses the necessary radio resources (typically on a statistical basis) and makes an admission decision. Thus, RAC has the responsibility to keep the overall load within the feasible region in which the RAN remains stable and is able to deliver the expected QoS. Subsequently, it is the task of the scheduler to assign resources dynamically to users so that the QoS commitments are indeed kept and radio resources remain highly utilized.

Radio admission control has the task to check the availability of radio resources when setting up a GBR radio bearer (upon an EPS bearer request from the core network). RAC may also be executed at (initial) RRC connection request from the UE—that is, when the UE attempts to enter connected mode [14]. Although at

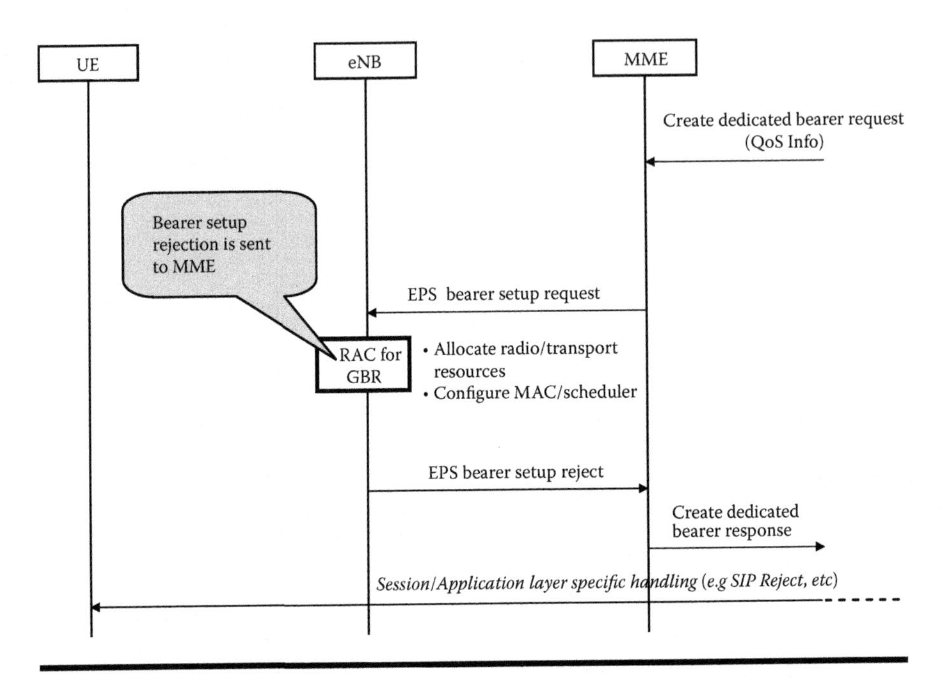

Figure 4.4 If the radio access network cannot support the requested EPS bearer, it rejects the bearer request. Note that UE is not involved in the EPS bearer setup procedure other than being notified by higher layer signaling—for example, using the session initiation protocol (SIP) between the core (service) network and the UE.

this stage the UE does not specify the requested service, the RAN may reject such a connection request due to a heavy load situation. The RAN may also reserve a *default bearer* to the UE so that as soon as the UE gets connected, an instant access to best effort services can be provided for the UE (as opposed to the service-specific bearers that need to configured and established according to the specific service requirements). RAC can be seen as part of the more general (overall) admission control procedure that also checks transport and hardware resources before admitting a new radio bearer or a radio bearer that is handed over from another eNode B. RAC is seen as a single cell RRM function and it does not require inter-eNode B communication (see Figures 4.3 and 4.4).

Although RAC is mainly outside the scope of standardization, some aspects can be expected to be common for various LTE implementations. The provided QoS for the EPS bearer service (and the associated physical radio bearer) and, specifically, the (average) bit rate are basically determined by:

A: the number of transferred bits while the bearer is accessing (using) the medium; and
B: how often and for how long the bearer gets access to the medium (see also Figure 4.5).

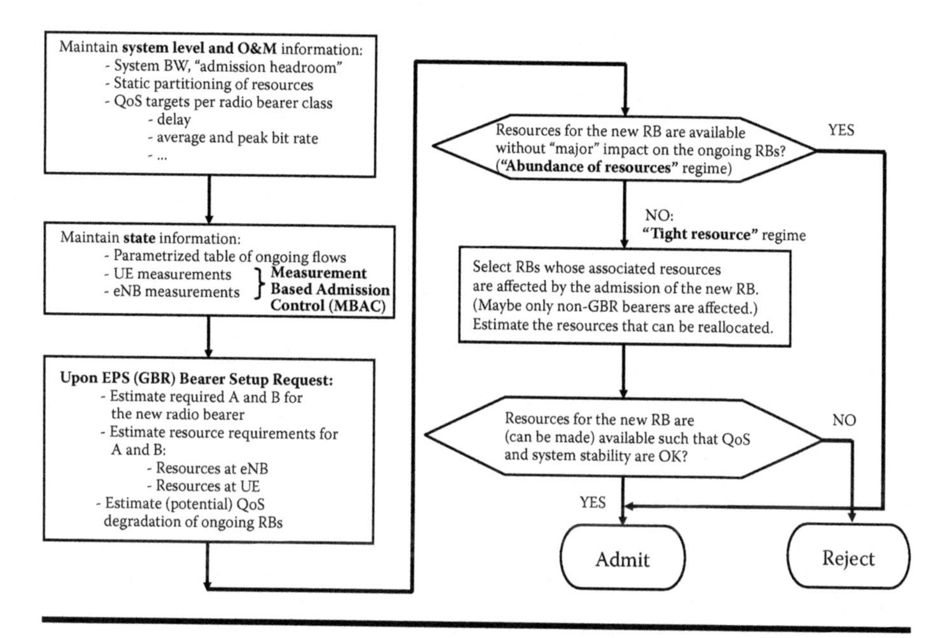

Figure 4.5 A high-level (schematic) view of a possible admission control algorithm. It is important to realize that because AC is not standardized, different realizations of LTE RANs will run different admission control algorithms. In this figure we depict an approach according to which the AC algorithm divides the RAN load situation into a lightly loaded and a heavily loaded regime in order to facilitate fast admission (accept/reject) decisions.

The first aspect (A) depends on the signal-to-interference-and-noise (SINR) ratio and the applied MCS on the scheduled resource blocks as well as on the number of scheduled resource blocks. The B aspect is determined by the load (in terms of ongoing EPS and physical layer [PHY] bearers) and the associated QoS requirements. It follows that the input parameters to the admission control algorithm must allow for the evaluation of both aspects. Obviously, aspect B has a major impact on the user perceived packet delay. Therefore, different combinations of A and B can be appropriate for different services. For instance, a voice service requires a low-delay, regular access to the wireless medium with a relatively low average bit rate requirement. In contrast, the perceived QoS of a file download service is largely determined by the overall bit rate rather than by the experienced delay of individual packets.

Regarding the operation of the admission control algorithm, it can be useful to distinguish two load regimes. In the "abundance of resources" regime, the resources for newly arriving bearer requests are available with a broad margin, so a thorough inspection of the current resource situation (including radio and transport resources) is not necessary. In the "tight resource" regime, the resource requirement of newly arriving radio bearers and the availability of the necessary resources must be checked before the admission decision.

4.2.2 Dynamic Packet Assignment—Scheduling

As it has been described in Section 4.1.2, the radio interface in LTE is used as one common channel, shared by all users in the cell, which are scheduled in the time-frequency domain, optionally extended with the antenna configuration as a third dimension of the resource space. The eNode B controls the assignment of resources on the uplink and on the downlink shared channels, called the PUSCH (physical uplink shared channel) and the PDSCH (physical downlink shared channel), respectively. Correspondingly, it is necessary to differentiate a downlink scheduler and an uplink scheduler function in the eNode B. The location of the DL scheduler in the eNode B is a straightforward choice; actually, it is the only feasible choice. However, in principle, the UL scheduler function could have been distributed into the UEs as well, resulting in a UE-controlled multiple access scheme for the uplink channel. The main reason to place the UL scheduling control (centrally) into the eNode B is to maintain intracell orthogonality.

Although the primary objective and the operation are essentially the same for the uplink and the downlink schedulers, there are a few important differences in terms of the available channel state information and buffer status information at the eNode B for the uplink and the downlink channels.

The scheduler can assign resources in units of pairs of resource blocks (RBs), where an RB consists of 12 subcarriers in the frequency domain and one slot (0.5 ms) in the time domain, as illustrated in Figure 4.2. To signal the scheduled RBs pertaining to a particular UE for both the UL and DL channels, the PDCCH (physical downlink control channel) is used. If the UE recognizes its identity on the PDCCH, it decodes the corresponding control information and identifies the DL RBs that carry data addressed to the UE and the UL RBs that have been granted for the UE to send UL data. The PDCCH is carried in the first one to three OFDM symbols in each subframe. The number of OFDM symbols used for the PDCCH can vary dynamically from one subframe to the other, depending on, for example, the number of UEs to be scheduled in the given transmission time interval (TTI) or the granularity of the allocations.

In order to limit the control signaling overhead associated with the dynamic signaling of the RB allocation to the UE, the so-called semipersistent scheduling is supported. Semipersistent scheduling allows one to assign resources ahead in time—typically in a periodic pattern, which can be especially useful in the case of applications that generate predictable amounts of data periodically, like voice over Internet protocol (VoIP).

The scheduler selects the UEs to be scheduled and the RBs to be assigned primarily based on two factors: the channel quality and the QoS requirements of the radio bearers of the UE combined with the amount of pending traffic in the transmit buffers. The availability and the accuracy of the link quality and buffer status information in the eNode B for the UL and for the DL directions are fundamentally different.

Moreover, the freedom of the scheduler in selecting RBs for the same UE is also different in the UL and DL directions. In the DL the scheduler can assign any arbitrary set of RBs for the UE, but in the UL the RBs assigned to a particular UE have to be adjacent in order to maintain the single carrier property. As a consequence, the downlink scheduler can take full advantage of frequency-dependent scheduling, exploiting multiuser diversity in the frequency domain as well as in the time domain. In the UL the single carrier property limits the possibility of fully utilizing frequency selective scheduling.

4.2.2.1 Obtaining Channel Quality Information

In order to be able to perform channel-dependent scheduling on the downlink channel, the eNode B has to obtain channel quality reports from the UEs—at least for those that have pending DL data. The CQI (channel quality indicator) reports are used by the UE to send information about the DL channel quality back to the eNode B. In order to enable the UE to measure the channel quality on a resource block, so-called reference signals are transmitted in each RB. Out of the 12×7 REs (assuming normal cyclic prefix), four resource elements in each RB (for single antenna transmission) are used to transmit reference symbols. The reference symbols are needed also for channel estimation to enable coherent reception.

The CQI reports can be sent either on the PUCCH (physical uplink control channel), if the UE has no UL assignment, or on the PUSCH if the UE has a valid UL grant. The granularity of the CQI report in the frequency domain can be configured ranging from wideband reporting to per-RB reporting. (At the time of writing, the formats and triggering criteria for sending CQI reports have not yet been settled in the 3GPP.) The UE can send channel rank and precoding matrix reports bundled together with the CQI report in order to support multiantenna transmission at the eNode B. More details on the RRM function controlling the multiantenna transmission are discussed in Section 4.2.7.

To obtain channel-quality information on the uplink channel at the eNode B is somewhat easier than for the downlink channel because the eNode B can perform measurements on the UL transmission of the UE. Reference symbols similar to those in the downlink are inserted in each RB in the uplink. Note, however, that the channel quality can be estimated only on RBs on which the UE is actually transmitting. Because the UE typically does not transmit in the full bandwidth, the eNode B gets (relative) channel quality information only on RBs that it has assigned to the UE, while the RB assignment should be done according to best RB quality selected from the full bandwidth. (To obtain precise *absolute* channel quality, including uplink path loss, is made difficult by the fact that the UE exercises power control for uplink transmissions.)

In order to allow the eNode B to estimate the channel quality on all RBs from the same UE, it is possible to transmit so-called channel sounding reference signals from the UE. The UE transmits the sounding signal for one symbol duration within a subframe occupying the entire bandwidth, if the eNode B instructs the UE to do so. Recall that the flexibility of the RB assignments in the UL is constrained by the single carrier property. Therefore, realizing a fully flexible channel-dependent scheduling in the uplink is difficult, even if full bandwidth channel quality information is available.

4.2.2.2 Obtaining Buffer Status Information

What was difficult in obtaining channel quality information for the downlink versus the uplink is now the opposite case for the buffer status information. Downlink buffer status information is naturally available for the downlink scheduler because data buffering is done in the eNode B. However, because the UL buffers are located in the UE, the UL scheduler in the eNode B can have some (approximate) knowledge of the UL buffer status only if the UE reports this information to the eNode B.

Regarding the UL buffer status reporting, it is useful to differentiate two cases depending on whether the UE has a valid UL grant (i.e., the UE is in the middle of a continuous UL transmission) or does not have a UL grant (i.e., its UL buffers were emptied in a previous scheduling instance and it needs to request new UL resources upon arrival of the first packet into the empty buffer). To request resources in this latter case, there are two possibilities in LTE: using the random access channel (RACH) or a dedicated scheduling request (SR) resource on the PUCCH, if there has been such a resource assigned to the UE by the eNode B. The SR sent on the PUCCH consists of only one bit of information, indicating only the arrival of new data. The main advantage of the dedicated SR as compared to RACH-based request is that it provides a faster access to UL resources, due to being contention free, and at the same time off-loads the RACH channel from frequent UL resource requests resulting from bursty packet data sources.

A prerequisite for having a dedicated SR resource is that the UE must have UL time synchronization; that is, it must have up-to-date time alignment value from the network, requiring that the UE had UL transmission within the past ×100 ms (the maximum time before the clocks in the UE and in the eNode B drift in the order of the cyclic prefix, 4.7 μs). However, not all UEs with UL time synchronization may have such a dedicated resource because the number of such resources is limited. The eNode B can assign and revoke SR resources via higher layer signaling, depending on the number of active UEs in the cell, the activity of the UE, etc. If the UE does not have such a dedicated SR resource, it has to rely on the normal RACH procedure to request a UL resource in a contention-based or contention-free manner. The RACH procedure can be made contention free if the eNode B assigns a dedicated preamble for the UE to perform the random access, which can be seen as a way of polling the UE.

Once the UE has a valid UL grant, it can send a detailed buffer status report via MAC control signaling, carried in the MAC header of UL user data. This means

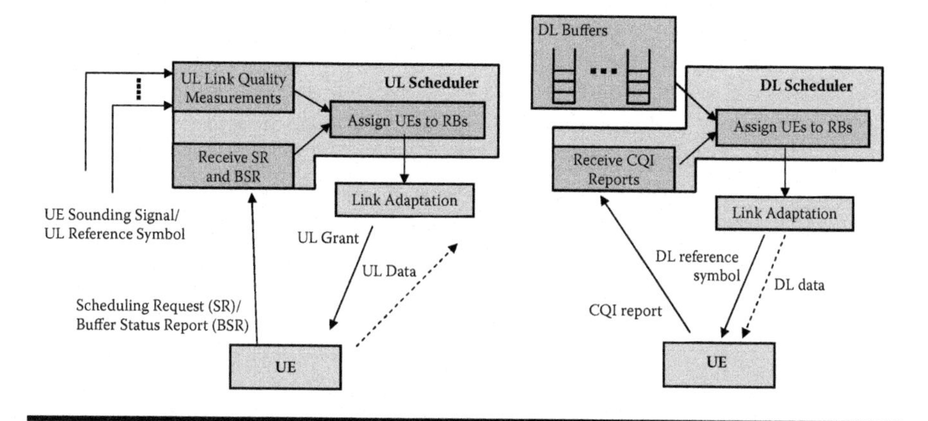

Figure 4.6 Illustration of UL and DL scheduler functions.

that during a continuous flow of data, the UE can send updated buffer status reports via in-band signaling and, in response, the eNode B will continuously assign new UL grants. We illustrate the operation of the UL and DL schedulers in Figure 4.6.

4.2.3 Link Adaptation and Power Allocation

Once the scheduler has selected the set of RBs to be assigned to a particular UE, the MCS and the power allocation have to be determined. This is done by the link adaptation function. Although, strictly speaking, the link adaptation is not part of the scheduler function, it is closely related to the resource assignment as such because the resource is fully defined not only by the time-frequency allocation but also by the format (i.e., modulation and coding) in which data should be transmitted on the given resource. The selection of MCS is done by the eNode B for the uplink and for the downlink. For the downlink direction, the selection is done based on CQI reports from the UE, taking into account the buffer content as well; for the uplink, it is selected based on the measured link quality at the eNode B and the buffer status report from the UE. Note that when selecting the MCS for the transmission in the next subframe, the eNode B has to predict the link quality—that is, the SINR based on measurements in previous subframes (UL) or based on previous CQI reports (DL).

Therefore, the predictability of the interference conditions has high importance from the MCS optimality point of view. Large and uncorrelated interference variations from one subframe to the other make the link prediction very difficult. From this aspect, the intercell interference coordination function can have a positive effect not only in decreasing the level of interference but also in decreasing the time variation of the interference and thereby making the link quality more predictable [29]. Finally, the selected MCS is signaled together with the downlink/uplink scheduling assignment to the UE on the DL control channel (PDCCH). This means that neither the UE nor the eNode B has to do blind decoding. The UE decodes the data received on PDSCH according to the MCS indicated on the PDCCH. In the UL, the eNode B decodes the UE transmission according to the MCS it has assigned to the UE associated with the UL grant.

The power allocation is also under eNode B control, and it is tightly coupled to the MCS selection. A given MCS is optimal only at a given SINR. Therefore, the selection of the MCS is always done with a target SINR in mind. Then it is the responsibility of the power control function to set the transmit power levels such that the target SINR is reached. For the DL transmission, the eNode B distributes its power on the RBs according to the corresponding target SINRs. In the simplest case, the DL power is distributed uniformly over the RBs (i.e., no downlink power control is employed). For a close to optimum power allocation, the so-called *water-filling power allocation* might be used [10,11], whereby higher power is allocated to subcarriers whose fading and interference are in favorable conditions. However, the downlink power allocation is fully controlled by the eNode B, so the power allocation algorithm does not need to be specified in the standard.

However, to control the UL transmission power, the eNode B needs to send power control commands to the UE, which needs to be specified in the standard. Similarly, the behavior of the UE in response to receiving such a power control command also needs to be specified. This means that the UL power control algorithm needs to be specified in the standard. According to this algorithm, the UE transmit power is set as follows:

$$min\ (P_{max},\ 10\log_{10}(M) + P_0 + \alpha \times PL + \Delta_{MCS} + f(\Delta_i))$$

where

P_{max} is the maximum allowed power;

M is the number of resource blocks assigned to the UE;

P_0 is a UE-specific parameter;

α is a cell-specific path loss compensation factor;

PL is the downlink path loss calculated in the UE based on reference power measurements;

Δ_{MCS} is a modulation- and coding scheme-specific parameter (the table of Δ_{MCS} values is configured in the UE via RRC signaling; MCS is signaled in the scheduling assignment); and

function $f()$ is also signaled via RRC, while Δ_i is the actual transmit power command signaled in each scheduling assignment.

4.2.4 Handover Control

Handover control is responsible for maintaining the radio link of a UE in active mode as the UE moves within the network from the coverage area of one cell to the coverage area of another. In LTE, the handovers are *hard handovers* (similarly to global system for mobile communications [GSM] systems) with preparation at the target cell. The handover is network controlled and UE assisted. Hard handover means that the switch from one cell to the other happens in a "break-before-make" fashion; that is, the UE has connectivity to only one cell at a time. This is in contrast to the 3G wideband code division multiple access (WCDMA) system, which employs soft handover and fast power control and in which the UE can be associated with multiple cells at the same time.

One of the most important reasons why soft handover is used in WCDMA is the interference sensitivity of the system (especially in the uplink), stemming from the nature of code division multiple access. When soft handover is used the transmit power of the UEs, the caused interference can be decreased because the diversity gain of soft handover compensates for the smaller transmit power. In LTE there is perfect intracell orthogonality owing to the OFDM multiple access scheme, so the system is not (intracell) interference sensitive in the same sense as a WCDMA system. Moreover, LTE can exploit diversity in a number of other ways than soft handover, such as multiantenna transmission modes; because of the fast link adaptation and channel-dependent scheduling functions, it can adapt to instantaneous channel conditions in the best possible way, which makes the importance of a soft handover solution negligible.

The network-controlled and UE-assisted property of the handover means that the decision to move the radio link connection of the UE from one cell to the other is made by the network—more specifically, by the eNode B serving the UE, assisted by measurement reports received from the UE. The eNode B can utilize a number of other information sources as well for making the handover decision, including its own measurements, the availability of radio resources in candidate cells, load distribution, etc. However, the most important aspect that should drive the handover decision is the UE path gain measure. In other words, the handover control should ensure that the UE is always connected to the cell with the best average path gain. This is especially important in reuse-1 systems like LTE, where a UE connected to a cell other than the best cell may cause substantial additional interference to neighbor cells, especially to the best cell.

Another consequence of the reuse-1 system is that the link quality (i.e., the SINR) may change rapidly due to intercell interference as the UE moves toward the cell edge. The use of ICIC techniques may mitigate the high cell edge interference effects (see Section 4.2.5 for more details on ICIC). Nevertheless, the fast deterioration of the link quality at the cell edge means that the system needs to act rather quickly upon the changing link conditions before the link gets lost, which requires a fast handover execution and signaling mechanisms that are robust with respect to intercell interference. More specifically, the overall handover procedure time has to be reliable and short enough, including the time elapsed until the handover situation is recognized, the time needed for preparing the handover at the target cell, and the time needed for executing the handover. Another consequence of the fast change of the link quality at the cell edge is that a large hysteresis in the source

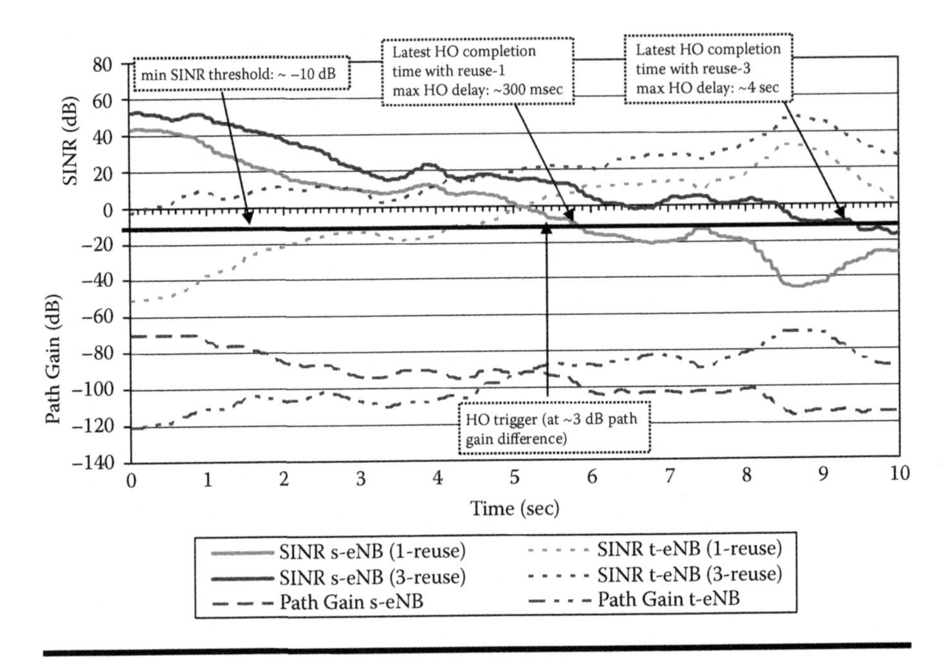

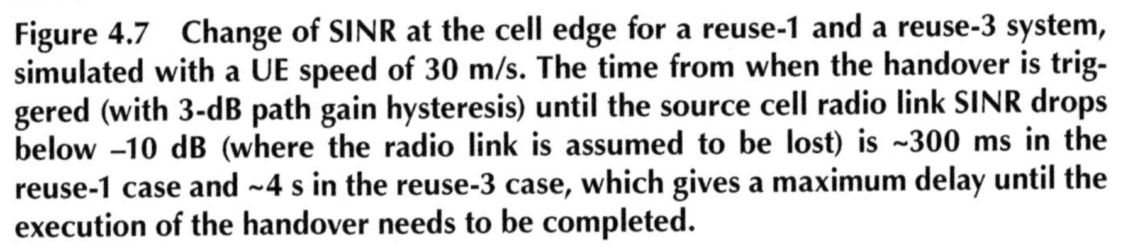

Figure 4.7 Change of SINR at the cell edge for a reuse-1 and a reuse-3 system, simulated with a UE speed of 30 m/s. The time from when the handover is triggered (with 3-dB path gain hysteresis) until the source cell radio link SINR drops below –10 dB (where the radio link is assumed to be lost) is ~300 ms in the reuse-1 case and ~4 s in the reuse-3 case, which gives a maximum delay until the execution of the handover needs to be completed.

and target cell path gain differences, used for the handover decision, may not be allowed. However, a smaller hysteresis may trigger more handovers.

Figure 4.7 shows the change of SINR at the cell edge (obtained from simulations) in the function of time as the UE moves with a speed of 30 m/s from one cell to the other in a reuse-1 and in a reuse-3 system, respectively. As it can be seen in the figure, the SINR deteriorates more rapidly in the reuse-1 system, leaving a shorter time for the execution of the handover (~300 ms in the reuse-1 case and several seconds in the reuse-3 case).

A fast handover execution is not only needed to combat the rapid change of link quality but also is important from the user-perceived performance point of view. This means that, in order to achieve good handover performance from a radio efficiency and also from a user-perceived quality point of view, it is required to have low interruption time and no user data packet losses during the handover. In order to meet these requirements, the following handover procedure is used in LTE, as illustrated in Figure 4.8.

Without discussing all details of the procedure, we would like to point out a few important aspects to observe. After the decision for a handover has been made in the

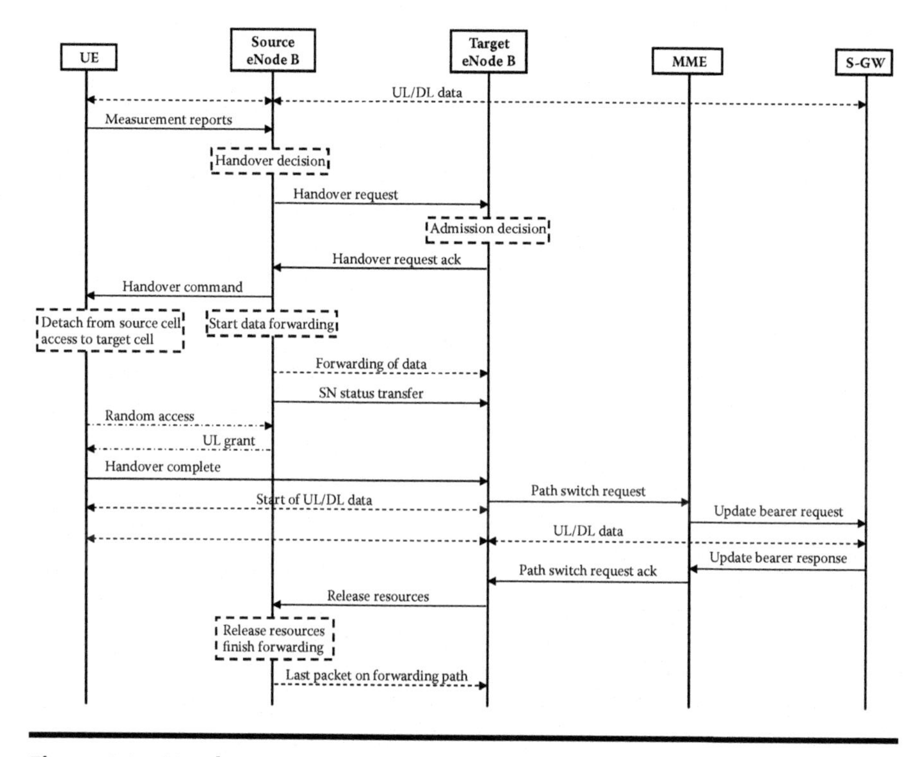

Figure 4.8 Handover message sequence.

source eNode B, it signals to the selected target eNode B and requests the reservation of resources. If the admission decision has passed successfully in the target, the target eNode B prepares a transparent container (which is for interpretation only for the UE, including necessary information for the UE to access the target cell) and sends it back to the source eNode B in the handover request acknowledge message. Next, the source eNode B can command the UE to execute the handover and at the same time start the forwarding of user data. The source eNode B forwards PDCP service data units (SDUs) (i.e., IP packets that could not be successfully sent in the source cell) toward the target eNode B. Note that the L2 protocols including RLC/MAC are reset in the target cell (i.e., no HARQ/ARQ status information is preserved); the header compression engine in the PDCP layer is also reset. For more details on the handover procedure, see 3GPP TS 36.300 [14].

When the UE arrives at the target cell, it accesses the cell via the RACH. However, in order to reduce the interruption time due to potential collisions on the RACH, it is possible to use a dedicated preamble for the access. The term *preamble* refers to the signature sequence that is sent by the UE on the RACH slot, and it is used to identify the access attempt. There are 64 preambles for contention-based access, and a separated set of preambles is used in a contention-free, dedicated manner—for example, for handover access (or for access to regain uplink time synchronization). The target eNode B can reserve a dedicated preamble for the particular handover instance of the UE and can signal this preamble to the UE via the transparent container in the handover command. Because the preamble is dedicated, no other UEs can use it at the same time, which ensures that the access attempt will be contention free. With the preceding handover scheme, an interruption time in the range of 15 ms can be ensured.

Finally, it is worth mentioning that LTE provides an efficient recovery mechanism for handover failure cases, when the handover could not have been commanded by the network due to the loss of the radio link. Although such radio link losses should be rare events in a well-planned network, their occurrence cannot be completely ruled out, especially due to the potential harsh interference conditions on the cell edge. If the UE loses the radio link, it reselects to a suitable cell and initiates a connection reestablishment. If the UE context is available at the selected eNode B (i.e., if the UE reselects to a cell belonging to the source eNode B or to a cell of an eNode B that has been prepared for a handover), the UE context can be recovered. In such cases, the interruption time and the user-perceived performance will be almost as good as in the nonfailure case. In all other cases, the UE has to reestablish connectivity via an idle to active state transition, which will take a somewhat longer time. Also, if the source eNode B wants to decrease the impact of a potential handover failure, it can prepare multiple target cells; later, after the handover has been successfully completed, it can cancel the preparation in the other cells.

4.2.5 Intercell Interference Coordination (ICIC)

Intercell interference coordination has the task to manage radio resources (notably the radio resource blocks) so that intercell interference is kept under control. The specific ICIC techniques that will be used in LTE are still in some key points that have already been agreed to [46]. ICIC is inherently a multicell RRM function that needs to take into account the resource usage status and traffic load situation of multiple cells. The presingle-antenna, as well as multiple-antenna systems, has been actively studied by the research community (see, for instance, references 24–41). In this chapter we focus on the process.

Within the 3GPP there is a fairly wide consensus that LTE should be a reuse-1 system in which all resource blocks should be used by each cell. In such systems, UEs served by neighboring cells may cause (uplink) interference to eNode Bs, as illustrated in Figure 4.9, while eNode Bs may cause downlink interference to served UEs. However, eNode Bs can employ scheduling strategies that allow them to reduce the probability for causing such intercell interference by carefully selecting the scheduled resource blocks.

A resource block collision between two cells, as described in Figure 4.9, can be reduced either by avoiding scheduling some of the resource blocks in some of the cells (reuse-n, $n > 1$, or fractional reuse; see Figure 4.10) or by coordinating the

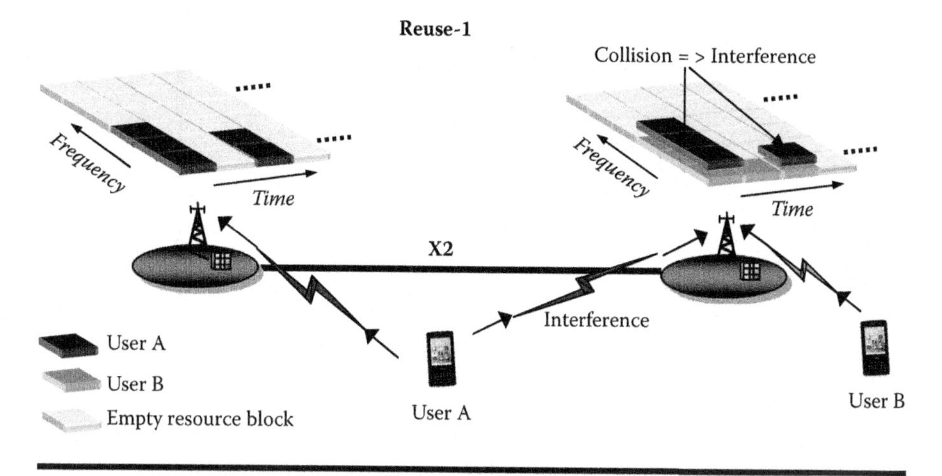

Figure 4.9 Intercell interference in a reuse-1 system (showing the uplink case).

allocation of the resource blocks in neighbor cells. In case of fractional reuse, certain resource blocks are barred for use on the cell edge, as illustrated in Figure 4.11. Because reuse-*n* systems tend to underutilize radio resources, the consensus within the 3GPP is that LTE should be a reuse-1 system [46]. Similar arguments can be raised against fractional reuse systems with certain resource blocks barred on the cell edge. Therefore, in LTE such (static) barring will not be employed, either. Building on the reuse-1 agreement, the coordination of the scheduler operation as an optional means to improve the cell edge SINR distribution is, however, supported.

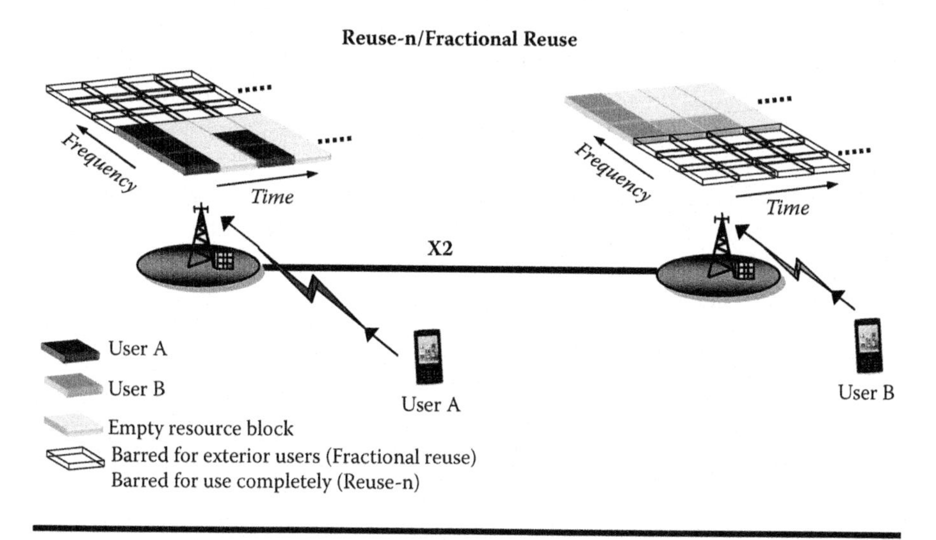

Figure 4.10 Intercell interference in a reuse-*n* or in a fractional reuse system.

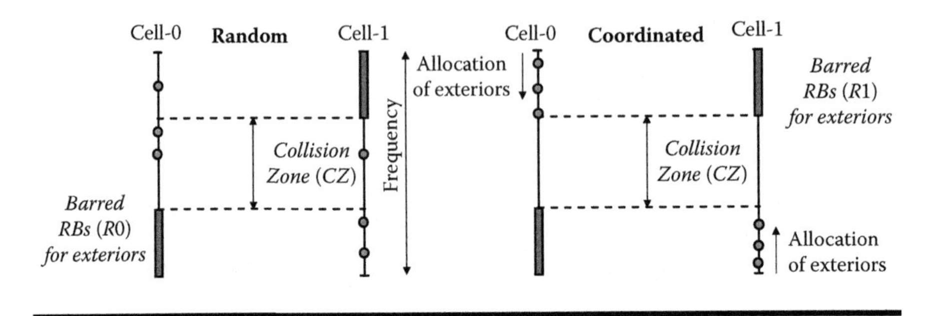

Figure 4.11 Example of fractional reuse with (static) barring of resource blocks on the cell edge. In a random allocation, both cells select the RBs to schedule UEs randomly in the frequency domain (or based on the frequency-dependent channel quality of RBs). In the coordinated allocation, the two cells start the allocation of RBs from the two ends of the frequency band, starting the allocation with the exterior UEs first. In addition, (static) barring of RBs to be used for exterior UEs may be employed in both cells; this provides a guaranteed zone where no exterior–exterior collisions can occur.

Intercell interference coordination (including the coordination of resource block scheduling and power allocation) can be thought of as a set of means that reduces the probability and mitigates the impact of intercell *collisions*. In fact, these types of "collision models" have been extensively studied in the literature [26,27,38–41] and have been the subject of system-level simulations within the 3GPP [42].

In principle, the coordination of RB allocation between the cells can be performed in the time or in the frequency domain. Because a time domain coordination performed on the scheduling time scale (1 ms) would be infeasible due to, among others, delay and generated signaling load on the X2 interface (while a time domain coordination on a longer time scale would imply increased radio interface delays and underutilization of cell resources), the primary approach adopted for ICIC in LTE is the frequency domain coordination.

Before the details of how to avoid resource block collisions are discussed and in order to better understand the expected gains of ICIC algorithms, it is worth investigating the impacts of potential collisions. We can observe the following consequences that a collision may have:

- Fewer user data bits can be carried in one RB because the link adaptation needs to select lower modulation order or lower coding rate to compensate the lower SINR.
- Fewer numbers of RBs can be allocated to the UE in one subframe due to hitting the UE power limit (resulting in higher UE power consumption as well).
- More HARQ retransmissions may be needed for successful data delivery.

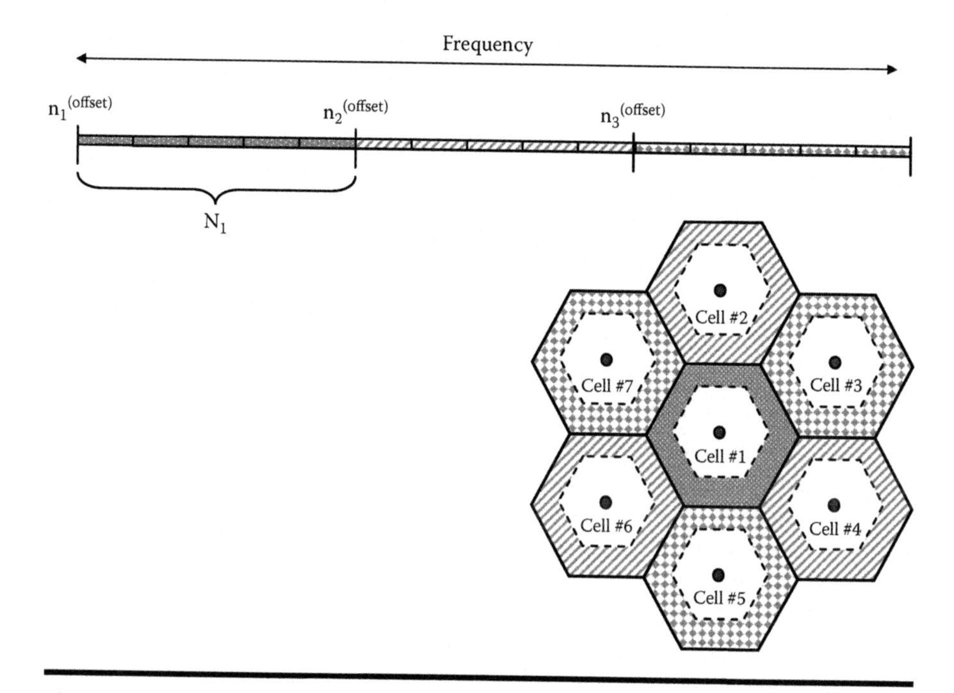

Figure 4.12 Illustration of frequency offset assignment to cells in the proposed ICIC algorithms.

If a collision occurs, the scheduler will need to assign more resources (RBs) to the UE to compensate the loss in the carried number of bits due to any of the preceding consequences (i.e., fewer bits per RB, more retransmission, etc.). This means that more resources (and maybe longer time) will be used to carry the same amount of user bits in the no-ICIC case than in the ICIC case. However, it is important to observe that this difference in the number of RBs used to carry the given amount of data will not necessarily appear in the most typical performance measure, in the user throughput. Whether the collisions will have an effect on the throughput or not depends to a large extent on the type of the traffic model.

We can differentiate full buffer and nonfull buffer traffic models, each of which can be further classified as peak rate limited or non-peak rate limited. The full buffer assumption means that an unlimited amount of data is waiting for transmission (i.e., basically one traffic source could utilize the full bandwidth in each subframe, unless the available power limits the number of RBs that can be assigned). In the nonfull buffer case, it is assumed that the traffic source generates finite amounts of data in bursts with certain interarrivals, where there is an idle period between the service time of two consecutive bursts (i.e., there is no continuous load on the system). Peak rate limitation means that the there is an upper limit on the maximum number of bits (or number of RBs) that can be assigned for transmission within a given time interval for the particular UE. In the simplest case, the peak rate limitation can be interpreted on the subframe time scale, imposing a limit on the maximum number of RBs that can be assigned to the UE in one subframe. In what follows, we show results for the *full buffer–peak rate limited* and for the *nonfull buffer–non-peak rate limited* scenarios.

Although the actual ICIC algorithm is not standardized, 3GPP R1-074444 [42] analyzes the performance of some alternative ICIC methods. Because of its relevance to the current 3GPP status, we discuss some of the results of this latter contribution.

In one of the proposed approaches for UL ICIC in LTE, (Figure 4.12) each cell is assigned a color that corresponds to a specific offset value (n_1, n_2, n_3) in the frequency domain. The offset value is an important parameter to the examined ICIC algorithms as described in the following. The assignment of the colors to the cells can be done via the operation and maintenance subsystem or dynamically between eNode Bs utilizing the X2 interface.

To realize the potential gain from avoiding resource block collisions, there is a need to distinguish UEs residing in the interior and exterior parts of the cell, basically similarly to proposed reuse partitioning techniques. This is because, typically, only collisions between cell edge (exterior) UEs cause noticeable SINR degradation. In the following we discuss simulation results that were obtained assuming a quasi-static preconfiguration of parameters using the following ICIC schemes:

First, the scheduler determines for each subframe which UEs will get scheduled and how many resource blocks they will get, based on the fairness and QoS criteria. Then, depending on the ICIC scheme, it will be selected which particular resource blocks in the frequency domain will be assigned to the UE:

- *No ICIC (reference case).* The eNode B scheduler does not employ restrictions on the schedulable resource blocks. That is, the scheduler in each cell works independently of the used resource blocks in the neighboring cells.
- *Start–stop index (SSI).* In this scheme, there is a start index ($n_i^{(offset)}$) and a stop index ($n_i^{(offset)} + N_i$) associated with the set of available resource blocks. The scheduler uses the resource blocks between the start and stop indexes for exterior UEs. If this pool of resource blocks is depleted, some exterior UEs will not get scheduled within a specific subframe. If resource blocks remain in this pool after exterior UEs have been scheduled, they can be utilized by interior UEs. Using disjoint subsets of resource blocks (defined by the start and stop indexes) in neighboring cells, exterior collisions can be completely avoided. Interior UEs are scheduled on the remaining resource blocks (i.e., after whatever resource blocks are allocated to the exterior UEs).
- *Start index (SI).* This scheme is similar to the SSI scheme, but there is no stop index. That is, the scheduler schedules exterior UEs starting from the resource block identified by the start index. Because there is no stop index, all resource blocks in this scheme may be assigned to exterior UEs (although such a situation is unlikely to happen, assuming proper dimensioning and a call admission control mechanism).

- *Random start index (RSI).* This scheme is similar to the SI scheme, except that the start indexes are defined without cell-wise coordination. This scheme is a fully distributed scheme in the sense that a central intelligence that assigns the start indexes is unnecessary.
- *Start index geometry weight (SIGW).* This scheme is similar to SI but uses a continuous measure (which we call the geometry index) to sort the UEs (rather than distinguishing exteriors and interiors). Thereafter, the scheduling algorithm is similar to that of the SI scheme, except that now the scheduler schedules the most exterior UEs first, starting from a preconfigured start index, and proceeds toward the interior UEs.
- *Random index geometry weight (RIGW).* This scheme is a combination of the RSI and SI schemes; that is, UEs are sorted in terms of their geometry index ("most exterior" being scheduled first) from a randomly chosen start index.

Numerical results for these schemes (Figure 4.13) indicate that geometry-based ICIC schemes provide the highest performance gains as compared to the other ICIC mechanisms. Therefore, it is important to obtain UE geometry information with sufficient accuracy by the scheduling eNode B. Recognizing this, the use of measurement reporting techniques to obtain this knowledge is currently under discussion in the 3GPP [45]. The ICIC gain is greatly dependent on the load of the system, which is in line with the findings of several other papers; see, for instance, Fodor [27] and Kiani, Øien, and Gesbert [33].

Figure 4.14 plots results for the *nonfull buffer–non-peak rate limited* traffic scenario as well (plotting only the no-ICIC and the best ICIC algorithms). As it can be seen from the figure, only a negligible difference is found in the fifth percentile throughput curves of the ICIC and the no-ICIC cases. Because there is no continuous traffic load on the system (nonfull buffer case), the occasional collisions that occur in the no-ICIC case can be compensated by additional RB allocation, which results in the same user-level throughput. However, the differences in terms of consumed UE power and lower number of bits carried per RB are clearly visible from the figure.

At the time of writing, the status of ICIC is captured by 3GPP R1-075014 [46]. According to this fairly broad consensus in RAN1, the release-8 LTE standard will not support ICIC in the downlink. Uplink intercell interference coordination consists of two interrelated mechanisms, the details of which are currently under discussion within the 3GPP.

The first part is a proactive ICIC mechanism supported with communication over the X2 interface between eNode Bs [46]. The basic idea of this scheme is that a potentially disturbing eNode B proactively sends an RB-specific indication to its potentially disturbed neighbor. This message indicates which resource blocks will be scheduled (with a high probability) with high power (i.e., by cell edge UEs). Thus, this message allows the receiving eNode B to try to avoid scheduling

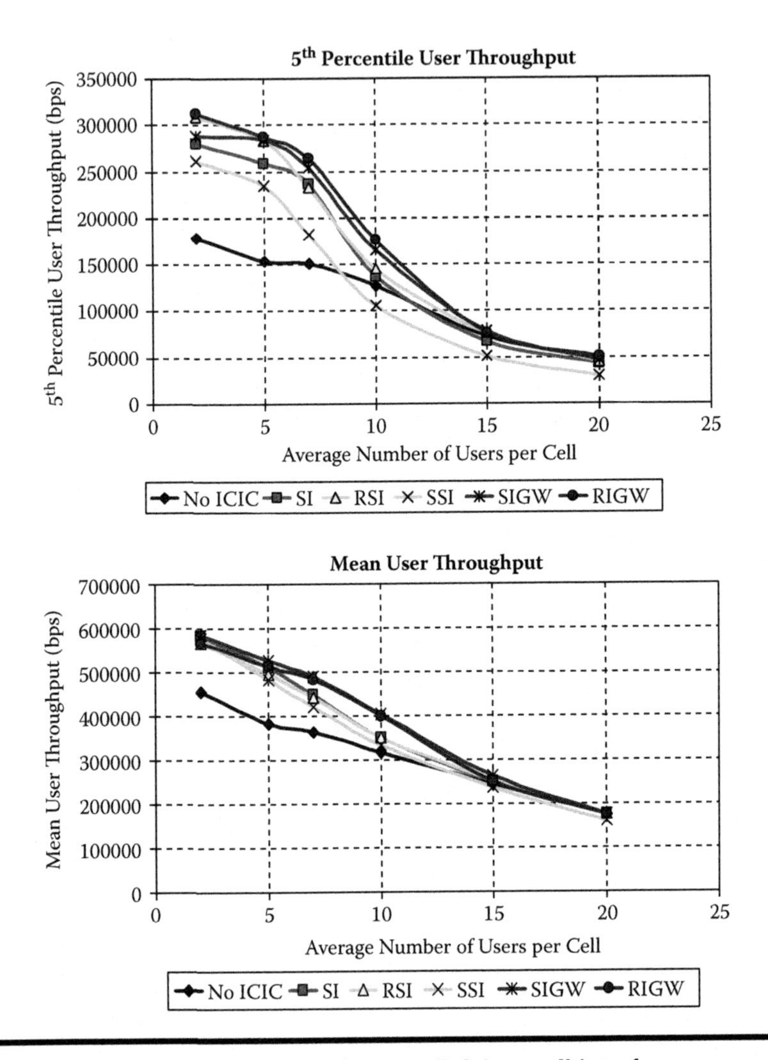

Figure 4.13 System simulation results on uplink intercell interference coordination; full buffer–peak rate limited traffic scenario.

the same resource blocks for its cell edge UEs. This way the proactive scheme allows neighbor eNode Bs to reduce the probabilities of "exterior–exterior" (i.e., cell edge) UEs simultaneously taking into use the same resource blocks. The avoidance of such exterior–exterior collisions has been found useful by system-level simulations. An important part of the proactive scheme is the identification of cell edge UEs. The 3GPP currently studies the use of UE measurement reporting (similar but not identical to that used for handovers) for this purpose (see references 43–48); these proposals are expected to be discussed more at future 3GPP meetings.

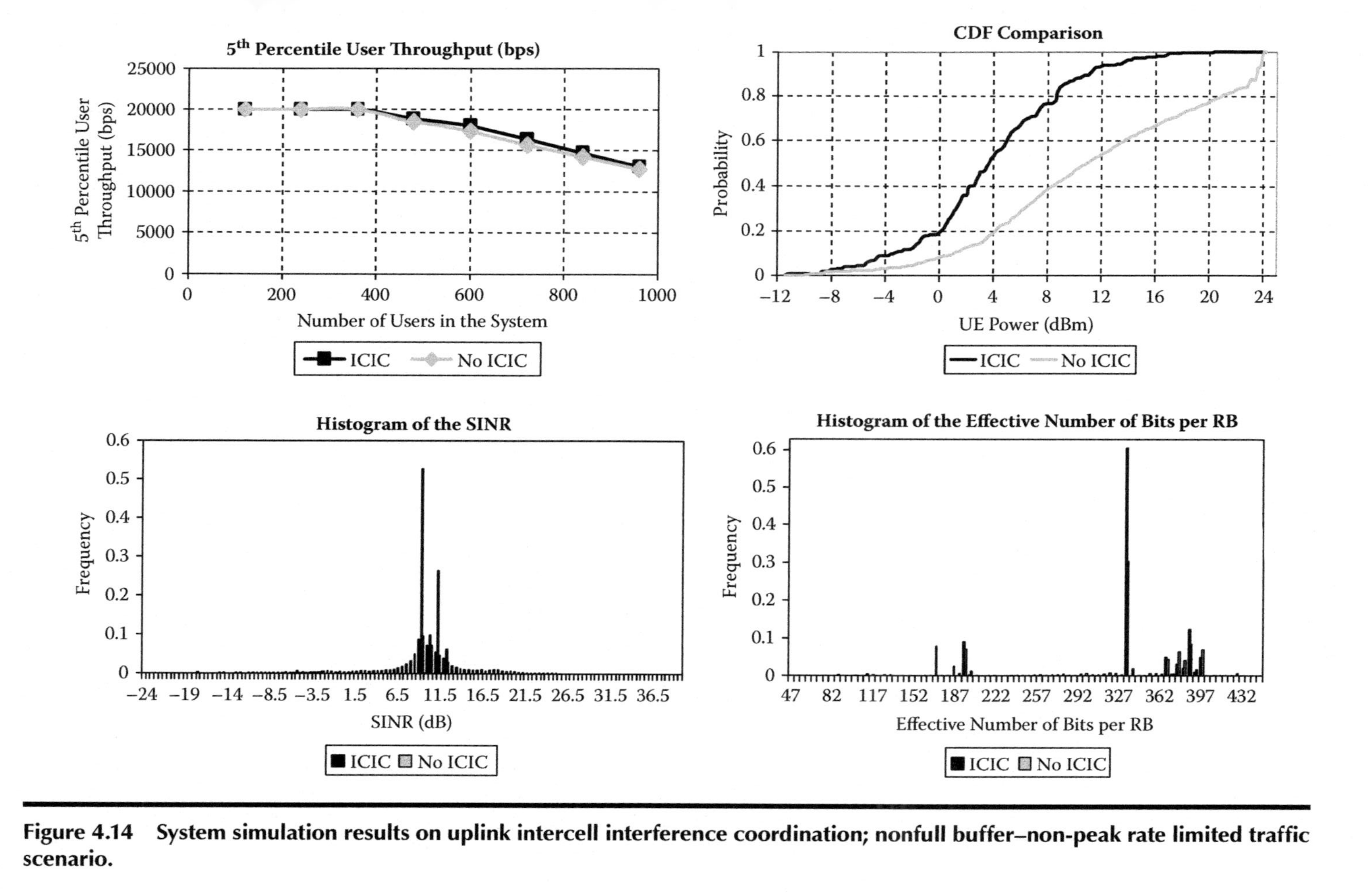

Figure 4.14 System simulation results on uplink intercell interference coordination; nonfull buffer–non-peak rate limited traffic scenario.

In addition, the 3GPP discusses the use of the overload indicator (OI) originally proposed for intercell power control purposes. It is currently agreed in 3GPP that the OI also carries information at the resource block granularity. As opposed to the proactive scheme, the overload indication is a reactive scheme that indicates a high detected interference level on a specific resource block to neighbor eNode Bs. The details of OI-based ICIC and its joint operation with the proactive scheme have yet to be defined in 3GPP.

4.2.6 Load Balancing

During the standardization process, it was early agreed that, for *intrafrequency* cells (for both idle and connected mode UEs), the best radio condition is the main mobility driver. Load balancing (LB) has the task to handle uneven distribution of the traffic load over multiple *interfrequency* and *inter-RAT (radio access technology)* cells. The purpose of LB is thus to influence the load distribution over multiple frequency and RAT layers in such a manner that radio resources remain highly utilized and the QoS of in-progress sessions is maintained to the largest possible extent while call dropping probabilities are kept sufficiently small. LB algorithms may result in handover and cell reselection decisions with the purpose of redistributing traffic from highly loaded cells to underutilized cells. Load balancing in idle mode (called camp load balancing) as well as in connected mode (often referred to as load balancing) has been identified in Appendix E of reference 14 as mobility drivers. Both camp and traffic load balancing are applicable in interfrequency and inter-RAT cases only.

4.2.7 MIMO Configuration Control

As it has been explained in Section 4.1.2, the radio resource domain in LTE can be basically interpreted as a three-dimensional domain of [time, frequency, antenna port], corresponding to the time multiplexing, frequency multiplexing, and spatial multiplexing possibilities in LTE. The availability of multiple antennas at the transmitter and receiver sides, also called MIMO systems, enables the use of various diversity methods (transmit or receiver diversity) and the use of spatial multiplexing. In the case of spatial multiplexing, the same time-frequency resource is used to transmit different data belonging to the same or different streams, provided that the spatial "channels" are separable enough. In this section, we first give an overview of the multiantenna solutions in general. Then we discuss the different MIMO variants in the context of LTE, also addressing the required resource management functions used for the antenna port configuration control.

A MIMO system can be generally characterized by a set of input and output antennas, where there are N antennas on the transmit side and M antennas on the receiver side, which yields an $N \times M$ MIMO system, as depicted in Figure 4.15. The input and the output should be interpreted from the radio channel perspective

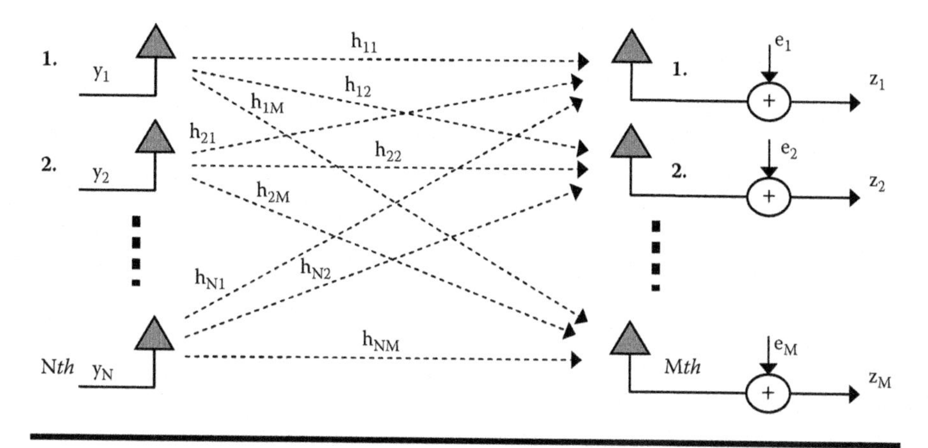

Figure 4.15 A MIMO antenna system.

(i.e., the N transmit antennas provide the input to the wireless channel, while the M receive antennas deliver the output of the channel). The transfer function of the multiantenna system can be characterized by the channel transfer matrix **H**, where the $h_{i,j}$ element of the matrix gives the channel coefficient between transmit antenna i and receive antenna j. This means that the system can be described by the following equation:

$$\mathbf{z} = \mathbf{H} \times \mathbf{y} + \mathbf{e};$$

that is,

$$\begin{pmatrix} z_1 \\ \vdots \\ z_M \end{pmatrix} = \begin{pmatrix} h_{1,1} & h_{1,2} & \cdots & h_{1,N} \\ h_{2,1} & h_{2,2} & \cdots & h_{2,N} \\ \vdots & \vdots & \ddots & \vdots \\ h_{M,1} & h_{M,2} & \cdots & h_{M,N} \end{pmatrix} \cdot \begin{pmatrix} y_1 \\ \vdots \\ y_N \end{pmatrix} + \begin{pmatrix} e_1 \\ \vdots \\ e_M \end{pmatrix}, \tag{4.1}$$

where

vector **z** contains the received signals on the M receive antennas (i.e., Z_i is the received signal on the ith receive antenna);

vector **y** contains the N transmitted signals on the different transmit antennas (i.e., y_j is the signal transmitted on the jth transmit antenna); and

e contains the noise plus interference received on the different receive antennas.

The preceding system equation is a frequency domain equation (i.e., **z** and **y** are the frequency domain representation of the received and transmitted signals, respectively). In the special case of an OFDM system, where the modulated signal is generated in the frequency domain, **z** and **y** correspond to the received and transmitted modulation symbols directly (i.e., the received signal after the fast Fourier transform (FFT) operation and the transmitted signal before the IFFT operation). For simplicity, we assume a non-frequency-selective channel, which means that the matrix **H** is independent of the OFDM carrier on which the signal is transmitted.

Obviously, the received signal on any of the receiver antenna ports will contain components from the transmitted signals of all the transmitter antenna ports. That is, the signals transmitted from the different transmit antennas will cause interference with each other and the receiver needs somehow to separate out the different spatial "channels" in order to be able to demodulate the different data streams transmitted from the different antennas. Although different receiver algorithms are used in practice, it is common to all methods that the receiver needs to estimate the channel transfer matrix **H** and apply an inverse of the channel transfer on the received signal in order to separate out the different streams transmitted from the different antennas. One of the methods often used for this purpose is the so-called zero-forcing MIMO technique, where the receiver tries to null out the channel from the direction of the interferer. In what follows, we illustrate MIMO spatial multiplexing for the zero-forcing technique, where the following operation is applied at the receiver:

$$\mathbf{v} := \mathbf{H}^{-1} \times \mathbf{z} = \mathbf{H}^{-1} \times \mathbf{H} \times \mathbf{y} + \mathbf{H}^{-1} \times \mathbf{e} \tag{4.2}$$

$$= \mathbf{y} + \mathbf{H}^{-1} \times \mathbf{e}. \tag{4.3}$$

where **v** denotes the received signals after performing the inverse channel operation. (In order to ensure that the matrix **H** is quadratic and thereby the inverse operation is meaningful, we assume an $N \times N$ MIMO system for illustration purposes.) After the different streams have been separated out via the inverse operation, it is possible to use one of the well-known receiver decision algorithms on the obtained vector **v**, such as the minimum mean square error (MMSE) algorithm (also called interference rejection combining [IRC] in the literature), to decide for the modulation symbols.

The channel estimation (i.e., the estimation of the matrix **H**) can be done based on the reception of antenna-specific reference symbols. As has been mentioned also in Section 4.2.2, certain time-frequency resource elements in each resource block are allocated for the transmission of reference symbols in order to enable channel estimation. Each antenna port has its own reference symbol, where the time-frequency resource element used to transmit the reference symbol differs for different antenna ports. This is actually what defines and identifies an antenna port.

By observing the receiver Equation (4.2), it becomes obvious that the inverse transform of the channel transfer matrix, and thereby the separation of the spatial channels, is possible only if the inverse of the matrix **H** exists. It is known from elementary matrix calculus that the inverse of a matrix **A** exists if and only if det **A** ≠ 0—that is, when the rows and also the columns of the matrix are independent. (The matrix analogy is used only for illustration purposes; in a real MIMO system, the perfect matching of the analogy may not be possible.) The number of independent rows and columns of a matrix is defined as the rank of the matrix and, for an arbitrary matrix **A** of size $N \times M$, it holds that $rank(\mathbf{A}) \leq min(N,M)$.

We define the *channel rank* (r_{ch}) of a MIMO system as the maximum number of independent streams that can be transmitted in parallel (i.e., using the same time-frequency resource element). The *transmission rank* (r_{tr}) is defined as the number of independent streams actually transmitted. (Obviously, in all feasible configurations it should always hold that $r_{tr} \leq r_{ch}$). Transferring the preceding matrix calculus analogy into the context of MIMO systems yields the intuitive interpretation of the channel rank as the rank of the channel transfer matrix **H**—that is,

$$r_{ch} = rank(\mathbf{H}) \leq min(N,M).$$

The MIMO use case that we have assumed in the preceding discussion is the *spatial multiplexing* case when different data streams are transmitted from the different antennas, using the same time-frequency resource block. However, other use cases of MIMO systems, when the same data stream is sent from the different antennas, are generally called *diversity schemes*. That is, we can differentiate the following two primary use cases of MIMO systems:

- *Spatial multiplexing* can be employed in cases when the channel rank $r_{ch} > 1$ (and likewise the transmission rank $r_{tr} > 1$) and the SINR is high enough. Because the transmission power has to be shared between the streams sent from the different antennas, the SINR per stream will be lower compared to a single antenna transmission. Therefore, the spatial multiplexing transmission mode is beneficial only if the original SINR was high enough so that the loss in terms of channel capacity due to the lower SINR per stream is compensated by the multiplication of the transmission streams. In other words, this means that the original SINR has to be on the flat part of the Shannon curve, where a 1/2, 1/3,… decrease of the SINR does not result in a 1/2, 1/3,… decrease of the channel capacity. It is also often said that the spatial multiplexing is most suitable to increase the peak rate in the inner part of the cell, where the channel conditions are favorable.
- In contrast to spatial multiplexing, *diversity schemes* are typically used for bad SINR scenarios in order to improve the SINR by exploiting diversity gains among the spatial channels of the multiantenna system. In the case of the diversity schemes, the transmission rank is $r_{tr} = 1$ (i.e., only one piece of data is transmitted in a given time-frequency resource block, meaning

that the one and the same data stream is transmitted from each transmit antenna port). There are many different variants of diversity schemes; the most relevant examples from an LTE system point of view include:

- transmit diversity
- beam forming
  - space–time block codes (STBCs) (not used in LTE)
  - space–frequency block codes (SFBCs)
  - cyclic delay diversity (CDD)

Common to all diversity schemes is that they perform some kind of transformation on the data stream prior to transmission in order to map the different transforms of the signal to the different antennas and send them out accordingly. This transformation operation is often called *precoding*. The spatial multiplexing and diversity schemes can also be combined, meaning that a precoding operation can be used in the case of multistream transmission as well. For example, it is also possible to use two transmit antennas for beam forming toward two receiving users, while two other antennas spatially multiplex two streams of data to a third receiving user equipment. The general structure of the channel transmission processing chain (applicable also for the LTE downlink) is shown in Figure 4.16.

In LTE, at most two code words can be transmitted at the same time; this means that even in the case of four MIMO streams, only two code words are involved (i.e., one code word is split into two parts to result in two streams). The bits of each code word are first mapped to modulation symbols according to the modulation scheme employed (QPSK [quadrature phase shift keying], 16 QAM [quadrature amplitude modulation], 64 QAM), and then the modulation symbols are mapped to layers by the layer mapper. The layers are the input to the precoder and the number of layers (L) is always less than or equal to the number of transmit antennas (i.e., $L \geq N$). The layer mapping differs depending on whether the transmission will be spatial multiplexing or transmit diversity.

Let us denote the first and second code words after the modulation operation with the row vectors of $\overline{\mathbf{d}}^{(1)}$ and $\overline{\mathbf{d}}^{(2)}$, respectively, each with length of D. The layer mapper generates R number of consecutive output vectors of length L from the code words as input. Let us denote the output of the layer mapper at the ith epoch

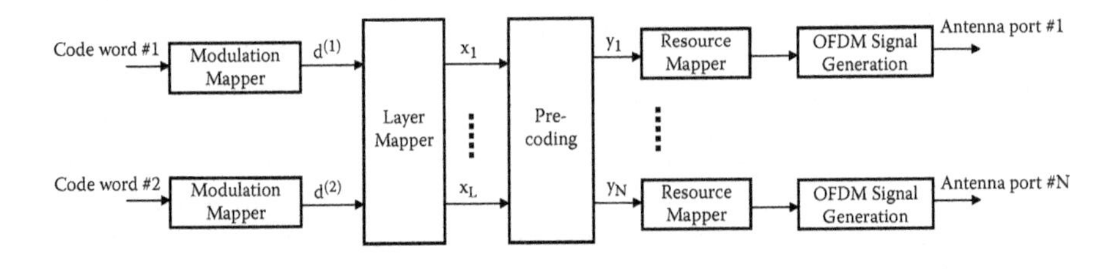

Figure 4.16 MIMO transmit processing chain.

with the column vector $\mathbf{x}(i)$. For example, the layer mapping in the case of spatial multiplexing, assuming four transmit antennas and two code words, looks like the following:

$$\mathbf{x}(i) = \begin{pmatrix} d_{2i}^{(1)} \\ d_{2i+1}^{(1)} \\ d_{2i}^{(2)} \\ d_{2i+1}^{(2)} \end{pmatrix}.$$

The precoder takes the output vectors of the layer mapper one by one (denoted by x from now on; for simplicity we omit the index i) and generates a matrix of $\mathbf{Y}$ with size of $N \times R$, where N is the number of transmit antennas and R is the number of resource elements on which the L layers are transmitted. That is, the precoder encodes the layers into a block of vectors, where the first dimension of the encoded block is the transmit antennas (spatial domain) and the second dimension can be the time or the frequency domain, resulting in STBC or SFBC, respectively. In other words, the columns of the matrix $\mathbf{Y}$ correspond to what is transmitted on the respective transmit antennas on one given time-frequency resource element. The resource elements to which the different columns of the matrix are mapped can be in the same subframe but on different carrier frequencies (SFBC case), or they can be in different subframes (STBC case). In LTE, only the SFBC type of precoding is applied. In the simplest case, the matrix $\mathbf{Y}$ can be a single vector, meaning that the layers are only spatially encoded but not in the time or the frequency domain. Next, we show examples for the precoding operation for the different spatial multiplexing and diversity schemes used in LTE.

The precoder operation for *transmit diversity* is defined for two and four antenna ports and has the following form (two-antenna-ports case):

$$\mathbf{Y} = \begin{pmatrix} x_1 & x_2 \\ x_2^c & -x_1^c \end{pmatrix},$$

where the notation c denotes the complex conjugate. This type of precoding is also called Alamouti coding and corresponds to SFBC.

Precoding for *spatial multiplexing* typically involves a set of precoder matrices,

$$\{\mathbf{W}(1), \mathbf{W}(2), \ldots, \mathbf{W}(K)\},$$

also called a codebook, where the actual precoder matrix to be used can be dynamically changed on a subframe basis according to the channel feedback reports from the UE. The precoder matrix describes the mapping of the different layers to the antenna ports (the relation between $\mathbf{Y}$ and $\mathbf{x}$)—that is,

$$\mathbf{Y} = \mathbf{W} \times \mathbf{x}.$$

The codebook-based precoding is one of the primary transmission modes in LTE. The standard defines one codebook for the two-transmit-antenna case and one for the four-transmit-antenna case. For more details on the actual codebooks, see 3GPP TS 36.211 [5].

The precoding can be further combined with CDD transmission, where the mapping of the symbols from the layers to the antenna ports is done by the precoder matrix and the CDD matrix together (i.e., $\mathbf{Y} = \mathbf{D}(k)\,\mathbf{W}(i)\,\mathbf{x}$, where $\mathbf{D}(k)$ is the transform matrix performing the CDD operation:

$$\mathbf{D}(k) = \begin{pmatrix} 1 & 0 & 0 & 0 \\ 0 & e^{-j2\pi k\delta} & 0 & 0 \\ 0 & 0 & e^{-j2\pi k2\delta} & 0 \\ 0 & 0 & 0 & e^{-j2\pi k3\delta} \end{pmatrix},$$

where k represents the frequency domain index of the resource element to which the transmission is mapped and δ is the delay shift. The idea with the CDD transformation is to employ an increasing phase shift on the antenna ports. Depending on the actual channel matrix, on some antenna ports the phase shift will match the actual channel and thereby result in an increased SINR; on other antenna ports, it may null out the transmission, constituting a source of diversity.

In *beam forming,* a single symbol is multiplied by different weight factors and sent on the different antenna elements; this introduces antenna-specific phase adjustments. As a result of the different phase adjustments on the different antennas, the transmitted signal can be steered in specific directions. That is, the "precoding" operation for beam forming, assuming four transmit antennas, is the following:

$$\mathbf{Y} = \begin{pmatrix} w_1 \\ w_2 \\ w_3 \\ w_4 \end{pmatrix} \cdot x$$

where x is the modulation symbol corresponding to the single layer.

The preceding transmit diversity, spatial multiplexing, and beam-forming schemes apply for the downlink directions only. In the uplink, there is only a limited set of multiantenna transmission capability. More specifically, the first version of LTE will support only two types of multiantenna schemes: closed-loop transmit antenna switching and multiuser MIMO (MU-MIMO). In the antenna switching solution, the eNode B can decide on a subframe level which of the two transmit antennas the UE should use for the next transmission. However, the gains with such an antenna switching solution are questionable. In the MU-MIMO case, multiple UEs, which have quasi-orthogonal channels, are scheduled on the same resource block, which realizes a way of spatial multiplexing in the uplink. Such a MU-MIMO setup can be seen as a case when the transmit antennas of the MIMO system are at separate UEs and the receive antennas are at the eNode B. The location of the receive antennas at the same node (at the eNode B) facilitates the processing of the multiple UE streams, and this is exactly what enables the use of MU-MIMO in the uplink. In the downlink, MU-MIMO would be more problematic due to the receiver processing being in separate nodes (i.e., in separate UEs).

For the operation of the downlink multiantenna schemes, various feedback reporting mechanisms are required from the UE. Recall that the UE sends CQI reports, which is one form of an SINR measure, to assist the channel-dependent scheduling in the eNode B. For supporting the multiantenna transmission, the UE sends, in addition to the CQI reports, *channel rank* and *precoding matrix* selection reports with a periodicity in the order of subframe length (1 ms). The UE performs a prediction of the expected throughput assuming different precoder matrices and transmission rank values and reports the recommended transmission rank and precoder matrix back to the eNode B that yields the highest expected throughput. The eNode B may or may not follow the recommendations of the UE. The precoder matrix that has been selected by the eNode B for the transmission is indicated as part of the transport format signaled in the scheduling assignment on PDCCH. The recommended transmission rank is an average over all the feasible set of sub-bands; the precoder matrix recommendation can be sent per sub-band (i.e., for different sub-bands, the UE may recommend different precoder matrices to use). The frequent reporting of CQI, precoder matrix, and transmission rank recommendations, commonly called channel feedback reporting (CFR), allows one to follow the fast fading link fluctuations with the adjustment of the multiantenna transmission parameters as well.

However, the MIMO transmission mode (i.e., whether transmit diversity, spatial multiplexing, or beam forming mode is used) is configured semistatically via RRC signaling. The different MIMO transmission modes also imply different configurations of the required fast time scale channel reports and thereby the resources that need to be reserved on the uplink control channels for the channel reporting. The codebook restriction is also configured via the RRC protocol. In the case of codebook restriction, the eNode B can restrict the set of precoder matrices that the UE can recommend in the channel feedback report.

4.2.8 MBMS Resource Control

The delivery of multicast broadcast multimedia services (MBMS) is supported in LTE via the following two transmission modes:

- *MBMS single frequency network (MBSFN) transmission mode.* In this mode of operation, multiple cells are transmitting exactly the same signal on the same resource at the same time. When the same OFDM signals (same waveforms) arrive at the UE with delay differences less than the cyclic prefix of the OFDM signal, the multiple signals will be constructively added by the UE receiver without any additional action in the receiver. Basically, the signals arriving from different cells will appear exactly the same for the UE receiver as multipath components of the same signal. This mode of operation is suitable for a large coverage area (e.g., a national TV channel).

- *Single cell transmission mode.* In this mode, the UE is receiving the MBMS signal from only one cell (i.e., the combination of multiple signals from different cells is not possible). The single cell transmission mode can be more resource efficient than the MBSFN transmission when only a few UEs are interested in receiving the considered MBMS service. There is also the possibility to improve the received signal quality of the single cell transmission by adapting the transmission based on feedback information from the UEs, such as HARQ or CQI feedback. Such an adaptation mechanism helps to compensate the efficiency loss due to the lack of multicell combination. Note that in MBSFN mode, the use of feedback information is not supported. The single cell mode allows transmission of different MBMS services on the same radio resources even in adjacent cells. Therefore, this mode of operation can be advantageous for a more localized broadcasting scenario (e.g., broadcasting in a sport arena or for public safety purposes).

The preceding MBMS transmission modes can be further classified depending on whether a dedicated carrier or a mixed carrier is used for the transmission. In a dedicated carrier case, the given carrier is reserved solely for MBMS transmission (i.e., no unicast transmission is present on the carrier). Therefore, the coverage area of a single transmitter is not limited by unicast capacity. This allows the deployment of dedicated carriers on a more spares but higher power/higher tower transmitter infrastructure (e.g., as an overlay to the unicast deployment). Additionally, the dedicated carrier mode enables saving some of the L1/L2 control signaling (e.g., HARQ feedback, scheduling assignments) associated with unicast transmissions only. The mixed carrier mode enables multiplexing of unicast and multicast services.

The MBSFN and the single cell transmission modes have different implications on the required RRM functions. In the MBSFN mode, the most important requirement is to ensure that exactly the same OFDM waveform is transmitted from all involved cells. This requires an accurate time synchronization among the cells; the allocation of the same time-frequency resource blocks for the MBMS transmission in all cells, including the same transport format (i.e., modulation and coding scheme); and the identical mapping of user data packets into RBs [14].

The method used for time synchronization is not the subject of the 3GPP standardization, and it is left for vendor selection to choose a method from the available legacy solutions, such as IEEE 1588 [9] or GPS-based solutions. The uniform allocation of time-frequency RBs and MCS in all cells involved in the MBSFN transmission is performed by a central coordination entity, called multicast/broadcast coordination entity (MCE). The MCE would typically be a separate physical

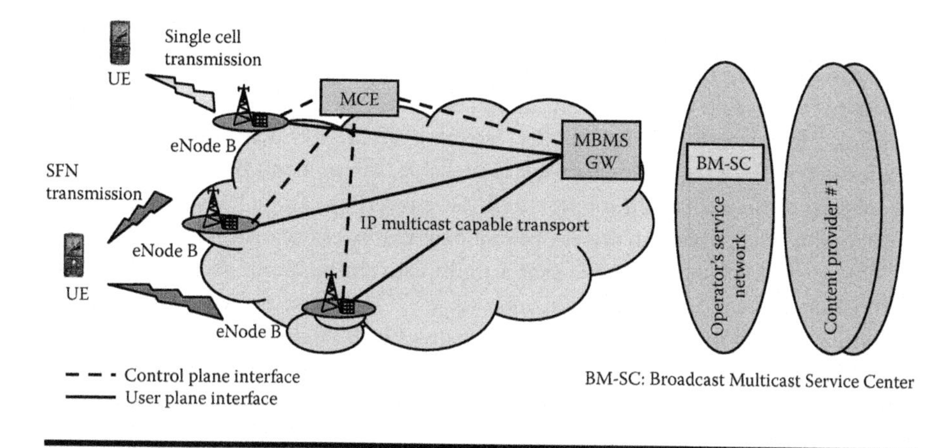

Figure 4.17 The E-UTRAN MBMS architecture.

node, part of the evolved universal terrestrial radio access network (E-UTRAN) and responsible for a certain set of eNode Bs. See Figure 4.17 for a complete picture of the E-UTRAN MBMS architecture.

The MCE would typically allocate time-frequency RBs in a periodic pattern at the start of the session, which would remain static (semistatic) throughout the lifetime of the MBMS session. The details of the MCE–eNode B resource configuration signaling has not been settled in 3GPP at the time of writing. When allocating the time-frequency resources to the MBMS transmission, the MCE may consider maintaining a guard zone around the border region of the MBMS area, where the same resource cannot be used in neighbor cells in order to avoid unicast traffic interfering with the MBMS transmission.

Finally, the identical mapping of user data into RBs is ensured by the content synchronization method, which operates in the user plane between the MBMS-GW and the eNode Bs [8,14]. It is based on adding byte counts and optionally also periodic time stamps to the packets sent between the MBMS-GW and eNode B as an indication of the time when the particular packet needs to be sent on the radio interface. Because the content synchronization scheme is purely a user plane procedure without RRM relevance and at the time of writing the exact method has not yet been selected by 3GPP, we do not address it in the rest of this chapter.

For single-cell transmission mode, the eNode B can decide autonomously and in a dynamic fashion which RBs it wants to use for sending MBMS data and which for unicast data. The selection of transport format can also be dynamic, much in the same way as for unicast traffic. This means that the scheduling of MBMS data can be done as part of the regular scheduling function in the eNode B. As it has been mentioned earlier, the single cell transmission can be employed with or without UE feedback. The UE feedback information helps to utilize the radio resources more efficiently by adapting the transmission to the actual radio conditions of the different UEs. For example, if no feedback information is available, the transmission power setting or the transport format selection has to assume the worst-case scenario: that is, that even a UE on the cell edge should be able to receive the transmission. However, if channel quality feedback or HARQ acknowledge/negative acknowledge (ACK/NACK) feedback is available, the transmit power or MCS can be adapted to the actual radio link conditions, and they can be less robust because unsuccessful receptions can always be compensated by a retransmission.

However, it still remains true that a multicast transmission has to adapt to the channel quality conditions of the worst-case UE. Therefore, a single cell multicast transmission will, on average, use more resources (e.g., higher transmission power) compared to a single unicast transmission. This means that the interference caused to neighbor cells by a multicast transmission will be higher and the tolerance to interference will be lower than in a unicast transmission. Therefore, the intercell interference coordination between neighbor cells can be equally, or even more, important for the multicast transmission case as compared to the unicast transmission.

One difference, though, between the unicast and the multicast intercell interference coordination is that, although in the unicast case the coordination is more easily done in the frequency domain than in the time domain, in the multicast case it is just the opposite. The MBMS transmissions are typically periodic in time (i.e., there is MBMS transmission in only every *N*th subframe and, when there is MBMS transmission in the given subframe, all RBs in the subframe are used for MBMS—that is, no unicast data are sent in the same subframe). This is especially true for MBSFN transmission where, according to the L1 specification, MBSFN and non-MBSFN transmission cannot occur in the same subframe. For single-cell transmission, multicast (non-MBSFN) and unicast transmission may occur in the same subframe. See also Section 4.2.5 for more details on intercell interference coordination.

4.3 Radio Resource Management Related Measurements

Because the operation of E-UTRA—including channel-dependent scheduling, power control, idle and connected mode mobility, admission control, and radio resource management in general—relies on measured values, it is natural that the various physical layer measurements are instrumental in LTE. Recognizing this, the 3GPP has defined the basic measurement-related requirements and the physical layer measurements in 3GPP TR 36.801 [19] and 3GPP TS 36.214 [20], respectively. The most important measurement aspects include the usefulness, accuracy, and complexity of a particular measurement as well its typical L1 measurement interval and the measurement's impact on UE power consumption. In this section, we discuss the most important measurements in LTE, grouping them into UE and eNode B measurements and, when appropriate, drawing an analogy with well-known WCDMA measurements.

4.3.1 Measurements Performed by the User Equipment

UE measurements are needed to serve the following purposes:

- *Intra-LTE (intra- and interfrequency) cell reselection and handovers as well as inter-RAT handovers (handovers to WCDMA and GSM/enhanced data rates for GSM evolution radio access networks [GERANs]).* This is because radio coverage is one of the most important mobility drivers in both idle and connected modes [14].
- *Admission and congestion control.* Measurement-based admission and congestion control play an important role in maintaining service quality for end users (based on single [local] eNode B measurements).
- *Uplink power control, scheduling, and link adaptation.* These essential radio network functions are inherently adaptive and rely on fast and accurate measurements [18].
- *Operation and maintenance.* This set of functions enables network operators to observe the performance and reliability of the network and to detect failure situations. Measurements are the primary input to these functions.

The UE measurement quantities follow. In addition, Figure 4.18 provides a brief overview of the required RRM measurements and their counterparts in

Measurement Type	Purpose	L1 Measurement Interval	Protocol to Report: RRC/MAC	Higher Layer Filtering: Mandatory/No	Analogy with WCDMA
Reference Symbol Received Power (RSRP)	HO, Cell Reselection	200 ms	RRC	M	CPICH RSCP
Reference Symbol Received Quality (RSRQ)	HO, Cell Reselection	200 ms	RRC	M	CPICH Ec/Io
Carrier Received Signal Strength Indicator (RSSI)	Inter-Frequency and Inter-RAT HO	200 ms	RRC	M	UTRA RSSI
Channel Quality Indicator (CQI)	Scheduling, DL Power Control, Link Adaptation	1 TTI (1 ms)	MAC	N	CQI

HO: Handover
DL: Downlink
TTI: Transmission Time Interval
RRC: Radio Resource Control
MAC: Medium Access Control
CPICH: Common Pilot Indicator Channel
RSCP: Received Signal Code Power
UTRA: Universal Terrestrial Radio Access

Figure 4.18 Measurements performed by the user equipment (UE). The UE measurements are instrumental for intra-LTE and inter-RAT mobility control and for channel-dependent (opportunistic) scheduling as well as other vital physical layer procedures such as power control and link adaptation. (See also reference 18.)

WCDMA. Positioning-related measurements are not listed because they depend upon the exact positioning method used in E-UTRAN [22]. Further details can be found in 3GPP TS 36.214 [20].

- Reference symbol received power (RSRP) is determined for the considered cell as the linear average over the power contributions (in [W]) of the resource elements that carry cell-specific reference signals within the considered measurement frequency bandwidth.
- Reference symbol received quality (RSRQ) is defined as the ratio N RSRP/(E-UTRA carrier RSSI), where N is the number of RBs of the E-UTRA carrier RSSI measurement bandwidth. The measurements in the numerator and denominator should be made over the same set of resource blocks.
- E-UTRA carrier received signal strength indicator (RSSI) comprises the total received wideband power observed by the UE from all sources, including co-channel serving and nonserving cells, adjacent channel interference, thermal noise, etc.
- CQI is per sub-band, per group of sub-bands, and over the entire bandwidth.

4.3.2 Measurements Performed by the eNode B

In E-UTRAN, certain types of measurements should be performed internally in the eNode B and will not be exchanged between the eNode Bs. These measurements do not need to be specified in the standard; rather, they will be implementation dependent. On the other hand, measurements that are to be exchanged between the eNode Bs over the X2 interface need to be standardized. The possible measurements should serve the following procedures (at the time of writing, under study by the 3GPP):

- intra-LTE and inter-RAT handovers;
- intercell interference coordination; and
- operation and maintenance.

The eNode B measurements are described next. The current description does not explicitly take into account the impact of multiple transmit and receive antennas on the measured quantities and measurement procedures (an issue still under discussion at the 3GPP). Most of the eNode B measurements are implementation specific and need not be specified in the standard:

- DL total Tx (transmit) power: transmitted carrier power measured over the entire cell transmission bandwidth;
- DL resource block Tx power: transmitted carrier power measured over a resource block;

- DL total Tx power per antenna branch: transmitted carrier power measured over the entire bandwidth per antenna branch;
- DL resource block Tx power per antenna branch: transmitted carrier power measured over a resource block;
- DL total resource block usage: ratio of downlink resource blocks used to total available downlink resource blocks (or simply the number of downlink resource blocks used);
- UL total resource block usage: ratio of uplink resource blocks used to total available uplink resource blocks (or simply the number of uplink resource blocks used);
- DL resource block activity: ratio of scheduled time of downlink resource block to the measurement period;
- UL resource block activity: ratio of scheduled time of uplink resource block to the measurement period;
- DL transport network loss rate: packet loss rate of GTP-U (GPRS tunneling protocol–user plane) packets sent by the access gateway on S1 user plane. The measurement should be done per traffic flow. The eNode B should use the sequence numbers of GTP-U packets to measure the downlink packet loss rate;
- UL transport network loss rate: packet loss rate of GTP-U packets sent by the eNode B on S1 user plane. The measurement should be done per traffic flow. The access gateway should use the sequence numbers of GTP-U packets to measure the downlink packet loss rate;
- UL RTWP: received total wideband power, including noise measured over the entire cell transmission bandwidth at the eNode B;
- UL received resource block power: total received power, including noise measured over one resource block at the eNode B;
- UL SIR (per UE): ratio of the received power of the reference signal transmitted by the UE to the total interference received by the eNode B over the UE occupied bandwidth;
- UL HARQ BLER: the block error ratio based on CRC check of each HARQ-level transport block;
- propagation delay: estimated one-way propagation delay measured during random access transmission;
- UE Tx time difference: time difference between the reception of the UE transmitted signal and the reference symbol transmission time instant; and
- DL RS Tx power: downlink reference signal transmit power determined for a considered cell as the linear average over the power contributions (in [W]) of the resource elements that carry cell-specific reference signals transmitted by the eNode B within its operating system bandwidth.

For intercell interference coordination purposes, it may be useful to measure the user plane load (e.g., in terms of number of sent user plane packets/bits per second). The definitions of such measurements and associated procedures are for further study.

4.4 User Equipment Behavior

A fundamental design principle of LTE as well as its predecessors has been to allow the network to control UEs in connected mode—that is, UEs that have a radio resource control connection to the network (often casually referred to as "active" mode UEs). This design principle has been useful for protocol design and, most importantly, for radio resource management purposes because the network is in the position of making network-wide near-optimal decisions, including intra-LTE and inter-RAT handovers, load balancing, and others.

In contrast, when the UE is not connected ("idle"), it has to act (much more) autonomously, although it is possible for the network to influence this behavior, as we shall see in this section. To this end, some issues of the UE behavior need to be standardized, which indeed has some RRM-related aspects.

The UE procedures in idle mode (including public land mobile network [PLMN]) and RAT selection, initial cell selection and cell reselection, cell reservations and access restrictions, and the receiving of broadcast information and paging) are specified by 3GPP TS 36.304 [15]. A schematic view of the UE behavior focusing on PLMN, RAT, and cell (re-)selection is depicted by Figure 4.19.

During the standardization process, it was recognized that the identification of idle mobility (basically, cell reselection) drivers for scenarios in which the UE moves within the same carrier frequency (intrafrequency mobility)—as well as when it can choose between different frequency layers and even various radio access technologies (interfrequency and inter-RAT mobility)—is important because it has an impact on the preferred idle mode mobility procedures.

The most important driver is, naturally, radio coverage; that is, the UE should always camp on the cell that provides the best signal strength characterized by the so-called *S criterion* defined by means of the RSRP (the *S value*) described in the measurement section of this chapter. According to 3GPP TS 36.304 [15], the *S* criterion is defined for cell selection as follows:

$$S_{rxlev} \triangleq Q_{rxlevmeas} - Q_{rxlevmin} - P_{compensation} > 0,$$

where $Q_{rxlevmeas}$ is the measured cell RSRP value, $Q_{rxlevmin}$ is the minimum required RSRP level in the cell in decibels of measured power (dBm), and the compensation power level ($P_{compensation}$) is an additional cell-specific offset value that can be set by the network operator (at the time of writing, discussed by the 3GPP).

In addition to this fundamental driver, the 3GPP has agreed that, for inter-frequency and inter-RAT mobility, other factors should also be supported—most

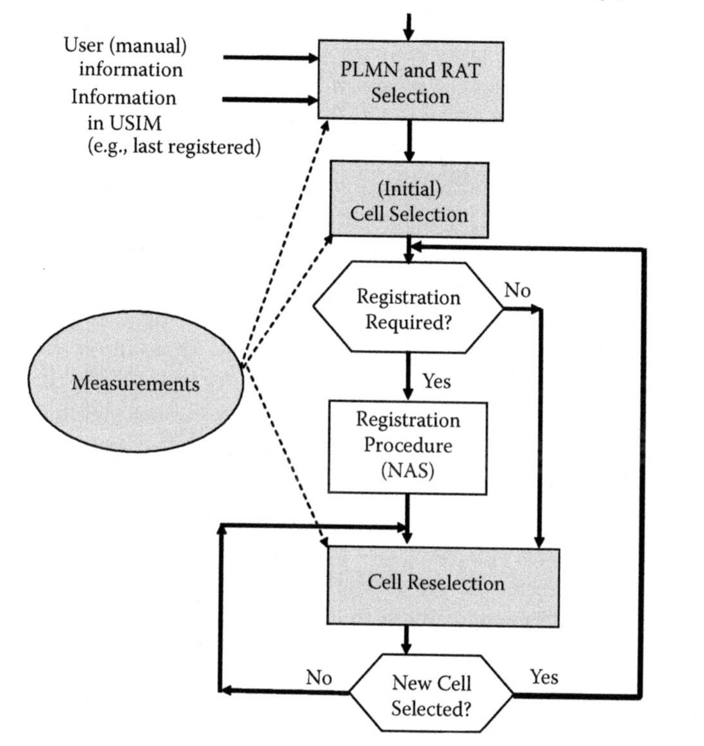

Figure 4.19 A schematic view of the UE behavior in LTE.

importantly, idle mode *load balancing* involving network control of idle mode UE distribution. Such load balancing has been motivated by two factors. First, the UE in idle mode does consume radio resources in the form of control plane signaling when responding to paging and when sending tracking area update messages to the network [14]. Second, some services (notably high bit rate point-to-point as well as multicast and broadcast services) can only be provided on LTE, which means that only UEs camping on LTE have immediate access to such services (i.e., without having to perform a handover).

Accordingly, the UE behavior is designed to meet the most important idle mobility drivers that are identified in 3GPP TS 36.300 [14]. When the UE is powered on or upon recovery from lack of coverage, the UE needs to select an appropriate PLMN and RAT. The main input to PLMN and RAT selection algorithms is provided manually by the human user (e.g., preferred RAT) or fetched from the user subscription identity module (USIM) (e.g., the most recently used PLMN).

At cell selection, the UE searches for a suitable cell of the selected PLMN and RAT and chooses that cell to (initially) obtain available services by tuning to its control channels. (We say that the UE *camps* on that cell.) According to 3GPP TS 36.304 [15], the UE will now, if necessary, *register* its presence in the tracking area of the chosen cell and RAT and, as an outcome of a successful *location registration,* the selected PLMN becomes the *registered* PLMN.

The UE now continuously searches for the best cell on which to camp. The particular cell reselection process to achieve this depends on the cell reselection scenario, UE capabilities, subscriber priority class, and other factors (for details, see 3GPP R2-080238 [47]). Although the baseline for UE behavior in LTE has been that in GSM and WCDMA, a new (LTE-specific) aspect is the handling of priorities between available RATs and frequency layers. According to this priority scheme, a set of broadcasted system information parameters configures, for each UE, a default RAT and frequency layer priority list. In addition, the LTE RAN uses RRC signaling to configure a UE-specific priority list for each UE. The setting of the priority list provides network control to steer camping mobiles to particular RATs and frequency layers facilitating the *camp load balancing* that is one of the idle mode mobility drivers in 3GPP TS 36.300 [14].

More specifically, in the priority scheme, interfrequency and inter-RAT cell reselection relies on two sets of parameters. The first set is the cell-specific (and thereby frequency- and RAT-specific) set of parameters common to all UEs camping on a specific RAT and frequency. The second set is the UE-specific set of parameters that provides the means to control UE behavior depending on UE class and, in general, operator-defined policy. Here, we focus on the cell parameters that need to be broadcasted in the cell (see Figure 4.20) and note that the UE-specific parameters are similar to the broadcasted ones listed in 3GPP TS 36.331 [21].

Figure 4.20 lists the parameters needed for interfrequency and inter-RAT cell reselection. The $\text{Thresh}_{serving,low}$ threshold is needed in order to compare the appropriate S value of the serving cell when deciding if the UE can reselect to a lower-priority cell than the currently serving cell. On the other hand, the $\text{Thresh}_{serving,high}$ parameter is needed in order for the UE to know when it should trigger interfrequency and inter-RAT measurements. That is, if the S parameter of the serving cell is above this threshold, the UE should not measure on intrafrequency and inter-RAT (unless it is not camping on its highest-priority RAT/frequency layer). This parameter is not needed when the UE is camping on a cell that is not of the highest-priority layer because, if the UE is camping on a lower-priority cell, it periodically searches for the highest-priority layer (see also Figure 4.21). The $\text{Thresh}_{x,low}$ parameter specifies the minimum level that must be fulfilled by a cell in order for it to be selectable. This value needs to be given for each RAT and frequency layer (x). $\text{Thresh}_{x,high}$ specifies the value that a higher-priority cell needs to fulfill in order for the UE to reselect to that cell. Thus, this parameter is not used when the UE is camping on its highest-priority layer.

The $T_{HigherPrioritySearch}$ parameter is needed by UEs currently camping on a cell that has lower than highest priority. These UEs need to look periodically for higher-priority cells. Because interfrequency and inter-RAT measurements

Parameter	Meaning	Comment
$Qoffset_{s,n}$	Offset between serving and neighbor cells in the R criterion.	There may be two Qoffset values: layer specific and cell specific. However, Qoffset is not needed in the priority-based scheme.
$Thresh_{serving, low}$	The minimum level that must be fulfilled for camping on a cell on the serving frequency layer (FFS: specific to the serving RAT?)	Assumes single (scalar) "S" value, for instance, only signal strength or only signal quality.
$Thresh_{serving, high}$	If the measured (S) value is under this threshold, the UE shall start non-intra-frequency measurements.	
$Thresh_{x, low}$	The minimum level that must be fulfilled for camping on a cell on a frequency/RAT layer. x specifies the freq./RAT.	An equivalent definition is: The minimum level that must be fulfilled for camping on a cell on a frequency/RAT layer *that has lower priority than the currently serving frequency/RAT layer.*
$Thresh_{x, high}$	If the measured (S) value of x exceeds this value and x is of higher priority than the currently serving cell, the UE shall reselect to x. x specifies the freq./RAT.	This means that even if the S value is higher than $Thresh_{x, low}$ and x is of higher priority than the current one, the UE does not necessarily reselect to x unless $Thresh_{x, low} > Thresh_{x, high}$
$T_{HigherPrioritySearch}$	Minimum periodicity for higher-priority frequency layer measurements.	
$T_{LowerPrioritySearch}$	Minimum periodicity for lower-priority frequency layer measurements.	
$T_{EqualPrioritySearch}$	Minimum periodicity for equal-priority frequency layer measurements.	
<RAT, Band, Priority>	**List of RATs and associated priorities that the UEs may select (in the neighborhood of this cell), that is a sort of "neighbor RAT list".**	

Figure 4.20 Broadcasted system information parameters to facilitate the priority-based scheme for interfrequency and inter-RAT cell reselection. The *R criterion* refers to the ranking criterion that is used to *rank* cells that fulfill their respective *S* criteria. This ranking criterion is defined in TS 36.304 [15] and includes the RSRP measurement quantity and an offset value that can be specified between two cells, as shown in the table.

have an impact on the UE discontinuous reception and transmission (DRX) performance, this parameter represents a trade-off between minimizing battery consumption and ensuring that the time it can take for the UE to detect the availability of its highest-priority RAT and frequency is short enough. When the measured *S* value in the currently serving cell is under $Thresh_{serving,low}$, the UE starts interfrequency and inter-RAT measurements including RATs and frequencies of lower priority. However, the measurement frequency for lower-priority cells may be set lower than for high-priority cells by setting the $T_{LowerPrioritySearch}$ parameter. Similarly, a separate value for equal priority cells can be set in the $T_{Equal\ Priority\ Search}$ parameter.

The *<RAT, band, priority>* information informs the UEs about the default (non-UE-specific) priority list that UEs can use as a base priority list. If the UE has

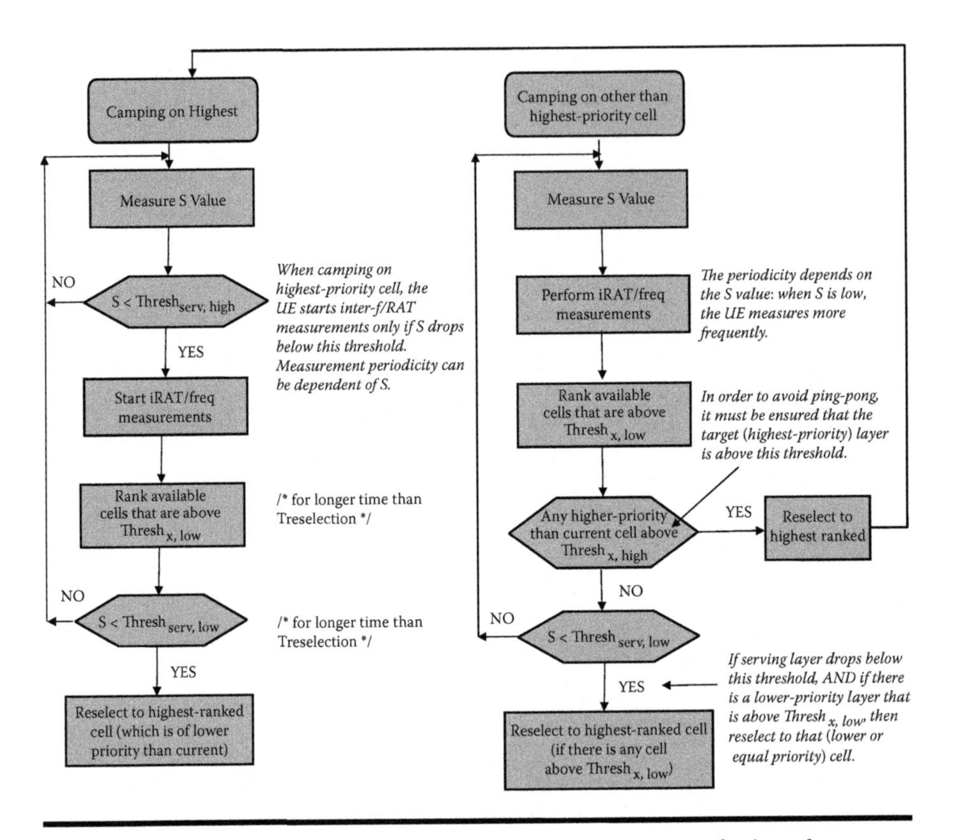

Figure 4.21 A high-level view and the usage of the parameters for interfrequency and inter-RAT cell reselection. The left-hand side flow diagram illustrates the UE behavior when the UE camps on its highest-priority RAT and frequency layer. The right-hand side diagram shows the case when the UE camps on a RAT/frequency layer that is not the UE's highest-priority layer.

already received a UE-specific priority list (by RRC or nonaccess stratum [NAS] signaling), then the UE ignores this parameter set.

Figure 4.21 summarizes the usage of the interfrequency and inter-RAT system information broadcast parameters. When the UE is camping on its highest-priority cell, it periodically measures the S value (e.g., the RSRP) of the currently serving cell and compares this value with $\text{Thresh}_{serving, high}$. If the S value is below this threshold, the UE starts to look for interfrequency and inter-RAT cells. The cells whose measured S value is above $\text{Thresh}_{x,low}$ will be ranked (taking into account their associated priorities). When the serving cell's S value is below the $\text{Thresh}_{serving, lowvalue,}$ the UE reselects to the highest-ranked such cell. The CR procedure is a bit different when the UE is camping on a cell that is not its highest-priority cell. In this case, the UE constantly looks for availability for higher-priority cells by performing inter-RAT and interfrequency measurements. If it finds any such (higher-priority) cell, it will reselect to that cell. In addition, similarly to the previous case, if the currently serving cell drops below $Thresh_{serving, low}$, the UE reselects to a lower-priority cell if that is above $Thresh_{x,low}$.

4.5 Radio Resource Management in Multi-RAT Networks

The drivers for inter-RAT radio resource management are captured in 3GPP TS 36.300 [14]. For idle mode, the methods and parameters are specified in 3GPP TS 36.304 [15]. The general working assumption in the 3GPP at the time of writing is that idle mode inter-RAT management is based on absolute priorities. For connected mode, the details of inter-RAT handover managements will be covered by 3GPP TS 36.331 [21], upon which the 3GPP is working.

A schematic overview of inter-RAT RRM is provided in Figure 4.22. A general principle for inter-RAT (intra-3GPP) radio resource management is that the UE is not connected to multiple RATs simultaneously. In particular, when a UE is using a service that is best delivered over LTE, all other services that are used simultaneously by this UE must also use LTE. Another general principle for inter-RAT handovers is that (1) triggering of inter-RAT measurements as well as the handover decision is made by the RAN that currently serves the UE, and (2) the target RAN gives guidance for the UE on how to make the radio access. The target RAN provides information on the target cell, including radio resource configuration, necessary identities, etc., in a transparent container. Although the handover command must be compiled and sent to the UE by the serving RAN, the target RAN assists in this by providing information about the target cell. For inter-RAT handovers, the serving RAN needs to be able to (1) trigger inter-RAT measurements, (2) make a comparison between different radio access technologies, and (3) make a handover decision and command the UE. The target RAN must be able to send information in a transparent container to the UE that guides the UE on how to make radio access in the target RAN. In the "from LTE to UTRAN/GERAN" direction, this is made possible by identifying the following events defined in terms of appropriate measurement quantities, associated threshold, offset, and timer values:

- The estimated quality of the currently used frequency (LTE) falls below a certain threshold. Essentially, this event triggers the start of UE measurements on UTRAN or GERAN. The threshold value that defines this event should be sufficiently low in order to avoid unnecessary inter-RAT measurements.
- The estimated quality of the currently used frequency (LTE) is above a certain threshold. This event should stop inter-RAT measurements.

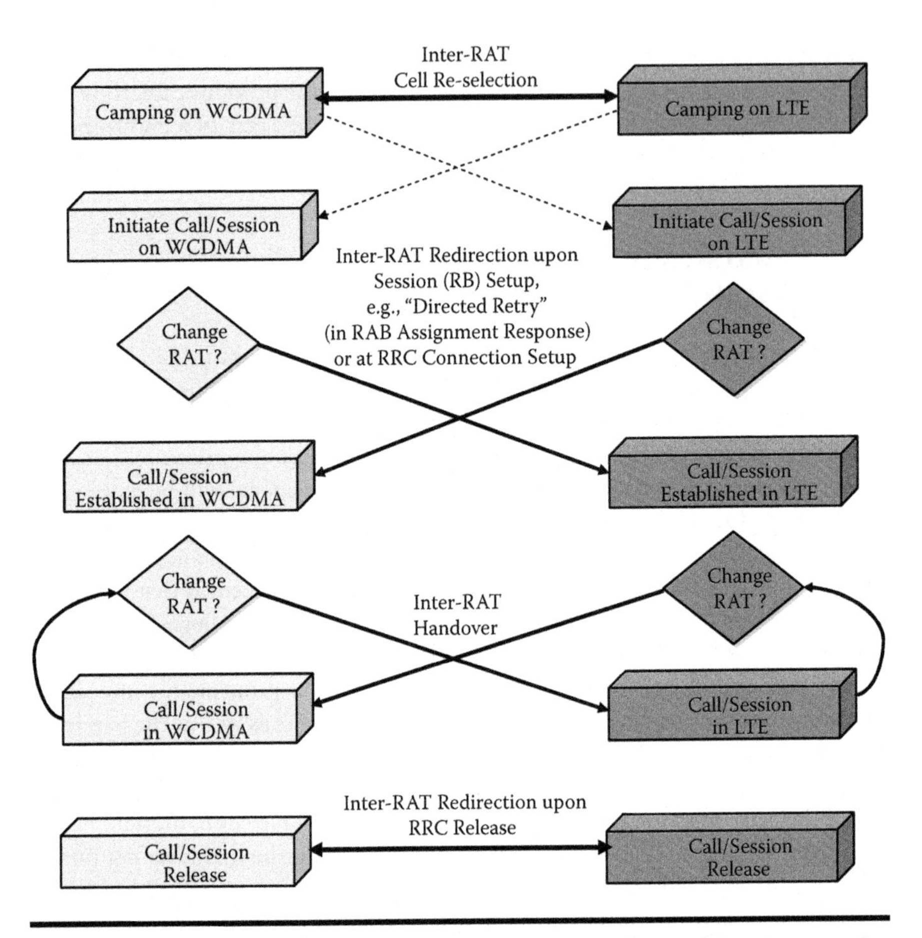

Figure 4.22 Radio resource management techniques for multi-RAT networks. Changing the serving RAT can happen at different phases during the lifetime of a connection. The figure illustrates this by indicating when inter-RAT cell reselection, redirection upon connection establishment, handover, and redirection upon connection release happen.

- The estimated quality of the other system (e.g., UTRAN) is above or below a certain threshold. This event allows the other RAT to be selectable.
- The estimated quality of the currently used system (LTE) is below a certain threshold and the estimated quality of the other system is above a certain threshold. When this event occurs, it should start the inter-RAT handover procedure. The exact specification of these events will be part of 3GPP TS 36.331 [21].

4.6 Summary

In this chapter, we have seen that a number of advanced RRM functions are needed in today's wireless systems, which are being developed and standardized in particular for the 3GPP LTE networks in order to fulfill the ever-increasing capacity demands by utilizing the radio interface more efficiently. Considering the facts that the available radio spectrum is a limited resource and the capacity of a single radio channel between the UE and the network is also limited by the well-known theoretical bounds of Shannon, the remaining possibilities to increase the capacity are to increase the number of such "independent" radio channels in addition to trying to approach the theoretical channel capacity limits on each of these individual channels. Advanced radio resource management techniques play a key role in achieving these goals.

A straightforward way of increasing the number of such "independent" channels is to increase the system bandwidth or the number of cells in a given deployment. In LTE the maximum supported system bandwidth size has been increased to 20 MHz and a variety of flexible system bandwidth configurations is possible; the number of cells in a network is more of a deployment issue than a system design principle. The other, and less straightforward, possibility for increasing the number of independent radio channels is to employ various spatial multiplexing techniques and advanced receiver structures that can better separate out the radio channels in the spatial domain. As we have seen in this chapter, LTE employs all of the preceding methods, including MIMO, beam forming, and advanced receiver methods, to increase system capacity and coverage.

The other component of increasing the capacity comes from better utilization of such single radio channels and trying to approach the theoretical limits of the channel capacity. This is primarily achieved by fast link adaptation and dynamic scheduling methods—all part of RRM functions of LTE, which try to follow the fast fluctuations of the radio link and to exploit time, frequency, or multiuser diversity of the radio channel. This chapter has shown methods and examples of how these advanced RRM functions are realized in LTE and how, together, they will make LTE a high-performance, competitive system for many years to come.

4.6.1 Outlook

Although the first release of the LTE standard (release 8 in 3GPP release numbering) has not yet been completed at the time of writing, the discussion about future enhancements of the LTE system, which will fulfill or even exceed requirements being set by ITU for IMT Advanced systems, have already started in the 3GPP. The work in ITU is targeted to set only the requirements that a system should fulfill in order to qualify as IMT-Advanced capable. At the time of writing, this work of defining the requirements is ongoing in ITU [12].

Such an enhanced version of the LTE system is also often referred to as LTE-Advanced. The LTE-Advanced system will be fundamentally based on the LTE technology but will include some additions that enable improvement of the system toward and beyond IMT Advanced requirements. The improvements are not pointing exclusively in the direction of achieving even higher spectral efficiencies with more advanced radio link algorithms, etc., but, rather, are focusing more on other aspects of the system. For example, such issues as spectrum aggregation, self-configuration, self-optimization, advanced repeater structures, and distributed antenna systems (DASs) are being discussed in the 3GPP [13] as candidate technologies for LTE-Advanced.

Acronyms

3GPP	3rd Generation Partnership Project
AC	admission control
ACK	acknowledgment
AMBR	aggregated maximum bit rate
AP	access point
App.	application
ARP	allocation retention priority
BLER	block error rate
BM-SC	broadcast multicast service center
BS	bearer service
BSR	buffer status report
CDD	cyclic delay diversity
CDMA	code division multiple access
CFR	channel feedback reporting
CPICH	common pilot indicator channel
CQI	channel quality indicator
CZ	collision zone
DAS	distributed antenna system
DFT	discrete Fourier transform
DL	downlink
eNB	eNode B
EPC	evolved packet core
EPS	evolved packet system
E-UTRAN	evolved universal terrestrial radio access network
FDD/TDD	frequency domain duplexing/time domain duplexing
FDMA	frequency division multiple access
GERAN	GSM/enhanced data rates for GSM evolution radio access networks
GBR	guaranteed bit rate

GPRS	general packet radio service
GPS	global positioning system
GSM	global system for mobile communications
GTP	GPRS tunneling protocol
GTP-U	GPRS tunneling protocol—user plane
GW	gateway
HARQ	hybrid automatic repeat request
HO	handover
HSPA	high-speed packet access
ICIC	intercell interference coordination
IEEE	Institute of Electrical and Electronics Engineers
IMT	international mobile telecommunications
IP	Internet protocol
IRC	interference rejection combining
ITU	International Telecommunication Union
L1	layer 1
L2	layer 2
LA	link adaptation
LB	load balancing
LTE	long term evolution
MAC	medium access control
MBMS	multicast broadcast multimedia services
MBR	maximum bit rate
MBSFM	MBMS single frequency network
MCE	multicast/broadcast coordination entity
MCS	modulation and coding scheme
MIMO	multiple input, multiple output
MME	mobility management entity
MMSE	minimum mean square error
MU-MIMO	multiuser MIMO
NACK	negative acknowledgment
NAS	nonaccess stratum
OFDM	orthogonal frequency division multiplexing
OI	overload indicator
PCRF	policy control and resource function
PDCP	packet data convergence protocol
PDN	packet data network
PDCCH	physical downlink control channel
PDSCH	physical downlink shared channel
PHY	physical layer
PL	path loss
PLMN	public land mobile network
PRACH	physical random access channel

PUCCH	physical uplink control channel
PUSCH	physical uplink shared channel
QAM	quadrature amplitude modulation
QoS	quality of service
QPSK	quadrature phase shift keying
RAC	radio admission control
RACH	random access channel
RAN	radio access network
RAT	radio access technology
RB	resource block
RBC	radio bearer control
RE	resource element
RIGW	random index geometry weight
RLC	radio link control
RRC	radio resource control
RRM	radio resource management
RSCP	received signal code power
RSI	random start index
RSRP	reference symbol received power
RSRQ	reference symbol received quality
RSSI	received signal strength indicator
RTWP	received total wideband power
s-eNB	source eNode B
S-GW	serving gateway
SAE	system architecture evolution
SC-FDMA	single-carrier FDMA
SCTP	stream control transmission protocol
SDP	session description protocol
SDU	service data unit
SFBC	space-frequency block codes
SFN	single-frequency network
SI	start index
SIGW	start index geometry weight
SINR	signal-to-interference-and-noise ratio
SIP	session initiation protocol
SIR	signal-to-interference ratio
SR	scheduling request
SSI	start–stop index
STBC	space–time block codes
TCP	transmission control protocol
TDMA	time division multiple access
t-eNB	target eNode B
TTI	transmission time interval

Tx	transmission
UDP	user datagram protocol
UE	user equipment
UL	uplink
USIM	user subscription identity module
UTRA	universal terrestrial radio access
VoIP	voice over Internet protocol
WCDMA	wideband code division multiple access

References

1. Dahlman, E., S. Parkvall, J. Skold, and P. Beming. 2007. *3G Evolution, HSPA and LTE for mobile broadband.* Orlando, FL: Academic Press.
2. 3GPP TS 36.322. E-UTRA radio link control (RLC) protocol specification. ftp://ftp.3gpp.org/Specs/archive/36_series/36.322/, 2008.
3. 3GPP TS 36.321. E-UTRA medium access control (MAC) protocol specification. ftp://ftp.3gpp.org/Specs/archive/36_series/36.321/, 2008.
4. 3GPP TS 36.323. E-UTRA packet data convergence protocol (PDCP) specification. ftp://ftp.3gpp.org/Specs/archive/36_series/36.323/, 2008.
5. 3GPP TS 36.211. E-UTRA physical channels and modulation. ftp://ftp.3gpp.org/Specs/archive/36_series/36.211/, 2008.
6. 3GPP TS 25.913. Requirements for evolved UTRA (E-UTRA) and evolved UTRAN (E-UTRAN). ftp://ftp.3gpp.org/Specs/archive/25_series/25.913/, 2006.
7. Oestges, C., and B. Clerckx. 2007. *MIMO wireless communications: From real-world propagation to space–time code design.* Orlando, FL: Academic Press.
8. 3GPP R3-070941. May 2007. Comparison of robust E-MBMS content synchronization protocols.
9. IEEE Std. 1588. IEEE standard for a precision clock synchronization protocol for networked measurement and control systems, 2008.
10. Ju, W., and W. M. Cioffi. 2001. On constant power water filling. *IEEE International Conference on Communications* (ICC), 1665–1669.
11. Lozano, A., A. M. Tulino, and S. Verdú. 2006. Optimum power allocation for parallel Gaussian channels with arbitrary input distributions. *IEEE Transactions on Information Theory* 52 (7), 3033–3051.
12. ITU-R Resolution 57. Principles for the process of development of IMT Advanced. http://www.itu.int/dms_pub/itu-r/opb/res/R-RES-R.57-2007-PDF-E.pdf, 2007.
13. Overview and process/roadmap of ITU-R IMT Advanced standards. http://www.3gpp.org/ftp/workshop/2007-11-26_RAN_IMT_Advanced/Docs/, 2007.
14. 3GPP TS 36.300. Evolved UTRA and evolved UTRAN, overall description. ftp://ftp.3gpp.org/Specs/archive/36_series/36.300/, 2008.
15. 3GPP TS 36.304. Evolved (UTRA) and evolved (UTRAN): User equipment procedures in idle mode. ftp://ftp.3gpp.org/Specs/archive/36_series/36.304/, 2008.
16. 3GPP TR R3.018. Evolved UTRA and UTRAN radio access architecture and interfaces. http://www.3gpp.org/ftp/tsg_ran/WG3_Iu/R3_internal_TRs/, 2007.
17. 3GPP TR 25.813. E-UTRA radio interface protocol aspects. http://www.3gpp.org/ftp/Specs/html-info/25813.htm, 2006.

18. 3GPP TS 36.213. E-UTRA physical layer procedures. ftp://ftp.3gpp.org/Specs/archive/36_series/36.213/, 2008.
19. 3GPP TR 36.801. E-UTRA measurement requirements. http://www.3gpp.org/ftp/Specs/html-info/36801.htm, 2008.
20. 3GPP TS 36.214. E-UTRA physical layer measurements. http://www.3gpp.org/ftp/Specs/html-info/36214.htm, 2008.
21. 3GPP TS 36.331. E-UTRA radio resource control (RRC) protocol specification. ftp://ftp.3gpp.org/Specs/archive/36_series/36.331/, 2008.
22. 3GPP TS 25.171. Requirements for support of assisted global positioning system. http://www.3gpp.org/ftp/Specs/html-info/25171.htm
23. Zander, J. 2001. *Radio resource management for wireless networks.* Boston: Artech House Publishers.
24. Katzela, I., and M. Naghshineh. 1996. Channel assignment schemes for cellular mobile telecommunication systems: A comprehensive survey. *IEEE Personal Communications* June:10–31.
25. Liu, G., and H. Li. 2003. Downlink dynamic resource allocation for multi-cell OFDMA system. *58th IEEE Vehicular Technology Conference, VTC 2003* 3:1698–1702.
26. Fodor, G. 2006. Performance analysis of a reuse partitioning technique for OFDM-based evolved UTRA. *IEEE International Workshop on Quality of Service*, New Haven, CT, pp. 112–120.
27. Fodor, G. 2007. On scheduling and interference coordination policies for multicell OFDMA networks. *IFIP Networking '07,* Atlanta, GA, *Springer Lecture Notes in Computer Science* 4479:488–499.
28. Koutsimanis, C., and G. Fodor. 2008. A dynamic resource allocation scheme for guaranteed bit rate services in OFDMA networks. *IEEE International Conference on Communications*, Beijing, May.
29. Fodor, G. 2008. A low intercell interference variation scheduler for OFDMA networks. *IEEE International Conference on Communications*, Beijing, May.
30. Reider, N. 2007. Intercell interference coordination techniques in mobile networks. Master' thesis, Budapest University of Technology and Economics (BUTE).
31. Koutsimanis, C. 2007. Resource allocation for narrow band and elastic services in OFDMA wireless networks. Master' thesis, Royal Institute of Technology (KTH), COS/RCS 2007-3, Stockholm, Sweden. Available at: http://www.cos.ict.kth.se/publications/publications/2007/2630.pdf
32. Simonsson, A. 2007. Frequency reuse and intercell interference coordination in E-UTRA. *IEEE VTC Spring,* 3091–3095.
33. Kiani, S. G., G. E. Øien, and D. Gesbert. 2007. Maximizing multicell capacity using distributed power allocation and scheduling. *IEEE Wireless Communications and Networking Conference*, 1692–1696.
34. Lee, N.-H., and S. Bahk. 2007. Dynamic channel allocation using the interference range in multicell downlink systems. *IEEE Wireless Communications and Networking Conference*, 1718–1723.
35. Koutsopoulos, I., and L. Tassiulas. 2006. Cross-layer adaptive techniques for throughput enhancement in wireless OFDM-based networks. *IEEE Transactions on Networking* 14 (5):1056–1066.

36. Thanabalasingham, T., S. Hanley, L. L. H. Andrew, and J. Papandripoulos. 2006. Joint allocation of subcarriers and transmit powers in a multiuser OFDM cellular network. *IEEE International Conference on Communications '06*, 269–274.
37. Abardo, A., A. Alessio, P. Detti, and M. Moretti. 2007. Centralized radio resource allocation for OFDMA cellular systems. *IEEE International Conference on Communications '07*, 269–274.
38. Bosisio, R., and U. Spagnolini. 2007. Collision model for the bit error rate analysis of multicell multiantenna OFDMA systems. *IEEE International Conference on Communications '07.*
39. Bosisio, R., and U. Spagnolini. 2008. Interference coordination versus interference randomization in multicell 3GPP LTE system. *IEEE Wireless Communications and Networking*, Las Vegas, NV.
40. Kurt, T., and H. Delic. 2005. Space-frequency coding reduces the collision rate in FH-OFDMA. *IEEE Transactions on Wireless Communications* 4 (5):2045–2050.
41. Elayoubi, S.-E., B. Fourestié, and X. Auffret. 2006. On the capacity of OFDMA 802.16 systems. *IEEE International Conference on Communications '06*, 1760–1765.
42. 3GPP R1-074444. November 2007. On intercell interference coordination with/without traffic load indication.
43. 3GPP R2-072577. June 2007. E-UTRA UE measurement reporting.
44. 3GPP TS 36.213. E-UTRA physical layer procedures. http://www.3gpp.org/ftp/Specs/html-info/36213.htm, 2008.
45. 3GPP R2-072577. June 2007. E-UTRA UE radio measurement reporting.
46. 3GPP R1-075014. November 2007. Way forward on UL ICIC/overload indicator for LTE. Telecom Italia, Ericsson, Alcatel-Lucent, Orange, Qualcomm Europe, Telefonica, Vodafone, KPN.
47. 3GPP R2-080238. January 2008. Update of parameters and flow charts for cell reselection.
48. 3GPP R1-074852. November 2007. Additional RSRP reporting trigger for ICIC.

Chapter 5

MIMO OFDM Schemes for 3GPP LTE

Gang Wu and Xun Fan

Contents

5.1 Introduction

Multiple-input, multiple-output (MIMO) techniques have been integrated as one of the key approaches to provide the peak data rate, average throughput, and system performance in 3rd Generation Partnership Project (3GPP) long term evolution (LTE). Based on the function of the multiple transmission symbol streams in MIMO, the operation mode of multiple transmit antennas at the cell site (denoted as MIMO mode) have spatial division multiplexing (SDM), precoding, and transmit diversity (TD). Base on the allocation of the multiple transmission streams in MIMO, the MIMO mode is denoted as single user (SU)-MIMO if the multiple transmission symbol stream is solely assigned to a single UE and multiuser (MU)-MIMO if the SDM of the modulation symbol streams for different UEs use the same time-frequency resource [1]. The LTE downlink is an orthogonal frequency division multiplexing (OFDM) system, so the MIMO modes proposed in LTE are MIMO OFDM schemes.

In 3GPP LTE, the MIMO mode is restricted by the UE capability (e.g., number of receive antennas) and is determined by taking into account the slow channel variation. The MIMO mode is adapted slowly (e.g., only at the beginning of communication or every several 100 ms), in order to reduce the control signaling (including feedback) required to support the MIMO mode adaptation. For the control channel, only a single stream using the multiple transmit antennas is supported. The function of the MIMO in the general downlink physical channels is shown in Figure 5.1. The baseband signal processing procedure is defined in terms of the following steps [3]:

- scrambling coded bits in each of the code words to be transmitted on a physical channel;
- modulating scrambled bits to generate complex-valued modulation symbols;
- mapping the complex-valued modulation symbols onto one or several transmission layers;

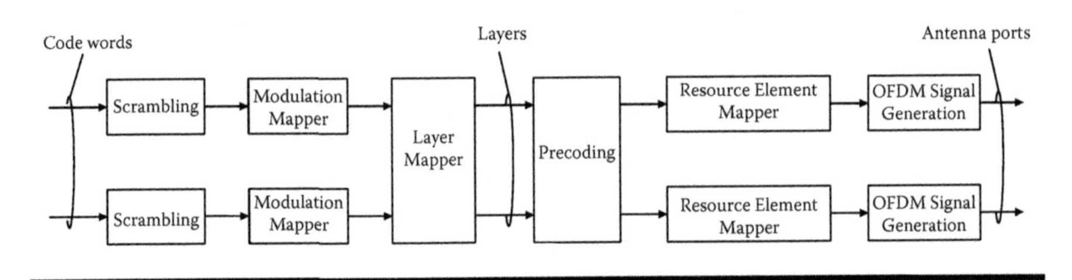

Figure 5.1 Overview of physical channel processing with MIMO.

- precoding the complex-valued modulation symbols on each layer for transmission on the antenna ports;
- mapping complex-valued modulation symbols for each antenna port to resource elements; and
- generating complex-valued time-domain OFDM signals for each antenna port.

The baseline antenna configuration for MIMO is two transmit antennas at the cell site and two receive antennas at the UE. The higher-order downlink MIMO and antenna diversity (four transmit [Tx] and two or four receive [Rx] antennas) are also supported. The baseline antenna configuration for uplink single-user MIMO is two transmit antennas at the UE and two receive antennas at the cell site. If the UE has only a single power amplifier and two transmit antennas, the antenna switching/selection is the only option that is supported for SU-MIMO. In addition, the following techniques are supported [2]:

- codebook-based precoding with a single precoding feedback per full system bandwidth when the system bandwidth (or subset of resource blocks [RBs]) is smaller than or equal to 12 RBs and per five adjacent resource blocks or the full system bandwidth (or subset of resource blocks) when the system bandwidth is larger than 12 RBs; and
- rank adaptation with single rank feedback referring to full system bandwidth; node B can override rank report.

5.2 SU-MIMO

5.2.1 Layer Mapping for Spatial Multiplexing

The complex-valued modulation symbols for each of the code words to be transmitted are mapped onto one or several layers. Complex-valued modulation symbols $d^{(q)}(0),\ldots,d^{(q)}(M_{\text{symb}}^{(q)}-1)$ for code word q are mapped onto the layers $x(i) = [x^{(0)}(i)\ldots x^{(\upsilon-1)}(i)]^T$, $i = 0,1,\ldots,M_{\text{symb}}^{\text{layer}}-1$, where υ is the number of layers and $M_{\text{symb}}^{\text{layer}}$ is the number of modulation symbols per layer.

For spatial multiplexing, the layer mapping is done according to Table 5.1. The number of layers υ is less than or equal to the number of antenna ports P used for transmission of the physical channel.

5.2.2 Layer Mapping for Transmit Diversity

For transmit diversity, the layer mapping is done according to Table 5.2. There is only one code word, and the number of layers υ is equal to the number of antenna ports P used for transmission of the physical channel.

Table 5.1 Code-Word-to-Layer Mapping for Spatial Multiplexing

Number of Layers	Number of Code Words	Code-Word-to-Layer Mapping $i=0,1,\ldots,M_{symb}^{layer}-1$	
1	1	$x^{(0)}(i)=d^{(0)}(i)$	$M_{symb}^{layer}=M_{symb}^{(0)}$
2	2	$x^{(0)}(i)=d^{(0)}(i)$ $x^{(1)}(i)=d^{(1)}(i)$	$M_{symb}^{layer}=M_{symb}^{(0)}=M_{symb}^{(1)}$
3	2	$x^{(0)}(i)=d^{(0)}(i)$ $x^{(1)}(i)=d^{(1)}(2i)$ $x^{(2)}(i)=d^{(1)}(2i+1)$	$M_{symb}^{layer}=M_{symb}^{(0)}=M_{symb}^{(1)}/2$
4	2	$x^{(0)}(i)=d^{(0)}(2i)$ $x^{(1)}(i)=d^{(0)}(2i+1)$ $x^{(2)}(i)=d^{(1)}(2i)$ $x^{(3)}(i)=d^{(1)}(2i+1)$	$M_{symb}^{layer}=M_{symb}^{(0)}/2=M_{symb}^{(1)}/2$

5.2.3 Precoding

The precoder takes as input a block of vectors $x(i) = [x^{(0)}(i)\ldots x^{(\upsilon-1)}(i)]^T$, $i=0,1,\ldots,M_{symb}^{layer}-1$ from the layer mapping and generates a block of vectors $y(i)=[\ldots y^{(p)}(i)\ldots]^T$, $i=0,1,\ldots,M_{symb}^{ap}-1$ to be mapped onto resources on each of the antenna ports, where $y^{(p)}(i)$ represents the signal for antenna port p.

5.2.3.1 Precoding for Transmission on a Single Antenna Port

For transmission on a single antenna port, precoding is defined by

$$y^{(p)}(i)=x^{(0)}(i)$$

Table 5.2 Code-Word-to-Layer Mapping for Transmit Diversity

Number of Layers	Number of Code Words	Code-Word-to-Layer Mapping $i=0,1,\ldots,M_{symb}^{layer}-1$	
2	1	$x^{(0)}(i)=d^{(0)}(2i)$ $x^{(1)}(i)=d^{(0)}(2i+1)$	$M_{symb}^{layer}=M_{symb}^{(0)}/2$
4	1	$x^{(0)}(i)=d^{(0)}(4i)$ $x^{(1)}(i)=d^{(0)}(4i+1)$ $x^{(2)}(i)=d^{(0)}(4i+2)$ $x^{(3)}(i)=d^{(0)}(4i+3)$	$M_{symb}^{layer}=M_{symb}^{(0)}/4$

where $p \in \{0,4,5\}$ is the number of the single antenna port used for transmission of the physical channel and $i = 0,1,\ldots,M_{\text{symb}}^{\text{ap}} - 1$, $M_{\text{symb}}^{\text{ap}} = M_{\text{symb}}^{\text{layer}}$.

5.2.3.2 Precoding for Spatial Multiplexing

Precoding for spatial multiplexing is only used in combination with layer mapping for spatial multiplexing, as described in Section 5.2.1. Spatial multiplexing supports two or four antenna ports, and the set of antenna ports used is $p \in \{0,1\}$ or $p \in \{0,1,2,3\}$, respectively.

5.2.3.3 Precoding for Zero- and Small-Delay CDD

For zero-delay and small-delay cyclic delay diversity (CDD), precoding for spatial multiplexing is defined by

$$\begin{bmatrix} y^{(0)}(i) \\ \vdots \\ y^{(P-1)}(i) \end{bmatrix} = D(k_i)W(i)\begin{bmatrix} x^{(0)}(i) \\ \vdots \\ x^{(\upsilon-1)}(i) \end{bmatrix}$$

where

the precoding matrix $W(i)$ is of size $P \times \upsilon$;
the quantity $D(k_i)$ is a diagonal matrix for support of cyclic delay diversity;
k_i represents the frequency-domain index of the resource element to which the complex-valued symbol $y(i)$ is mapped; and
$i = 0,1,\ldots,M_{\text{symb}}^{\text{ap}} - 1$, $M_{\text{symb}}^{\text{ap}} = M_{\text{symb}}^{\text{layer}}$.

The matrix $D(k_i)$ should be selected from Table 5.3, where a UE-specific value of δ is semistatically configured in the UE and the eNodeB by higher-layer signaling. The quantity η in Table 5.3 is the smallest number from the set $\{128,256,512,1024,2048\}$ such that $\eta \geq N_{\text{RB}}^{\text{DL}} N_{\text{sc}}^{\text{RB}}$.

For spatial multiplexing, the values of $W(i)$ should be selected among the precoder elements in the codebook configured in the eNodeB and the UE. The eNodeB can further confine the precoder selection in the UE to a subset of the elements in the codebook using codebook subset restrictions. The configured codebook should be selected from Table 5.5 or Table 5.6.

5.2.3.4 Precoding for Large-Delay CDD

For large-delay CDD, precoding for spatial multiplexing is defined by

$$\begin{bmatrix} y^{(0)}(i) \\ \vdots \\ y^{(P-1)}(i) \end{bmatrix} = W(i)D(i)U\begin{bmatrix} x^{(0)}(i) \\ \vdots \\ x^{(\upsilon-1)}(i) \end{bmatrix}$$

Table 5.3 Zero- and Small-Delay Cyclic Delay Diversity

Set of Antenna Ports Used	Number of Layers υ	$D(k_i)$	d: No CDD	d: Small Delay
{0,1}	1 2	$\begin{bmatrix} 1 & 0 \\ 0 & e^{-j2\pi \cdot k_i \cdot \delta} \end{bmatrix}$	0	$2/\eta$
{0,1,2,3}	1 2 3 4	$\begin{bmatrix} 1 & 0 & 0 & 0 \\ 0 & e^{-j2\pi \cdot k_i \cdot \delta} & 0 & 0 \\ 0 & 0 & e^{-j2\pi \cdot k_i \cdot 2\delta} & 0 \\ 0 & 0 & 0 & e^{-j2\pi \cdot k_i \cdot 3\delta} \end{bmatrix}$	0	$1/\eta$

where the precoding matrix $W(i)$ is of size $P \times \upsilon$ and $i = 0,1,\ldots,M^{ap}_{symb} - 1$, $M^{ap}_{symb} = M^{layer}_{symb}$. The diagonal size $\upsilon \times \upsilon$ matrix $D(i)$ supporting cyclic delay diversity and the size $\upsilon \times \upsilon$ matrix U are both given in Table 5.1 for different numbers of layers υ.

The values of the precoding matrix $W(i)$ should be selected from among the precoder elements in the codebook configured in the eNodeB and the UE. The eNodeB can further confine the precoder selection in the UE to a subset of the elements in the codebook using codebook subset restriction. The configured codebook should be selected from Table 5.5 or Table 5.6.

Table 5.4 Large-Delay Cyclic Delay Diversity

Number of Layers υ	U	$D(i)$
1	[1]	[1]
2	$\begin{bmatrix} 1 & 1 \\ 1 & e^{-j2\pi/2} \end{bmatrix}$	$\begin{bmatrix} 1 & 0 \\ 0 & e^{-j2\pi i/2} \end{bmatrix}$
3	$\begin{bmatrix} 1 & 1 & 1 \\ 1 & e^{-j2\pi/3} & e^{-j4\pi/3} \\ 1 & e^{-j4\pi/3} & e^{-j8\pi/3} \end{bmatrix}$	$\begin{bmatrix} 1 & 0 & 0 \\ 0 & e^{-j2\pi i/3} & 0 \\ 0 & 0 & e^{-j4\pi i/3} \end{bmatrix}$
4	$\begin{bmatrix} 1 & 1 & 1 & 1 \\ 1 & e^{-j2\pi/4} & e^{-j4\pi/4} & e^{-j6\pi/4} \\ 1 & e^{-j4\pi/4} & e^{-j8\pi/4} & e^{-j12\pi/4} \\ 1 & e^{-j6\pi/4} & e^{-j12\pi/4} & e^{-j18\pi/4} \end{bmatrix}$	$\begin{bmatrix} 1 & 0 & 0 & 0 \\ 0 & e^{-j2\pi i/4} & 0 & 0 \\ 0 & 0 & e^{-j4\pi i/4} & 0 \\ 0 & 0 & 0 & e^{-j6\pi i/4} \end{bmatrix}$

Table 5.5 Codebook for Transmission on Antenna Ports {0,1}

Codebook Index	Number of Layers υ	
	1	2
0	$\begin{bmatrix} 1 \\ 0 \end{bmatrix}$	$\frac{1}{\sqrt{2}}\begin{bmatrix} 1 & 0 \\ 0 & 1 \end{bmatrix}$
1	$\begin{bmatrix} 0 \\ 1 \end{bmatrix}$	$\frac{1}{2}\begin{bmatrix} 1 & 1 \\ 1 & -1 \end{bmatrix}$
2	$\frac{1}{\sqrt{2}}\begin{bmatrix} 1 \\ 1 \end{bmatrix}$	$\frac{1}{2}\begin{bmatrix} 1 & 1 \\ j & -j \end{bmatrix}$
3	$\frac{1}{\sqrt{2}}\begin{bmatrix} 1 \\ -1 \end{bmatrix}$	–
4	$\frac{1}{\sqrt{2}}\begin{bmatrix} 1 \\ j \end{bmatrix}$	–
5	$\frac{1}{\sqrt{2}}\begin{bmatrix} 1 \\ -j \end{bmatrix}$	–

5.2.3.5 Codebook for Precoding

The precoding is used to convert the antenna-domain MIMO signal processing into the beam-domain processing. For two transmit antennas, it was decided that the unitary precoding matrix is to be used for support of SU-MIMO. Also, UE feedback is based on the codebook-based precoding for frequency-domain duplexing (FDD).

For transmission on four antenna ports, the quantity $W_n^{\{s\}}$ denotes the matrix defined by the columns given by the set $\{s\}$ from the expression $W_n = I - 2u_n u_n^H / u_n^H u_n$, where I is the 4×4 identity matrix and the vector $\mathbf{u_n}$ is given by Table 5.6.

The codebooks for various ranks for four antennas are constructed as follows [5]:

1. *Choose rank-one codebook.* The rank-one codebook consisting of some N unit-norm 4×1 complex vectors is chosen. Note that the codebook can be chosen to satisfy any metric. We evaluate two codebooks here: one is a Grassmanian-based codebook optimized to achieve high throughput for correlation antennas; the other is a constant-modulus Grassmanian codebook optimized to achieve high chordal distance.
2. *Obtain householder vectors.* From each vector in the rank-one codebook, obtain a set of householder-basis vectors. Suppose $\mathbf{x}$ is a valid precoding

Table 5.6 Codebook for Transmission on Antenna Ports {0,1,2,3}

Codebook index	$\mathbf{u}_n$	*Number of layers* υ			
		1	*2*	*3*	*4*
0	$u_0 = [1 \quad -1 \quad -1 \quad -1]^T$	$W_0^{\{1\}}$	$W_0^{\{14\}}/\sqrt{2}$	$W_0^{\{124\}}/\sqrt{3}$	$W_0^{\{1234\}}/2$
1	$u_1 = [1 \quad -j \quad 1 \quad j]^T$	$W_1^{\{1\}}$	$W_1^{\{12\}}/\sqrt{2}$	$W_1^{\{123\}}/\sqrt{3}$	$W_1^{\{1234\}}/2$
2	$u_2 = [1 \quad 1 \quad -1 \quad 1]^T$	$W_2^{\{1\}}$	$W_2^{\{12\}}/\sqrt{2}$	$W_2^{\{123\}}/\sqrt{3}$	$W_2^{\{3214\}}/2$
3	$u_3 = [1 \quad j \quad 1 \quad -j]^T$	$W_3^{\{1\}}$	$W_3^{\{12\}}/\sqrt{2}$	$W_3^{\{123\}}/\sqrt{3}$	$W_3^{\{3214\}}/2$
4	$u_4 = [1 (-1-j)/\sqrt{2} - j (1-j)/\sqrt{2}]^T$	$W_4^{\{1\}}$	$W_4^{\{14\}}/\sqrt{2}$	$W_4^{\{124\}}/\sqrt{3}$	$W_4^{\{1234\}}/2$
5	$u_5 = [1 (1-j)/\sqrt{2}\, j (-1-j)/\sqrt{2}]^T$	$W_5^{\{1\}}$	$W_5^{\{14\}}/\sqrt{2}$	$W_5^{\{124\}}/\sqrt{3}$	$W_5^{\{1234\}}/2$
6	$u_6 = [1 (1+j)/\sqrt{2} - j (-1+j)/\sqrt{2}]^T$	$W_6^{\{1\}}$	$W_6^{\{13\}}/\sqrt{2}$	$W_6^{\{134\}}/\sqrt{3}$	$W_6^{\{1324\}}/2$
7	$u_7 = [1 (-1+j)/\sqrt{2}\, j (1+j)/\sqrt{2}]^T$	$W_7^{\{1\}}$	$W_7^{\{13\}}/\sqrt{2}$	$W_7^{\{134\}}/\sqrt{3}$	$W_7^{\{1324\}}/2$
8	$u_8 = [1 \quad -1 \quad 1 \quad 1]^T$	$W_8^{\{1\}}$	$W_8^{\{12\}}/\sqrt{2}$	$W_8^{\{124\}}/\sqrt{3}$	$W_8^{\{1234\}}/2$
9	$u_9 = [1 \quad -j \quad -1 \quad -j]^T$	$W_9^{\{1\}}$	$W_9^{\{14\}}/\sqrt{2}$	$W_9^{\{134\}}/\sqrt{3}$	$W_9^{\{1234\}}/2$
10	$u_{10} = [1 \quad 1 \quad 1 \quad -1]^T$	$W_{10}^{\{1\}}$	$W_{10}^{\{13\}}/\sqrt{2}$	$W_{10}^{\{123\}}/\sqrt{3}$	$W_{10}^{\{1324\}}/2$
11	$u_{11} = [1 \quad j \quad -1 \quad j]^T$	$W_{11}^{\{1\}}$	$W_{11}^{\{13\}}/\sqrt{2}$	$W_{11}^{\{134\}}/\sqrt{3}$	$W_{11}^{\{1324\}}/2$
12	$u_{12} = [1 \quad -1 \quad -1 \quad 1]^T$	$W_{12}^{\{1\}}$	$W_{12}^{\{12\}}/\sqrt{2}$	$W_{12}^{\{123\}}/\sqrt{3}$	$W_{12}^{\{1234\}}/2$
13	$u_{13} = [1 \quad -1 \quad 1 \quad 1]^T$	$W_{13}^{\{1\}}$	$W_{13}^{\{13\}}/\sqrt{2}$	$W_{13}^{\{123\}}/\sqrt{3}$	$W_{13}^{\{1324\}}/2$
14	$u_{14} = [1 \quad 1 \quad -1 \quad -1]^T$	$W_{14}^{\{1\}}$	$W_{14}^{\{13\}}/\sqrt{2}$	$W_{14}^{\{123\}}/\sqrt{3}$	$W_{14}^{\{3214\}}/2$
15	$u_{15} = [1 \quad 1 \quad 1 \quad 1]^T$	$W_{15}^{\{1\}}$	$W_{15}^{\{12\}}/\sqrt{2}$	$W_{15}^{\{123\}}/\sqrt{3}$	$W_{15}^{\{1234\}}/2$

vector; the corresponding householder vector **u** is obtained so that the first column of $(\mathbf{I} - 2\mathbf{u}\mathbf{u}^\mathbf{H})$ is equal to **x.** Note that this is always possible for all unit norms **x.** Further, the resulting **u** is unique up to a phase rotation and is guaranteed to have norm one.

3. *Obtain rank-four codebook.* From the set of householder vectors **u,** construct the 4×4 unitary matrices $\mathbf{H}(\mathbf{u}) = (\mathbf{I} - 2\mathbf{u}\mathbf{u}^\mathbf{H})$. These matrices form the rank-four codebook.

4. *Obtain rank-two and rank-three codebooks by column selection.* Codebooks for ranks two and three are obtained by selecting submatrices from the rank-four codebook. One exemplary construction is that the rank-R codebook just contains the first R columns of each matrix in the rank-four codebook. However, more sophisticated construction techniques would pick the submatrices in order to optimize the chordal distance or some other performance-related criterion.

Some salient features of the preceding construction include:

- *Optimum performance for rank one.* This feature ensures optimum cell-edge throughput performance. In addition, multiple rank-one codebooks can be defined to fit different channel scenarios.
- *Optimum chordal distance for rank three.* One possible construction of rank-three codebooks is to pick the *last* three columns of the 4×4 matrices $\mathbf{H}(\mathbf{u})$. It can be shown that, in this case, the chordal distance is exactly equal to that of the rank-one codebook. Thus, if the rank-one codebook has the optimum chordal distance, so does the proposed rank-three codebook.
- *Near-optimum performance for ranks two and four.* The proposed construction achieves high throughput for ranks two and four.
- *Nested structure.* Clearly, the rank-one codebook just consists of the first columns of the matrices in the rank-four codebook. Further, by construction, codebooks of ranks two and three are also submatrices of the rank-four codebook.
- *Efficient computation.* The householder structure greatly simplifies signal-to-interference-plus-noise ratio (SINR) computation for rank-two and rank-three precoding matrices.

Link-level simulation results comparing the proposed codebook to other codebook structures are shown in the following subsections.

5.2.3.5.1 Four Node-B, Two UE Antennas, TU6 Channel Model

Figure 5.2 plots the percentage throughput gains versus geometry when each codebook is compared to an antenna selection codebook. Gains are shown for antenna correlations of 0.1 and 0.5, respectively. As seen from the figure,

- Precoding provides significant gains over antenna selection even at high geometries. In particular, impressive gains are obtained at high antenna correlations.
- The proposed householder-based codebook constructions consistently outperform discrete Fourier transform (DFT)-based codebook constructions. Even when the DFT-based codebooks are optimized, they suffer a 3–5% loss when compared to the proposed constructions at low geometry. At higher geometries, the performance of all codebooks is comparable.

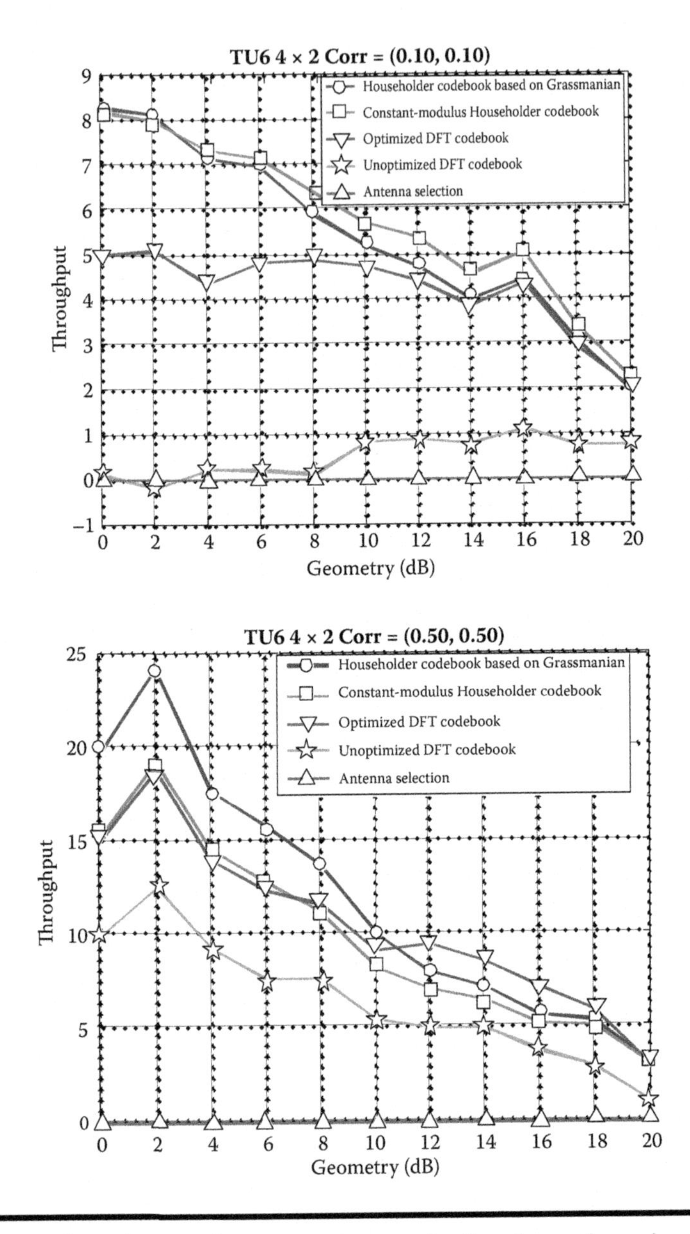

Figure 5.2 Throughput gains over antenna selection for various size-16 codebooks, with two UE antennas over a TU6 channel.

5.2.3.5.2 Four Node-B, Four UE Antennas, TU6 Channel Model

Figure 5.3 plots the percentage throughput gains versus geometry when each codebook is compared to an unoptimized DFT codebook with no ordering. Gains

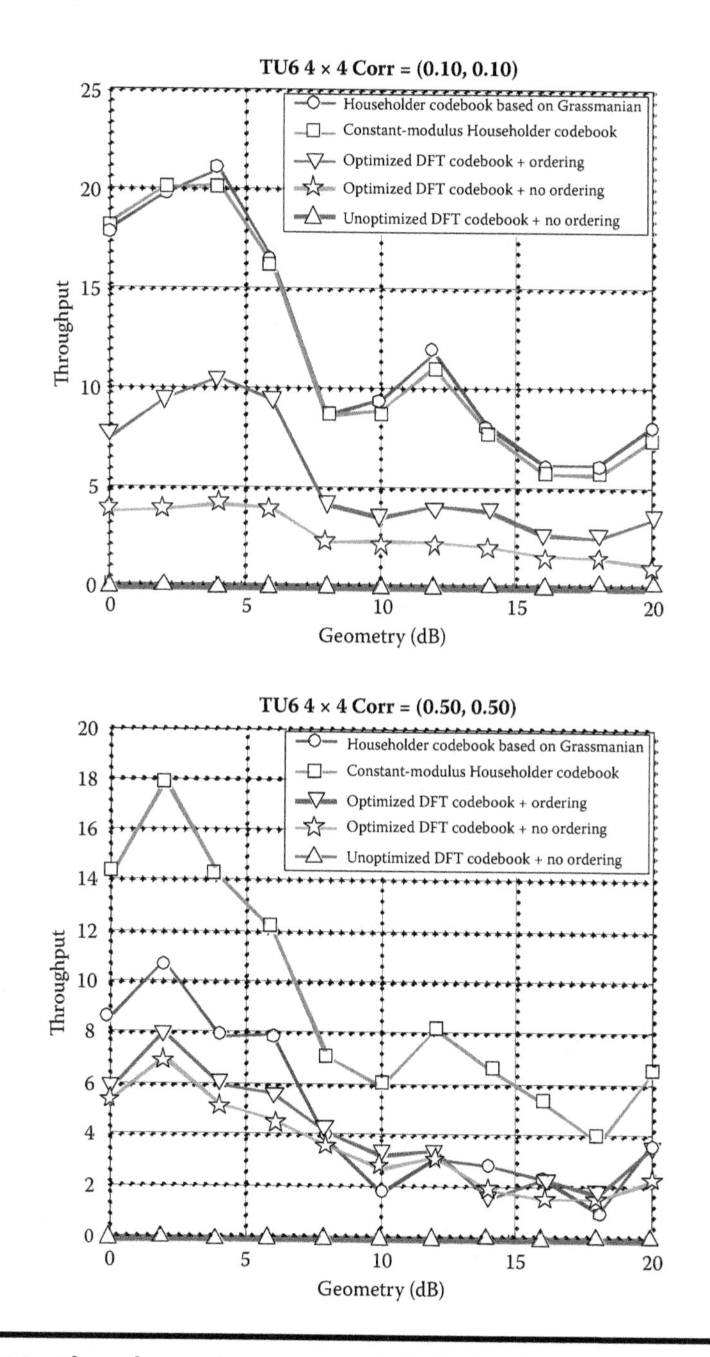

Figure 5.3 Throughput gains over unoptimized, unordered DFT codebook for various size-16 codebooks, with four UE antennas over a TU6 channel.

are shown for antenna correlations of 0.1 and 0.5, respectively. As seen from the figure,

- Ordering results in negligible, very small throughput gain over no ordering.
- The proposed constant-modulus codebook consistently outperforms DFT-based precoders at all geometries, for both low and high antenna correlations. In particular, even when the DFT-based codebook is optimized and ordered for optimum throughput, it still suffers an 8% throughput loss when compared to the proposed codebooks at low geometries. Further, even at high geometries, the DFT-based codebook suffers 2% throughput loss when compared to the proposed constant-modulus codebook.

5.2.3.5.3 Four Node-B, Four UE Antennas, SCME-C and -D Channel Models

Figure 5.4 plots the percentage throughput gains versus geometry when each codebook is compared to an unoptimized DFT codebook with no ordering. The first plot uses the SCME-C (Spatial Channel Model Extension) channel model and the second uses the SCME-D channel model, which has slightly higher correlation. As seen from the figure,

- The proposed constant-modulus codebook still outperforms DFT-based precoders at all geometries for both channel models. This holds even when the DFT-based codebook is optimized and ordered for maximum sum throughput.
- Using a householder codebook based on Grassmanian rank-one vectors further increases the throughput gains by 2–4%.
- It should be noted that the two householder-based codebooks outperform each other under different channel conditions. However, with proper choice of the base rank-one codebook, the householder structure itself appears to achieve good performance when compared to DFT-like codebooks.
- With the SCME channel model, ordering seems to provide significant gain for DFT-based codebooks. However, it should be noted that ordering adds five additional bits on top of the 4-b precoding indicator. However, this 9-b composite DFT-based codebook still performs worse compared to the two 4-b householder-based designs.

5.2.3.6 Precoding for Transmit Diversity

The precoding operation for transmit diversity is defined for two and four antenna ports. For transmission on two antenna ports, $p \in \{0,1\}$, the output $y(i) = [y^{(0)}(i)\ y^{(1)}(i)]^T$

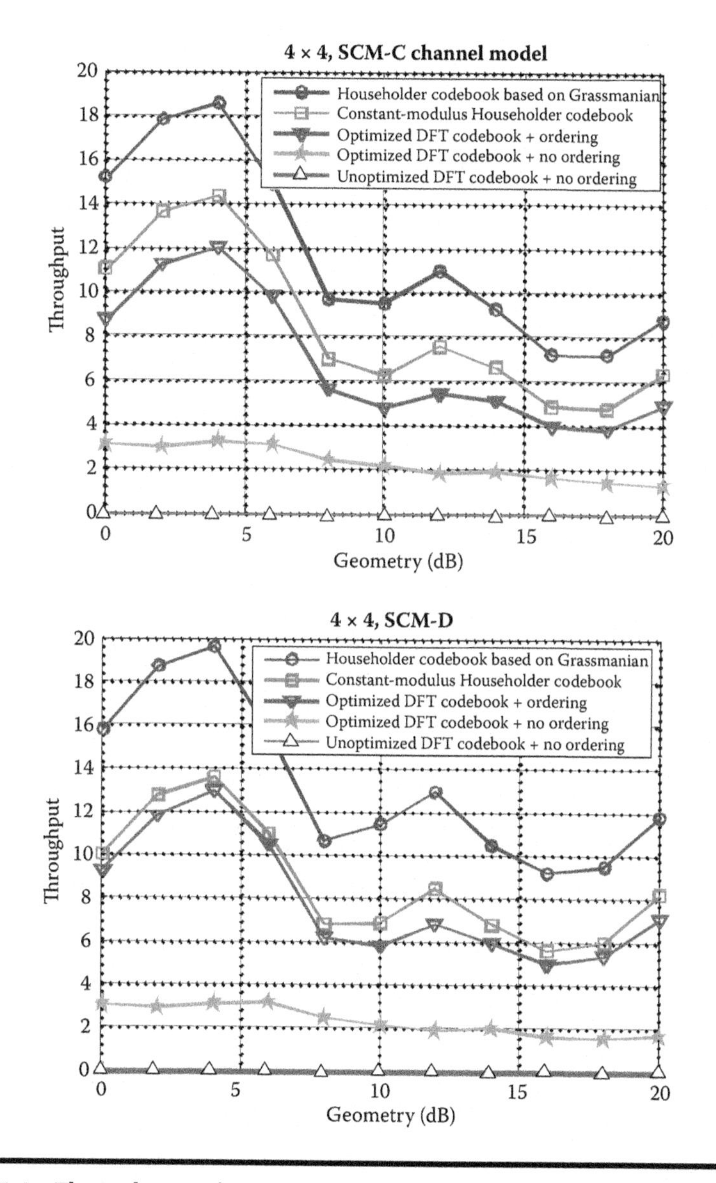

Figure 5.4 Throughput gains over unoptimized, unordered DFT codebook for various size-16 codebooks, with four UE antennas and SCM channel models.

of the precoding operation is defined by

$$\begin{bmatrix} y^{(0)}(2i) \\ y^{(1)}(2i) \\ y^{(0)}(2i+1) \\ y^{(1)}(2i+1) \end{bmatrix} = \frac{1}{\sqrt{2}} \begin{bmatrix} 1 & 0 & j & 0 \\ 0 & -1 & 0 & j \\ 0 & 1 & 0 & j \\ 1 & 0 & -j & 0 \end{bmatrix} \begin{bmatrix} \mathrm{Re}(x^{(0)}(i)) \\ \mathrm{Re}(x^{(1)}(i)) \\ \mathrm{Im}(x^{(0)}(i)) \\ \mathrm{Im}(x^{(1)}(i)) \end{bmatrix}$$

for $i = 0,1,\ldots,M_{\text{symb}}^{\text{layer}} - 1$ with $M_{\text{symb}}^{\text{ap}} = 2M_{\text{symb}}^{\text{layer}}$.

For transmission on four antenna ports, $p \in \{0,1,2,3\}$, the output $y(i) = [y^{(0)}(i)\; y^{(1)}(i)\; y^{(2)}(i)\; y^{(3)}(i)]^T$ of the precoding operation is defined by

$$\begin{bmatrix} y^{(0)}(4i) \\ y^{(1)}(4i) \\ y^{(2)}(4i) \\ y^{(3)}(4i) \\ y^{(0)}(4i+1) \\ y^{(1)}(4i+1) \\ y^{(2)}(4i+1) \\ y^{(3)}(4i+1) \\ y^{(0)}(4i+2) \\ y^{(1)}(4i+2) \\ y^{(2)}(4i+2) \\ y^{(3)}(4i+2) \\ y^{(0)}(4i+3) \\ y^{(1)}(4i+3) \\ y^{(2)}(4i+3) \\ y^{(3)}(4i+3) \end{bmatrix} = \frac{1}{\sqrt{2}} \begin{bmatrix} 1 & 0 & 0 & 0 & j & 0 & 0 & 0 \\ 0 & 0 & 0 & 0 & 0 & 0 & 0 & 0 \\ 0 & -1 & 0 & 0 & 0 & j & 0 & 0 \\ 0 & 0 & 0 & 0 & 0 & 0 & 0 & 0 \\ 0 & 1 & 0 & 0 & 0 & j & 0 & 0 \\ 0 & 0 & 0 & 0 & 0 & 0 & 0 & 0 \\ 1 & 0 & 0 & 0 & -j & 0 & 0 & 0 \\ 0 & 0 & 0 & 0 & 0 & 0 & 0 & 0 \\ 0 & 0 & 0 & 0 & 0 & 0 & 0 & 0 \\ 0 & 0 & 1 & 0 & 0 & 0 & j & 0 \\ 0 & 0 & 0 & 0 & 0 & 0 & 0 & 0 \\ 0 & 0 & 0 & -1 & 0 & 0 & 0 & j \\ 0 & 0 & 0 & 0 & 0 & 0 & 0 & 0 \\ 0 & 0 & 0 & 1 & 0 & 0 & 0 & j \\ 0 & 0 & 0 & 0 & 0 & 0 & 0 & 0 \\ 0 & 0 & 1 & 0 & 0 & 0 & -j & 0 \end{bmatrix} \begin{bmatrix} \mathrm{Re}(x^{(0)}(i)) \\ \mathrm{Re}(x^{(1)}(i)) \\ \mathrm{Re}(x^{(2)}(i)) \\ \mathrm{Re}(x^{(3)}(i)) \\ \mathrm{Im}(x^{(0)}(i)) \\ \mathrm{Im}(x^{(1)}(i)) \\ \mathrm{Im}(x^{(2)}(i)) \\ \mathrm{Im}(x^{(3)}(i)) \end{bmatrix}$$

for $i = 0,1,\ldots,M_{\text{symb}}^{\text{layer}} - 1$ with $M_{\text{symb}}^{\text{ap}} = 4M_{\text{symb}}^{\text{layer}}$.

5.3 MU-MIMO

In 3GPP LTE, MU-MIMO is still a topic in discussion. Per-user unitary and rate control (PU²RC) is a typical proposed MU-MIMO scheme [6,7]. Figure 5.5 shows the block diagram for PU²RC. In such a system, a set of precoder matrices, $\mathbf{E} = \{\mathbf{E}^{(0)} \ldots \mathbf{E}^{(G-1)}\}$, is used, where $\mathbf{E}^{(g)} = \left[\mathbf{e}_0^{(g)} \;\cdots\; \mathbf{e}_{M-1}^{(g)}\right]$ is the gth precoding

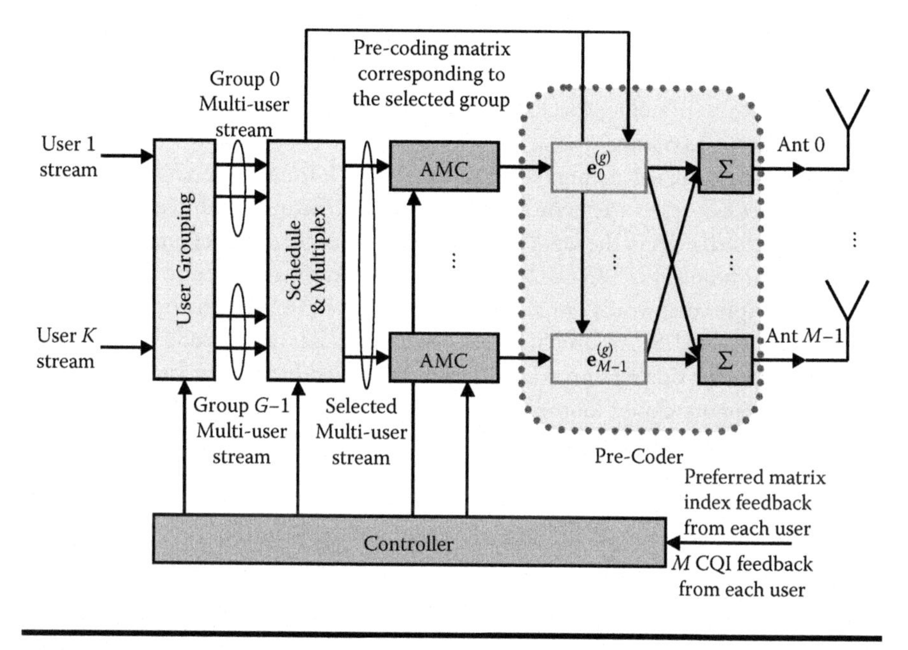

Figure 5.5 PU²RC transmitter chain.

matrix, and $\mathbf{e}_m^{(g)}$ is the *m*th precoding vector in the set. Each UE will calculate a channel quality indicator (CQI) value for every vector in every matrix in the set E. The system designer can trade off the amount of feedback overhead with the scheduling flexibility at the node B by choosing an appropriate value of *G* and deciding the amount of information that the UE needs to feed back to the node B. For most flexibility, the UE can feed back every CQI value for every matrix in the set E and thus a total of *GM* CQI values for every UE.

For lowest overhead, the UE needs to feed back only the best CQI value and the corresponding vector index, which would require $\log_2(GM)$ bits per UE in addition to the actual CQI value. This supports the restricted space division multiple access (SDMA)-only case because each user can be scheduled on only one specific beam. This also supports the transmit beam-forming-only case when a user is scheduled on the time-frequency unit. A trade-off of these two is that each UE feeds back the *M* CQI values corresponding to its preferred/best group. The node B can now select a group *g*, with an accompanying basis $\mathbf{E}^{(g)}$, and schedule up to *M* code words of independent data to different UEs, depending on the scheduler design.

The following are the main features of PU²RC:

- Multiple sets of unitary precoding are predetermined and the degree of freedom in the unitary precoding selection provides multiuser diversity gain in the space domain.

- PU²RC with partial feedback: UE reports the preferred unitary precoding (i.e., preferred matrix index).
- PU²RC with full feedback: UE reports all the CQIs corresponding to all the unitary precoding.
- Using the spatial channel information fed back from the receivers, the transmitter can select a relevant unitary precoding. Because multiple users can be scheduled even in the same time-frequency resource and they are assigned different beams, PU²RC–MIMO realizes SDMA. In fact, because a single user's multiple code words can also be scheduled even in the same time-frequency resource, PU²RC realizes SDM. When a single user's single code word is transmitted in a single beam, PU²RC performs closed-loop beam forming.
- To support closed-loop beam forming, UE needs to report the preferred beam in the preferred unitary precoding (i.e., preferred vector index as well as preferred matrix index).

Given a matrix in the set E, the node B will use the following procedure:

- PU²RC with partial feedback;
- gather the feedback information, which indicates "a preferred precoding matrix" and the CQI values for all the precoding vectors in the matrix;
- group users who declare the same preferred precoding matrix;
- select a group with the highest group priority; how to define the group priority depends on the scheduling policy;
- select code words of multiple users with the highest priority in the selected group; how to define the code word priority (or user priority) depends on the scheduling policy;
- apply appropriate AMC schemes to the selected code words; and
- apply a precoding scheme corresponding to the selected group.

The codebook of PU²RC is considered as a unitary matrix. One example is the Fourier basis—that is,

$$\mathbf{e}_m^{(g)} = \frac{1}{\sqrt{M}}\left[w_{0m}^{(g)} \quad \cdots \quad w_{(M-1)m}^{(g)} \right]^{\mathrm{T}}$$

$$w_{nm}^{(g)} = \exp\left\{ j\frac{2\pi n}{M}\left(m + \frac{g}{G} \right) \right\}$$

Consider the case of two transmit antennas ($M = 2$) and two possible groups ($G = 2$). Here we will have the following precoder sets:

$$\left\{ \mathbf{E}_0 = \frac{1}{\sqrt{2}}\begin{bmatrix} 1 & 1 \\ 1 & -1 \end{bmatrix}, \ \mathbf{E}_1 = \frac{1}{\sqrt{2}}\begin{bmatrix} 1 & 1 \\ j & -j \end{bmatrix} \right\}$$

Consider now four transmit antennas ($M = 4$) and still only two groups ($G = 2$). Here we will have the following precoder sets:

$$\mathbf{E}_0 = \frac{1}{\sqrt{4}}\begin{bmatrix} 1 & 1 & 1 & 1 \\ 1 & e^{j\pi/2} & e^{j\pi} & e^{j3\pi/2} \\ 1 & e^{j\pi} & e^{j2\pi} & e^{j3\pi} \\ 1 & e^{j3\pi/2} & e^{j3\pi} & e^{j9\pi/2} \end{bmatrix} = \frac{1}{\sqrt{4}}\begin{bmatrix} 1 & 1 & 1 & 1 \\ 1 & j & -1 & -j \\ 1 & -1 & 1 & -1 \\ 1 & -j & -1 & j \end{bmatrix}$$

$$\mathbf{E}_1 = \frac{1}{\sqrt{4}}\begin{bmatrix} 1 & 1 & 1 & 1 \\ e^{j\pi/4} & e^{j3\pi/4} & e^{j5\pi/4} & e^{j7\pi/4} \\ e^{j\pi/2} & e^{j3\pi/2} & e^{j5\pi/4} & e^{j7\pi/2} \\ e^{j3\pi/4} & e^{j9\pi/4} & e^{j15\pi/4} & e^{j21\pi/4} \end{bmatrix}$$

$$= \frac{1}{\sqrt{4}}\begin{bmatrix} 1 & 1 & 1 & 1 \\ \frac{1}{\sqrt{2}}(1+j) & \frac{1}{\sqrt{2}}(-1+j) & \frac{1}{\sqrt{2}}(-1-j) & \frac{1}{\sqrt{2}}(1-j) \\ j & -j & j & -j \\ \frac{1}{\sqrt{2}}(-1+j) & \frac{1}{\sqrt{2}}(1+j) & \frac{1}{\sqrt{2}}(1-j) & \frac{1}{\sqrt{2}}(-1-j) \end{bmatrix}$$

Throughput performance is carried out to verify the advantage of the PU^2RC. Figure 5.6 compares performance results of the SU-MIMO, MU-MIMO, and combined schemes in a 4×4 MIMO system. For the sake of simplicity and because the channel is i.i.d. (independent and identically distributed) fading, no precoding is included in this simulation. As can be noted from the figure, the SU-MIMO scheme has a very large gain over the MU-MIMO scheme when there are few users in a cell; this is especially true when the signal-to-noise ratio (SNR) is high. The MU-MIMO scheme achieves multiuser diversity gain (MUDG) on a per-code-word basis, which accounts for the big performance gap over the SU-MIMO scheme. The combined scheme simply follows the envelope of the other two schemes. Here, large gains are achieved with the combined scheme; note how the crossover point between SU- and MU-MIMO changes with different SNRs.

Figure 5.7 shows the results of the 2×2 MIMO case where we have correlated fading and precoding. Here, the MU-MIMO has more gain over the SU-MIMO because the combined effect of precoding and a correlated channel means that there is a big difference between the two stream qualities. Namely, the SU-MIMO UE derives very little benefit from having a second stream because the CQI on that stream is very low. The MU-MIMO, on the other hand, gains from this fact because the second stream interferes very little with the first stream, and it still gets the full MUDG.

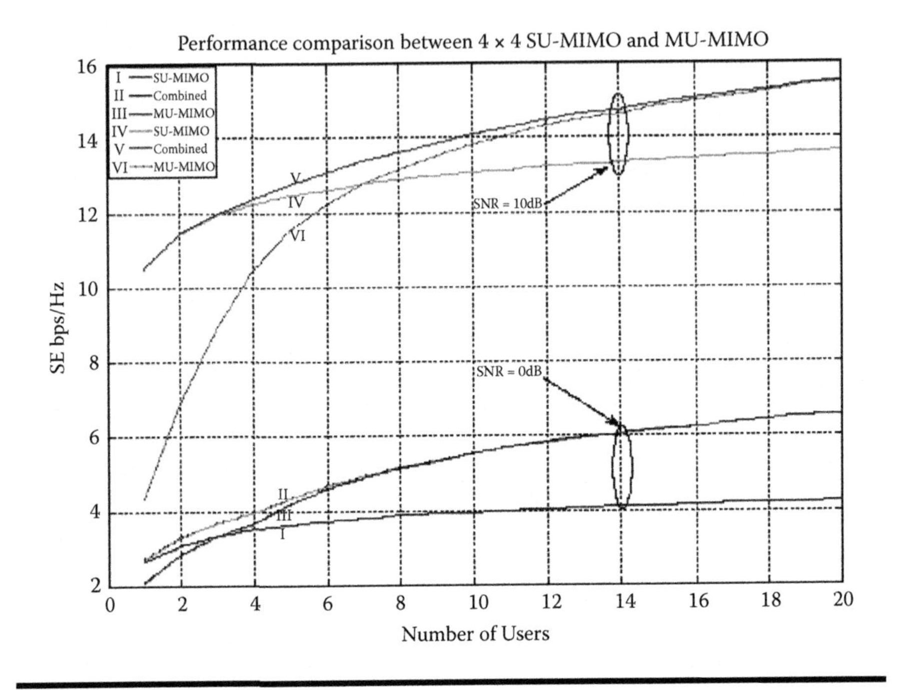

Figure 5.6 Performance for a 4 × 4 MIMO. SU-MIMO uses MMSE-SIC and four code words per UE. MU-MIMO uses MMSE and one code word per UE.

5.4 Physical Channel Procedures Supporting MIMO

UE is semistatically configured via higher-layer signaling to receive the physical downlink shared channel based on one of the following transmission modes [4]:

- single-antenna port;
- transmit diversity;
- open-loop spatial multiplexing;
- closed-loop spatial multiplexing; and
- multiuser MIMO.

5.4.1 UE Procedure for Reporting Channel Quality Indication, Precoding Matrix Indicator, and Rank

The time and frequency resources that can be used by the UE to report CQI, precoding matrix indicator (PMI), and rank indication (RI) are controlled by the eNB. For spatial multiplexing, as given in reference 3, the UE determines an RI corresponding to the number of useful transmission layers. For transmit diversity, as given in reference 3, RI = 1.

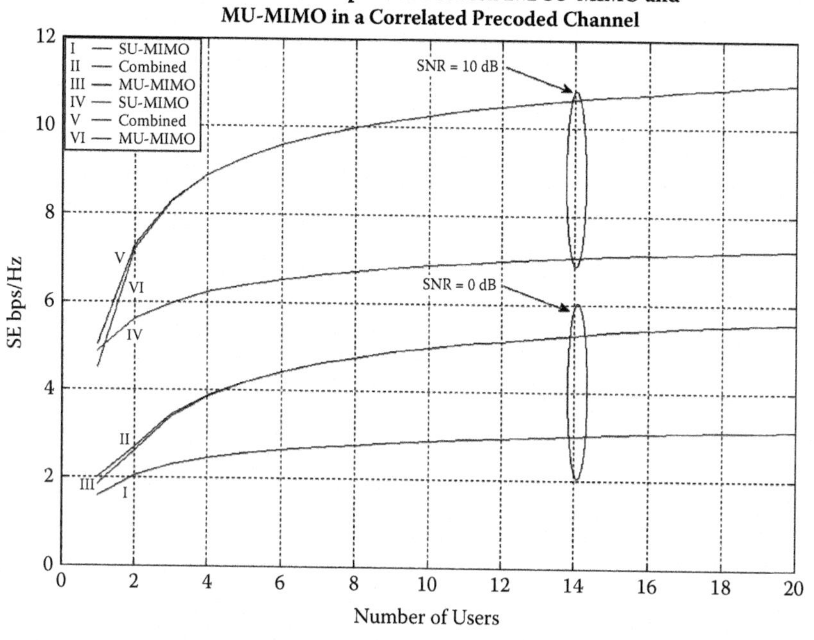

Figure 5.7 Performance for a 2 × 2 MIMO. SU-MIMO uses MMSE-SIC and two code words per UE. MU-MIMO uses MMSE and one code word per UE. The channel is correlated and precoded.

CQI, PMI, and RI reporting is periodic or aperiodic. A UE transmits CQI, PMI, and RI reporting on a physical uplink control channel (PUCCH) for subframes with no physical uplink shared channel (PUSCH) allocation. A UE transmits CQI, PMI, and RI reporting on a PUSCH for those subframes with PUSCH allocation for (a) scheduled PUSCH transmissions with or without an associated scheduling grant or (b) PUSCH transmissions with no uplink shared channel (UL-SCH). The CQI transmissions on PUCCH and PUSCH for various scheduling modes are summarized in Table 5.7. If both periodic and aperiodic reporting occur in the same subframe, the UE transmits only the aperiodic report in that subframe.

When reporting RI, the UE reports a single instance of the number of useful transmission layers. For each RI reporting interval during closed-loop spatial multiplexing, a UE determines an RI from the supported set of RI values for the corresponding eNodeB and UE antenna configuration and reports the number in each RI report. For each RI reporting interval during open-loop spatial multiplexing, a UE determines an RI for the corresponding eNodeB and UE antenna configuration

Table 5.7 Physical Channels for Aperiodic or Periodic CQI Reporting

Scheduling Mode	*Periodic CQI Reporting Channels*	*Aperiodic CQI Reporting Channel*
Frequency nonselective	**PUCCH; PUSCH**	**PUSCH**
Frequency selective	**PUCCH; PUSCH**	**PUSCH**

in each reporting interval and reports the detected number in each RI report to support selection between RI = 1 transmit diversity and RI > 1 large-delay CDD open-loop spatial multiplexing.

When reporting PMI, the UE reports either a single or a multiple PMI report. The number of RBs represented by a single UE PMI report can be N_{RB}^{DL} or a smaller subset of RBs. The number of RBs represented by a single PMI report is semi-statically configured by higher-layer signaling. A UE is restricted to reporting PMI and RI within a precoder codebook subset specified by a bitmap configured by higher-layer signaling. For a specific precoder codebook and associated transmission mode, the bitmap can specify all possible precoder codebook subsets that the UE can assume the eNB may be using when the UE is configured in the relevant transmission mode.

The set of sub-bands (S) that a UE evaluates for CQI reporting is semistatically configured by higher layers. A sub-band is a set of k contiguous PRBs, where k is also semistatically configured by higher layers. Note that the last sub-band in set S may have fewer than k contiguous PRBs, depending on N_{RB}^{DL}. The number of sub-bands for system bandwidth given by N_{RB}^{DL} is defined by $N = \lceil N_{RB}^{DL}/k \rceil$. The term "wideband CQI" denotes a CQI value obtained over the set S.

For single-antenna port and transmit diversity, as well as open-loop spatial multiplexing and closed-loop spatial multiplexing with RI = 1, a single 4-b wideband CQI is reported. For RI > 1, closed-loop spatial multiplexing PUSCH-based triggered reporting includes reporting a wideband CQI that comprises.

5.4.2 Aperiodic/Periodic CQI Reporting Using PUSCH

A UE performs aperiodic CQI reporting using the PUSCH upon receiving an indication sent in the scheduling grant. The aperiodic CQI report size and message format are given by radio resource control (RRC). The minimum reporting interval for aperiodic reporting of CQI and PMI is one subframe. In the case of frequency-selective CQI and PMI reports, the sub-band size for CQI and PMI is the same. The sub-band size for CQI and PMI is the same for transmitter–receiver configurations with and without precoding.

Table 5.8 Sub-Band Size versus DL System Bandwidth

System Bandwidth	*Sub-Band Size*
N_{RB}^{DL}	(k)
6–7	(Wideband CQI only)
8–10	4
11–26	4
27–64	6
64–110	4, 8

A UE is semistatically configured by higher layers to feed back CQI on the PUSCH using one of the following CQI reporting types:

- wideband feedback
  - A UE should report one wideband CQI value.
- higher-layer-configured sub-band feedback
  - A UE should report a wideband CQI and one CQI for each sub-band in the set of sub-bands (S) semistatically configured by higher layers. Sub-band CQIs are encoded differentially with respect to the wideband CQIs using 3 b.
  - Supported sub-band size (k) includes those given in Table 5.8. The k values are semistatically configured by higher layers as a function of system bandwidth.
- UE-selected sub-band feedback
  - The UE should select the M best sub-bands of size k (where k and M are given in Table 5.9 for each system bandwidth range) within the set of sub-bands S semistatically configured by higher layers and report the positions of these M sub-bands using $\lceil \log_2 (\lceil N_{RB}^{DL}/k \rceil) \rceil$ bits.
  - The UE should also report one CQI value reflecting transmission only over the selected M best sub-bands determined in the previous step.
  - Additionally, the UE should report one wideband CQI value.
  - The best-M CQI value is encoded differentially using 3 b relative to the wideband CQI.
  - Supported sub-band size k and M values include those shown in Table 5.9. The k and M values are a function of system bandwidth.

5.4.3 Periodic CQI Reporting Using PUCCH

A UE is semistatically configured by higher layers to feed back different CQI types on the PUCCH periodically, using the following periodicity parameters:

Table 5.9 Sub-Band Size (k) and M Values versus DL System Bandwidth

System Bandwidth N_{RB}^{DL}	Sub-Band Size k [RBs]	M
6–7	(Wideband CQI only)	(Wideband CQI only)
8–10	2	1
11–26	2	3
27–64	3	5
64–110	4	6

- N_P is the periodicity of the subframe pattern allocated for the CQI reports in terms of subframes where the minimum reporting interval is N_{PMIN}.
- N_{OFFSET} is the subframe offset.

Support for different CQI report types includes wideband CQI on set S and frequency-selective CQI. The supported sub-band size (k) values include those given in Table 5.8, which are a function of system bandwidth.

For the frequency-selective CQI, a CQI report in a certain subframe describes the channel quality in a particular part or in particular parts of the bandwidth (a part is frequency consecutive and an integer multiple of the sub-band size). Which bandwidth parts to use varies deterministically from one CQI report subframe to another, covering the entire set of sub-bands (S) after a finite period.

A UE with a scheduled PUSCH transmission in the same subframe as its CQI report uses the same PUCCH-based reporting format when reporting CQI on the PUSCH unless an associated PDCCH with scheduling grant format indicates an aperiodic report is required.

5.5 Summary

3GPP LTE has considered MIMO OFDM as one of the key techniques to achieve high throughput and performance. This chapter has introduced the two main categories of MIMO OFDM schemes in LTE—SU-MIMO and MU-MIMO—and the related physical channel procedures. The performance of the selected codebooks has been provided to show their advantages.

References

1. 3GPP TR25.814. Physical layer aspects for evolved UTRA.
2. 3GPP TR36.913. Requirements for evolved UTRA (E-UTRA) and evolved UTRAN (E-UTRAN).
3. 3GPP TR36.211. Physical channels and modulation (release 8).
4. 3GPP TR36.213. Physical layer procedures (release 8).

5. R1-070730. Precoding codebook design for four-node-B antenna.
6. R1-060335. Downlink MIMO for EUTRA.
7. R1-060912. PU2RC performance evaluation.

Link

http://www.3gpp.org/

Symbols

(k,l)	resource element with frequency-domain index k and time-domain index l
$a_{k,l}^{(p)}$	value of resource element (k,l) (for antenna port p)
D	matrix for supporting cyclic delay diversity
f_0	carrier frequency
M_{sc}^{PUSCH}	scheduled bandwidth for uplink transmission, expressed as a number of subcarriers
M_{RB}^{PUSCH}	scheduled bandwidth for uplink transmission, expressed as a number of resource blocks
$M_{bit}^{(q)}$	number of coded bits to transmit on a physical channel (for code word q)
$M_{symb}^{(q)}$	number of modulation symbols to transmit on a physical channel [for code word q]
M_{symb}^{layer}	number of modulation symbols to transmit per layer for a physical channel
M_{symb}^{layer}	number of modulation symbols to transmit per antenna port for a physical channel
N	a constant equal to 2,048 for $\Delta f = 15$ kHz and 4,096 for $\Delta f = 7.5$ kHz
$N_{CP,l}$	downlink cyclic prefix length for OFDM symbol l in a slot
$N_{cs}^{(1)}$	number of cyclic shifts used for PUCCH formats 1/1a/1b in a resource block with a mix of formats 1/1a/1b and 2/2a/2b
$N_{RB}^{(2)}$	bandwidth reserved for PUCCH formats 2/2a/2b, expressed in multiples of N_{sc}^{RB}
N_{RB}^{PUCCH}	number of resource blocks in a slot used for PUCCH transmission (set by higher layers)
N_{ID}^{cell}	physical layer cell identity
N_{ID}^{MBSFN}	MBSFN area identity
N_{RB}^{DL}	downlink bandwidth configuration, expressed in multiples of N_{sc}^{RB}

$N_{RB}^{min, DL}$ smallest downlink bandwidth configuration, expressed in multiples of N_{sc}^{RB}

$N_{RB}^{max, DL}$ largest downlink bandwidth configuration, expressed in multiples of N_{sc}^{RB}

N_{RB}^{UL} uplink bandwidth configuration, expressed in multiples of N_{sc}^{RB}

$N_{RB}^{min, UL}$ smallest uplink bandwidth configuration, expressed in multiples of N_{sc}^{RB}

$N_{RB}^{max, UL}$ largest uplink bandwidth configuration, expressed in multiples of N_{sc}^{RB}

N_{symb}^{DL} number of OFDM symbols in a downlink slot

N_{symb}^{UL} number of SC-FDMA symbols in an uplink slot

N_{sc}^{RB} resource block size in the frequency domain, expressed as a number of subcarriers

N_{RS}^{PUCCH} number of reference symbols per slot for PUCCH

N_{TA} timing offset between uplink and downlink radio frames at the UE, expressed in units of T_s

$n_{PUCCH}^{(1)}$ resource index for PUCCH formats 1/1a/1b

$n_{PUCCH}^{(2)}$ resource index for PUCCH formats 2/2a/2b

n_{PDCCH} number of PDCCHs present in a subframe

n_{PRB} physical resource block number

n_{VRB} virtual resource block number

n_{RNTI} radio network temporary identifier

n_f system frame number

n_s slot number within a radio frame

P number of cell-specific antenna ports

p antenna port number

q code-word number

$s_l^{(p)}(t)$ time-continuous baseband signal for antenna port p and OFDM symbol l in a slot

T_f radio frame duration

T_s basic time unit

T_{slot} slot duration

W precoding matrix for downlink spatial multiplexing

β_{PRACH} amplitude scaling for PRACH

β_{PUCCH} amplitude scaling for PUCCH

β_{PUSCH} amplitude scaling for PUSCH

β_{SRS}	amplitude scaling for sounding reference symbols
Δf	subcarrier spacing
Δf_{RA}	subcarrier spacing for the random access preamble
υ	number of transmission layers

Abbreviations

ACK	acknowledgment
BCH	broadcast channel
CCE	control channel element
CDD	cyclic delay diversity
CQI	channel quality indicator
CRC	cyclic redundancy check
DL	downlink
DTX	discontinuous transmission
EPRE	energy per resource element
MCS	modulation and coding scheme
NACK	negative acknowledgment
PBCH	physical broadcast channel
PCFICH	physical control format indicator channel
PDCCH	physical downlink control channel
PDSCH	physical downlink shared channel
PHICH	physical hybrid ARQ indicator channel
PRACH	physical random access channel
PRB	physical resource block
PUCCH	physical uplink control channel
PUSCH	physical uplink shared channel
QoS	quality of service
RBG	resource block group
RE	resource element
RPF	repetition factor
RS	reference signal
SINR	signal-to-interference-plus-noise ratio
SIR	signal-to-interference ratio
SRS	sounding reference symbol
TA	time alignment
TTI	transmission time interval
UE	user equipment
UL	uplink
UL-SCH	uplink shared channel
VRB	virtual resource block

Chapter 6

Single-Carrier Transmission for UTRA LTE Uplink

Basuki E. Priyanto and Troels B. Sorensen

Contents

6.1 Introduction

Uplink transmission in the long term evolution (LTE) of the UMTS (universal mobile telecommunications system) terrestrial radio access system (UTRA LTE) has numerous physical layer advances in comparison to UTRA WCDMA (wideband code division multiple access), mainly to achieve two to three times better spectral efficiency. These advances include flexible channel bandwidth up to 20 MHz, flexible user resource allocation in both time and frequency domains, and a shorter time transmission interval (TTI) of 1 ms. Specifically challenging for the uplink is that these enhancements are to be achieved, preferably, with reduced power consumption to extend the battery life and cell coverage.

The radio access technique is one of the key issues in the LTE uplink air interface. In LTE, orthogonal frequency division multiple access (OFDMA) has been selected as the multiple access scheme for downlink, and single-carrier frequency division multiple access (SC-FDMA) has been selected for uplink. OFDM is an attractive modulation technique in a cellular environment to combat a frequency selective fading channel with a relatively low-complexity receiver [1]. However, OFDM requires an expensive and inherently inefficient power amplifier in the transmitter due to the high peak-to-average power ratio (PAPR) of the multicarrier signal.

An alternative transmission scheme with the same attractive multipath interference mitigation property as OFDM is SC-CP (single-carrier transmission with cyclic prefix) [2–4]. Therefore, SC-CP can achieve a link-level performance comparable to OFDM for the same complexity, but at reduced PAPR [2]. In addition, the performance of SC-CP can be further improved by using a turbo equalization receiver [5]. The choice of single-carrier transmission in a discrete Fourier transform (DFT)-spread OFDM form allows for a relatively high degree of commonality with the downlink OFDM scheme and the possibility to use the same system parameters [6]. Moreover, it enables direct access to the frequency domain to perform frequency domain equalization [2].

This chapter is organized as follows. LTE uplink numerology is introduced in Section 6.2. The description of both radio access techniques is given in Section 6.3. The PAPR evaluation for each radio access scheme is discussed in Section 6.4. Section 6.5 describes the link-level model used for the evaluations in this work, including the specific parameter settings. The LTE uplink performance with various key techniques is also presented, including link adaptation, in Section 6.6, fast hybrid automatic repeat request (HARQ) in Section 6.7, antenna configuration in Section 6.8, flexible frequency allocation in Section 6.9, typical channel estimation in Section 6.10, and, finally, turbo equalization in Section 6.11 as a performance enhancement technique for SC-FDMA transmission. We conclude the work by showing the impact of the nonlinear power amplifier to both in-band and out-of-band performance in Section 6.12 and summarize the chapter in Section 6.13.

6.2 LTE Uplink Numerology

The basic physical layer parameters that were originally designed for the uplink are shown in Table 6.1 [6]. The transmission scheme is SC-FDMA in a DFT-spread OFDM form. The table shows that different transmission bandwidths have different parameter settings. The transmission is frame based with a radio frame duration of 10 ms and consists of 20 subframes of 0.5 ms. In technical specification TR 25.814 V7.1.0 [6], a subframe is equivalent to a TTI. An example of subframe structure for uplink transmission is shown in Figure 6.1.

Based on technical specification TR 25.814 V7.1.0 [6], a subframe consists of two short blocks and six long blocks. The short block is a reference symbol and transmitted in a time-multiplex format with the long block. It has the following purposes:

- uplink channel estimation for uplink coherent demodulation/detection; and
- possible uplink channel quality estimation for uplink frequency or time-domain channel-dependent scheduling.

The short block duration is half the long block duration. The long block is primarily used to transmit data.

In the work item phase, the 3rd Generation Partnership Project (3GPP) has modified certain physical layer parameters and hence has changed the assumptions that were initially applied for the evaluations in this chapter. In particular, the subframe or TTI duration has changed to 1 ms to include two 0.5-ms slots, each with a structure as shown in Figure 6.2. As it can be seen, the modified short-block duration is equal to the long block and placed in the middle of the slot (as of 2007, the final position within the slot had not been decided). Initially, the supported modulation schemes have been $\pi/2$ BPSK (binary phase key shifting), QPSK (quadrature phase shift keying), 8 PSK (phase shift keying), and 16 QAM (quadrature amplitude modulation). In the work item phase, the supported modulation scheme for data transmission was changed to QPSK, 16 QAM, and 64 QAM (optional).

6.3 Radio Access Techniques

The basic uplink transmission technique is single-carrier transmission with cyclic prefix to achieve uplink interuser orthogonality and to enable efficient frequency domain equalization at the receiver side. The transmitter and receiver structure for SC-FDMA transmission is shown in Figure 6.3. An SC-FDMA structure is identified by the insertion of DFT spreading and inverse discrete Fourier transform (IDFT) despreading at the transmitter and receiver, respectively. From this

Table 6.1 Parameters for LTE Uplink Based on the Study Item

Transmission BW (MHz)	1.25	2.5	5	10	15	20
Subframe duration	0.5 ms	0.5 ms	0.5 ms	0.5 ms	0.5 ms	0.5 ms
Subcarrier spacing	15 kHz	15 kHz	15 kHz	15 kHz	15 kHz	15 kHz
Sampling frequency (MHz)	1.92	3.84	7.68	15.36	23.04	30.72
Long block FFT size	128	256	512	1024	1536	2048
Long block number of occupied subcarriers	75	150	300	600	900	1200
Short block FFT size	64	128	256	512	768	1024
Short block number of occupied subcarriers	38	75	150	300	450	600
CP length (μs/sample)	(3.65/7) × 7, (7.81/15) × 1[a]	(3.91/15) × 7, (5.99/23) × 1	(4.04/31) × 7, (5.08/39) × 1	(4.1/63) × 7, (4.62/71) × 1	(4.12/95) × 7, (4.47/103) × 1	(4.13/127) × 7, (4.39/135) × 1

[a] (*x*1/*y*1) × *n*1, (*x*2/*y*2) × *n*2 means (*x*1/*y*1) for *n*1 reference signal or data blocks and (*x*2/*y*2) for *n*2 reference signal or data blocks.

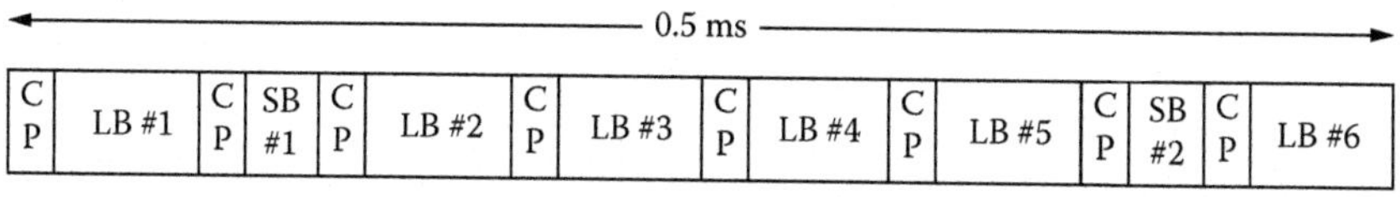

0.5 ms
CP
LB #1
CP
SB #1
CP
LB #2
CP
LB #3
CP
LB #4
CP
LB #5
CP
SB #2
CP
LB #6

Figure 6.1 LTE uplink subframe structure with two short blocks [6].

implementation structure, SC-FDMA is also known as DFT-spread (DFT-S) OFDM, which is a form of the single-carrier transmission technique where the signal is generated in the frequency domain. The DFT spreading combines parallel M-PSK/M-QAM symbols to form an SC-FDMA symbol.

To formulate the DFT-S system, we can start by defining s_m as the mth transmitted symbol at the output of an equivalent OFDM system. The modulator converts the random bit stream input to the M-QAM/M-PSK symbols represented by the vector $\mathbf{x_m}$ of length N_d. The OFDM system then constructs s_m as

$$\mathbf{s_m} = \mathbf{F_m} \times \mathbf{x_m} \tag{6.1}$$

where $\mathbf{F_m}$ is a matrix that performs an N_s-point inverse fast Fourier transform (IFFT) operation. N_s is the number of IFFT output samples. N_g is the length of the guard interval samples, and $N_{gs} = N_g + N_s$ is the length of the total time domain symbol.

$$\mathbf{D_m} = \begin{pmatrix} D_{m,0}[0] & \cdots & D_{m,N_d-1}[0] \\ \vdots & \ddots \quad D_{m,k}[n] \quad \ddots & \vdots \\ D_{m,0}[N_d-1] & \cdots & D_{m,N_d-1}[N_d-1] \end{pmatrix} \tag{6.3}$$

The mth transmitted symbol in a DFT-S OFDM system can be expressed as a vector of length N_{gs} samples, defined by

$$s_m = \mathbf{F_m} \times \mathbf{T_m} \times \mathbf{D_m} \times \mathbf{x_m} \tag{6.2}$$

0.5 ms
CP
LB #1
CP
LB #2
CP
LB #3
CP
SB #1
CP
LB #4
CP
LB #5
CP
LB #6

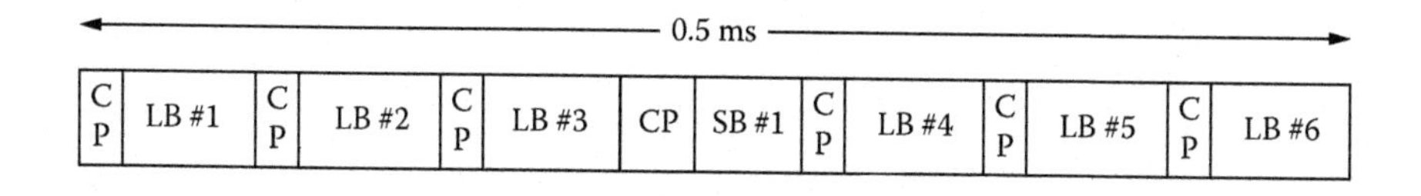

Figure 6.2 LTE uplink slot structure with one short block.

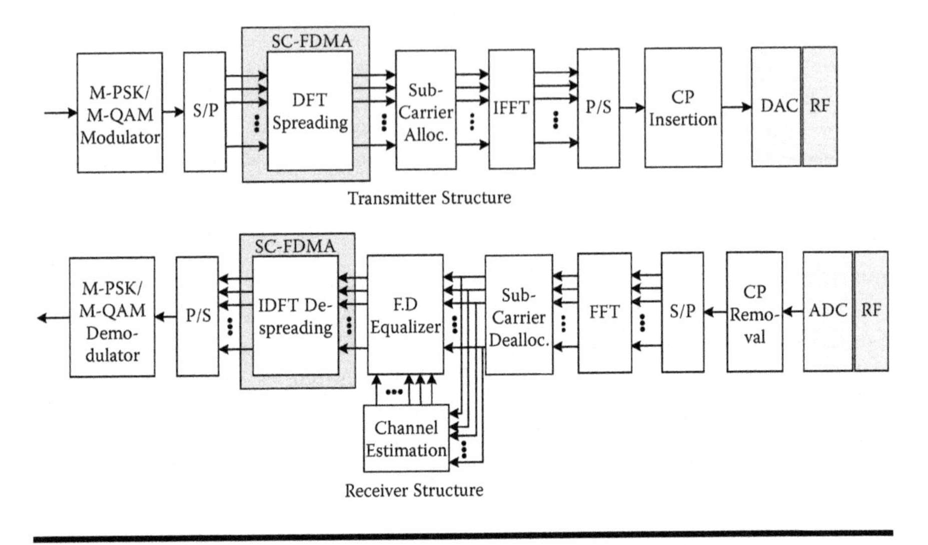

Figure 6.3 OFDMA and SC-FDMA basic transmitter and receiver structures.

$\mathbf{D_m}$ is a matrix which performs an N_d-point DFT operation. The DFT-S matrix element is expressed as

$$D_{m,k}[n] = \frac{1}{\sqrt{N_d}}.e^{-j2\pi k\left(\frac{n}{N_d}\right)} \tag{6.4}$$

where k is the M-QAM/PSK symbol index. $\mathbf{T_m}$ is an $N_s \times N_d$ mapping matrix for the subcarrier allocation. The subcarrier allocation can be classified in two allocation modes: distributed and localized allocation [6] (see also Figure 6.16). In distributed mode, $N_s = S_f \times N_d$, with S_f defined as the spreading factor. The matrix of subcarrier allocation in localized mode is defined as

$$\mathbf{F_m} = \begin{pmatrix} F_{m,0}[0] & \cdots & F_{m,N_s-1}[0] \\ & \ddots & \\ \vdots & F_{m,l}[n] & \vdots \\ & & \ddots \\ F_{m,0}[N_{gs}-1] & \cdots & F_{m,N_s-1}[N_{gs}-1] \end{pmatrix} \tag{6.7}$$

$$\mathbf{T_m} = [\mathbf{0}_{\mathbf{q \times S_f,1}} \mathbf{J}_{\mathbf{N_d,1}} \mathbf{0}_{\mathbf{N_s - N_d - q \times S_f}} ; \mathbf{0}_{\mathbf{N_s,N_d-1}}] \tag{6.5}$$

where q is the localized chunk index. $J_{a,b}$ is an $a \times b$ matrix of ones. For the distributed mode, the subcarrier allocation matrix is defined as

$$\mathbf{T_m} = [\mathbf{J}_{1,1} \mathbf{0}_{\mathbf{S_f-1,1}} \cdots \mathbf{J}_{\mathbf{N_d,1}} 0_{\mathbf{S_f-1,1}} ; 0_{\mathbf{N_s,N_d-1}}] \tag{6.6}$$

Finally, F_m has a matrix element given by

$$F_{m,l}[n] = \frac{1}{\sqrt{N_s}} . e^{j2\pi l\left(\frac{n-N_g}{N_s}\right)} \tag{6.8}$$

where l is the subcarrier index.

After the multipath channel and adding the additive white Gaussian noise (AWGN), removing the cyclic prefix, and going through the N point FFT at the receiver, the received signal vector in the frequency domain can be expressed as

$$\mathbf{z_m} = \mathbf{H_m} \cdot \mathbf{T_m} \cdot \mathbf{D_m} \cdot \mathbf{x_m} + \mathbf{w_m} \tag{6.9}$$

where $\mathbf{H_m}$ is the diagonal matrix of channel response and $\mathbf{w_m}$ is the noise vector. In this system, the maximum excess delay of the channel is assumed to be shorter than the cyclic prefix duration; therefore, the intersymbol interference (ISI) can easily be eliminated by removing the cyclic prefix.

The linear amplitude and phase distortion in the received signal as a result of the multipath channel is compensated by a frequency domain equalizer (FDE). After FDE, the output signal is

$$\mathbf{v_m} = \mathbf{C_m} \times \mathbf{x_m} \tag{6.10}$$

where $\mathbf{C_m}$ is the matrix diagonal of FDE coefficients:

$$\mathbf{C_m} = \mathbf{diag}(\mathbf{C_{m,1}}, \mathbf{C_{m,2}}, \ldots, \mathbf{C_{m,N_d}}) \tag{6.11}$$

The FDE coefficient generation is described in Section 6.5, where we describe the additional link-level processing required for a coded and more practical system. The steps presented in Equations (6.5) to (6.11) form the basis for the link-level model in Section 6.5.

6.4 Peak-to-Average Power Ratio Evaluation

The envelope variation of the transmitted signal can be represented in a PAPR. In this work, the PAPR of both SC-FDMA and OFDMA is evaluated. The PAPR is an essential parameter that can affect the radio frame (RF) part because a transmitter signal with high PAPR is sensitive to the nonlinear distortion at the power amplifier. The PAPR of the transmitted signal can be expressed as

$$CCDF(P_{ins}) = Prob\left\{\frac{|v_i|^2}{P_{avg}} > P_T\right\} \tag{6.12}$$

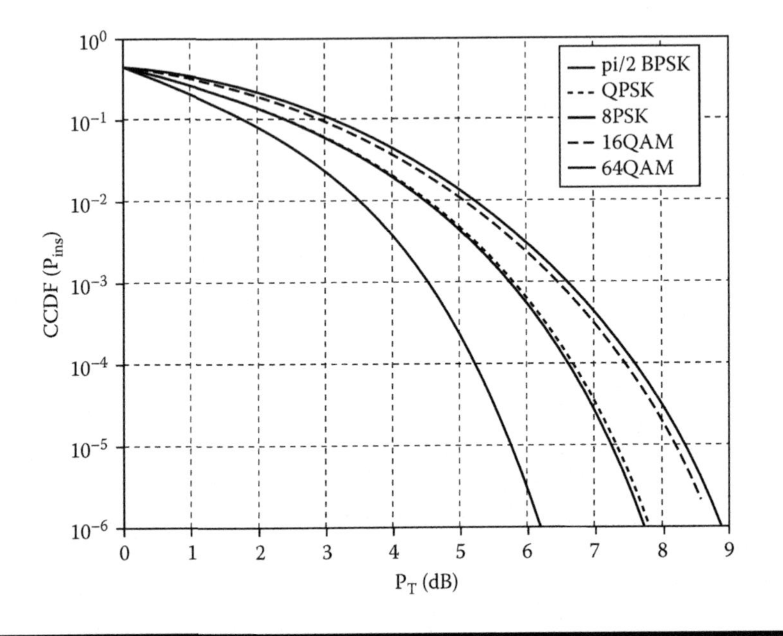

Figure 6.4 PAPR of SC-FDMA with various modulation schemes.

where P_{ins} is the instantaneous power of the signal, v_i is the ith signal samples, P_{avg} is the long-term average power, and P_T is the power threshold of the signal. For simplicity, P_{avg} has been normalized to one. In order to obtain accurate results, the transmitted signal has been oversampled four times according to the suggestion given in Han and Lee [7].

Figure 6.4 shows the PAPR results for SC-FDMA with various modulation schemes. The PAPR results vary depending on the modulation scheme, so higher-order modulation schemes exhibit a higher PAPR compared to lower-order modulation schemes. In SC-FDMA, as a single carrier transmission, the PAPR is mainly affected by the envelope variations in each modulation scheme; thus, the higher-order modulations lead to larger dynamic range and higher PAPR.

In comparison, the PAPR results for OFDMA with various modulation schemes are shown in Figure 6.5. It is observed that the PAPR is not affected by the type of modulation scheme, and all the modulation schemes have the same PAPR results. After the IFFT modulation in an OFDM system, the modulated symbols are added together to form a multicarrier signal. The addition increases the PAPR of the OFDM signal but diminishes the impact of the modulation scheme.

The evaluation of SC-FDMA and OFDMA with regard to PAPR is summarized in Figure 6.6. SC-FDMA has an advantage in obtaining a small PAPR for the lower-order modulation schemes. For $\pi/2$ BPSK, the PAPR of SC-FDMA is about

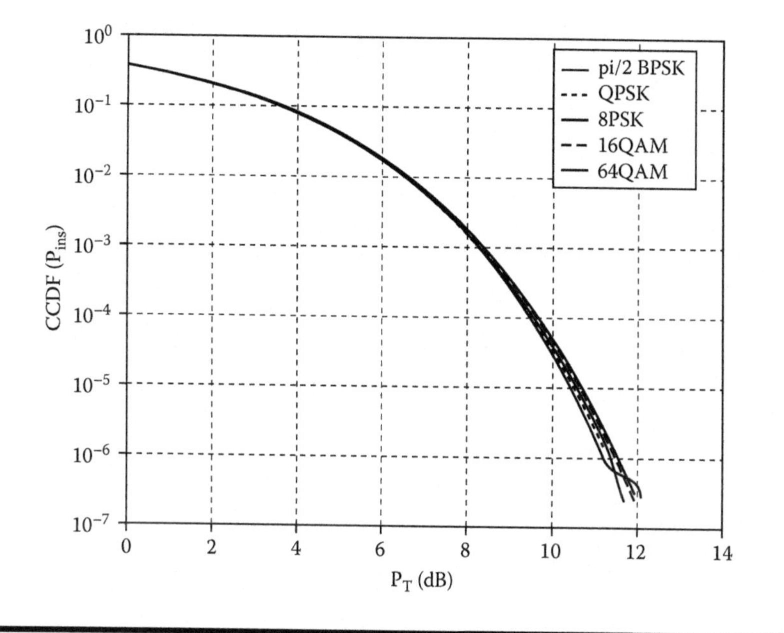

Figure 6.5 PAPR of OFDM for various modulation schemes.

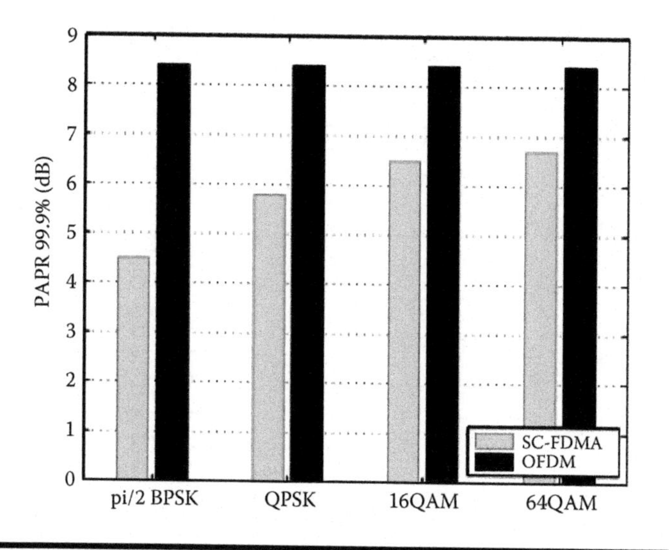

Figure 6.6 PAPR comparison of SC-FDMA and OFDMA.

4 dB lower than the same modulation format in OFDM. For 64 QAM, the difference is smaller than 1.6 dB in comparison to OFDMA.

6.5 Link-Level Model

The link-level model for the LTE uplink is shown in Figure 6.7. The left side is the transmitter, corresponding to the user equipment (UE), and the right side is the receiver, corresponding to the base station (BS). The transmission from the UE starts

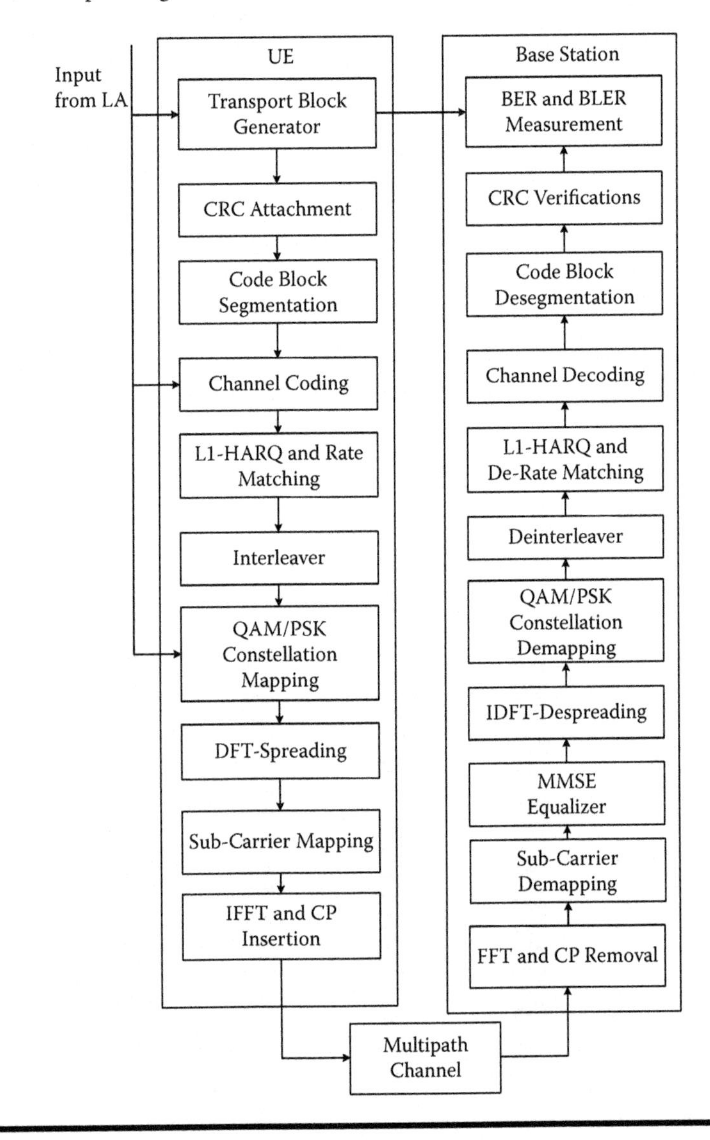

Figure 6.7 Layer-1 UTRA LTE uplink link-level simulator.

with the transport format and resource combination (TFRC) information from the link adaptation (LA) module to the transport block generator. The transport block generator generates the packet based on the requested size. Then the cyclic redundancy check (CRC)-based error-detecting code is added to the corresponding packet.

According to the packet size and the maximum code block size, the packet is segmented into several blocks. Each block is coded; the coding used for this study is the turbo coding defined in UTRA release 6 [8]. The rate matching block adjusts the output bit rate according to the requirements by puncturing or repeating the coded bits. The rate matching implementation follows the description in UTRA release 6 [8]. Afterward, the output coded bits are interleaved and are passed to the modulator. The modulator is implemented based on the supported modulation schemes defined in technical specification TR 25.814 V7.1.0 [6]: $\pi/2$ BPSK, QPSK, 8 PSK, and 16 QAM. For QAM modulation, a QAM remapping is performed to set the systematic bits at more reliable constellation points because this improves the decoder performance.

Single-carrier transmission is formed by the DFT spreading. The DFT spreading block precodes the PSK/QAM data symbols. Thus, all constellation symbols are mixed together to form an SC-FDMA symbol. Afterward, several SC-FDMA symbols are combined with pilot symbols in a time division multiplex (TDM) and mapped to the proper subcarriers. The 3GPP study item report [6] specifies both localized and distributed allocation mapping as discussed in Section 6.3. Finally, the basic OFDM processing discussed in Section 6.3 is performed. The data symbols are placed on orthogonal subcarriers by the IFFT and the cyclic prefix is added.

At the receiver side, the base station basically performs the reverse operations with respect to the process in the transmitter. The amplitude and phase variation due to the transmission in a frequency-selective fading are compensated by an equalizer. The equalizer is basically a one-tap frequency domain equalizer located at the output of the FFT in the receiver. The equalizer coefficients in Equation (6.10) can be generated based on either zero forcing (ZF) or minimum mean square error (MMSE). The zero forcing applies channel inversion, and the equalization coefficient is given by

$$\mathbf{C}_{\mathbf{m,k}} = \frac{\mathbf{H}^{*}_{\mathbf{m,k}}}{|\mathbf{H}_{\mathbf{m,k}}|^{2}} \tag{6.13}$$

The ZF has a disadvantage in terms of noise enhancement for small amplitudes of $H_{m,k}$ [9]. To overcome the noise enhancement problem in the ZF technique, the FDE complex coefficient, $C_{m,k}$, can also be obtained under the MMSE criterion. This is given by [10]

$$\mathbf{C}_{\mathbf{m,k}} = \frac{\mathbf{H}^{*}_{\mathbf{m,k}}}{|\mathbf{H}_{\mathbf{m,k}}|^{2} + \frac{\sigma_{n}^{2}}{\sigma_{s}^{2}}} \tag{6.14}$$

Table 6.2 UTRA LTE Uplink Simulation Parameters

Parameter	Value
Carrier frequency	2 GHz
Transmission BW	10 MHz
Subframe duration	0.5 ms
Subcarrier spacing	15 kHz
SC-FDM symbols/TTI	6 LBs, 2 SBs[a]
CP duration	4.1 μs
FFT size/useful subcarriers	1024/600
MCS settings	π/2 BPSK: 1/6, 1/3 QPSK: 1/2, 2/3, 3/4 8 PSK: 1/3, 1/2, 2/3, 3/4 16 QAM: 1/2, 2/3, 3/4, 4/5
Channel code	3GPP rel. 6 compliant turbo code with basic rate 1/3
Rate matching, interleaver	3GPP rel. 6 compliant
Channel estimation	Ideal
Antenna schemes	SISO
Channel model	Typical urban 6 paths [6]
Speed	3 km/h

[a] LBs = long blocks; SBs = short blocks.

where k denotes the subcarrier index, σ_n^2 is the variance of the additive noise, and σ_s^2 is the variance of the transmitted pilot symbol.

The performance of ZF and MMSE in an LTE scenario is investigated in both the SC-FDMA and OFDMA case. The simulation parameters are shown in Table 6.2.

Figure 6.8 shows the performance of a lower-order modulation and coding scheme (MCS), BPSK with coding rate 1/6. This result is to represent the impact to the low MCSs, which typically operate at lower signal-to-noise ratios (SNRs). SC-FDMA with ZF suffers from high block error rate (BLER), and the MMSE significantly improves the performance over ZF. OFDM with MMSE is slightly better than OFDM with ZF. Basically, in this case SC-FDMA and OFDMA achieve the same performance when MMSE is selected.

The results for the higher modulation schemes, 16 QAM rate 4/5, are shown in Figure 6.9, valid when the system is operated at higher SNRs. It is shown that the performance gap between SC-FDMA ZF and MMSE is getting smaller.

The results in Figures 6.8 and 6.9 show that different equalization techniques do not significantly affect the OFDM system. However, they become a critical issue for SC-FDMA. The forward error correction (FEC) code used in this analysis is the

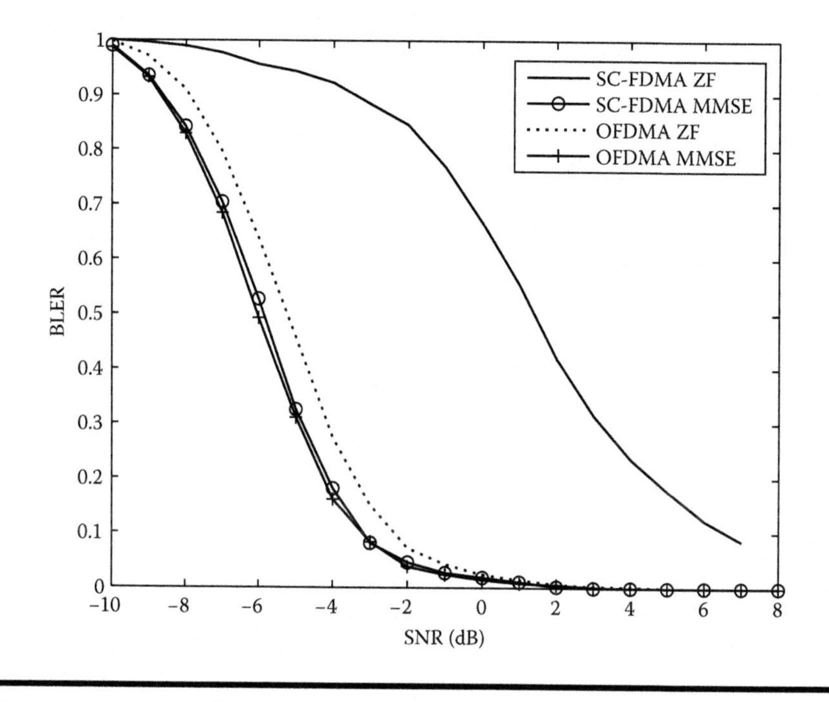

Figure 6.8 BLER performance of BPSK 1/6 SC-FDMA and OFDMA with different equalization.

turbo code [8]. In OFDM transmission, the FEC plays an important role in providing additional diversity, especially for the transmission over frequency-selective fading. Although in SC-FDMA the FEC is still important, it does not provide much additional diversity since on top of the DFT-S.

Figures 6.10 and 6.11 show the SC-FDMA spectral efficiency results for both MMSE and ZF equalization. The spectral efficiency, throughput, is calculated according to

$$\text{TP} = \frac{(1 - BLER) \times P}{T \times B} \tag{6.15}$$

where

BLER is the block error rate;
P is the subframe (block) size in bits;
T is the subframe duration in seconds; and
B is the bandwidth in hertz.

The results are presented for all modulation schemes suitable for LTE uplink. MMSE is significantly better than ZF, especially in the low SNRs, where SC-FDMA

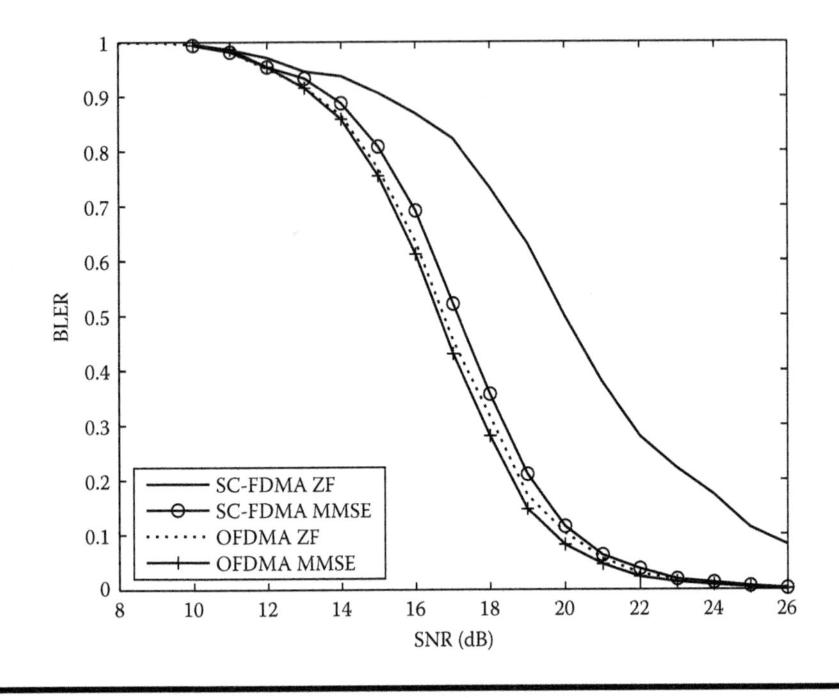

Figure 6.9 BLER performance of 16 QAM 4/5 SC-FDMA and OFDMA with different equalization.

with ZF suffers from noise enhancement. It is clearly shown that the SC-FDMA MMSE is always preferred to ZF in all simulated conditions. Hence, we will assume MMSE equalization for the following results.

6.6 Impact of Link Adaptation

A simple link adaptation technique is adopted in this work [11] in which the link adaptation is defined as the maximum achievable throughput/spectral efficiency for various modulations and coding schemes. In this work, the selection of modulation and coding scheme is based on average SNR. The measured SNR at the base station, generally referred to as channel state information (CSI), varies depending on the channel condition. Here, the CSI measurement is assumed ideal, whereas in practice it will be influenced by measurement inaccuracy and quantization noise.

The spectral efficiency of different MCSs in the single-input, single-output (SISO) case without HARQ is plotted versus the average SNR in Figure 6.12. Variable-rate turbo encoding and modulation from $\pi/2$ BPSK to 16 QAM provide peak data rates from 1.2 to 23 Mb/s for the system with 10-MHz bandwidth.

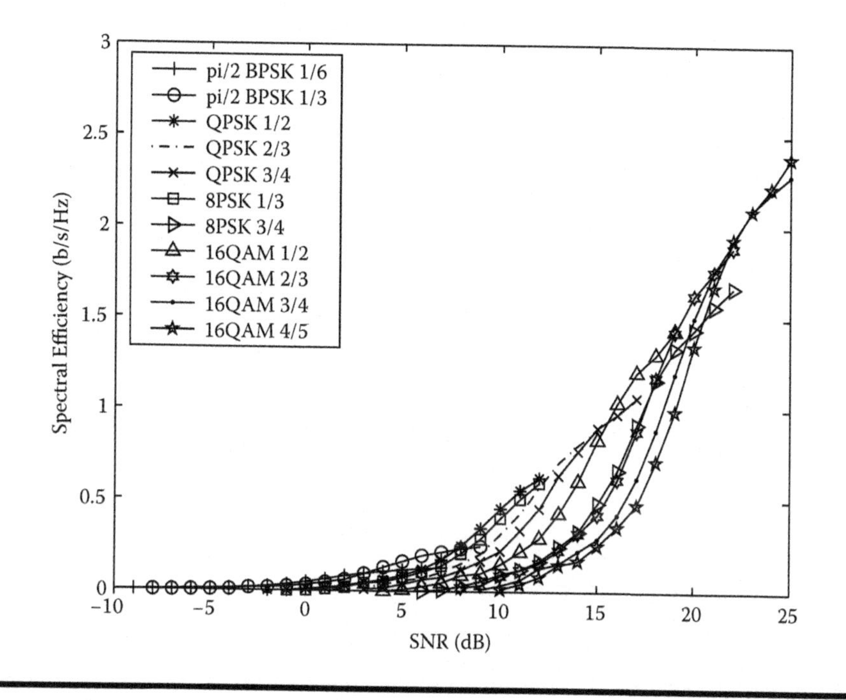

Figure 6.10 Spectral efficiency of SC-FDMA with zero forcing equalization.

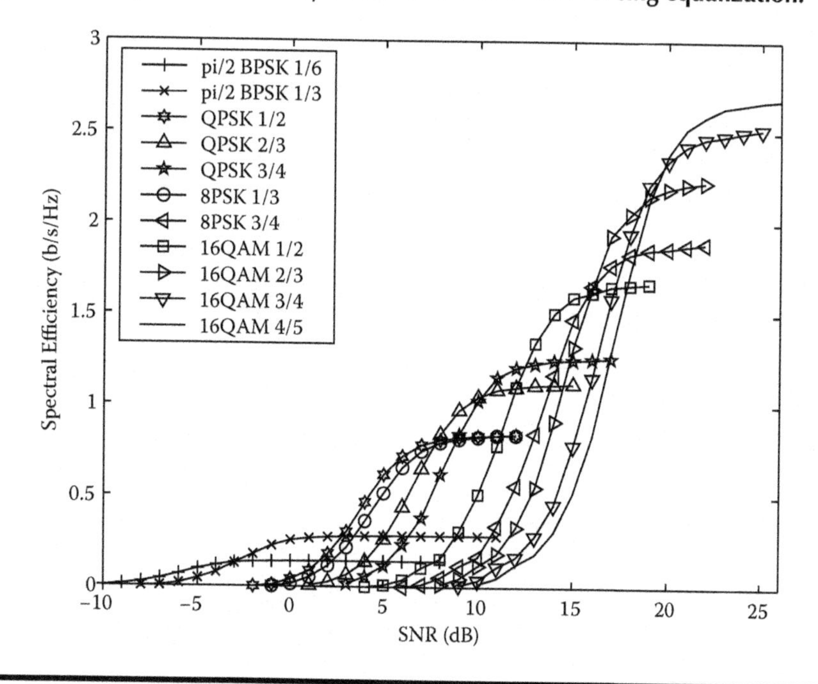

Figure 6.11 Spectral efficiency of SC-FDMA with MMSE equalization.

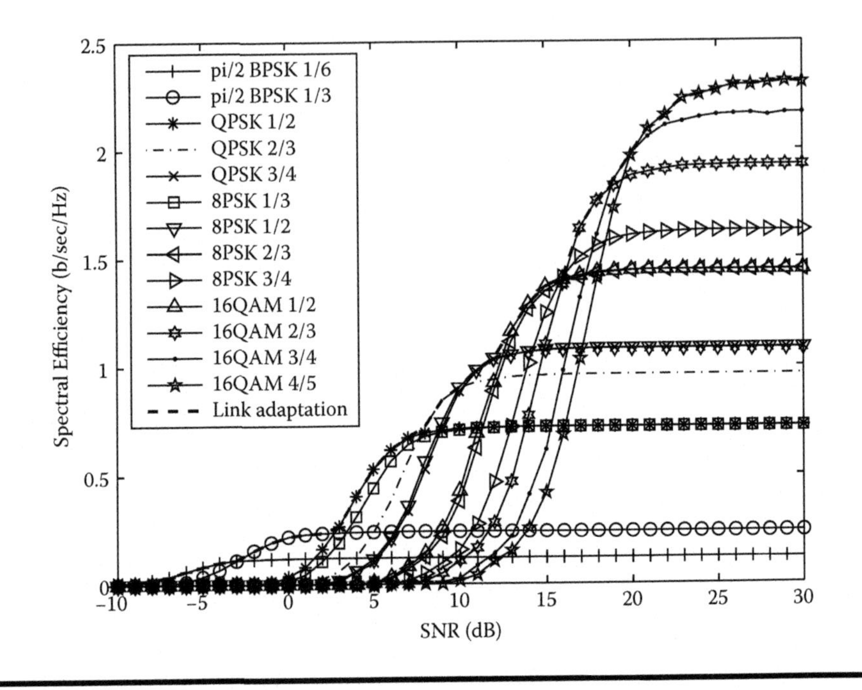

Figure 6.12 Spectral efficiency of SISO SC-FDMA with various MCS schemes.

In these peak data rates, the 14% pilot overhead described previously has been taken into account. The MCS formats used and corresponding data rate values are summarized in Table 6.3. It can be observed that 8 PSK does not contribute to the link adaptation curve.

Table 6.3 Peak Data Rate for 10-MHz System

MCS	*Data Rate*
$\pi/2$ BPSK 1/6	1.2 Mb/s
$\pi/2$ BPSK 1/3	2.4 Mb/s
QPSK 1/2 8 PSK 1/3	7.2 Mb/s
QPSK 2/3	9.6 Mb/s
QPSK 3/4, 8 PSK 1/2	10.8 Mb/s
16QAM 1/2, 8 PSK 2/3	14.4 Mb/s
8 PSK 3/4	16.2 Mb/s
16 QAM 2/3	19.2 Mb/s
16 QAM 3/4	21.6 Mb/s
16 QAM 4/5	23 Mb/s

6.7 Impact of Fast HARQ

A fast level-1 (L1) HARQ retransmission concept is used with the aim of facilitating more aggressive and spectrally efficient packet scheduling [12]. Two soft-combining HARQ retransmission mechanisms are considered: chase combining and incremental redundancy. Both retransmission mechanisms ensure that the past retransmission is fully utilized. The major difference between these two retransmission mechanisms is that chase combining requires the same transmitted signal in each retransmission, whereas incremental redundancy transmits different parity bits in each retransmission.

The HARQ within the medium access control (MAC) sublayer has the following characteristics:

- *N*-process stop-and-wait (SAW) HARQ is used. The HARQ processes are transmitted over *N* parallel time channels in order to ensure the continuous transmission to the UE.
- The HARQ is based on acknowledges/negative acknowledges (ACKs/NACKs). The received data packets are acknowledged after each transmission. NACK implies a retransmission request for either an additional redundancy for HARQ incremental redundancy or an identical packet for the HARQ chase combining case.
- For the uplink, the HARQ is based on synchronous retransmissions.

In the following simulations, we have assumed no error in the transmission of ACK/NACK from the BS to the UE, six SAW parallel channels, and a maximum of four retransmissions. In practice, four to six SAW parallel channels are used [13].

Figure 6.13 shows the performance results of 16 QAM for lower coding rate (1/2) and higher coding rate (4/5). The HARQ can improve the performance, as can be observed in the low SNR region for each coding rate. The benefit of incremental redundancy over chase combining can also be seen, especially for the higher coding rate. As a consequence, the general assumption for this study is to use incremental redundancy for coding rates exceeding 1/2 and chase combining for coding rates equal to or less than 1/2.

The spectral efficiency with LA including all of the MCSs is shown in Figure 6.14. The combining gain of HARQ can be observed in the smoothing out of the maximum achievable spectral efficiency results, when compared to Figure 6.12. It can also be observed that, at low SNR, most of the spectral efficiency curves are overlapped. This condition is beneficial when there is an error in the link adaptation mechanism. Although the UE transmits with a suboptimal MCS, it can still obtain the same spectral efficiency due to HARQ.

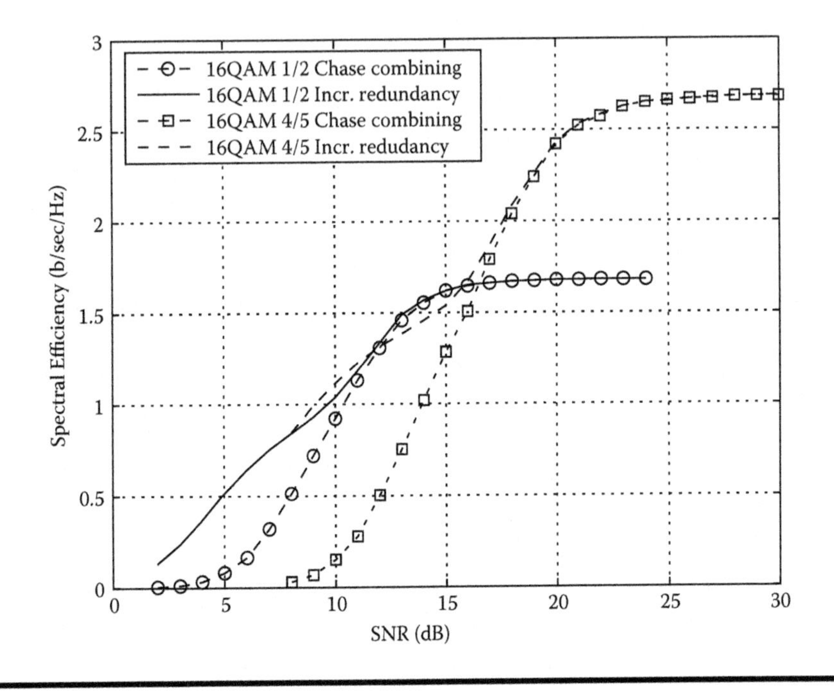

Figure 6.13 Spectral efficiency of SC-FDMA SISO case for chase combining (CC) and incremental redundancy (IR).

6.8 Antenna Configuration

Multiple antennas are essential for UTRA LTE in order to reach the targeted peak data rates—in particular, with the aim of improving the performance at the lower SNR operating range. In this work the combination of SC-FDMA with SISO and the single-input, multiple-output (SIMO) antenna configurations has been studied. In LTE, the baseline for uplink is to use one transmit antenna at the UE and two receive antennas with maximal ratio combining (MRC) at the base station [14].

After MRC, the output signal is represented as follows:

$$V_{m,k} = \frac{z1_{m,k} H1^*_{m,k} + z2_{m,k} H2^*_{m,k}}{H1_{m,k} H1^*_{m,k} + H2_{m,k} H2^*_{m,k}} \tag{6.16}$$

where $z1$ and $z1$ are the received signals in the frequency domain representation of each antenna branch.

In Figure 6.15, the spectral efficiency for both SISO and SIMO antenna schemes is presented. The spectral efficiency of OFDMA is also presented for comparison. In the SISO case, OFDMA outperforms SC-FDMA with an MMSE

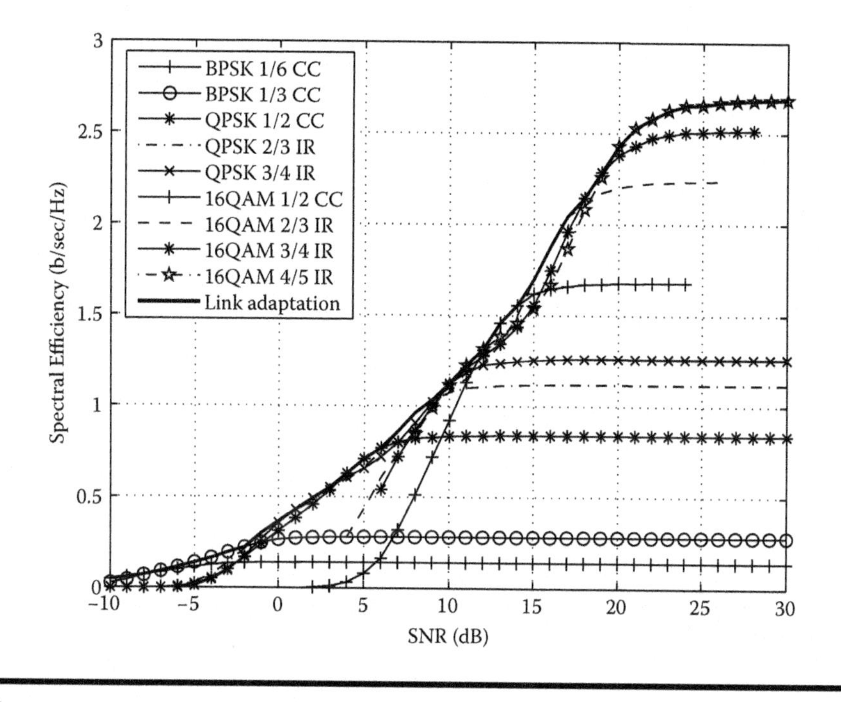

Figure 6.14 Spectral efficiency of SC-FDMA SISO case with HARQ.

receiver. The demodulation for OFDMA is performed in the frequency domain, whereas for SC-FDMA, demodulation is performed in the time domain. A deep fading may affect the whole data symbol in SC-FDMA; thus, the performance of SC-FDMA is worse for a frequency-selective fading channel. For the SIMO case, the SC-FDMA and OFDMA performance is improved due to the diversity gain, and the spectral efficiency gap between OFDMA and SC-FDMA becomes unnoticeable. The MMSE receiver in SIMO with MRC further reduces the noise in comparison to the SISO case, as well as greatly increases the spectral efficiency of SC-FDMA.

6.9 Impact of Frequency Allocation

LTE uplink transmission allows a flexible frequency allocation within the channel bandwidth. In the study item phase [6], the 3GPP has defined the localized and distributed frequency/subcarrier allocation as shown in Figure 6.16. The distributed case is used to gain frequency diversity, whereas the localized SC-FDMA can be used to simplify subcarrier allocation and benefit from opportunistic scheduling. Lately, the 3GPP has only considered the localized case and not the distributed case because the localized case with frequency hopping can

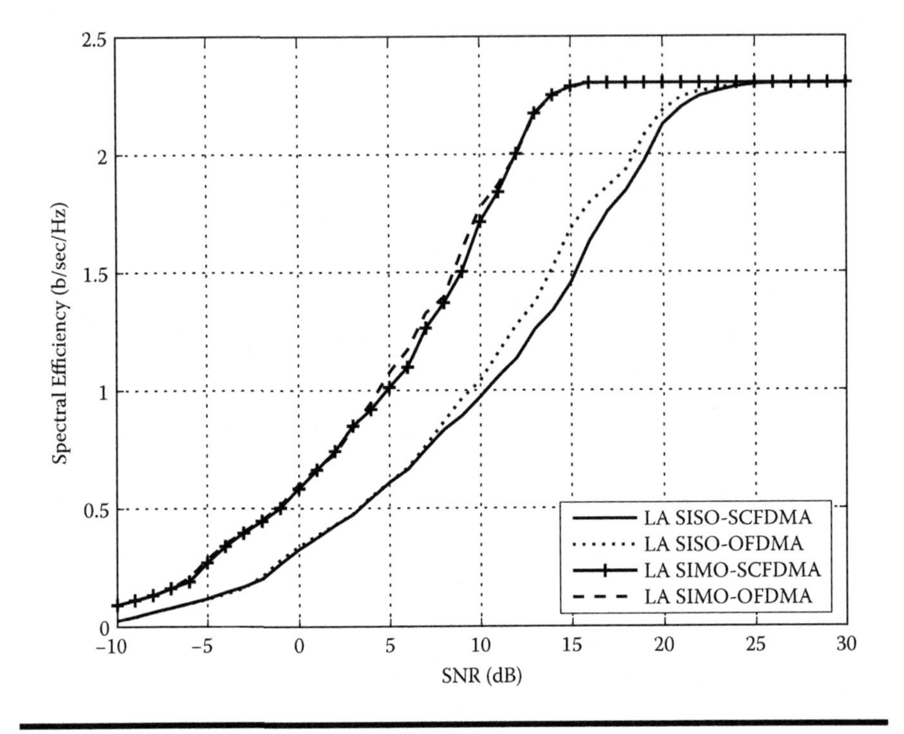

Figure 6.15 Spectral efficiency for SC-FDMA and OFDMA with two antenna configurations (SISO and SIMO).

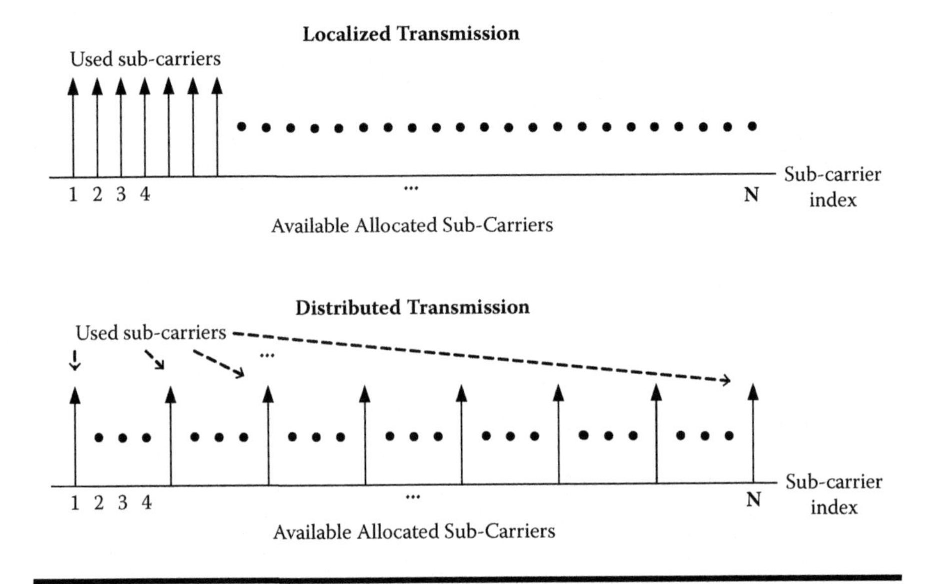

Figure 6.16 Localized and distributed frequency allocation in LTE uplink.

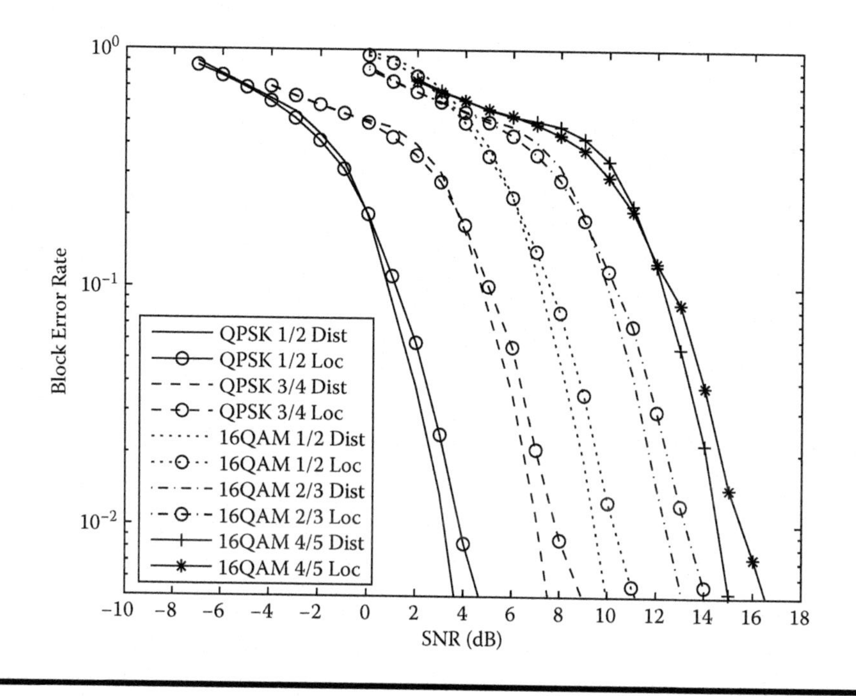

Figure 6.17 BLER performance of distributed versus localized cases in SC-FDMA.

also gain the frequency diversity. Moreover, the distributed case is sensitive to the frequency error.

The BLER performance of the distributed and localized case in SC-FDMA with HARQ is presented in Figure 6.17. The SIMO case is assumed, and the channel bandwidth is 10 MHz. A 2.5-MHz bandwidth is allocated for the transmission. For the distributed case, this configuration will allow a repetition factor of four because all the subcarriers carrying data are spread over the full 10-MHz bandwidth.

It is shown that the distributed cases perform better than the localized one in most cases. For BLER 1%, it is clear that the distributed case is superior to the localized. It has about a 1-dB gain over localized in 16 QAM, rate 4/5. The distributed case can maximize the frequency diversity by distributing the data in the whole bandwidth. In the real situation, when real channel estimation is taken into account, the localized case is expected to perform better than the distributed case. The reason is that, considering the limited amount of pilot signals, the distributed case has to estimate the channel transfer function (CTF) in the full bandwidth, whereas the localized case estimates the channel transfer function only in the allocated channel bandwidth. Thus, the distributed case is more sensitive to the error introduced by channel estimation.

6.10 Channel Estimation

For the transmission over time-varying and frequency-selective fading channels, the receiver requires an estimation of the CTF in both time and frequency domains. The receiver utilizes the received pilot signals as described in Section 6.2 to perform channel estimation. In the following, we consider a practical one-dimensional estimator that operates independently in time and frequency domains. First, the CTF is estimated from the received short blocks. Second, time domain channel interpolation is considered. For the simplicity of derivation of the channel estimation algorithm, the channel impulse response is considered constant over the duration of an OFDM symbol.

6.10.1 *Frequency-Domain Channel Estimation*

For frequency-domain channel estimation, first an estimate of the CTF at the pilot subcarriers is obtained. Then the full channel transfer function is calculated using an interpolation method.

1. Estimate at pilot position: Let p_i $i = 0,\ldots, N_p - 1$ be a set of indexes containing the subcarrier indexes that carry pilot symbols, where N_p is the indexes of pilot symbols in the short block. A least-squares (LS) estimate of the CTF at these pilot positions can be calculated as

$$\tilde{\mathbf{h}}_p[i] = \frac{\mathbf{z}[\mathbf{p}_i]}{\mathbf{d}[\mathbf{p}_i]} \tag{6.17}$$

   where $\mathbf{z}[\mathbf{p}_i]$ and $\mathbf{z}[\mathbf{p}_i]$ are, respectively, the received short block after the FFT and the transmitted short block in the *i*th pilot subcarrier. $\mathbf{h}_\mathbf{p}$ is an N_p long column vector containing the LS estimates of the channel at pilot subcarriers.
2. Wiener filtering interpolation: Wiener filtering is the optimum interpolation method in terms of MSE. Using the statistics of the channel and noise, it performs an MMSE interpolation of the estimates at pilot subcarriers, optimally reducing the effects of noise and channel distortion. The full CTF is obtained by

$$\tilde{\mathbf{h}} = \mathbf{R}_{\mathbf{hh_p}} \cdot \left(\mathbf{R}_{\mathbf{h_p h_p}} + \sigma_\mathbf{w}^2 \mathbf{I}_{\mathbf{N_p}}\right)^{-1} \tilde{\mathbf{h}}_\mathbf{p} \tag{6.18}$$

   where $\mathbf{R}_{\mathbf{hh_p}}$ is the cross-correlation matrix of the true CTF and the true CTF at pilot subcarriers, $\mathbf{R}_{\mathbf{h_p h_p}}$ is the autocorrelation matrix of the true CTF at pilot subcarriers, $\sigma_\mathbf{w}^2$ is the noise power, and $\mathbf{I}_{\mathbf{N_p}}$ is the $N_p \times N_p$ identity matrix. Note that the CTF coefficients are assumed to be uncorrelated to the noise process.

6.10.2 Time-Domain Channel Estimation

A simple time-domain channel estimation based on linear interpolation is assumed. In a subframe, two short blocks are transmitted. The channel estimation for each subframe is a linear interpolation between these two short blocks.

Figure 6.18 shows the link-level simulation results with the preceding channel estimation techniques and at a UE speed of 120 km/h. It is shown that the OFDMA performance is more affected by the UE speed than SC-FDMA, especially at higher modulation and coding schemes (16 QAM rate 4/5). The UE at high speed has a higher Doppler frequency and creates intercarrier interference (ICI). In OFDM, each subcarrier carries different symbols and therefore OFDM is more sensitive to the ICI.

Figure 6.19 illustrates the results for a speed of 200 km/h. It is shown that 16 QAM rates of 3/4 and above do not contribute to the link adaptation curve and hence the system operates at low spectral efficiency. At high speed, the subcarrier orthogonality is destroyed by the ICI.

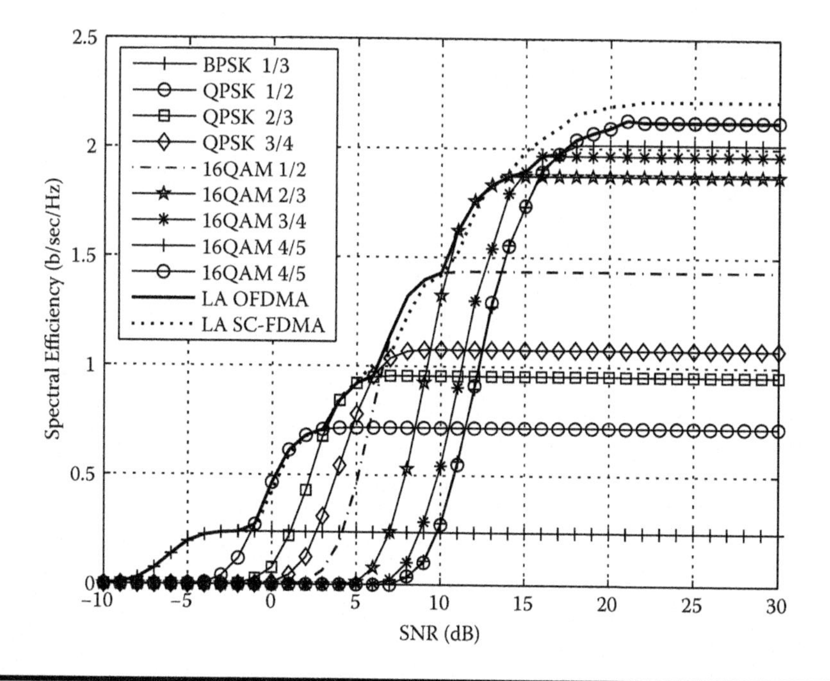

Figure 6.18 SC-FDMA and OFDMA with real channel estimation at a speed of 120 km/h.

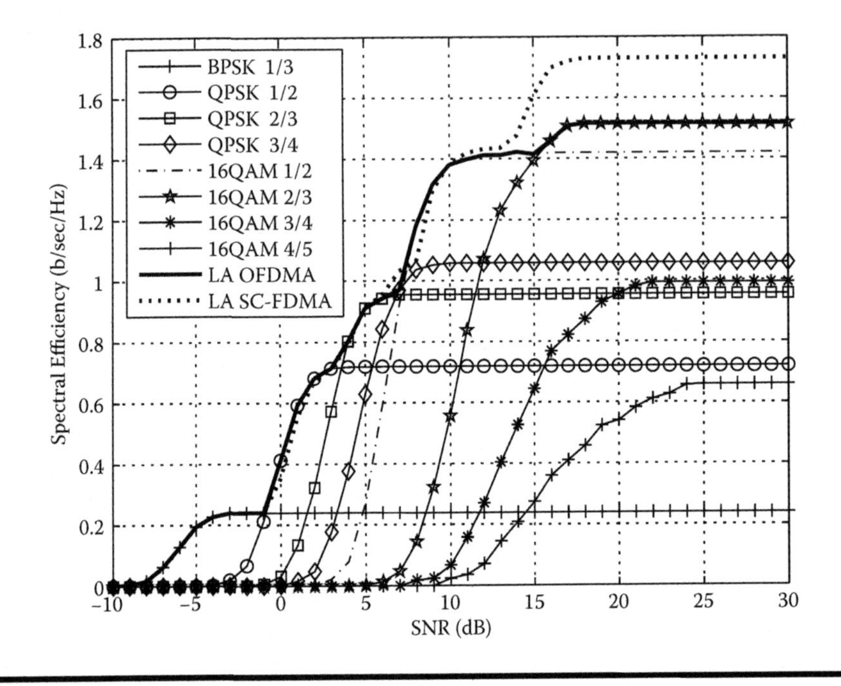

Figure 6.19 SC-FDMA and OFDMA with real channel estimation at a speed of 200 km/h.

6.11 SC-FDMA with Turbo Equalization

The turbo equalizer can improve SC-FDMA performance. Turbo equalization (TEQ) in the context of LTE uplink was proposed and investigated in Priyanto [11] and Berardinelli et al. [15]. Turbo equalization is an advanced iterative equalization and decoding technique that allows enhancement of the performance for the transmission over a frequency-selective fading channel. The turbo equalizer, however, increases the receiver complexity. Nevertheless, the implementation is performed in the base station, which does not have physical constraints and power consumption as stringent as those of the UE.

The block diagram of the proposed turbo equalizer receiver is shown in Figure 6.20. It uses a soft interference cancellation (SIC) equalizer as described in Laot, Glavieux, and Labat [16] and frequency domain equalization. This structure results in more efficient operation than the traditional time domain equalization [17] because the equalization is performed on one data block at a time. Moreover, the operations within this block involve an efficient FFT operation and a simple channel inversion operation.

In the following, the performance of the receiver with turbo equalizer is evaluated using a 1 × 2 SIMO antenna scheme based on the simulation parameters in Table 6.2.

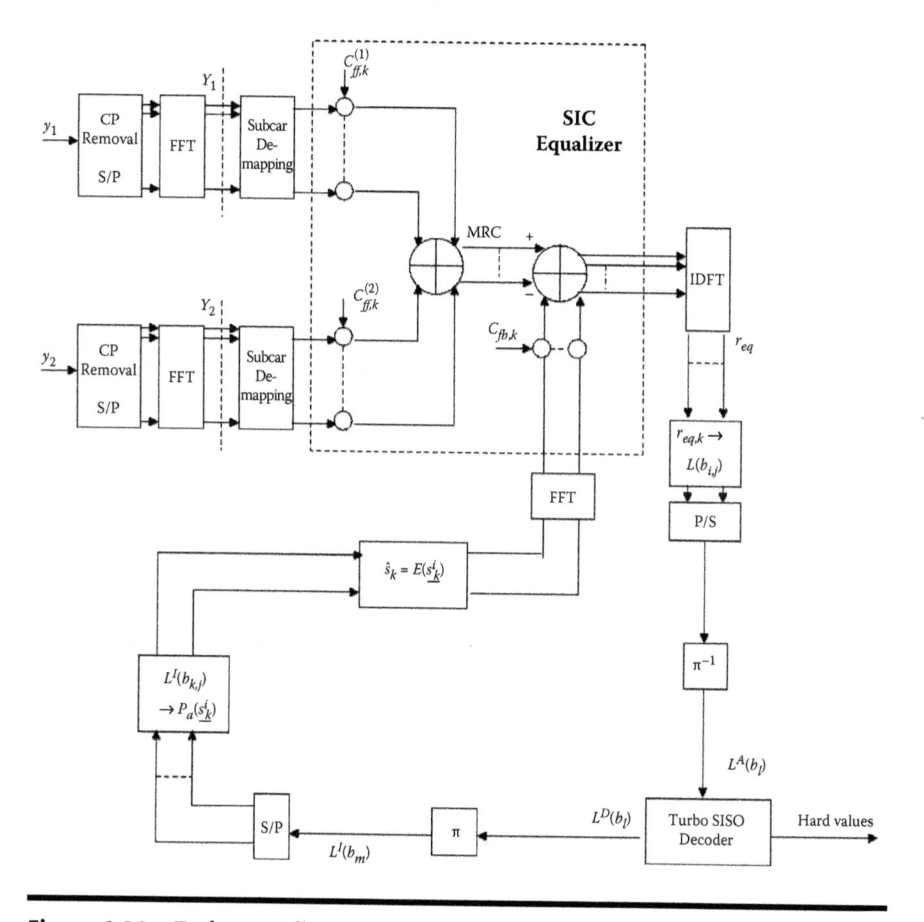

Figure 6.20 Turbo equalizer structure.

Figure 6.21 shows the BLER performance of the SC-FDMA 16 QAM system with three different coding rates. The results of TEQ with different iterations and SC-FDMA with simple MMSE are given. Two different results for turbo equalization are plotted with two and four iterations of the algorithm, respectively. Compared with the MMSE receiver, the turbo equalizer shows a gain of about 1 dB. In our assumptions, the cyclic prefix allows complete removal of the ISI; hence, the gain is exclusively due to the reduction of the noise component. The significant gain is already obtained from two iterations of the algorithm.

In Figure 6.22 the performance of OFDMA has also been included, and the results are shown for both QPSK and 16 QAM modulations in comparison with the SC-FDMA TEQ receiver performance for two iterations of the algorithm. Note that for QPSK 2/3 and QPSK 3/4, the SC-FDMA with MMSE system performs better than OFDMA. This is due to the fact that low-order modulation schemes are robust to the noise enhancement of SC-FDMA transmission. Moreover, OFDM

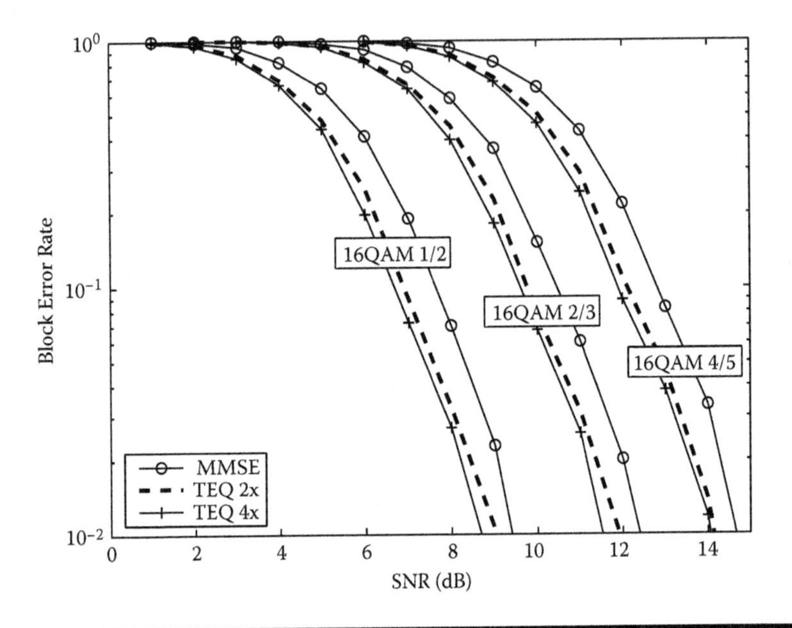

Figure 6.21 BLER performance for SC-FDMA 1 × 2 MRC with and without turbo equalizer in TU 06 channel.

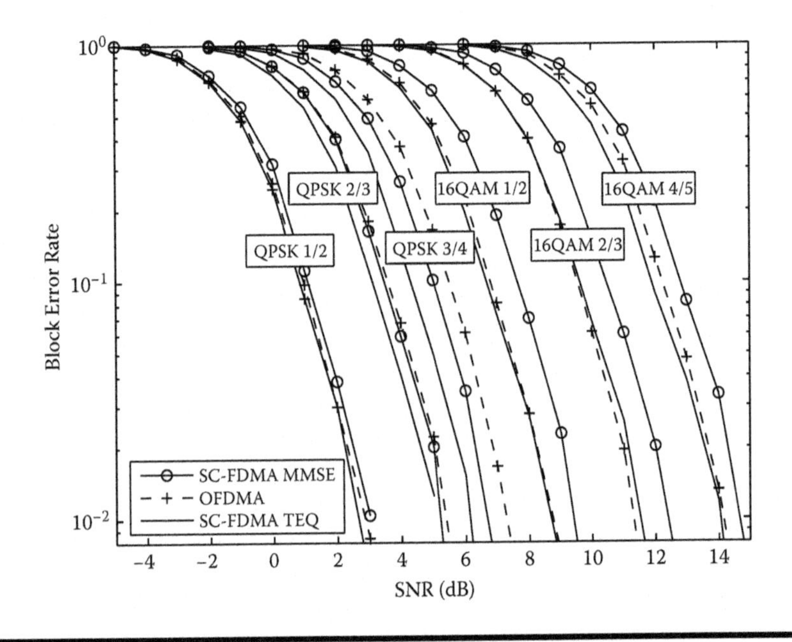

Figure 6.22 BLER performance for SC-FDMA versus OFDMA in TU 06 channel.

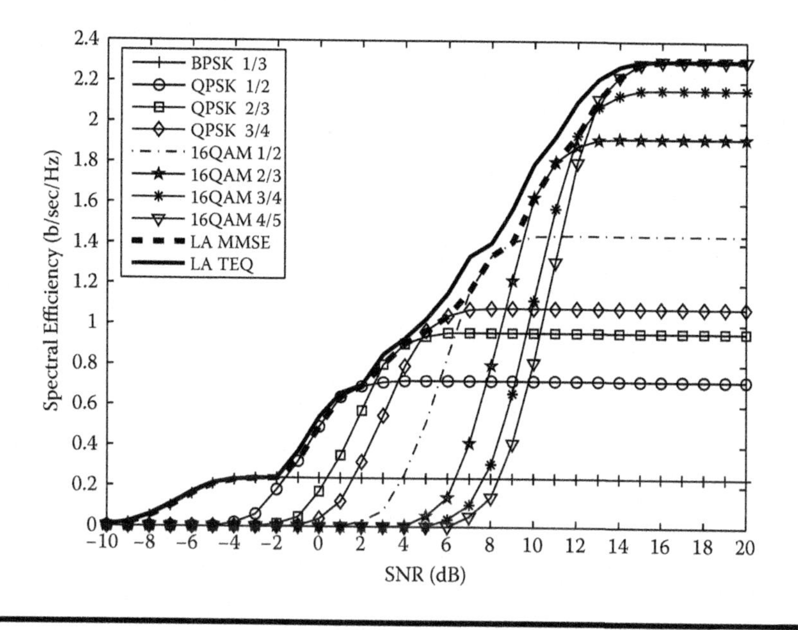

Figure 6.23 Spectral efficiency for SC-FDMA in TU 06 channel.

transmission is more sensitive to the coding rate than SC-FDMA [12]. SC-FDMA takes advantage of spreading the data symbols over the transmission bandwidth, where the effect of a deep fade affecting few subcarriers will be diminished in the receiver combining all the subcarrier information.

The receiver with turbo equalizer gives better performance in most cases, and in the worst case it shows the same performance as OFDMA. Therefore, the turbo equalizer makes it possible to reduce the noise enhancement due to the time domain transmission in the SC-FDMA system and makes performance the same as or better than that of OFDMA.

The performance results for several MCSs in terms of spectral efficiency are shown in Figure 6.23. The figure shows the link adaptation curve for both the MMSE and the TEQ receiver. It is shown that the TEQ is effective for higher-order MCSs, whereas for BPSK it does not show any performance improvement. Therefore, in order to exploit its performance gain, the turbo equalizer receiver could be used for high-order modulation schemes, whereas the simple MMSE receiver can still be used for BPSK. In this way, the computational load for low-order modulations can be reduced.

6.12 Impact of Nonlinear Power Amplifier

The selection of an air interface technique is also influenced by the RF transmitter constraints, especially the nonlinear power amplifier. The RF transmitter imperfections create in-band performance degradation and increase the out-of-band

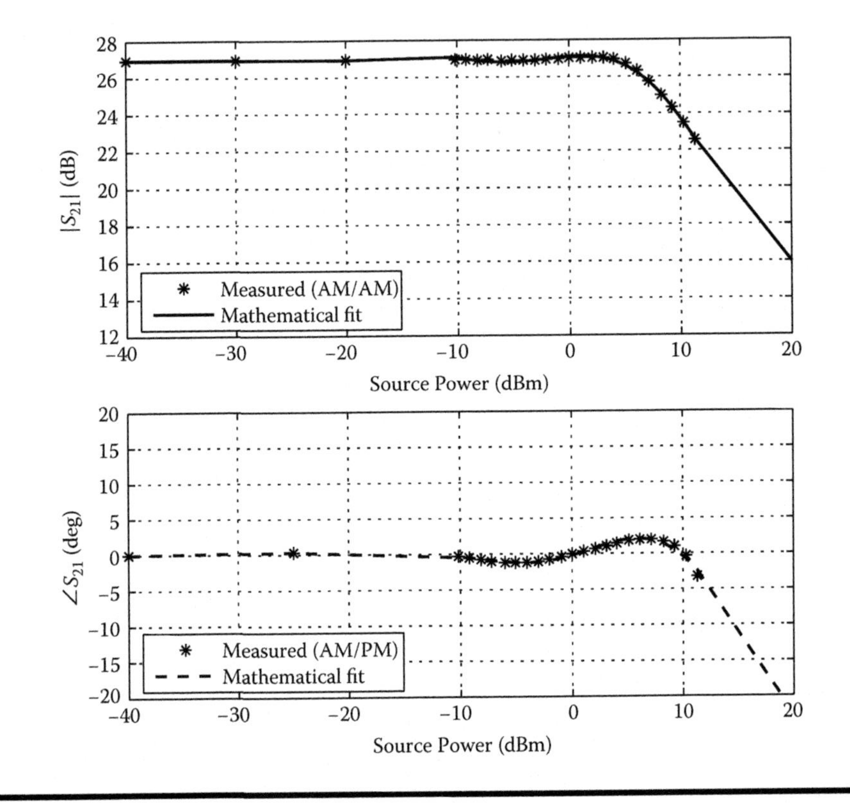

Figure 6.24 Measured AM/AM, AM/PM (asterisks), and model fit (solid) versus source power (decibels of measured power [dBm]) [19].

emissions. The latter can lead to violation of the spectral emission mask (SEM) and adjacent channel leakage ratio (ACLR) requirements. The trade-off among the implementation costs, the impact of the imperfections on achieving high performance, and meeting the requirements is the prominent issue.

The nonlinear power amplifier is usually characterized by AM/AM (amplitude to amplitude) and AM/PM (amplitude to phase) distortion curves [18]. The AM/AM and AM/PM characteristics used in this study, represented as $|S_{21}|$ and $\angle S_{21}$, are shown in Figure 6.24. These characteristics are based on the measurement result of a UE wideband CDMA power amplifier (PA) module [19]. This PA has a good compromise between linearity and power efficiency for varying envelope input signals. From Figure 6 24, the power amplifier has a 27-dB linear gain and 1-dB compression point at +6.1-dBm input power.

Error vector magnitude (EVM) is commonly used as a direct measure of the transmitter signal quality, in particular when related to any RF hardware imperfections in the system. EVM, however, does not cover the disturbance caused to other users, where ACLR must be used as a measure [20]. The EVM is a measure of the

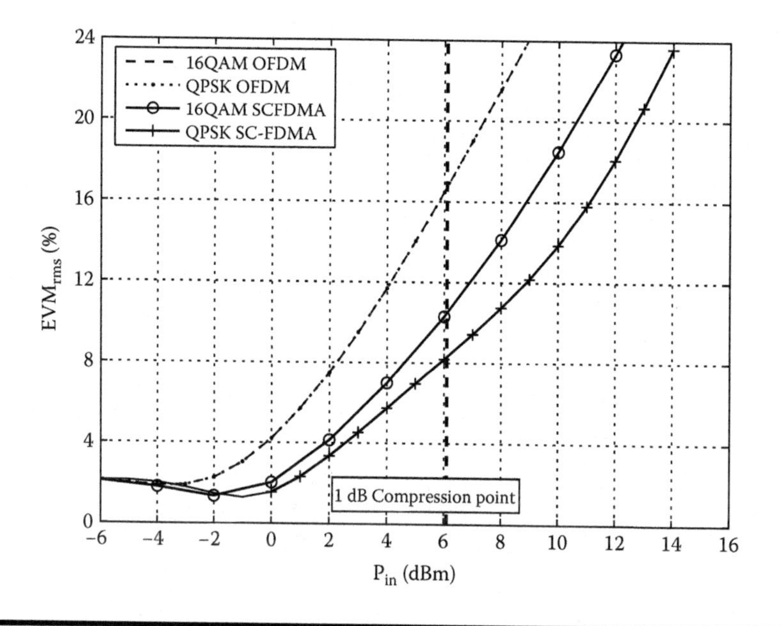

Figure 6.25 EVM results for nonlinear PA.

difference between the measured and the reference signals. In LTE, the EVM is measured in the constellation domain [21–23] rather than in the time domain as for the previous UTRA specifications [24]. This approach allows pre- and post-FFT time/frequency synchronization and amplitude/phase correction to be included in the frequency domain equalization block at the receiver.

Figure 6.25 shows the EVM results for both SC-FDMA and OFDMA air interfaces. The results for both QPSK and 16 QAM modulation are presented. SC-FDMA has better EVM performance compared to OFDMA. A change to lower modulation, QPSK, further improves the SC-FDMA performance, but this is not the case for OFDMA. These results are due to the fact that SC-FDMA has lower PAPR than OFDMA, as shown in Figures 6.4 and 6.5.

The impact of the nonlinear power amplifier on SC-FDMA and OFDMA spectral efficiency is shown in Figure 6.26. These results are obtained when an input back-off (IBO) of 4 dB is assumed to maintain high-efficiency power amplification and measured at a BLER target of 10%. For 16 QAM, rate 4/5, SC-FDMA has a gain of 0.7 dB over OFDMA, although only a small gain of 0.2 dB is obtained at the lower coding rate. These results confirm that the SC-FDMA is the preferred solution over OFDMA as well as in the presence of nonlinear power amplifiers.

Figure 6.27 shows the impact of the nonlinear amplifier on the spectrum emission of SC-FDMA and OFDMA. An input back-off of 4 dB and 16 QAM

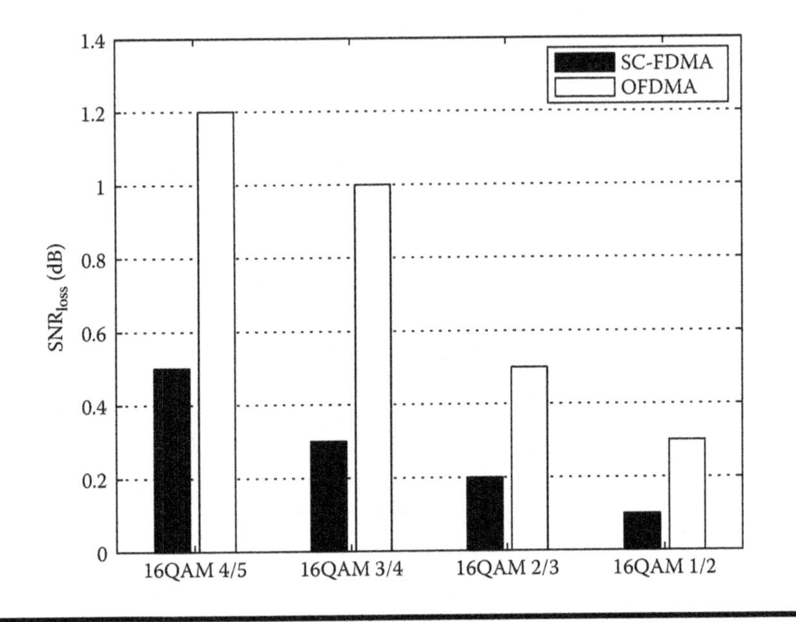

Figure 6.26 Performance loss at 10% BLER due to the nonlinear PA.

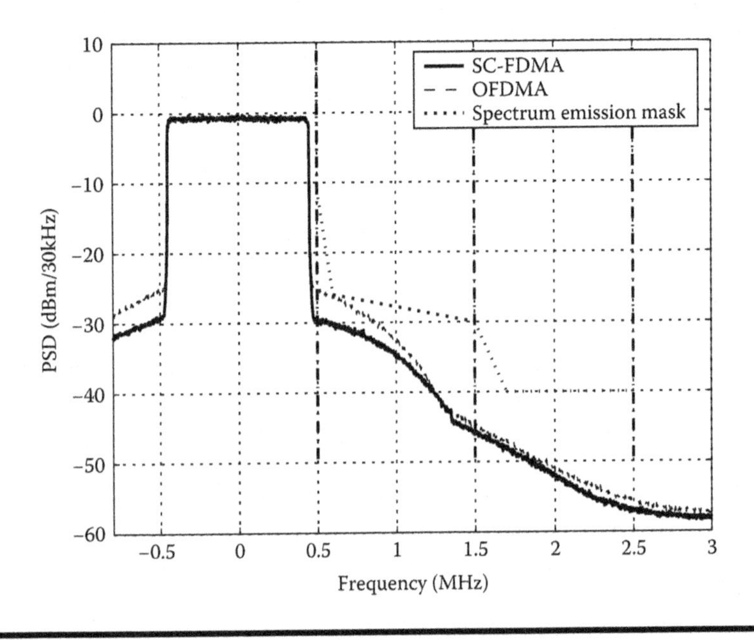

Figure 6.27 Spectrum emission of SC-FDMA and OFDMA.

modulation are assumed. It is shown that the spectrum emission of OFDMA is marginally below the spectrum emission mask, whereas a better result is obtained for the SC-FDMA transmitter, in agreement with the PAPR results.

6.13 Summary

In this chapter, we have presented the key techniques for LTE uplink as well as presented the baseline performance. Radio access technology is the key aspect in LTE uplink, and two radio access schemes, SC-FDMA and OFDMA, have been studied. The performance results are obtained from a detailed UTRA LTE uplink link-level simulator. The simulation results show that both SC-FDMA and OFDMA can achieve high spectral efficiency; however, SC-FDMA benefits in obtaining lower PAPR than OFDMA, especially for low-order modulation schemes. A 1×2 SIMO antenna configuration highly increases the spectral efficiency of SC-FDMA, making the performance of SC-FDMA with the MMSE receiver comparable to OFDMA, especially at high coding rates. The peak spectral efficiency results for SC-FDMA confirm that it meets the requirement to achieve a spectral efficiency improvement of two or three times that of 3GPP release 6.

Despite an increase in base station receiver complexity, the turbo equalizer can reduce the noise enhancement in the SC-FDMA system. Results show that a turbo equalizer receiver for a SIMO system can improve the BLER performance by almost 1 dB compared to a traditional MMSE receiver. Most of the performance gain is obtained with two iterations of the algorithm. Furthermore, the simulation results show that the SC-FDMA turbo equalizer performance is better than or equal to that of OFDMA for all the modulation and coding schemes.

A nonlinear power amplifier introduces in-band performance degradation and increases the out-of-band emission. The simulation results confirm that SC-FDMA is more robust against the nonlinear power amplifier because it achieves high spectral efficiency also in the presence of nonlinearities. This leads to better power amplifier efficiency, which can be used to increase the cell coverage.

References

1. Bingham, J. 1990. Multicarrier modulation for data transmission: An idea whose time has come. *IEEE Communications Magazine* May: 5–14.
2. Falconer, D., S. Ariyavisitakul, A. Benyamin-Seeyar, and B. Edison. 2002. Frequency domain equalization for single-carrier broadband wireless systems. *IEEE Communications Magazine* 40 (4): 58–66.
3. Sari, H., G. Karam, and I. Jeanclaude. 1995. Transmission techniques for digital terrestrial TV broadcasting. *IEEE Communications Magazine* 33 (2): 100–109.
4. Czylwik, A. 1997. Comparison between adaptive OFDM and single carrier modulation with frequency domain equalization. *Proceedings of Vehicular Technology Conference '97*, Phoenix, AZ, May 1997, 2: 865–869.

5. Kansanen, K. 2005. Wireless broadband single-carrier systems with MMSE turbo equalization receivers, PhD dissertation, University of Oulu, Oulu, Finland.
6. 3GPP. 2006. Physical layer aspects for evolved universal terrestrial radio access (UTRA) (release 8). Technical specification TR 25.814 V7.1.0, September 2006.
7. Han, S., and J. Lee. 2005. An overview of peak-to-average power ratio reduction techniques for multicarrier transmission. *IEEE Wireless Communications* 12 (2): 56–65.
8. 3GPP. June 2005. Multiplexing and channel coding. TS 25.212 V6.5.0 (2005-06).
9. Fazel, K., and S. Kaiser. 2003. *Multi-carrier and spread spectrum systems*. Chichester, England: John Wiley & Sons.
10. Sari, H., G. Karam, and I. Jeanclaude. 1994. Frequency-domain equalization of mobile radio and terrestrial broadcast channels. *Proceedings of IEEE Global Telecommunications Conference,* vol. 1, November 1994.
11. Priyanto, B. 2008. Air interfaces of Beyond 3G systems with user equipment hardware imperfections: Performance and requirements aspects, PhD dissertation, Aalborg University, Aalborg, Denmark.
12. Kolding, T., F. Frederiksen, and P. Mogensen. 2002. Performance aspects of WCDMA systems with high-speed downlink packet access (HSDPA). *Proceedings of IEEE VTC Fall 2002,* Vancouver, Canada, July 2002, 477–481.
13. Holma, H., and A. Toskala. 2007. *WCDMA for UMTS, HSPA evolution and LTE,* 4th ed. Chichester, England: John Wiley & Sons.
14. 3GPP. March 2006. Requirements for evolved UTRA (E-UTRA) and evolved UTRAN (E-UTRAN). TR 25.913.
15. Berardinelli, G., B. Priyanto, T. Sørensen, and P. Mogensen. 2008. Improving SC-FDMA performance by turbo equalizer in UTRA LTE uplink. *Proceedings of IEEE VTC Spring 2008,* May 2008.
16. Laot, C., A. Glavieux, and J. Labat. 2001. Turbo equalization: Adaptive equalization and channel decoding jointly optimized. *IEEE Journal on Selected Areas in Communications* 19 (9): 1744–1752.
17. Tuchler, M., and J. Hagenauer. 2000. Turbo equalization using frequency domain equalizers. *Proceedings of Allerton Conference,* Monticello, AR, October 2000, 1:144–153.
18. Razavi, B. 1998. *RF microelectronics,* 1st ed. Upper Saddle River, NJ: Prentice Hall PTR.
19. Staudinger, J. 2002. An overview of efficiency enhancements with application to linear handset power amplifier. *Proceedings of IEEE Radio Frequency Integrated Circuits Symposium.*
20. Gu, Q. 2006. *RF system design of transceivers for wireless communications.* New York: Springer.
21. Nokia. September 2006. Definition of E-UTRA EVM for the BS. 3GPP, Tallinn, Estonia. Technical Report R4-060737.
22. Nokia. September 2006. Methodology for deriving E-UTRA EVM BS requirements. 3GPP, Tallinn, Estonia. Technical report R4-060738.
23. Siemens. November 2006. EVM measure for LTE. 3GPP, Riga, Latvia. Technical report R4-061227.
24. 3GPP. December 2006. Base station (BS) conformance testing (FDD). Technical specification TS 25.141 V7.6.0.

Chapter 7

Cooperative Transmission Schemes

Wolfgang Zirwas, Wolfgang Mennerich, Martin Schubert, Lars Thiele, Volker Jungnickel, and Egon Schulz

Contents

Abstract

The 3rd Generation Partnership Project (3GPP) long tem evolution (LTE) release 8 specification defines the basic functionality of a new high-performance air interface providing high user data rates in combination with low latency based on MIMO (multiple input, multiple output), OFDMA (orthogonal frequency division multiple access), and an optimized system architecture evolution (SAE) as main enablers. At the same time, in the near future, increasing numbers of users will request mobile broadband data access everywhere—for example, for synchronization of e-mails, Internet access, specific applications, and file downloads to mobile devices like personal digital assistants (PDAs) or notebooks. In the future, a 100-fold increase in mobile data traffic is expected [1], making further improvements beyond LTE release 8 necessary and possibly ending up in new LTE releases or in a so-called international mobile telecommunications (IMT) advanced system [2–4]. IMT-advanced is a concept from the ITU (International Telecommunication Union) for mobile communication systems with capabilities that go further than that of IMT-2000. IMT-advanced was previously known as *systems beyond IMT-2000.*

The main enemy in today's cellular radio systems is interference. Although intracell interference is of minor importance for OFDMA systems like LTE, intersector as well as intersite interference reduces achievable spectral efficiency by factors. Multiuser MIMO (MU-MIMO) and OFDMA partly help to combat interference by beam forming gains and due to selection of resources less affected by interference. In the interference-limited case—meaning small cell sizes like pico or urban macrocells and high-transmission power—the performance of multicell systems is remarkably smaller than that of a single interference-free site. At the same time, it is well known [6] that, in cooperative antenna systems, several distributed radio stations simultaneously transmit suitably coherently precoded data signals by common beam forming over all radio stations, and antenna elements are able to overcome interference successfully. Cooperative antenna systems even exceed the performance of a single isolated cell due to rank enhancement effects [18].

For real systems, cooperation over large areas is not feasible because accurate channel state information (CSI) for many base stations (BSs) and much user equipment (UE) is needed. Therefore, practical solutions for cooperative antenna systems, generally termed in the following as COOPA (cooperative antenna) systems, have to been found. In addition to a classification of COOPA implementations, main challenges like channel estimation, feedback overhead, and backbone traffic overhead and possible solutions like cooperative relaying will be given in this chapter. A further important topic will be the implementation of a basic cooperation area (CA) based on LTE release 8 features with minimum adaptations to the standard as well as backward compatibility for release 8 conforming UEs and simultaneously robust operation.

7.1 Introduction

In the future, mobile network operators (MNOs) will have to provide broadband data rates to an increasing number of users with lowest cost per bit and probably as flat rates, such as is known from fixed network providers. Intercell interference is the most limiting factor in current cellular radio networks, which means that any type of practical feasible interference mitigation will be of highest importance to tackle the expected 100-fold traffic challenge. Cooperation is the only means known to really overcome interference and is therefore a very likely candidate for integration into an enhanced LTE system.

In the beginning of 2008, the first 3GPP LTE release 8 standard emerged, defining most of its fundamental parameters and procedures as described in the previous chapters. In the context of next-generation mobile networks (NGMNs) and the LTE SAE trial initiative (LSTI) [8], the single-link performance of LTE for 2 × 2 MIMO systems has been validated based on existing real-time hardware (HW) [5]. Outdoor multiuser measurement campaigns for single sites as described in this book more than proven the high benefits of MIMO and OFDMA, providing a user data rate of more than 100 Mb/s in a significant part of the radio cell.

The logical next step will be the optimization of the multicell network, which includes the transmission from more than one sector as well as from more than one site. As known very well, cellular radio systems in general (and specifically LTE) suffer significantly from interference in urban areas, reducing overall spectral efficiency from an ideal of about 10 to a few bits per second per hertz per cell. System-level simulations as well as the NOMOR [9] LTE emulator proved that MIMO in combination with OFDMA as an early form of adaptive resource coordination helps to overcome interference issues better than 1 × 1 or 1 × 2 OFDM systems like wireless local area networks (WLANs). Other common techniques, like frequency reuse greater than one [10], are already known from the global system for mobile communications (GSM) or from 3G systems like the universal terrestrial radio access (UTRA) frequency division duplex (FDD) system [11] and are quite easy to implement. But the resulting spectral efficiencies are in the order of 1 b/s/Hz/cell and thus far below the theoretical optimum of a COOPA system.

Downlink (DL) LTE baseline (2,2) systems with two antenna ports (APs) at the narrowband (eNB) and two receive antennas at the UE achieve spectral efficiencies in the order of 2–3 b/s/Hz/cell according to system-level (SL) simulations; this is significantly smaller than what is predicted in the theory for full cooperation among all eNBs of a network. A factor of 10–15 in terms of spectral efficiency can be expected [6], if one compares a (1,1) single-input, single-output (SISO) system without cooperation with a cooperative (4,4) MIMO system. Thus, in spite of the tremendous technical challenges described in this chapter, it will be important to define and standardize the most suitable type of cooperation in one of the next LTE releases.

In early 2008, a study item was launched at 3GPP [12] searching for useful enhancements of LTE release 8, which might end in a new LTE release R9. In parallel, there are strong global activities for the definition of the successor of IMT-2000, the so-called IMT-advanced systems [2], where it is likely that some form of cooperation will be included. Accordingly, research on cooperative antenna systems has become one of the hottest topics; it will form the input of ongoing standardization activities and has led already to proposals for the LTE evolution [14].

In IEEE 802.16j [15]—the relaying task group defining mobile multihop relay (MMR) solutions for WiMAX (worldwide interoperability for microwave access) systems—cooperative relaying has been adopted as one type of relaying. Although theoretical and simulation results are very promising, at the same time it is clear that full cooperation in a cellular radio system is not feasible because this would require accurate channel knowledge from all eNBs to all UEs, processing of very large channel matrices at a central processing unit (CPU), and the exchange of tremendous amounts of data between all eNBs involved in a cooperative network with extremely short delays (in the order of a few milliseconds).

In LTE release 8, the final decision about DL MU-MIMO has been shifted from LTE release 8 to future releases due to a limited time schedule. MU-MIMO is tightly related to cooperative systems, and it is widely anticipated as a precursor of cooperation. Note that relevant LTE specifications for open- and closed-loop MIMO can be found in reference 16. Very simple MU-MIMO schemes had been evaluated for release 8, where only one single PMI (precoding matrix indicator) for the full frequency band is fed back per UE to the eNB, resulting in a simpler implementation with very small feedback overhead as well as in very small performance gains due to inaccurate adaptation to frequency-selective radio channels. In contrast, results under the assumption of accurate channel knowledge have led to significantly higher MU-MIMO performance gains [17]. In this way, for the first time, it has been demonstrated that the spectral efficiency scales linearly with the overall number of antennas in the interference-limited network also. Under practical system considerations, a compromise has to be found between full network-wide cooperation and LTE release 8 SU-MIMO precoding based on restricted size LTE codebooks, with the same precoding matrix applied to the full bandwidth.

Throughout this chapter, different COOPA concepts will be introduced, such as intra- and inter-eNB cooperation, distributed antenna systems (DASs) based on remote radio heads (RRHs), or as self-organizing networks [18–20], addressing different aspects of the previously mentioned issues. Cooperative relaying seems to be especially promising because it combines multihop coverage gains with MIMO capacity gains. As a further application of cooperative relaying, the integration of backbone functionality into the air interface will be analyzed. Rising backhaul traffic is one of the main challenges for MNOs for the introduction of LTE with its increased data traffic compared to GSM and even high-speed downlink packet access (HSDPA). For that purpose, adjacent NBs are directly interconnected over the air in the same licensed spectrum and served in this way from a central eNB. The central eNB is the only one that has a fixed broadband backhaul connection, reducing the overall infrastructure effort.

The combination of cooperative antenna systems with different other techniques, such as antenna tilting, interference rejection combining (IRC), or interference (IF) coordination, is very promising because it allows restricting cooperation to a few cells in the order of three to five. The small size of these interference-free zones—denoted as CAs—limits the processing as well as the feedback effort to these zones. The combination with the previously mentioned techniques guarantees maximum benefit in spite of this reduced effort. By damping the general interference floor stemming from far-off eNBs, canceling of a limited number of residual strongest interferers within the CAs boosts overall performance, thereby justifying the relative high overhead for organizing the CAs. Interference between CAs should be handled and controlled by other means, such as resource coordination.

This leaves the task to organize a CA comprising only two transmitting and two receiving radio stations in the simplest case. For example, two NBs at one site may cooperate to overcome intersector interference between two UEs. The possible implementation of this fundamental building block of any cooperative system within the framework of LTE and its implications will be the main topic of this chapter. The chapter is organized into five sections. Section 7.2 provides a basic analysis of cooperative antenna systems, Section 7.3 classifies the basic types of cooperation systems, Section 7.4 is concerned with implementation issues, and Section 7.5 concludes the chapter.

Note that basic concepts of LTE, such as resource blocks, codebook-based closed-loop MIMO precoding with unitary precoding matrices, and single-stream and dual-stream transmission will not be explained in detail in this chapter. The interested reader is referred to the previous chapters.

7.2 Basic Analysis of Cooperative Antenna Systems

This section will first give main rationales for cooperative antenna systems. A theoretical analysis will be given in the second part based on interference functions.

7.2.1 The Benefits of Cooperation

Figure 7.1 provides main motivations for the application of cooperative transmission techniques for outdoor cellular mobile radio systems like LTE:

- *Reuse of sites and HW.* For incumbent mobile network operators, which already have a large installed base, it is important to reuse existing sites as well as installed HW as far and as effectively as possible. However, higher data rates, together with higher RF, will typically shrink possible cell sizes.

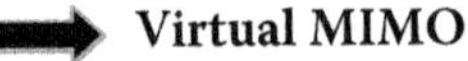

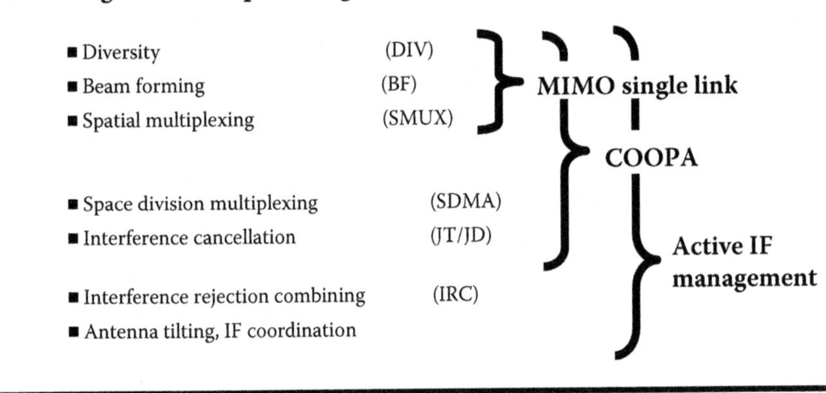

Figure 7.1 Rationale of COOPA systems.

Cooperation between the antenna ports of different sites and sectors will enhance the existing cell sites so that the original site structures might be reusable. Virtual MU-MIMO systems—sometimes also called spatial division multiple access (SDMA) [19]—allow for MIMO gains without requiring eNBs or UEs with large, costly, and space- and power-consuming antenna arrays as well as RF front ends (RF-FEs). Note that a significant part of HW costs of an eNB is contributed by the RF part, while, from a business perspective, a doubling of RF chains does not translate into a doubling of spectral efficiency as can be seen from SL simulations for LTE release 8.

- *Low SINR, low rank.* For indoor single-link MIMO systems, typically the received power will be quite large. Rich scattering environments ensure high-rank radio channels, allowing for simple exploitation of the spatial dimension by transmission of multiple data streams in parallel [21] over different spatial layers. In certain outdoor scenarios with strong line of sight (LOS), however, the radio channel from a UE to a single site might be rank deficient, thus limiting the achievable spatial multiplexing gains. More severe are frequent coverage holes due to strong shadowing or penetration losses for indoor users served by an outdoor eNB. In such holes, the signal-to-noise ratios (SNRs) are small. Because LTE is a cellular radio system, interference from adjacent radio cells is even more detrimental. From multicell simulations, it is known that about 40% of the UEs will achieve wideband signal-to-interference-and-noise ratios (SINR) below 0 dB. It is well known from projects like the EU-founded project WINNER [22] that significant spatial multiplexing gains require—depending on the scenario—SINRs in the order of 5–10 dB or more. SINRs larger than 5 (10) dB will be achieved for only 30 (15)% of best-case UEs. With frequency-flat single-user transmission, as in HSDPA, the major MIMO gains will stem from diversity realized by optimum combining or beam forming.

Advanced proposals for frequency-selective MU-MIMO transmission, which go beyond the scope of LTE release 8, cope better with this situation by exploiting frequency-selective multiuser diversity. For this purpose, each UE resource block (RB) with SINRs above average is scheduled with its best-fitting MIMO mode. In combination with codebook-based LTE like beam forming, spatial multiplexing to one or multiple users (SDMA) will be possible on more than 90% of the RBs, even though the SINR of many UEs is below 0 dB due to multicell interference [17]. As will be shown in more detail later, cooperation now turns detrimental interference power into useful signal energy, thus providing macrodiversity and beam-forming gains. Moreover, the channel rank is increased (often called rank enhancement) because radio channels from multiple sites are less correlated than that from a single site, leading to SINRs beyond those of isolated radio cells [18]. Beam forming and diversity gains, in combination with rank enhancement and interference cancellation, are the reason for the huge performance gains, promised by cooperative antenna systems from theory [6] as well as in many SL simulations.

The lower part of Figure 7.1 provides a differentiation of single-link MIMO, COOPA systems, and active interference management as seen by our group:

- *MIMO.* Conventional open- and closed-loop MIMO systems provide diversity (DIV), beam-forming (BF), and spatial multiplexing (SMUX) gain.
- *COOPA.* MU-MIMO such as SDMA (sometimes called virtual MIMO because, for example, several UEs with one AP each form a virtual antenna array) is a first step into the area of cooperation (see earlier discussion). The main focus of COOPA systems is on distributed MU-MIMO systems. For COOPA, several eNBs form so-called CAs of limited size with the goal to cancel interference within the CA completely—called zero forcing (ZF)—or to a certain level—called minimum mean square error (MMSE)—based on algorithms like joint transmission or detection [23].

- Active interference management. Cooperation areas require large overhead in terms of CSI estimation and feedback of CSI from UEs to eNBs over uplink (UL) radio channels. Moreover, they generate significant backbone traffic for accurate and timely exchange of data samples and CSI information between the involved NBs. Therefore, the size of cooperation areas will have to be restricted to very few cells and be supplemented by other techniques. In the case of OFDMA, a coordination of radio resources in adjacent cells might avoid collision of worst-case users at the cell edge or antenna tilting reduces far off interferers. Although antenna tilting is a well- known, proven, and often-applied technique for cellular radio systems (e.g., WCDMA), it is even more important and more effective in combination with COOPA systems [18]. For conventional cellular systems, antenna tilting requires a difficult trade-off between IF reduction into adjacent radio cells and reduced signal power for UEs of its own radio cell at the cell edge. As CAs form IF-free areas comprising several radio cells, antenna tilting can concentrate on reducing second- instead of first-tier IF.

  The effect on the overall interference floor can be seen in Figure 7.2, where antenna tilting increases the power ratios between the strongest interferers of the interference floor. As a result, the performance gain due to partial IF cancellation (i.e., a limited number of strongest interferers) is significantly increased. A further, very effective method in this context is

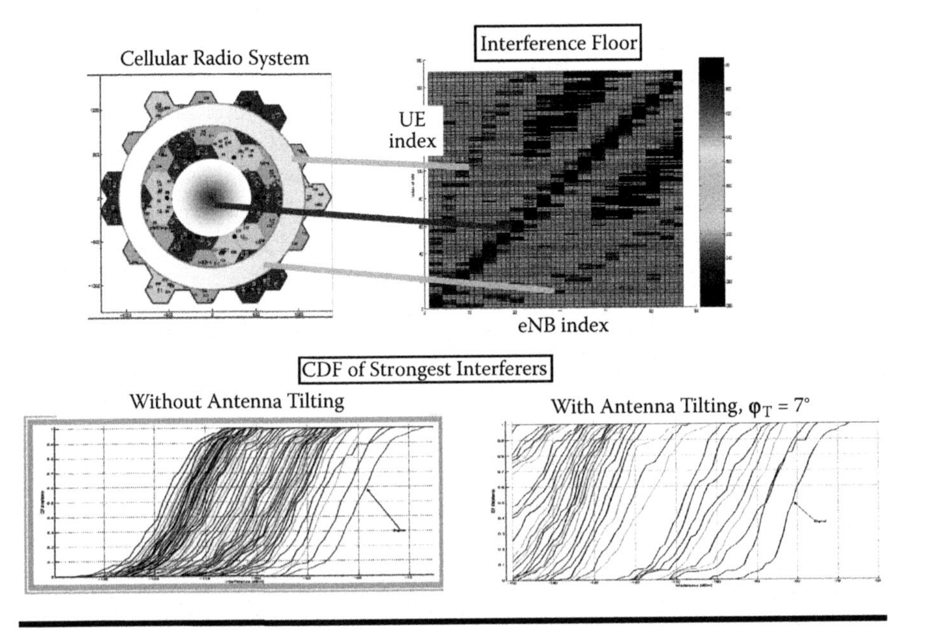

Figure 7.2 Interference floor with and without antenna tilting and simulation results for active interference management. At the top, the comparison of UE and eNB indices reveals that strongest IF comes from adjacent cells or second-tier eNBs, where sectors are directed to the center cell.

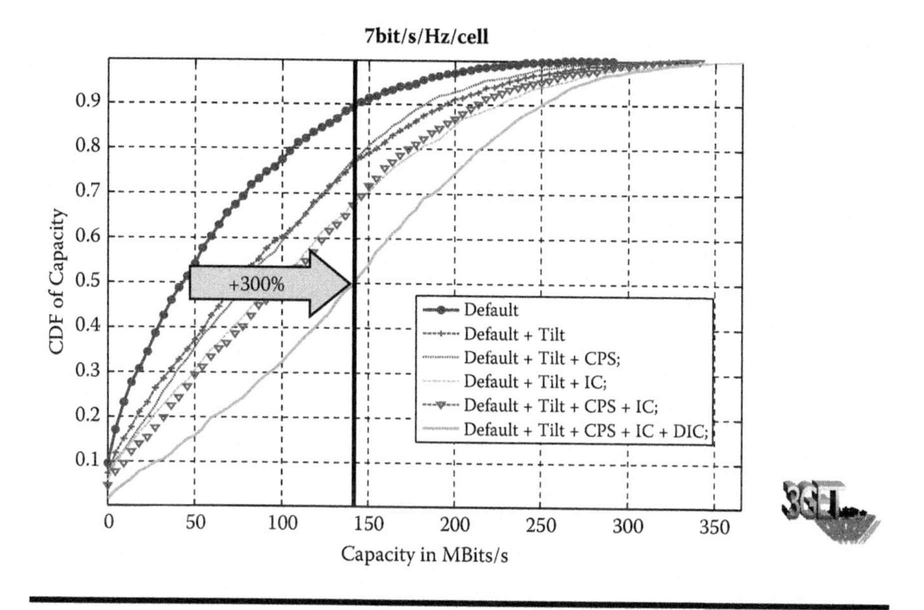

Figure 7.3 Gain for active IF management. A CA consists of two sectors and two adjacent NBs. Default = pure SISO; Tilt = antenna tilting; CPS = coordination; IC = IRC; DIC = dynamic IRC of worst-case interferer.

> advanced IRC at the UEs. When a UE is mobile, the eNBs leading to the strongest interference change quite often [24]. Typical large-scale parameter (LSP) correlation is in the order of 50 m, which means that, as soon as a UE has moved by about 50 m, another eNB might become the strongest interferer. IRC delays the need to reorganize the CA in such cases because the UE can adapt to different local interference conditions. Zirwas et al. [18] achieved overall performance gains of several hundred percent in certain environments by combining all techniques properly, as can be seen from the simulation results in Figure 7.3.

The main and most important message of Figure 7.1 is that the newly introduced spatial dimension for LTE systems may be used for conventional MIMO techniques like DIV, BF, or SM; however, in the context of cellular radio systems with strong intercell interference (ICI), often suppression of a few of the strongest interferers (e.g., by IRC or by means of cooperation) will be far more beneficial. In this context, accurate closed-loop MIMO precoding will save valuable spatial degrees of freedom (SDF) for IF suppression of other cell interference (OCI) based on IRC at the UE side.

With an active overall interference management scheme in mind, cooperation becomes much more effective because the IF management shapes the IF floor so that only one interferer or a few of the strongest interferers remain for local IF cancellation. Thus, the size of the CAs can be quite small—in the order of a few cells.

7.2.2 Theoretical Background and Motivation for Cooperative Strategies in Wireless Networks

The single-user point-to-point link is relatively well understood, both in theory and in practice. With modern signal processing and coding, a large portion of the capacity can be achieved [44,52,54]. However, these results and techniques cannot always be transferred to communication networks, which possibly have several source–destination pairs and simultaneous transmission over multiple network nodes. There are various ways in which nodes can interact and cooperate. Due to the high dimensionality of the problem, it is typically not possible to arrive at closed-form expressions or even to handle the problem analytically. There are many open questions, even for relatively simple channels.

In this section we will discuss some principles for multiuser communication in a network. One focus will be the modeling and analysis of interference between users. Without interference, one could apply theoretical results from wireline communications, where the network is often modeled as a collection of point-to-point links. However, neglecting the impact of interference can lead to significant performance losses. In general, the performance of one link can depend on the chosen strategy of other links. All users are potentially coupled by interference. Therefore, a joint optimization approach involving cooperation between users is generally favorable, although it also complicates the analysis of wireless networks. In the following, some basic strategies for cooperative network communication, which are far from comprehensive, will be discussed.

Relaying. The distance from source to destination can be bridged by using a relay. Compared to a single link, less transmission power is needed per transmitter. This advantage can be exploited in different ways. For given power limits, relaying increases the coverage of the system. For a given coverage area, relaying reduces the peak power and therefore the interference radiated to other users.

A possible relaying strategy is "amplify and forward," where the received signal is amplified and retransmitted. This strategy is simple but has the disadvantage that noise is amplified. This is avoided by decoding the information before retransmission. This is known as "decode and forward."

In Figure 7.4, a simple relaying channel is depicted, where the relay assists the direct link. Even for this basic channel, the information-theoretical capacity is not fully known (e.g., Cover and Gamal [43]).

Network coding. The core notion of network coding is to allow mixing of data at intermediate network nodes. A receiver sees these data packets and deduces from them the messages that were originally intended for that data sink [39,45,56,57].

Interference coordination and filtering. If certain knowledge about the occurrence of interference is available, then appropriate countermeasures can be taken (e.g., interference filtering by signal processing or interference avoidance by proper resource allocation; see Figure 7.5). Coordination strategies can be centralized or decentralized. For example, a central control unit can assign orthogonal resources such that interference is avoided. But the nodes can also act independently. In this

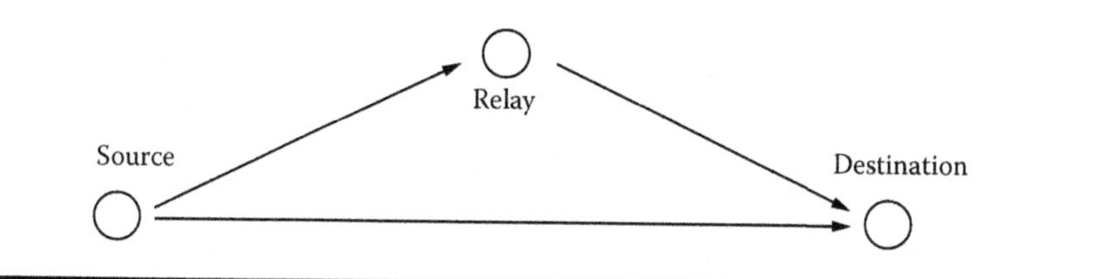

Figure 7.4 A basic relay channel; the relay assists the direct link.

case, they need to observe the interference from neighboring nodes. Strategies based on observation are often categorized as "cognitive radio." One can further distinguish strategies according to the degree of channel knowledge. Sometimes, efficient coordination is possible based on medium-term statistical channel knowledge.

Game theory can be used for analyzing such networks. The users can be regarded as players competing for the available resources.

Coherent base station cooperation. Information can be distributed to several network nodes, from which it is simultaneously transmitted to the destination. Conversely, a received signal can be reconstructed by combining the outputs of multiple distributed receivers. If this is done coherently, then the network terminals form a distributed "antenna array." This is sometimes referred to as "joint transmission" or "network MIMO" (see Figure 7.6).

Cooperation can be demanding in terms of infrastructure and synchronization, but it is nevertheless interesting because it provides a benchmark for achievable performance. The cooperation (combining) gain can be significant (see, for example, Section 7.2.2.4). Partial cooperation with reasonable implementation and overhead

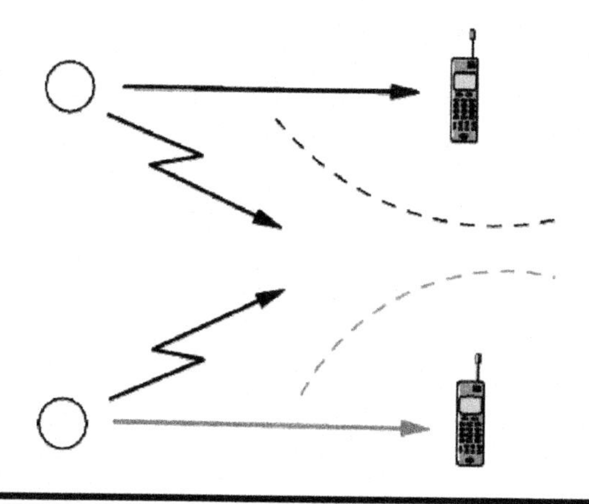

Figure 7.5 Interference can be avoided or reduced by *interference coordination*, either by properly allocating the resources or by interference filtering.

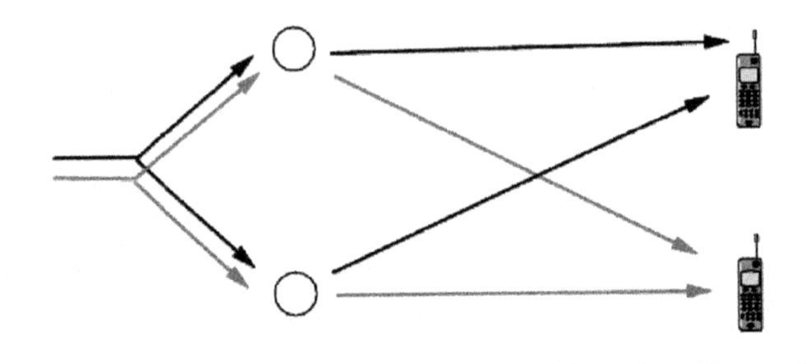

Figure 7.6 Combining signals are transmitted over multiple nodes simultaneously. This provides additional diversity and SNR improvement.

complexity can be a viable technique (e.g., cooperation between sectors; see, for example, Zirwas et al. [18]). The first advantage of cooperation is the improvement of the SNR, especially in areas between the transmitters (the "cell edge"). This can be viewed as a form of coherent macrodiversity. Another advantage is the possibility of effectively reducing interference. In this context, this strategy is sometimes referred to as "network MIMO" because data streams can be spatially multiplexed analogously to a classical point-to-point MIMO system.

7.2.2.1 Axiomatic Approach to Interference Modeling

Future wireless networks are expected to provide high-rate services for densely populated user environments. The application content is moving from audio to high-speed data transfer. The number of communication links is increasing. Therefore, interference is becoming an increasingly limiting factor. Interference puts a limit on how many users per cell can be served at a certain data rate. It therefore makes sense to ask for theoretical performance bounds, the region of achievable performances, and optimal transmission strategies. These results can serve as guidelines for the development of algorithms and implementations.

A general framework for interference modeling was introduced by Yates in Yeung et al. [55], with extensions in references 46, 49, and 51. It was proposed to model interference as a function of the transmit powers. Consider K users, $\mathcal{K} = \{1,2,\ldots, K\}$, whose transmit powers $p_1,\ldots, p_K$ are collected in a vector

$$\mathbf{p} = [p_1,\ldots, p_K]^T. \tag{7.1}$$

The resulting interference at user k is given by $I_k(\mathbf{p})$. The set of all possible power vectors is denoted by $\mathcal{P}$.

The behavior of the system depends on *how* the interference functions $I_1(\mathbf{p}),\ldots, I_K(\mathbf{p})$ depend on the available transmission powers. This dependency can be complicated and nonlinear because the interference can adapt to the current power allocation. Therefore, instead of developing a model for each particular system layout, we can focus on some core properties, summarized by the following framework of axioms.

Definition 7.1

We say that $I:\mathcal{P}\mapsto\mathbb{R}_+$ is an *interference function* if the following axioms are fulfilled:

$$\textbf{A1}\text{ (non-negativity) } I(\mathbf{p})\geq 0$$

$$\textbf{A2}\text{ (scale invariance) } I(\alpha\mathbf{p})=\alpha I(\mathbf{p})\quad\forall\alpha\in\mathbb{R}_{++}$$

$$\textbf{A3}\text{ (monotonicity) } I(\mathbf{p})\geq I(\mathbf{p}')\quad\text{if } \mathbf{p}\geq\mathbf{p}'$$

This definition differs slightly from Yates's original definition [55]. The difference is that the model [18] was developed for a particular power control problem, whereas the preceding model can be used in a more flexible and versatile way. We use the term "interference" in a very broad sense that is not limited to physical interference power [51]. This will become clear with following examples.

Example 7.1

In power control theory (e.g., references [49] and [59]), it is customary to model the interference coupling between users by a $K\times K$ *link gain matrix*

$$\mathbf{V}=[\mathbf{v}_1,\ldots,\mathbf{v}_K]^T. \tag{7.2}$$

The vector $\mathbf{v}_k\in\mathbb{R}_+^K$ contains the interference coupling coefficients of the kth user. Introducing an extended power vector

$$\mathbf{p}=[\mathrm{p}_1,\ldots,\mathrm{p}_K,\sigma^2]^{\mathrm{T}}, \tag{7.3}$$

where σ^2 is the noise variance and the interference plus noise experienced by this user is

$$I_k(\underline{\mathbf{p}})=\mathbf{p}^T\mathbf{v}_k+\sigma^2. \tag{7.4}$$

Example 7.2

Consider K users with signal-to-interference ratios

$$SIR_k(\mathbf{p})=\frac{p_k}{I_k(\mathbf{p})},\quad\forall k \tag{7.5}$$

where $I_1,\ldots,I_K$ are arbitrary interference functions. A vector

$$\gamma=[\gamma_1,\ldots,\gamma_K]^T>0 \tag{7.6}$$

is said to be feasible if and only if [52]

$$C(\gamma)=\inf_{\mathbf{p}\in\mathcal{P}}\left(\max_{1\leq k\leq K}\frac{\gamma_k}{SIR_k(\mathbf{p})}\right)\leq 1. \tag{7.7}$$

> Here, $C(\gamma)$ is a single measure for the joint performance of the K-user channel. The function $C(\gamma)$ is an interference function. The properties A1–A3 of the underlying interference functions $I_1, \ldots, I_K$ are preserved by the min–max optimization.

These two examples show that the theory of interference functions can be exploited in quite different ways and is not limited to power allocation problems. The advantage of the axiomatic approach is its generality. It provides a simple framework for optimizing interference-coupled multiuser systems. Such simplified modeling helps one to study the system in an analytical way. Analyzing the core properties of a system facilitates a rigorous framework, which is of theoretical value, but it also provides useful guidelines for the development of algorithms. Some examples will be discussed in the next section.

7.2.2.2 Resource Allocation Algorithms for Interference-Coupled Multiuser Systems

In the previous section we discussed interference models. Now it will be shown how these models can be used for the optimization of interference-coupled networks.

Unlike single-user systems, the optimization of coupled multiuser systems typically involves a trade-off between the users' performances. Increasing one user's performance comes at the cost of other users. Increasing performance often means increasing interference to other users or occupying shared resources. Therefore, there is no such thing as "the" optimal strategy. The systems performance is characterized by a region, the so-called *feasible set,* which is the set of all possible user performances. This is illustrated in Figure 7.7.

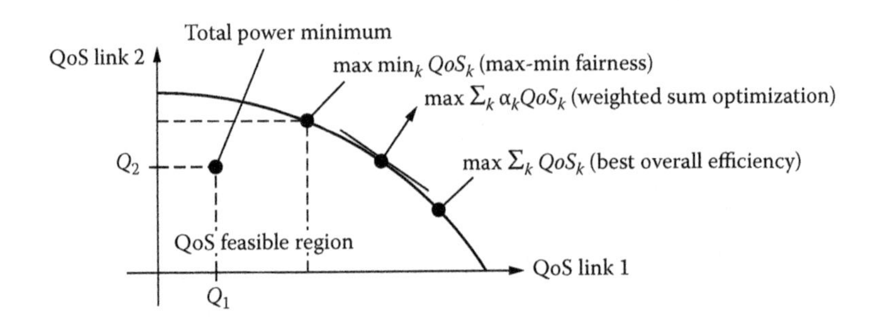

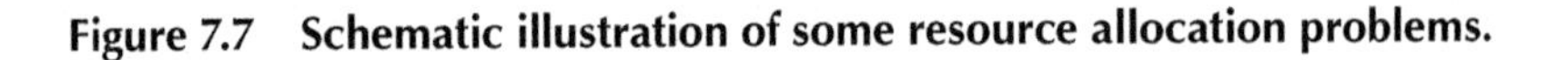

Figure 7.7 Schematic illustration of some resource allocation problems.

Assume that the actual performance measure (the quality of service [QoS]) is a continuous, strictly monotonic function ϕ of the signal-to-interference ratio (SIR)—that is,

$$QoS_k(\mathbf{p}) = \phi(SIR_k(\mathbf{p})), \quad 1 \le k \le K. \tag{7.8}$$

Some examples are BER: $\phi(x) = Q(\sqrt{x})$; MMSE: $\phi(x) = 1/(1 + x)$; BER-slope for α-fold diversity: $\phi(x) = x^{-\alpha}$; or capacity: $\phi(x) = \log(1 + x)$.

Let γ be the inverse function of ϕ; then

$$\gamma_k = \gamma(Q_k), \quad 1 \le k \le K, \tag{7.9}$$

is the minimum SIR level needed by the kth user to satisfy the QoS target Q_k. Thus, the problem of achieving certain QoS requirements carries over to the problem of achieving SIR targets $\gamma_k > 0$, $\forall k$.

The problem of achieving SIR targets $\gamma_1,\ldots,\gamma_K$ with minimum total transmission power was addressed by Yates in Yeung et al. [55]. The problem can be written as

$$\min_{\mathbf{p}\in\mathcal{P}} \sum_k p_k \quad s.t. \frac{p_k}{I_k(\mathbf{p})} \ge \gamma_k, \quad \forall k \in \mathcal{K}, \tag{7.10}$$

where $\mathcal{P} \subset \mathbb{R}_+^K$ is a compact set of possible power vectors. The set $\mathcal{P}$ depends on the power constraints of the system under consideration. Typically, each user has a maximum power limit $p_k^{\max}$ that must not be exceeded (i.e., $\mathcal{P} = \{\mathbf{p} : 0 \le p_k \le p_k^{\max}, \forall k\}$). However, other types of power constraints are possible (e.g., a sum-power constraint, which is typical for broadcast transmission).

Note that the problem (7.10) can be *infeasible* if the targets $\gamma_1,\ldots,\gamma_K$ cannot be achieved. We need to assume feasibility. Also, the interference function $I(\mathbf{p})$ must be strictly monotonic, with respect to the noise component, in addition to the properties A1–A3. Consider the general power set

$$\mathcal{P} = \{\mathbf{p} > 0 : p_{K+1} = \sigma^2\}. \tag{7.11}$$

The presence of a constant noise component is important to ensure that the infimum (7.10) is actually attained. In this case, it follows from Yates [55] (see also the alternative proof in Schubert/Boche [51]) that problem (7.10) is solved by the following fixed-point iteration:

$$p_k^{(n+1)} = \gamma_k I_k(\mathbf{p}(n)) \quad \forall k \in \mathcal{K}. \tag{7.12}$$

In each iteration, the users update their transmission powers depending on the observed interference power. This iteration converges component-wise to the unique global optimum of problem (7.10), regardless of the chosen initialization $\mathbf{p}^{(0)} \in \mathcal{P}$.

This interference framework is quite basic, and it holds (sometimes in approximation) for many interference scenarios. Despite its simplicity, a globally convergent algorithm (7.12), with a nice analytical structure and a known convergence behavior, exists [40,46].

7.2.2.3 Exploiting Convexity

The fixed-point iteration (7.12) has the advantage of being applicable to general interference functions. However, a better convergence behavior can be expected when convexity or concavity can be exploited. As an example, consider the interference function

$$I_k(\underline{\mathbf{p}}) = \min_{\mathbf{v}_k \in V_k} \mathbf{p}^T \mathbf{v}_k + \sigma^2. \tag{7.13}$$

$\mathbf{v}_k$ is a vector of interference coupling coefficients, chosen from a set $V_k \subset \mathbb{R}_+^K$. For each $\mathbf{p}$, the vector $\mathbf{v}_k$ is adjusted so that interference is minimized. This is a typical task of a receiver; therefore, we refer to $\mathbf{v}_k$ as the *receive strategy*.

Note that the function in Equation (7.13) fulfills A1–A3, so it is an interference function. It is also concave because the point-wise minimum over linear functions is concave. Boche and Schubert [40] showed that the power minimization problem (7.10) can be solved with superlinear convergence, whereas the fixed-point iteration (7.12) has only linear convergence [40,46]. This improved convergence speed is facilitated by exploiting the concavity of the interference function (7.13).

Boche and Schubert [41] also showed that every concave interference function can be expressed in the form of Equation (7.13); thus, the power minimization problem (7.10) can be solved with superlinear convergence. Similar results can be shown for convex interference functions. Superlinear convergence can be achieved for any convex or concave interference function.

Another interesting operating point is obtained by the principle of *proportional fairness*. The motivation for proportional fair resource allocation [47] is the observed utility inefficiency of min–max fairness. A min–max fair scheduler allocates the resources in such a way that all communication links achieve the same QoS. This strategy does not perform well in the presence of bottleneck links. If one link is very weak, then it may require all of the available resources in order to achieve an acceptable QoS. Min–max fairness is therefore known for its bad overall system efficiency.

Proportional fairness avoids this effect by putting more emphasis on the links with good channel conditions. This was first studied in the context of wireline networks [47]. Later, it was also successfully applied in the wireless context (see, for example, Viswanth et al. [54]). Proportional fairness is especially useful for elastic traffic because it favors good channel states and avoids the bad ones.

In Schubert/Boche [51], this problem was studied within the framework of log-convex interference functions. A function is said to be log-convex if the logarithm of the function is convex. An interference function $I(\mathbf{p})$ is said to be a *log-convex interference function* if A1–A3 are fulfilled and if $I(\exp(\mathbf{s}))$ is convex on $\mathbb{R}^K$ after a change of variable $\mathbf{p} = \exp(\mathbf{s})$ (component-wise). For this special class of interference function, efficient algorithms can be derived for certain performance measures [51].

Log convexity can be regarded as a generalization of convexity. It was shown in Schubert/Boche [51] that every convex interference function is a log-convex interference function. However, the converse is generally not true. This kind of convexity is sometimes referred to as "hidden convexity" because it facilitates efficient solutions for problems that are otherwise nonconvex [42].

7.2.2.4 Simulated Cooperation Gains and Benchmarks

Consider the interference function (7.13). We will now discuss how the theoretic framework of interference functions can be used for the optimization of a cooperative network, as illustrated in Figure 7.8.

Consider K single-antenna transmitters and M distributed receiving antennas, which can (partly) cooperate. Vector $\mathbf{h}_k$ contains the complex channel gains from user k to the M distributed antennas. Independent signals $s_1,\ldots, s_K$ are transmitted over vector-valued channels $\mathbf{h}_1,\ldots, \mathbf{h}_K \in \mathbb{C}^M$, with spatial covariance matrices

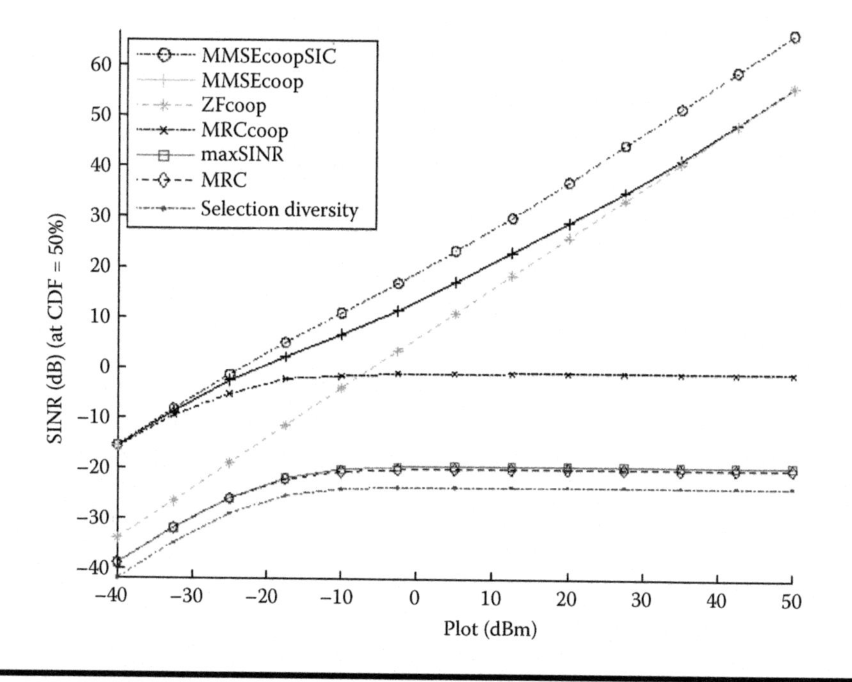

Figure 7.8 SINR versus transmission power in an interference-saturated multicellular system (reuse 1) with base station cooperation and SDMA.

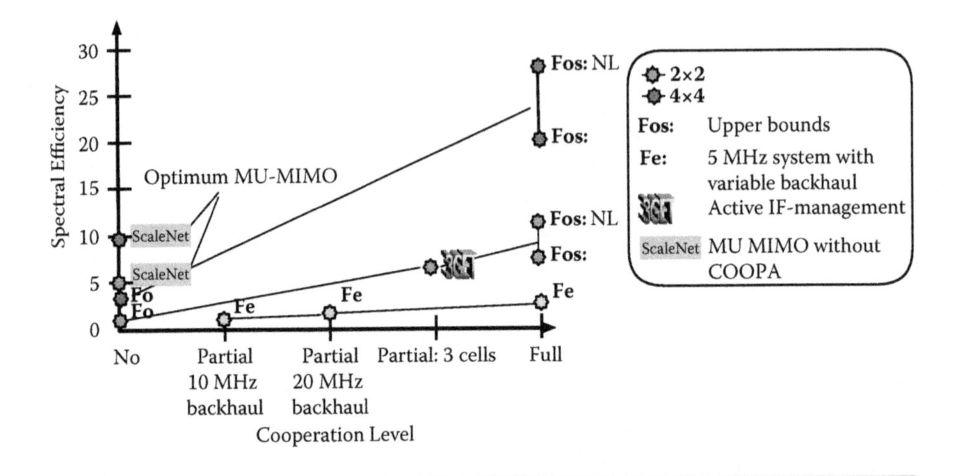

Figure 7.9 Potential gains for different degrees of cooperation. Fos: theoretical upper bounds [6]; FE: 5-MHz system with varying inter-NB backhaul traffic [7]; 3GET: active IF management with CAs of size 3 [38]; ScaleNet: optimum MU-MIMO without cooperation, but full CSI knowledge [17].

$\mathbf{R}_k = E[\mathbf{h}_k\mathbf{h}_k^H]$. The antenna outputs are combined by linear filters $\mathbf{u}_1,\ldots,\mathbf{u}_K$ (the distributed "beam formers"). The output of the kth beam former is

$$y_k = \mathbf{u}_k^H\left(\sum_{l=1}^{K}\mathbf{h}_l s_l + \mathbf{n}\right), \tag{7.14}$$

where $\mathbf{n} \in \mathbb{C}^M$ is an additive white Gaussian noise (AWGN) vector, with $E[\mathbf{n}\mathbf{n}^H] = \sigma^2\mathbf{I}$. The interference power coupling coefficients of the kth user are

$$[\mathbf{v}_k(\mathbf{u}_k)]_l = \begin{cases} \dfrac{\mathbf{u}_k^H\mathbf{R}_l\mathbf{u}_k}{\mathbf{u}_k^H\mathbf{R}_k\mathbf{u}_k} & 1 \le l \le K, l \ne k \\ \dfrac{\sigma^2\,\|\mathbf{u}_k\|^2}{\mathbf{u}_k^H\mathbf{R}_k\mathbf{u}_k} & l = K+1 \\ 0 & l = k \end{cases} \tag{7.15}$$

Assuming a normalization $\|\mathbf{u}_k\|^2 = 1$, the interference function for the beamforming case is

$$\begin{aligned}\mathcal{I}_k(\underline{\mathbf{p}}) &= \left[\max_{\|\mathbf{u}_k\|_2=1}\frac{\mathbf{u}_k^H\mathbf{R}_k\mathbf{u}_k}{\mathbf{u}_k^H\left(\sum_{l\ne k}p_l\mathbf{R}_l + \sigma^2\mathbf{I}\right)\mathbf{u}_k}\right]^{-1} \\ &= \min_{\|\mathbf{u}_k\|_2=1}\underline{\mathbf{p}}^T\mathbf{v}_k(\mathbf{u}_k)\end{aligned} \tag{7.16}$$

It can be observed that the interference coupling is not constant. For any power allocation $\mathbf{p} > 0$, the beam former $\mathbf{u}_k$ adapts to the interference in such a way that the SINR is maximized. For deterministic channels $\mathbf{h}_1,\ldots,\mathbf{h}_K$, we have $\mathbf{R}_l = \mathbf{h}_l\mathbf{h}_l^H$, so the interference resulting from optimum beam formers is obtained in closed form [50]:

$$I_k(\underline{\mathbf{p}}) = \frac{1}{\mathbf{h}_k^H\left(\sigma^2\mathbf{I} + \sum_{l\neq k} p_l\mathbf{h}_l\mathbf{h}_l^H\right)^{-1}\mathbf{h}_k}. \tag{7.17}$$

This interference function is a special case of the generic interference function (7.13).

Using this framework, we compare different combining strategies in a multicell scenario. In Figure 7.8, the achievable SINR is plotted over the total transmission power in the system. By achievable SINR, we mean the value at which the CDF equals 0.5. The system consists of seven cell sites, with three sectors each and two antennas per sector. The cell radius is 300 m and the SCME (spatial channel model extended) from the 1st winner project. Each sector contains two randomly placed users with a single antenna each. Thus, we have a system with a total number of 42 transmit antennas and 42 users. The total transmit power is the sum of all the users' individual powers. No power allocation is used, so each user transmits at full power. Full frequency reuse is assumed, and all users share the same frequency band and time slot. The results are averaged over 500 random experiments.

The following space diversity techniques are compared:

- *SD (selection diversity).* At each base station, the antenna with the strongest path gain is chosen.
- *MRC (maximum ratio combining).* At each base station, MRC is performed with respect to the users' channels.
- *MMSE (minimum mean square error).* For each user, the associated base station performs antenna combining such that MMSE is minimized. This corresponds to maximizing the individual SINR. In contrast to the MRC strategy, the interferers' channels are assumed to be known.
- *MRCcoop.* Maximum ratio combining is performed jointly for all base stations. This means that all antennas are coherently cooperating.
- *ZFcoop (zeroforcing with full base station cooperation).* The difference from MRCcoop is the full knowledge of the interferers' channels. Thus, interference can be completely eliminated by transmitting orthogonally to the interferers' channels.
- *MMSEcoop.* All base station antennas are combined coherently, based on the MMSE criterion. For large SNRs, this is asymptotically equivalent to ZFcoop.
- *MMSEcoopSIC.* This is the same as MMSEcoop, but, in addition, nonlinear interference cancellation (uplink) or dirty paper coding (downlink) is performed.

The purpose of this simulation is a qualitative comparison between some cooperative and noncooperative transmission strategies. It demonstrates that the theoretical gain by BS cooperation can be huge, depending on the SNR range and the degree of cooperation. Although the basic simulation parameters were chosen according to the LTE specifications, the simulation is not meant to replace a full LTE system-level simulation, and important aspects, like scheduling or frequency diversity, are not included. Therefore, the resulting SINR values are rather low. We are mainly interested in the differences between the curves.

It can be observed that, if full cooperation is possible, the system is no longer interference limited. For the case of MMSEcoopSIC, this even holds for an arbitrarily large number of users. This has an interesting consequence because it means that no matter how many users there are and how large their requested SINR is, it can always be supported as long as the power constraints are not violated. The cooperative system (MFcoop, ZFcoop, MMSEcoop, MMSEcoopSIC) is not interference limited, whereas conventional noncooperative systems are interference limited (like SD, MRC, and MMSE). That is, the interference puts a hard limit on the achievable SINR. Increasing the transmit power does not help in this case because the interference is increased. Finally, note that MMSE and MRC perform almost the same. This is explained by the large number of interferers; thus, the resulting interference is similar to isotropic noise.

7.3 Classification of Cooperation Schemes

Cooperative antenna systems may be implemented in many different ways; the following is a short classification of the most important COOPA proposals that is given based on the schematics in Figure 7.10. All cooperation schemes have in common that they form a distributed MU-MIMO or virtual MIMO system, where antenna elements (AEs) from different radio stations transmit simultaneously to different UEs.

a. *Intra-NB or intersector cooperation* is the simplest form of cooperation in a cellular network because all cells are located at the same site, avoiding any backbone interconnections between NBs. One of the channel cards can easily perform the CU functionality, given that it has enough processing capability. Intra-NB cooperation combats intersector IF, which typically reduces overall performance by about 20%. At a first glance, 20% performance gain seems to be small, but one has to keep in mind that, in conventional cellular systems, very narrow sector antennas—for example, 60° beam width for 120° sectors—have to be used to minimize intersector IF. In the case of COOPA systems, we face a paradigm shift as IF is turned into useful signal energy. Therefore, with wider antenna patterns, greater overlap areas and larger performance gains will be possible. This opens the door for an interesting first system deployment of LTE, where, with only one AP per sector, UEs may be

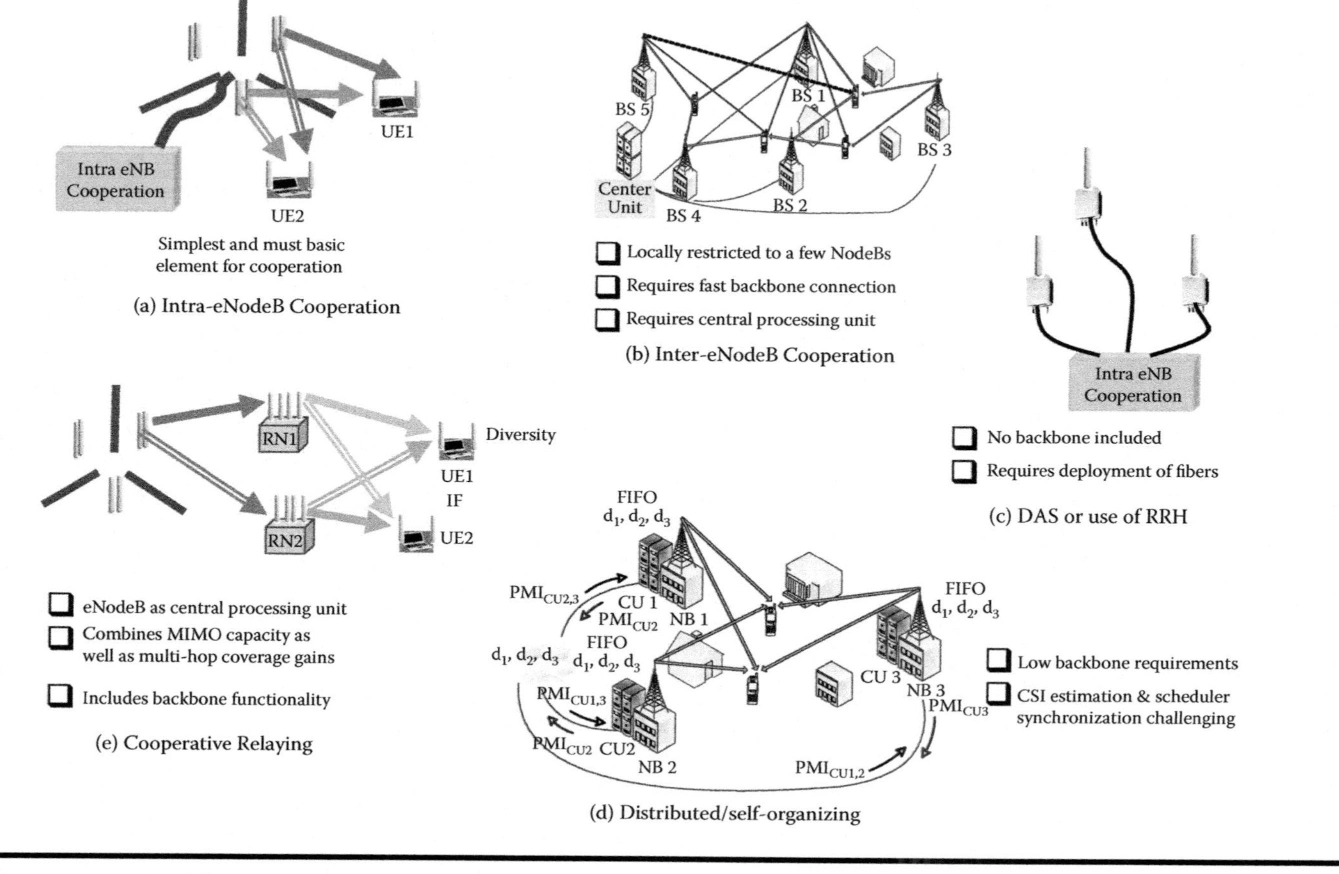

Figure 7.10 Main COOPA proposals.

served with LTE baseline 2 × 2 MIMO simply by combining the single APs of two adjacent sectors forming virtual MIMO arrays.

b. *Inter-eNB cooperation* is the conventional solution for building a CA between adjacent sites based, for example, on fast and broadband fiber interconnections among all involved eNBs. Typically, the CA will be limited to a few eNBs. Specifically for the UL, from time to time replacement of the eNBs forming the CA might be useful to adapt to varying IF constellations. For moving UEs, different eNBs might appear and disappear as strongest interferers due to varying shadowing conditions, with a typical large-scale parameter correlation of about 50 m in urban environments. Especially useful is intersite cooperation in countries like South Korea or China, where fiber deployment has reached a quite mature status. But one should keep in mind that, even if fibers are available, there might be significant leasing fees for using these fibers, generating a lot of unwanted operational expenditure (OPEX) for an MNO.

c. *Distributed antenna systems (DASs)* are based on RRHs that might be located at the edges of a building, at different locations in a shopping mall, or on different nearby buildings. The eNB with the CU serving the RRHs is located at one site, so the CAs are formed without any backbone network. The RRHs are typically connected to the baseband processing unit (BBU) over the open base station architecture initiative (OBSAI) [25] or common public radio interface (CPRI) [26], which carry data samples, resulting in very high data rates. For this reason, small distances will require fiber connections between BBU and RRH. Larger DAS deployments might face similar infrastructure challenges, as has been known for inter-eNB cooperation, if already existing fibers cannot be relied upon.

d. *Distributed/self-organizing (DSO) CAs* with or without inclusion of relay nodes (RNs) have been developed to overcome backbone restrictions and specifically to relax latency requirements for the X2 interface between cooperating eNBs [20]. The main idea is to do the same processing at each eNB in parallel at so-called mirror CUs. If each mirror CU selects its precoding matrix **W** for the same PMIs of all involved UEs and performs precoding for all data on the same RBs, the result on the air interface will be the same as for conventional inter-eNB cooperation. To enable distributed processing at mirror CUs, all UEs broadcast their PMIs to all involved eNBs. For precoding all involved eNBs, N_{NB} need all data d_1 to d_n of all supported UEs, N_{UE}, which can be easily fulfilled over the backbone network by IP multicast protocols [27]. Additionally, the medium access control (MAC) decisions at the eNBs have to be exchanged over the air on eNB–UE–eNB links. This is one of the challenges because over-the-air transmission is naturally less robust than a conventional X2 inter-eNB connection. At the same time, self-organizing networks can react very quickly, if properly implemented; by applying semistatic scheduling decisions as well as a suitable combination with data exchange over the X2-interface, robust implementations are possible. In addition to moderately higher data traffic on the S1-interface, fast data exchange in the order of a few milliseconds is the main advantage of DSOs over conventional inter-eNB cooperation because it allows faster adaptation to varying radio channel conditions.

e. *Cooperative relaying (CR),* in its most advanced form, not only provides diversity gains but also forms a CA including the AEs of one sector and one or more RNs. Similar to DSOs, the cooperation is organized without direct involvement of the backbone network, which is replaced by in-band relaying. The CU is located at the central eNB (requiring fast exchange of CSI and PMI over the relay links) or is replaced by two mirror CUs at the RNs, similarly to DSOs. A very interesting application of CR is the combination of cooperation with in-band relaying of backhaul traffic, which means that neighboring eNBs are served by one central serving eNB NB_S with fiber backbone connection, while the served eNBs NB_R have only a small data pipe to the backbone network. Thus, NB_Rs have relay functionality but "look" to UEs like conventional eNBs. This way, the advantage of reduced backbone infrastructure is combined with the benefit of IF cancellation within a CA, formed by one NB_S and one or two NB_Rs.

All proposals are different versions of a cooperative antenna system (in the following, called the COOPA system). Depending on side conditions and scenarios, one or another cooperation scheme out of solutions (a)–(e) might be selected. For those MNOs having easy access to fast low-latency fiber connections—as might be the case in South Korea or China—DAS or conventional inter-eNB cooperation is very promising. MNOs with low data rate digital subscriber line (DSL) connections per eNB will prefer CR, thus overcoming backbone limitations and improving overall system performance in one step.

In Figure 7.10, the potential gains of different system configurations are compared with respect to achievable spectral efficiencies, depending on the degree of cooperation—thus giving a rough idea about the potential of cooperative antenna systems. Different 2 × 2 as well as 4 × 4 MIMO system concepts with varying degrees of cooperation and with linear and NL precoding and different limitations for the backbone network are given. It is interesting that, for 2 × 2 MIMO and CAs with advanced interference management with spectral efficiencies of about 7 b/s, performance similar to that for full cooperation can be achieved.

7.3.1 Cooperation and the Backbone Network

One important issue in the context of CAs, which might be surprising at first glance, is the typically increased data traffic over the backbone network. One might assume that backbone traffic is of lower relevance compared to data transmission over the air, but increasing data traffic due to HSDPA or LTE is already a significant part of overall CAPEX and OPEX of a mobile radio system. T1/E1 lines with a data rate of a few megabits per second are quite common in current radio networks; required enhancements in this area are now a big challenge for MNOs. Generally, there are several options for providing broadband data connections to eNB sites over so-called S1 or between eNBs over so-called X2 interfaces as defined by 3GPP for LTE:

- Fibers are the most powerful technology and provide, in combination with GB Ethernet, almost 1 Gb/s or, in conjunction with so-called GPON systems, even several gigabits per second. A drawback is the typically high construction costs if fibers are not available.
- μ-wave links are commonly used for NB site connections. Today, several megabits per second are typical, but, given enough spectra, several 100 Mb/s would be feasible. There are mainly two possible drawbacks: increasing HW expenditures for higher data rate links and relatively large delays for μ-wave systems.
- An interesting alternative is optical free space feeder links, which might provide for distances well below the 1-km data rates of more than 1 Gb/s. Up to now, HW costs have been very high; however, given a larger usage of these systems, this might be changing in the future.
- For very high digital subscriber lines (VDSLs), digital spectrum management (DSM) level-3 techniques have been investigated to exploit existing wireline connections more effectively. Over short distances of a few hundred meters, several hundreds of megabits per second might be possible.
- Cooperative relaying combines cooperation gains with in-band wireless backbone functionality. This assumes optimum channel conditions for the first links from the serving NB_s to neighboring NB_Rs so that overall system performance loss due to in-band relaying is minimal. For this reason, LOS connections are assumed, which are further enhanced by directional antennas and MIMO techniques; in the simplest case, this means polarization multiplex. For LTE baseline, we can assume a spectral efficiency of roughly $R_{LTE} = 2$ b/s/Hz, compared to a peak data rate on the relay link of about $R_R = 8$ b/s/Hz. Therefore, by this simple calculation, a moderate relaying overhead of about 20% can be achieved because one fifth of the time is required for relaying of user data.

As far as possible, one should minimize the additional backbone data traffic for organization of the cooperation. Different approaches may help to achieve this goal. First, because data traffic will increase with increasing size of a CA, the number of participating NBs in a CA should be as small as possible. A combination with other techniques like antenna tilting, IF coordination, and advanced IRC helps to achieve high performance gains in spite of this smaller size of the CAs. Additionally, cooperation should be restricted to UEs with strong mutual interference and those that have large amounts of traffic to send. UEs with small data rates will result in a lot of overhead for organization of the cooperation while providing only result in performance gains. Here, voice over Internet protocol (VoIP) users should be mentioned specifically; they should be served conventionally without cooperation. Another helpful technique is advanced quantization schemes, which adapt channel quantization levels to the actual achievable cooperation gain. UEs with a modulation and coding scheme (MCS) support (e.g., 1 b/s) require less accurate precoding information than UEs with 6 b/s.

In the following, the required data rates on the S1, X2 interface and the over-the-air data rates are estimated for different implementations of the previously described cooperation concepts. The S1 interface provides the data connection to the backbone network, while the X2 interface has been defined for inter-eNB connections, which are essential for classical inter-eNB cooperation. As it is defined, X2 provides latencies of less than 20 ms and allows for fast tunneling of user data over GTPs (GPRS tunneling protocols; for more information, see 3GPP 36 series of specifications). The main targets are gigabit Ethernet lines, so data rates in the order of 1 Gb/s should be possible.

For comparison of the different COOPA concepts, some simplifying basic assumptions have been made:

- Each cooperation area consists of three eNBs.
- The two strongest out of three eNBs cooperate ⇒ (2 × 2, 2 × 2) MIMO.
- RF bandwidth is 20 MHz or 100 RBs.
- Mean data rate per eNB is $d_{NB} = 100$ Mb/s.
- Three sectors with two antenna elements each per site ⇒ data rate per site: $d_{site} = 300$ Mb/s.
- S1 = gigabit Ethernet preferred solution.
- X2 interface = <20 ms latency; GTP tunnel over gigabit Ethernet.
- Typical scheduling overhead is based on rough estimation (10–30% for one to three control symbols per subframe). For a more accurate estimation, see Chapter 8.

The following data rates will be used for comparison of the different proposals:

Data rate d_{OBSAI} is the data rate on the OBSAI interface between the baseband processing unit and RF–FE. The sampling rate is 20 MSps with four times oversampling and a quantization Q_{OBSAI} of 16–20 bit per data stream or per antenna ⇒ $d_{OBSAI} \approx 1.3$ Gb/s per antenna element.

CSI data rate d_{CSI} is the data rate for the feedback of CSI estimates from UEs to the CU. Similar signal and IF power for all scheduled UEs is assumed, allowing for a reduced CSI quantization. Note that, together with path loss information, this restriction can always be fulfilled. Full-band CSI for all RSs is assumed:

$\Rightarrow$ 200 RSs in four symbols per antenna and subframe (APs 1 and 2) and 200 RSs in two symbols for APs 3 and 4

subframe length of 1 ms

quantization $Q_{CSI} = 2–6…10$ bit (depends on MCS and intended SINR)

$\Rightarrow d_{CSI} \approx 200 * 4 * Q_{CSI}/1\text{ ms} = 1.6…8$ Mb/s per antenna

CB-based precoding data rate d_{CB} is the data rate for feeding back PMI values for precoding from UEs to the CU. A baseline 2 × 2 (two UEs and two eNBs) system leading to an overall 4 × 4 distributed MU-MIMO system is assumed:

each UE feeds back PMIs

full band, highest granularity $\Rightarrow$ 100 RBs per UE and subframe

quantization/CB size $Q_{CB} = \leq 5$ bit/10 bit (rank + PMI/increased SINR)

$\Rightarrow d_{CB} \approx 100 * Q_{CB}/1\text{ ms} = 0.5…1$ Mb/s per UE

Control data rate $d_{Control}$ is the data rate for exchange of scheduling information between different mirror CUs if distributed schedulers have to be synchronized:

10–30% control overhead (see Chapter 8)

$d_{Control} \approx 8…30$ Mb/s per cell (24…90 Mb/s per site)

CQI data rate d_{CQI} is the data rate for feeding back CQI information from UEs to the CU and informing the NBs about possible MCSs for different data streams. The assumption is a baseline 2 × 2 system so that, at maximum, two data streams can be supported. Adaptation and highest CQI feedback per subframe and for each RB:

$\Rightarrow$ 100 RBs per data stream and subframe

quantization $Q_{cqi} = 5$ bit (see LTE R8: 5-bit entry to CQI table)

$d_{CQI} \approx 100 * Q_{CQI}/1\text{ms} = 0.5$ Mb/s per data stream

QAM data rate d_{QAM} is the data rate in case of a subcarrier (SC)-wise quantization of the baseband Tx-signals. These signals have to be sent from the CU to each cooperating radio station:

full band $\Rightarrow$ 1,200 SCs for 14 symbols per antenna port and subframe (APs 1 and 2)

quantization $Q_{QAM} = 2–6…10$ bit (Q_{QAM} depends on MCS and intended SINR)

$\Rightarrow d_{QAM} \approx 1200 * 14 * Q_{QAM}/1\text{ ms} = 33.6…168$ Mb/s per antenna

In Figure 7.11a–f, six different network and COOPA concepts are depicted. For each solution, the data traffic on S1, X2 and the traffic that can be sent wirelessly (WL) have been calculated based on the preceding data rates for d_{OBSAI}, d_{CSI}, d_{CB}, $d_{Control}$, d_{CQI}, and d_{QAM}. For each solution, data and control traffic (e.g., for CSI information) are given. For some scenarios, upper bounds and achievable data rates, which can be expected after proper optimization, are given as $X2_{min}$- or WL_{opt}-values. In this case, lower and upper values for quantization levels have been used. The data rate of 300 Mb/s per site or 100 Mb/s per sector is a rough estimate for a baseline LTE implementation with 2 AEs per sector. In DL, peak data rates of almost 170 Mb/s per sector would be possible, but the averaging of cell center and cell edge users will reduce the mean data rate per cell significantly; 100 Mb/s is just a rough estimate allowing for easy calculation.

In Figure 7.11a, a conventional network as benchmark for data traffic on S1, X2 and WL links is given. In this simple case, each site with three eNBs has to transmit 300 Mb/s over the S1 interface and no traffic over X2 or WL.

Figure 7.11b assumes availability of fiber connections, over which RRHs are attached to the CU. In this case, it might even be possible to support the very high data rates of more than 10 Gb/s per site, which result from the OBSAI interface for several AEs.

Figure 7.11c is a solution where the CU receives all data for all eNBs over the S1 interface. Over the X2 interface, the CU receives CSI information from all eNBs, calculates proper precoding matrices, performs the precoding, and quantizes the resulting Tx (transmit) signals before these are transmitted to the neighboring sites. With optimum quantization and selection of cooperating UEs, the data rate over the X2 interface might be reduced significantly compared to solution 7.11b. In the case of 7.11b, RF signals have to be transmitted to the RRHs, but here just the quantized baseband signals are used.

In Figure 7.11d, the benefits of mirror CUs are being evaluated. This concept reduces data traffic for the X2 significantly. All data for all UEs are multicasted to all sites so that overall data traffic on S1 interfaces is tripled.

In Figure 7.11e, the concept of mirror CUs is extended to the previously described DSO system, where control and channel information is also exchanged over the air so that there is no data traffic on X2. At the same time, UEs must exchange and broadcast control and CSI information in the form of PMI values over the air on the WL link.

The solution in Figure 7.11f comprises different embodiments of cooperative relaying solutions. It can be seen at once that traffic on the X2 interface is zero and traffic from S1 interface is shifted to WL.

(f1) uses decode and forward (D&F) relays with the CU located at the serving eNB so that quantized and encoded Tx signals are being transmitted from eNB to RNs.

(f2) is basically the same solution as (f1) with the main difference that amplify and store and forward (A&S&F) relays in combination with remodulation are being used. The "store" operation is required to avoid feedback loops over the air, while remodulation allows adapting the

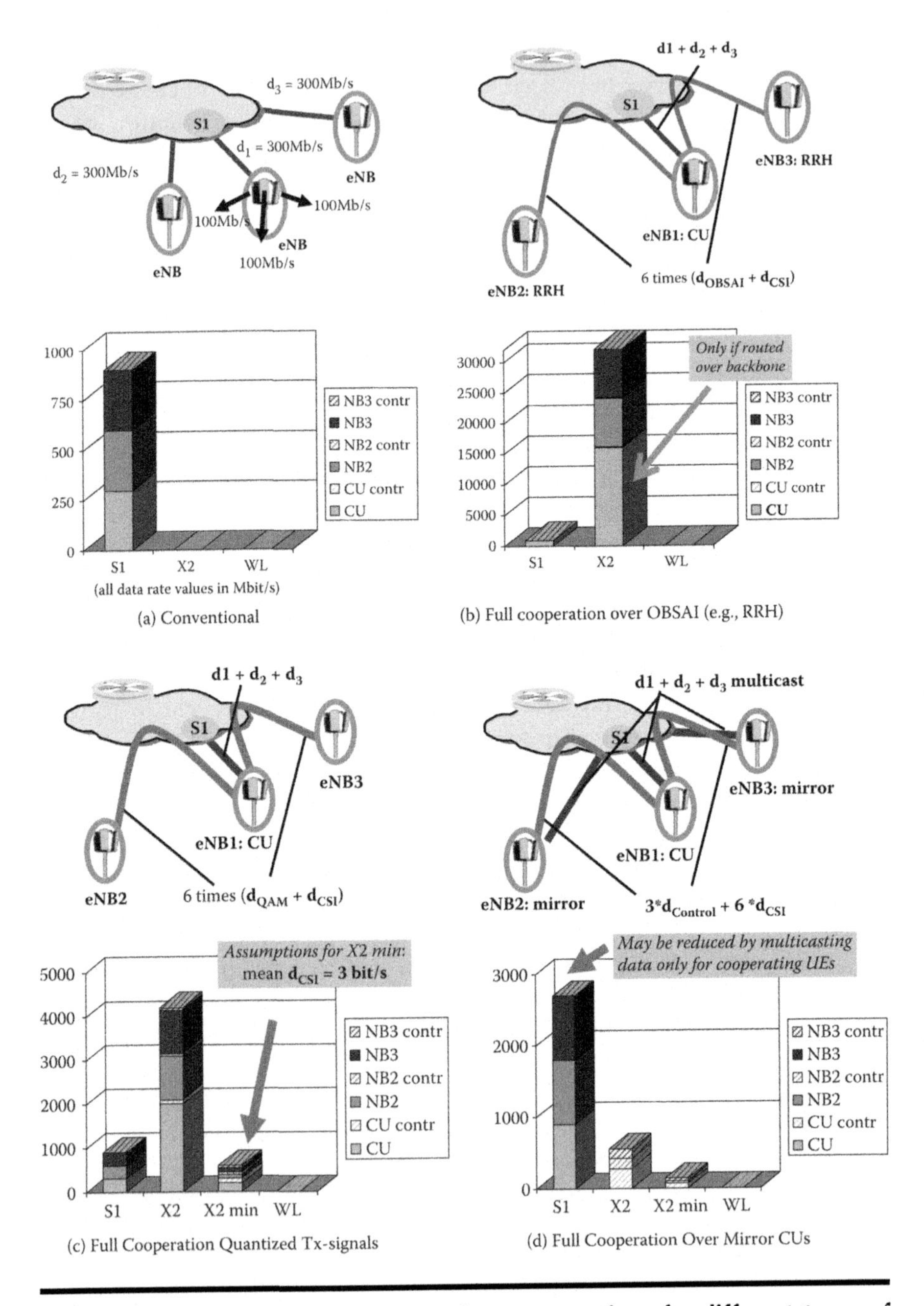

Figure 7.11 a–f: traffic on S1, X2, and WL connections for different types of cooperation.

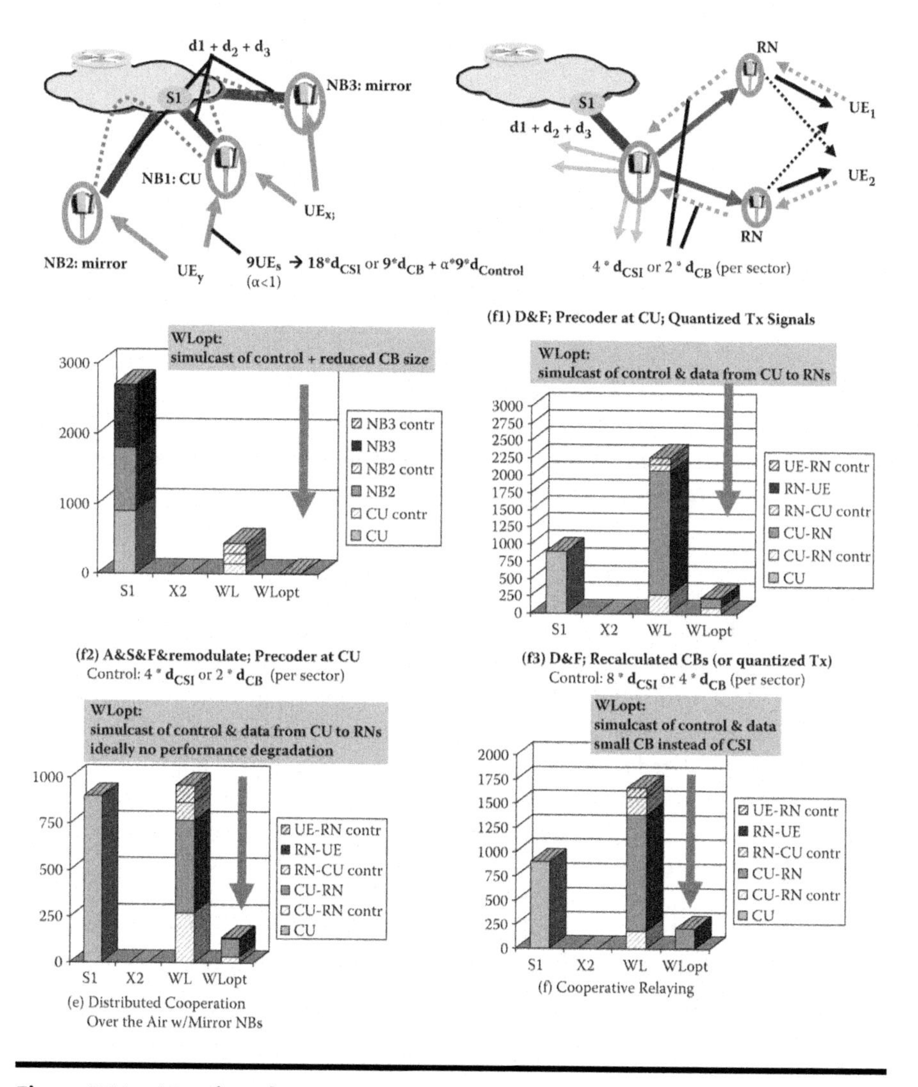

Figure 7.11 (Continued).

MCSs on NB–RN and RN–UE links separately (see also reference 15), which is important to avoid unnecessary overhead. For A&S&F instead of quantized data values, as for D&F, analogue Tx signals are sent to the RNs. For this reason, there will always be one channel use per Tx sample compared to the transmission of several bits for the D&F case. The inherent assumption is that the radio channels from NB to RNs have very high SNRs avoiding noise enhancement effects and that polarization multiplexing can be used, thus allowing for efficient data transmission on eNB–RN links.

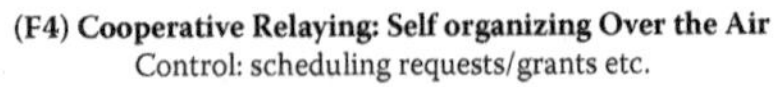

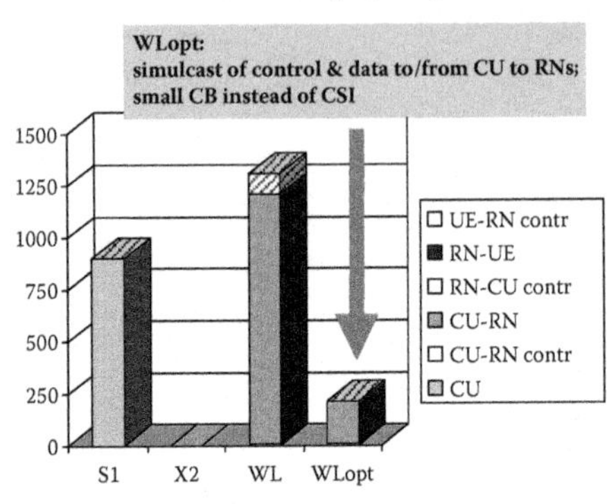

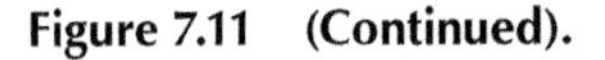
Figure 7.11 (Continued).

For (f3), the CU at the serving eNB has been replaced by two mirror CUs at both RNs per sector. D&F relays are assumed, but instead of precoded data, the "pure" data signals for all UEs are broadcasted to both RNs. Control and CSI information is exchanged between both RNs wirelessly over the serving eNB. Therefore, channel information might be outdated to some extent for this solution, which might limit maximum supportable mobile speed.

To overcome this limitation, in (f4) CSI or PMIs as well as other control information are exchanged between RNs over the air with help from the UEs. Because wireless links are less reliable, a hardened UL channel for this control information should be used to achieve a robust system concept. The main advantage here is the small delay and relatively simple implementation.

Sophisticated cooperative relaying solutions are very promising and combine in-band backhaul functionality, support of active interference management, and very fast adaptation to varying channel conditions. The values for WL_{opt} are rough estimates if all optimization potential is exploited and are subject to further research.

7.4 Basic Cooperation Area

It is important to analyze cooperative antenna systems from a very high level, as has been done in the preceding, to understand the impact on system level; however, at the same time, one has to look from the bottom up to understand all implications for a real-world realization. Starting from analysis of a single CA (see Figure 7.12),

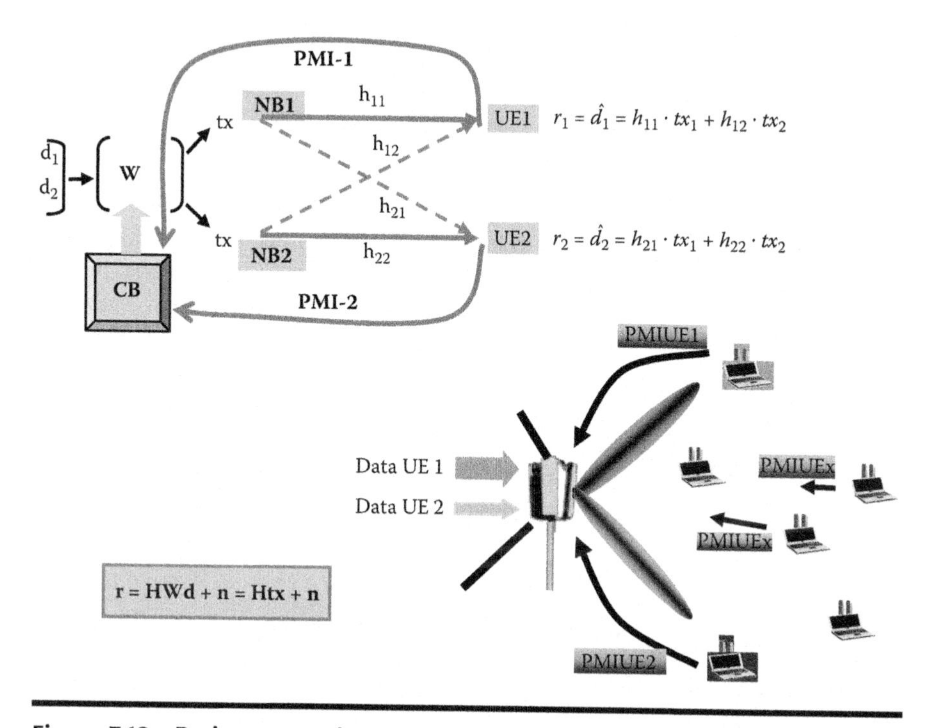

Figure 7.12 Basic cooperation area consisting of two eNBs and two UEs with one AE each.

this section will end with a likely LTE-conform implementation or give some indications about required enhancements of LTE R8.

The general closed-loop or precoded MU-MIMO system forming a CA consists of a group of N_{NB} eNBs and N_{UE} UEs having N_{tx} and N_{rx} antennas, respectively. As defined for LTE R8, codebook-based precoding is assumed.

The data symbol block, $d = [d_1, \cdot \cdot, d_{Ntr}]^T$ with $N_{tr} = N_{MS}N_{rx}$, is precoded by an $N_{tt} \times N_{tr}$ matrix **W** with $N_{tt} = N_{NB}N_{tx}$, in a case in which the number of data streams for each user n_d ($n_d \leq N_{rx}$) is N_{rx}. Here, the first N_{rx} data symbols are intended for the first user, the next N_{rx} symbols for the second user, and so on. When denoting i_{NB}/i_{UE} as the eNB/UE index and i_t/i_r as the transmit/receive antenna index, respectively, we can denote h_{ij}—where $i = N_{rx}(i_{UE} - 1) + i_r$, $j = N_{tx}(i_{NB} - 1) + i_t$—as the channel coefficient between the i_rth receive antenna of the i_{UE}th UE and the i_tth transmit antenna of the i_{NB}th eNB. The $N_{tr}N_{tt}$ channel coefficients can be expressed as the $N_{tr} \times N_{tt}$ channel matrix **H** with $[\mathbf{H}]_{ij} = h_{ij}$. The received signals on N_{tr} receive antennas, which are collected in the vector **r,** can be formulated as

$$r = \mathbf{HW}s + n, \tag{7.18}$$

where **n** is AWGN. The overall channel matrix **H** is an $N_{MS}N_{rx} \times N_{BS}N_{tx}$ matrix and is composed of the channel matrices for each user, which are of size $N_{rx} \times N_{BS}N_{tx}$. The following equation depicts this relationship:

$$\mathbf{H} = [\mathbf{H}_1, \cdot\cdot, \mathbf{H}_j, \cdot\ , \mathbf{H}_{\mathrm{NMS}}]^T, j\text{: user index.} \tag{7.19}$$

Here, $\mathbf{H}_j$ is the transpose of the channel matrix for user *j*, which is an $N_{NB}N_{tx} \times N_{rx}$ matrix. The N_{NB} eNBs need overall downlink channel state information **H** for the calculation of the precoding matrix **W** so as to form multiple spatial beams, which enable independent and decoupled data streams for N_{UE} users. The individual user *j* estimates its portion of the channel $\mathbf{H}_j$ and quantizes it by finding the best candidate from the predefined set of codes $\mathbf{C}_i$. The index of the chosen code *ij* is sent back to the eNB with the CU through a limited feedback channel. The eNB reconstructs the channel matrix $\hat{H}$ by looking up the codebook, which is shared by transmitters and receivers. This reconstructed channel matrix is used for the calculation of the precoding matrix **W**. Note that the notation of the general system is $(N_{NB} \times N_{tx}, N_{MS} \times N_{rx})$.

7.4.1 Challenges, or "Is Cooperation Feasible?"

In the preceding sections, many good reasons for cooperation have been given. Theoretically large capacity, as well as coverage gains in the order of several factors, has been predicted. Measurements in real-world outdoor scenarios have verified the potential cooperation gains by offline simulations based on the measured radio channels [28]. At the same time, simultaneous transmission to more than one single user on the same frequency-, time-, and, possibly, code-resource is challenging for a cellular mobile radio system because mutual time-varying interference may easily destroy any cooperation gain. The main issues in the context of cooperation and possible implementation strategies for LTE to overcome these topics include:

- *Backbone.* Some of the cooperation concepts have tough requirements regarding overall data traffic as well as delay constraints, which have to be fulfilled from the supporting backbone network, as analyzed in detail earlier. With respect to a viable business case of cooperation, this might be a big issue. From a technical point of view, fiber connections might provide any data rate and, in the case of transparent data lines or tunnels between the NBs, very small latencies will also be possible.
- *Tight synchronization in time and frequency* is one of the first issues that must be solved. Time synchronization for OFDMA systems has to ensure that receive signals from two different radio stations are received well within the guard interval (GI) for LTE of an OFDM symbol of length 4.68 μs; otherwise, intersymbol interference will occur. Analysis of measurements in the area of Berlin, Germany, has shown that typical delay spreads (DSs) of the combined radio channel for two cooperating UEs is about 1.5 times

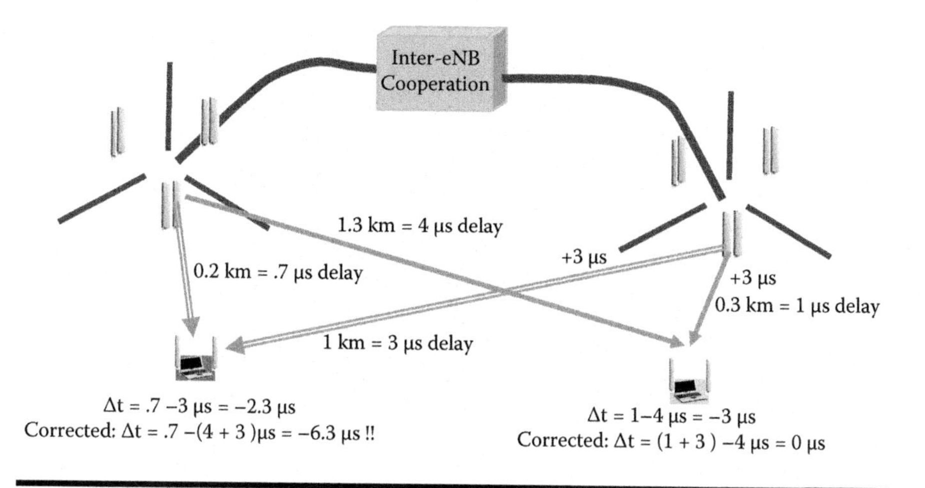

Figure 7.13 Timing advance for inter-eNB cooperation.

that of the DS to a single UE []. This is an encouraging result because different time delays for different eNB–UE links cannot be compensated by simple timing advance mechanisms known for single links, as is clear from Figure 7.13.

- *Frequency synchronization* errors cause OFDMA intercarrier interference or, if seen from a time domain perspective, different phase rotations for each transmitting radio station. For LTE, a quite narrow SC spacing of only 15 kHz has been chosen, so, without appropriate countermeasures, frequency stability of local oscillators (LOs) has to be extremely high. Keep in mind that for an RF frequency of 2.6 GHz, even a very challenging frequency stability of 1 ppm results in a frequency error of 2.6 kHz, which is 17% of the SC spacing. In early 2007, our group demonstrated feasibility of tight time and frequency synchronization by an SDMA demo on our LTE flexible HW platform. Two UEs were synchronized to one eNB over the air. For simultaneous transmission in UL, both UEs estimated the frequency offset of their LOs and prerotated their Tx signals by the opposite frequency offset, which has been called frequency advance and is easily possible using numerically controlled oscillators (NCOs) in the baseband unit. Residual phase offsets between both receive signals were estimated and corrected independently from each other at the NB [19] by a specific channel estimation algorithm. The demodulated QAM constellations for a two UE UL SDMA experiment can be found in Figure 7.14.
- With the topic of *channel estimation*, a much more severe and far-reaching issue is addressed compared to the topics discussed earlier. The channel matrix **H** consists of $i \times j$ channel elements, $[\mathbf{H}]_{ij} = h_{ij}$. Each h_{ij} represents a radio channel from one Tx- to one Rx (receive)-AE. Due to multipath reflections and moving

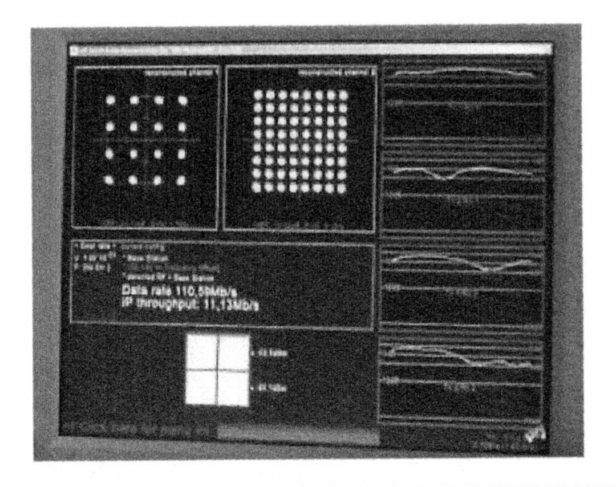

Figure 7.14 Separated user constellations (left) and channel coefficients from the two UEs to the two NB antennas after proper frequency synchronization.

mobile users, **H** is time varying and frequency selective so, in principle, each SC k ($k = 1 \ldots N_K$) and each symbol l ($l = 1 \ldots N_l$) has another channel matrix $\mathbf{H}_{\mathbf{k,l}}$ leading to a very high dimension of $N_{tt} \times N_{tr} \times N_K \times N_L$ complex channel coefficients. Ideally, these would have to be estimated with high accuracy. Luckily, the real dimension D of the radio channel—sometimes also called subspace dimension [29]—is much smaller than the full dimension. Assuming a 2×2 MIMO system with $N_K = 128$ SC—corresponding to a channel bandwidth (BW) of 1.25 MHz—and $N_L = 100$ symbols, the channel dimension D has been analyzed for an urban macroscenario with mobile speed of 10 m/s based on higher-order singular-value decomposition (HOSVD), which is basically a tensor-based approach [30,60] (Figure 7.15). As a result, the relevant dimension D_r is much smaller than the full dimension D and, in this case, for an HOSVD threshold of 10^{-5} by a factor of 240—that is, $(N_{tt} \times N_{tr} \times N_K \times N_l) = (2,2,128,100)$ compared to $(r_1, r_2, r_3, r_4) = (2,2,9,6)$, where r_1 to r_4 are the reduced dimensions for each parameter N_{tt} to N_L. This lower dimension has been partly exploited for LTE channel estimation based on the RS grids as defined in Chapter xxx for one to four antenna ports. The reduced dimension D_r in typical radio channels allows interpolation from estimates at REs with RSs for REs without RSs. It has been shown [33] that for lower SINRs and low channel dimensions D_r, several decibels of interpolation gain might be achieved.

This gain is important because cooperative JT techniques are more sensitive to interpolation errors than single-link MIMO transmissions. Figure 7.16 depicts the far-field radio channel for maximum ratio combining (MRC) of the intended signal as well as the IF null for JT based on ZF. Compared to MRC of the data signals, the IF nulls are located in an extremely small area so that even slowly moving users will leave this area

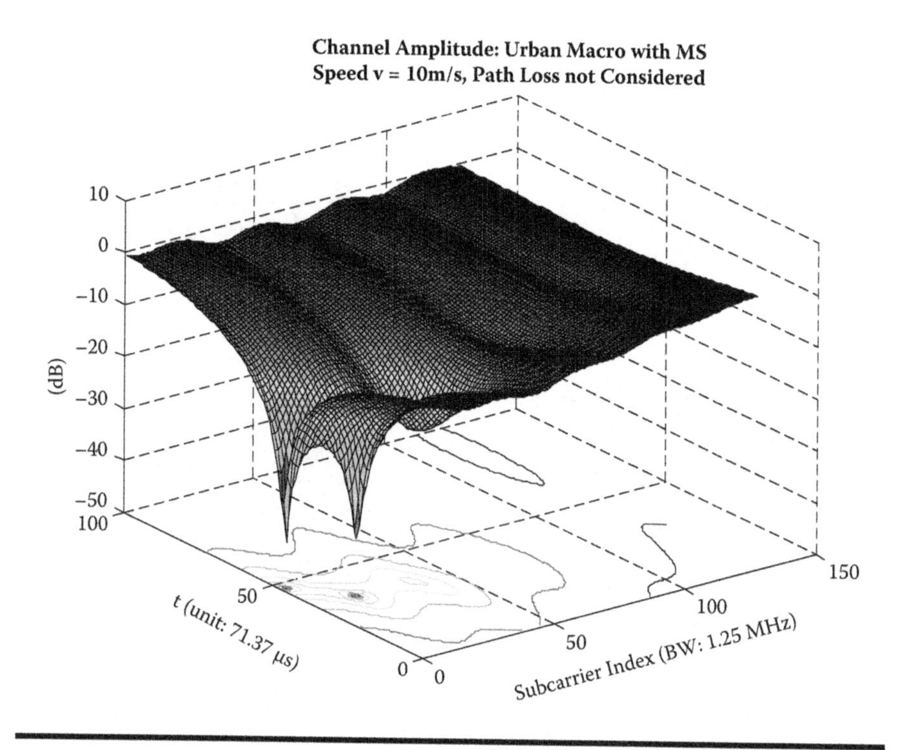

Figure 7.15 Typical SCMe radio channel.

very soon, resulting in significant IF. The same effect is well known from the time domain, where the normalized duration of fades for notches 20 dB below mean power might be smaller by a factor of 100 compared to that of power peaks (Figure 7.17).

- *Multicell channel estimation.* For cellular radio systems, there is a further hurdle for high-performance channel estimation because RSs cannot be received interference free but, rather, will overlap with RSs from other cells. In Thiele et al. [31], optimized allocation of RSs, together with specific RS sequences, has been developed. Depending on time variance of the radio channels, more or less intercell IF suppression is possible. In the 3G-LTE specification, reference symbols are intended for the purpose of intracell channel estimation; Frequency shifts sequences of length three enable the separate estimation of all three channels belonging to the sectors of the same NB. Furthermore, pilot symbols for the different transmit antennas are defined and located on different time and frequency positions (i.e., orthogonal in a resource block). In addition, pilots are scrambled using gold sequences (PRSs) to be defined by the network operators. Schiffermüller and Jungnickel [33] propose applying in addition block-orthogonal sequences

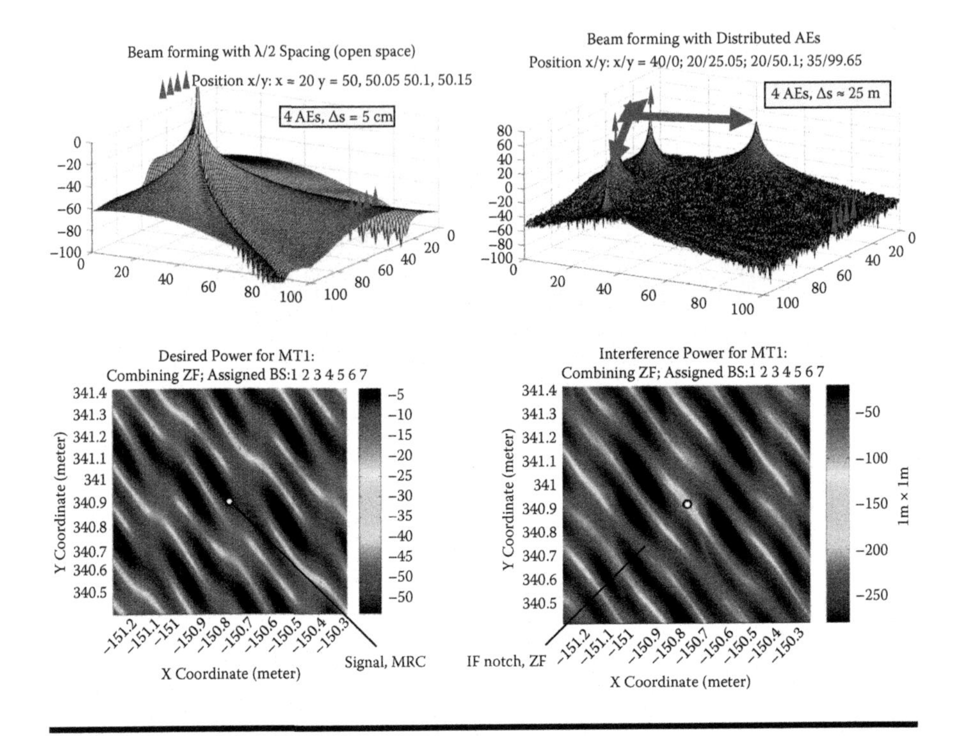

Figure 7.16 Far-field beam-forming pattern: (a) for central and distributed AEs; (b) for MRC versus ZF IF notches.

(e.g., Hadamard sequences) HS spread over the time domain from slot to slot with a maximum sequence length of 16.

Definition 7.2 (Block-Orthogonal Sequence)

Each row of a block-orthogonal sequence matrix HS is orthogonal to all other rows of the same matrix with full correlation length (i.e., $\mathbf{CC}^H = \mathbf{I}$). Reducing the correlation length to n yields a matrix with block-wise orthogonal properties, where each block is of size $n \times n$. Furthermore, each nth row should be identical for a given correlation length n. Figure 7.18 visualizes the suggested pilot grid, where the number (hex base) indicates the code $\mathbf{c}_v$ chosen from the sequence matrix $\mathbf{C}$ (i.e., the row of the Hadamard matrix of maximum length $N = 16$). The decimal numbers indicate the sequence length over the time domain. Note that the suggested scheme covers different sequence length $n = \{1,2,4,8,16\}$ because the sequence pattern repeats itself every n rows. Thus, the system may benefit from more precise channel estimation for increasing sequence length n. The suggested scheme assigning virtual pilot sequences to the multicell system is translational invariant with respect to the estimation error. Its block orthogonality is sustained even after a cyclic shift. Note that the suggested scheme can be easily extended

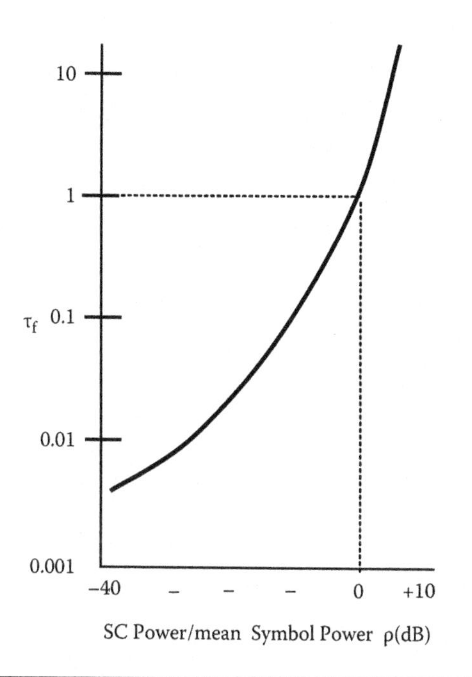

Figure 7.17 Normalized duration of fades.

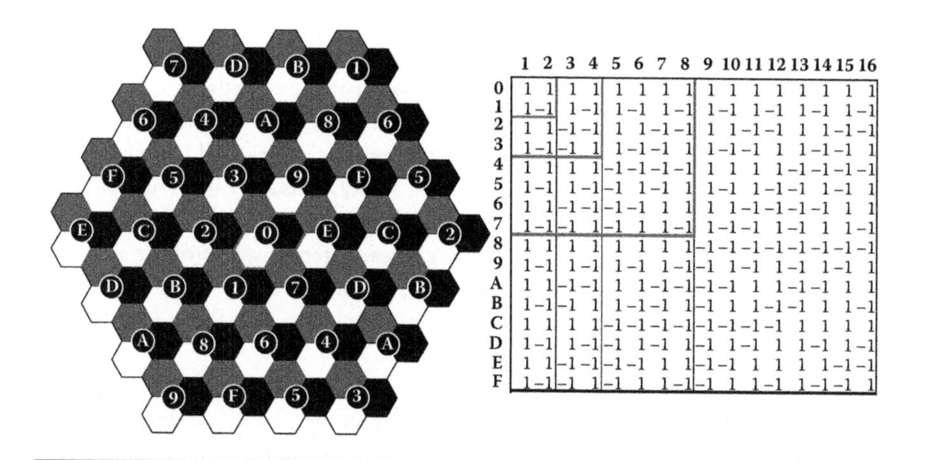

	1	2	3	4	5	6	7	8	9	10	11	12	13	14	15	16
0	1	1	1	1	1	1	1	1	1	1	1	1	1	1	1	1
1	1	−1	1	−1	1	−1	1	−1	1	−1	1	−1	1	−1	1	−1
2	1	1	−1	−1	1	1	−1	−1	1	1	−1	−1	1	1	−1	−1
3	1	−1	−1	1	1	−1	−1	1	1	−1	−1	1	1	−1	−1	1
4	1	1	1	1	−1	−1	−1	−1	1	1	1	1	−1	−1	−1	−1
5	1	−1	1	−1	−1	1	−1	1	1	−1	1	−1	−1	1	−1	1
6	1	1	−1	−1	−1	−1	1	1	1	1	−1	−1	−1	−1	1	1
7	1	−1	−1	1	−1	1	1	−1	1	−1	−1	1	−1	1	1	−1
8	1	1	1	1	1	1	1	1	−1	−1	−1	−1	−1	−1	−1	−1
9	1	−1	1	−1	1	−1	1	−1	−1	1	−1	1	−1	1	−1	1
A	1	1	−1	−1	1	1	−1	−1	−1	−1	1	1	−1	−1	1	1
B	1	−1	−1	1	1	−1	−1	1	−1	1	1	−1	−1	1	1	−1
C	1	1	1	1	−1	−1	−1	−1	−1	−1	−1	−1	1	1	1	1
D	1	−1	1	−1	−1	1	−1	1	−1	1	−1	1	1	−1	1	−1
E	1	1	−1	−1	−1	−1	1	1	−1	−1	1	1	1	1	−1	−1
F	1	−1	−1	1	−1	1	1	−1	−1	1	1	−1	1	−1	−1	1

Figure 7.18 Left: pilot reuse pattern based on orthogonal code sequence (e.g., Hadamard) in a triple-sectorized cellular system. Right: Hadamard sequences spread over space (rows) and time (columns) domains. Hex-base numbers indicate sites with the same virtual pilot sequence [37].

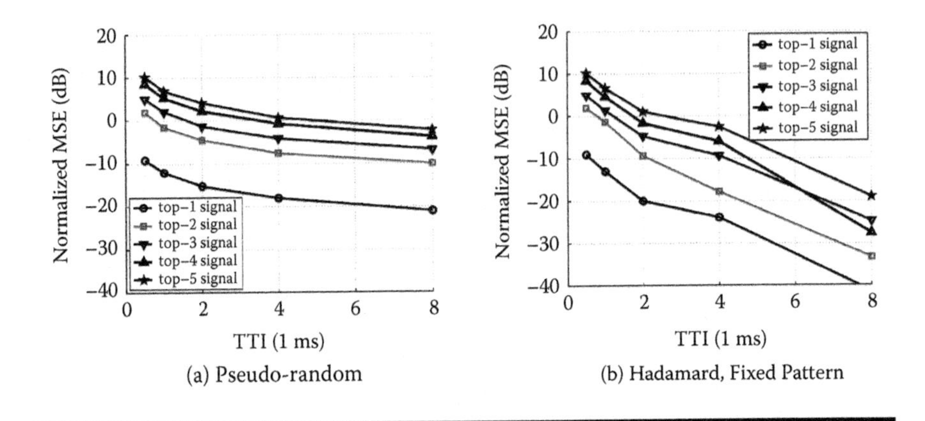

Figure 7.19 Normalized MSE obtained for the correlation estimator, in case of a static channel, for the five strongest sectors.

to the case of larger correlation length. Resulting performance improvements of channel estimation for the proposed scheme over PRS reference signals are given in Figure 7.19.

- *Feedback overhead.* Closely related to optimum channel estimation is the challenge to feed back CSI from all UEs to all eNBs involved in the CA so that the eNBs can do proper precoding. For TDD systems, one can partly rely on channel reciprocity, but for FDD systems one has to assume basically uncorrelated DL and UL radio channels. UL feedback in cellular systems is very expensive because it costs valuable UE power, thus reducing battery lifetime. At the same time, UEs have smaller Tx power and will have only limited UL capacity, in particular at the cell edge.
- *Codebook-based precoding* as defined in LTE release 8 is already a big step in reducing feedback; at the same time, more feedback will be required for high-performance precoding solutions. Many ideas exist for reducing feedback, such as the subspace approach [23], hierarchical codebooks, tracking CBs [34], or other compression techniques. Later we will propose a further technique for robust and self-adapting feedback overhead in the combination with advanced hybrid automatic repeat request (HARQ; in the following called COOPA HARQ) techniques.
- *Feedback delay* between measurement of the radio channel and availability at the CU is even more challenging. Based on the LTE RS and frame structure, it will take several 1-ms subframes to estimate the radio channel, prepare the UL message, decode the message at the eNB, and react by updating the precoder at the CU. The timing of these actions adds up easily to 10 ms and might be even higher if backbone delay is also involved. During this time duration, the radio channels might have changed significantly, even in

the case of relatively low mobile speed. COOPA HARQ also addresses this issue with the goal for a robust transmission mode.

- An organizational issue is the *coordination of different scheduling units,* especially in cases of distributed cooperation. Thus, one unit has to be defined as CU, which will control other scheduler units. This has to be done in a self-organizing manner as being investigated in 3GPP as self-optimizing networks (SONs).

7.4.2 Cooperation Area of Minimum Size as LTE Enhancement

In the following, the simplest type of CA is analyzed—namely a (2 × 1, 2 × 1) CA consisting of two different eNBs or RNs serving two UEs simultaneously, where each radio station has only one single antenna port. This can be seen as the core of any CA but at the same time allows analysis of most of the relevant issues addressed previously. Figure 7.12 (bottom) is a schematic representation of this basic CA for the application of intersector cooperation. It is assumed that a set of active UEs feed back their optimum PMIs and the CU of the CA selects two suitable UEs out of the set of active UE for cooperation on one, several, or all RBs. Suitable criteria for user grouping are similar path loss (PL), low channel correlation, and large mutual interference. Similar PL minimizes power rise for ZF or MMSE and low channel correlation helps to minimize the feedback overhead or maximizes performance for a given feedback overhead; large mutual interference guarantees high performance gains due to cooperation. In the following, only the two selected UEs will be considered further.

The CA is organized for the selected two UEs based on LTE release 8 according to the following parameters:

downlink/uplink, full duplex	2.53/2.68 GHz
antennas eNB/UE	2 TRx/2 Rx
sampling rate	30.72 MHz
used bandwidth	20 MHz
symbol period downlink/uplink long (short) block	71.4/70.1 (37.5) ns
cyclic prefix downlink/uplink	4.6875/4.13 ns
total number of subcarriers	2.048
number of used subcarriers/RB	up to 1.200/100
transmission scheme downlink	OFDMA
modulation in downlink	4, 16, 64 QAM
radio frame	10 ms
subframe or TTI	1 ms
symbols/subframe (DL)	14
resource block size	12 SCs in one subframe

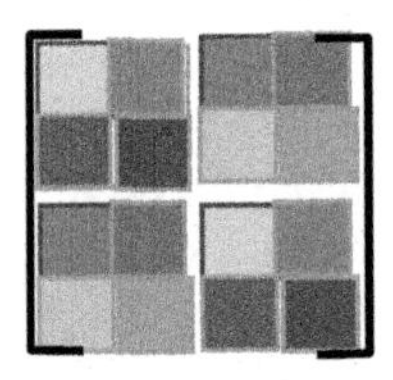

Full Channel Matrix **H**

Reduced Matrix
$\mathbf{H_R} = \mathbf{HR}$ of Size **2 × 2**
after IRC IF Suppression by **R**

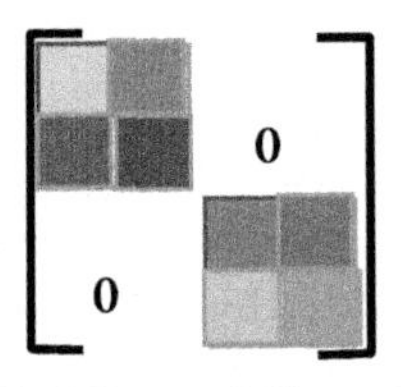

Block Diagonal Channel
Matrix $\mathbf{W_B H}$

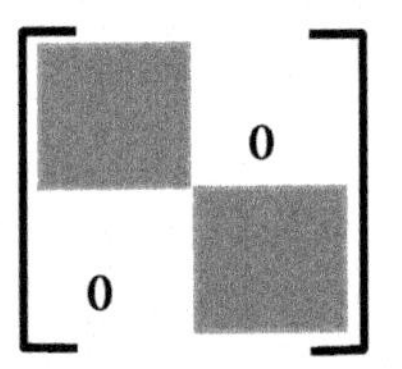

Block Diagonal
Matrix $\mathbf{WH_R}$ After **ZF**

Figure 7.20 Reduced size 2 × 2 $\mathbf{H_R} = \mathbf{HR}$ channel matrix for analysis of CAs.

In this list, two Tx and two Rx antenna ports per eNB and UE are stated, and the overall simulated system is of type (2 × 1, 2 × 1). The background is the overall system concept in mind, where the SDF of the two UE AEs is used for IRC by the receive matrix $\mathbf{R}_i$ of size 2 × 2 for $I = 1, 2$, as illustrated in Figure 7.20. The IRC will form the UE beam former so that one or several of the strongest interferers from other eNBs can be suppressed. If we use the SDF in terms of additional Rx antennas for IRC of intracell interference, this SDF will be missing for intercell interference cancellation. In mobile systems, the IRC of the strongest other cell interference has been proven to be quite effective because the UE can adapt if different eNBs become the strongest interferers [18].

As a result, the UEs feed back the CSI for the composite radio channel $\mathbf{HR}_i$ of size 2 × 2 instead of 2 × 4 or 4 × 4; thus, SDF is saved and feedback is reduced as well. The reduced model can be simply written as

$$\mathbf{r} = \mathbf{H}\,\mathbf{W}\,\mathbf{d} + \mathbf{n},$$

where **n** is AWGN and **H, W** are of size 2 × 2. In the following, the CA will be organized without taking AWGN into account (i.e., the purely interference-limited case will be analyzed). Thus, the comparison of different implementations will be independent of the cell size or Tx power. Codebook-based precoding is assumed, but for different types of codebooks.

7.4.3 Channel Estimation

Generally, a 2 × 2 CA for intersector cooperation can be implemented by two different strategies:

- One might "hide" the intersector cooperation as a conventional MU-MIMO transmission from a single sector. Each sector sends two different RS grids for AP 1, 3 and AP 2, 4, respectively. This allows the UEs to estimate their radio channels on orthogonal RS grids without knowing that the RSs are sent from different sectors. The advantage is that UEs conforming to release 8 can be supported. A common physical broadcast channel (PBCH) might have to be sent from both sectors to avoid confusion at the UEs, which will have some effect on cell planning. Performance gains will be restricted due to the limited PMI feedback, and four APs per sector are also not possible.
- The orthogonal frequency shifts between RSs of different sectors allow for quite accurate channel estimation; if limitations result from simultaneous data and/or RS transmission in other cells. The UEs should be attached logically to a single sector, which will deliver all required control information to this UE. Compared to release 8, a new MIMO mode in addition to the existing ones (i.e., a COOPA mode) has to be defined, allowing combination of two UEs semistatically into a CA. In COOPA mode, the UEs must feed back PMIs based on CSI estimates for their own and the neighboring sectors so that the NB can perform proper precoding.

7.4.4 Precoding Strategy

The general precoding strategy can be seen in Figure 7.12 (top), where two eNBs cooperate over a CU doing common precoding based on the PMI feedback (PMI 1 and PMI2) from two different UEs with one virtual AE each.

For LTE release 8, unitary precoding has been defined with a limited set of only three precoding matrices according to Figure 7.21 for dual stream transmission. For a (2 × 1, 2 × 1) MU-MIMO system, we always have dual stream transmission. It will be shown that performance for this limited LTE release 8 codebook (CB) size is poor in the general case. For that reason, we will allow for larger CB sizes and specifically nonunitary CBs so that the condition $\mathbf{W} \times \mathbf{W}^H = \mathbf{I}$, with $\mathbf{I}$ the identity matrix, might be violated.

Here we propose that each UE (UE1 and UE2) feeds back quantized amplitude and phase values of its estimated radio channels, h_{11}, h_{12}/h_{21}, and h_{22}. For this purpose, one might use the UL control-shared channel, which allows each UE to send in UL (in addition to PMIs) up to 200-b control information per subframe to the eNB.

Codebook Index	Number of Layers υ	
	1	2
0	$\begin{bmatrix}1\\0\end{bmatrix}$	$\frac{1}{\sqrt{2}}\begin{bmatrix}1 & 0\\0 & 1\end{bmatrix}$
1	$\begin{bmatrix}0\\1\end{bmatrix}$	$\frac{1}{2}\begin{bmatrix}1 & 1\\1 & -1\end{bmatrix}$
2	$\frac{1}{\sqrt{2}}\begin{bmatrix}1\\1\end{bmatrix}$	$\frac{1}{2}\begin{bmatrix}1 & 1\\j & -j\end{bmatrix}$
3	$\frac{1}{\sqrt{2}}\begin{bmatrix}1\\-1\end{bmatrix}$	-
4	$\frac{1}{\sqrt{2}}\begin{bmatrix}1\\j\end{bmatrix}$	-
5	$\frac{1}{\sqrt{2}}\begin{bmatrix}1\\-j\end{bmatrix}$	-

Figure 7.21 Codebook for transmission on antenna ports {0,1}.

In the case of LTE release 8, there are some good reasons for the limited CB size:

- Feedback overhead from UEs to eNBs in the UL is expensive in terms of resources and power consumption and therefore it should be minimized.
- For LTE baseline single-user (1×2, 1×2) or multiuser (1×2, 2×2) systems and the assumption of $\lambda/2$ spaced AEs (with λ as the RF-wavelength), beam patterns will be quite broad. At the same time, in cellular networks we can expect a high number of users per cell, so we should exploit multiuser diversity by applying user grouping. In combination with the broad beam pattern, there is a good chance to find UEs feeding back mutual orthogonal PMIs so that user grouping will be simple and effective.
- For DL MU-MIMO, we have a 2×4 system, so we can apply receive diversity or interference rejection combining at the UE side, which allows suppressing residual intracell interference caused by improper precoding.

At the same time, there are a number of good reasons for more advanced precoding schemes for future LTE enhancements:

- The goal of active IF management, based on interference-free CAs, is to increase spectral DL efficiencies by several hundred percent. Therefore, the additional (hopefully moderate) UL feedback overhead can be justified by large performance gains.

- Even in the case of $\lambda/2$ spacing, user grouping will grow increasingly difficult with an increasing number of AEs because resulting beam widths will be smaller. Therefore, even for large numbers of UEs, there will be much fewer UEs within a certain beam, so larger CB sizes will be required.
- Up to now in the LTE, single-sector MU-MIMO or SDMA has been investigated. In that case, there is a good chance that one beam selection fits for the full frequency band or all RBs. If we take a look at Figure 7.12 (bottom), there are two beams. As it is depicted, it fits in an open-loop scenario because beam separation is kept at all locations. In reality, multipath effects will intermix both beams, generating some mutual beam interference. From a cooperation perspective, it is more important that in the case of, for example, inter-eNB cooperation, AEs will be placed at different locations, generating generalized beam patterns. These are by no means orthogonal and show fast fluctuation from RB to RB, so feedback per RB will be required.

For our approach, the precoding matrix $\mathbf{W}$ is selected from a ZF or MMSE codebook, which means that $\mathbf{W}$ is the Morse Penrose pseudoinverse of the quantized channel matrix $\mathbf{H_Q} = [\ \cdots\ \text{PMI-}A_{i,j} \times e^{j \times \text{PMI-}\phi_{i,j}} \cdots\ ]$:

$$\mathbf{W}_{\text{A,PHI}} = \mathbf{H_Q}^{+};$$

$\text{PMI-}A_{i,j} = \text{quant}(\text{abs}(h_{i,j}))$; $\text{PMI-}\phi_{i,j} = \text{quant}(\text{angle}\ (h_{i,j}))$; $i = 1,\ldots,2$; $j = 1,\ldots,2$.

The quantize operation is "quant," which uses A bits for the amplitude value $\text{PMI-}A_{i,j}$ and phi bits for the angle $\text{PMI-}\phi_{i,j}$, so the accuracy under the assumption of normalized radio channels $h_{i,j}$ is for $\text{PMI-}A_{i,j}$ equal to $1/2^{\wedge A}$ and of the angle $\text{PMI-}\phi_{i,j}$ equal to $1/2^{\wedge PHI}$. Normalization includes the adaptation to different path loss values on each UE link, which can be estimated and corrected from wideband received signal strength indication (RSSI) measurements. The pertinent codebook, $C_{\text{A,PHI}}$, consists of $(2^{\wedge A} * 2^{\wedge PHI})^{\wedge 2}$ complex valued 2×2 matrices. The squaring is due to the fact that UE1 and UE2 feed back one value each, which leads to a combined codebook that is of higher dimension.

Amplitude as well as phase values are fed back separately. Our own investigations as well as those of other research groups have shown the importance of amplitude in addition to that of phase information for high-performance precoding solutions. Even so, quantization of amplitude $\text{PMI-}A_{i,j}$ can be done with a relatively smaller number of bits compared to $\text{PMI-}\phi_{i,j}$.

To compare the different precoding strategies, the SIR for two UEs with the same PL has been simulated for LTE release 8-conforming CBs as well as the previously described codebook $C_{A,PHI}$. The interference is the result of "improper" precoding. Two UEs are using the full bandwidth of 96 RBs. For each RB, the optimum precoder has been selected based on per-RB PMI feedback. Figure 7.22

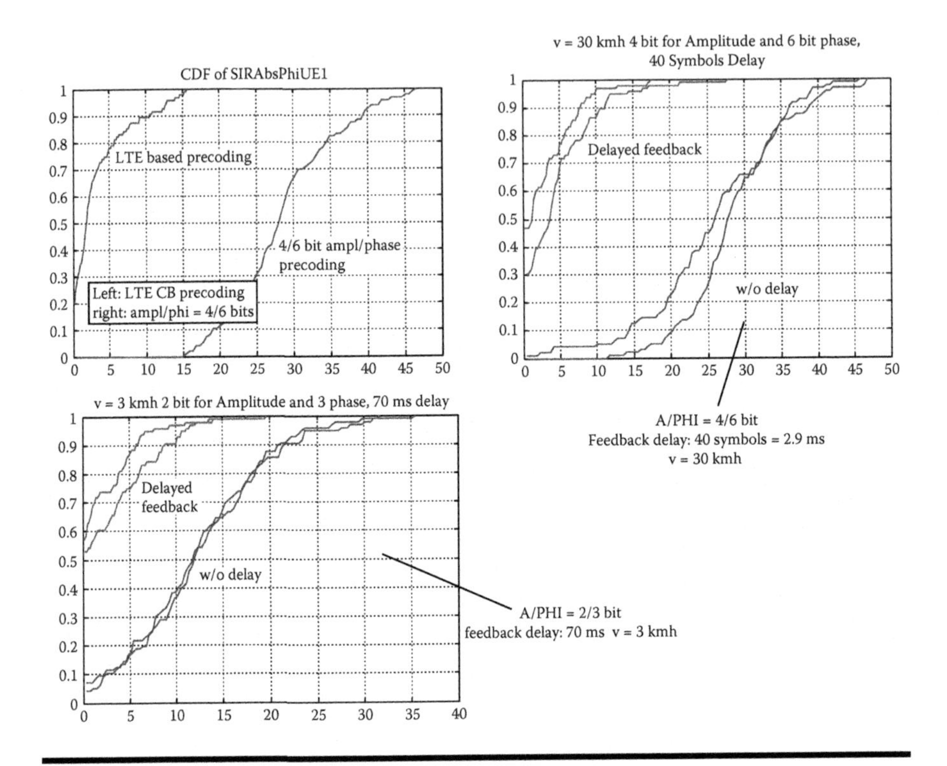

Figure 7.22 Simulation results for different precoding strategies and for different delays.

depicts several simulation results in the form of CDFs of the achievable SIR (note that the two curves are for UE1 and UE2):

- At the top left-hand side of the figure, LTE-based precoding is compared with $C_{A,PHI}$, where A and PHI are four and six, respectively. Thus, each PMI for each UE has 10 bit and the resulting SIR is, with about 25 dB at 50% CDF value, accordingly high. Compared to this LTE, release 8-based precoding achieves only for about 15% more than 10-dB SIR.
- On the bottom left-hand side is a simulation with a smaller CB size for A/PHI equal to 2/3. At the 50% CDF value, this results naturally in a smaller SIR compared to the 4/6 case of only 13 dB. Additionally, the effect of delayed feedback is analyzed, which, even for small mobile speeds of only 3 km/h, can be quite detrimental, if the feedback delay is large (here, 70 ms).
- On the top right-hand side, the same is shown for a mobile speed of 30 km/h, 40-symbol = 2.9 ms feedback delay, and 4/6 bit for A and PHI. The high performance gain without delay is almost completely lost for the delayed feedback case. (For a further reduction of feedback, see [61].)

7.4.5 COOPA HARQ

The high sensitivity to feedback delay calls for some specific solutions to come to a robust system concept and additionally keep overall overhead small. For this purpose, the COOPA HARQ concept has been developed. It integrates the conventional LTE HARQ concept with optimized CB-based precoding. The main goal of HARQ is to adapt optimally to varying link conditions. Thus, in a first transmission, only low redundancy is transmitted, but in cases in which the receiver cannot decode a data packet, a nonacknowledge (NACK) message is fed back to the transmitter. After reception of a NACK message, the transmitter sends additional redundancy for the first data packet. In contrast to ARQ schemes, in the case of HARQ, the received data from the first transmission are stored and will be combined suitably with the retransmitted receive signal [36].

For COOPA HARQ, the concept is extended to precoding. Instead of additional redundancy for the data packet, the retransmission of the data packet is done with an optimized precoding matrix $\mathbf{W}_{Q,2}$. Optimization can be done with respect to higher quantization levels A and PHI or with respect to a time-corrected channel matrix $\mathbf{H}_{\mathbf{delay}}$. In the first case, the first PMI feedback of each UE is, for example, only 3 bit, to conform to LTE release 8. As can be seen from Figure 7.22, only a few UEs will succeed if higher CIR (such as >10 dB) is requested; therefore, it will be necessary to send a modified NACK message, COOPA NACK. In addition to the NACK information, COOPA NACK contains some further PMI information, so the overall PMI information is, for example, 5 bit, long. With this additional PMI information, according to Figure 7.22 (bottom), a much higher SIR can be achieved and decoding probability is significantly increased. This helps to reduce overall feedback because a high number of feedback bits is generated only in those cases where high precoding accuracy is required also.

In case of significant feedback delay $\Delta\tau$ between the time t_1 of measuring $\mathbf{H}(t_1)$ and transmitting the data packet over the radio channel at time $t_2 = t_1 + \Delta\tau$, the radio channel will have evolved into the radio channel $\mathbf{H}_{\mathbf{delay}}(t_2)$. Thus, precoding with $\mathbf{W}(t_1)\mathbf{H}_{\mathbf{delay}}(t_2)$ will go wrong, as can be seen from the preceding simulation results. For COOPA HARQ (see schematic representation in Figure 7.23), it is important that the UE can measure $\mathbf{H}_{\mathbf{delay}}(t_2)$, which has been the radio channel for the first transmission and can therefore adapt its PMI feedback accordingly—either by calculating the difference of $\Delta\mathbf{W} = \mathbf{W}(t_2) - \mathbf{W}(t_1)$ itself or by feeding back $\mathbf{W}(t_2)$ so that the NB can calculate the difference. At the retransmission, the CU transmits $\text{Tx}(t_3) = \Delta\mathbf{W} \times \mathbf{H}_{\mathbf{delay}}\mathbf{d}$ ($\mathbf{d}$ is the same data vector as for the first transmission). The UEs combine both receive signals from t_1 and t_2 to $r_{\text{sum}} = r(t_1) + r(t_2) = (\mathbf{W}(t_1)$ $\mathbf{H}_{\mathbf{delay}}(t_2) + \Delta\mathbf{W} \times \mathbf{H}_{\mathbf{delay}})\ \mathbf{d} = \mathbf{W}(t_2)\mathbf{H}_{\mathbf{delay}}(t_2)\mathbf{d}$.

Note that the equation is true if $\mathbf{H}_{\mathbf{delay}}$ and $\mathbf{H}_{\mathbf{delay}}(t_2)$ are equal; otherwise, there will be only partial correction. This technique is nonetheless very important because it allows correcting precoding errors after transmissions have already taken place. Moreover, it allows for interpolation between both time instances,

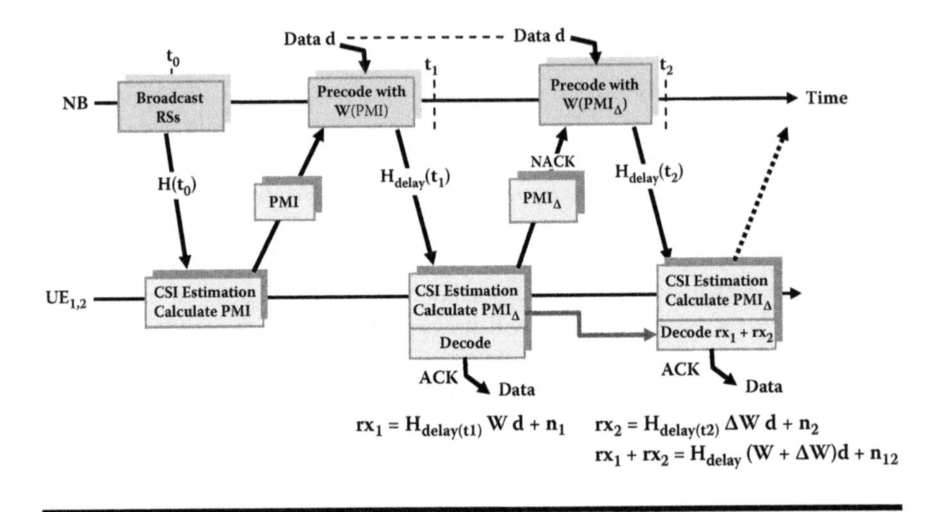

Figure 7.23 Schematic representation of COOPA HARQ concept.

and, accordingly, evaluations have revealed that even very simple linear interpolation over a full frame length of 10 ms is possible, while performance degradations compared to a constant radio channel are relatively small, even for the middle of the frame. Note that interpolation accuracy will be smallest at the center of the frame, if correct precoding is assumed for the beginning and end of the frame. Figures 7.24 and 7.25 contain simulation results highlighting achievable SIR with and without the COOPA HARQ concept.

The relative overall Tx power of $\Delta\mathbf{W}$ compared to $\mathbf{W}$ has been analyzed with respect to $\Delta\tau$ and mobile speed as well as feedback delay. For moderate scenarios, values of $\Delta\mathbf{W}$ will be small and in the order of –20 to –10 dB because the radio channel typically evolves slowly. For this reason, we can retransmit $Tx(t_3) = \Delta\mathbf{W} \times \mathbf{H}_{\mathbf{delay}}\mathbf{d} \times SCd$, where SCd is a proper scaling factor, so that, at the receiver,

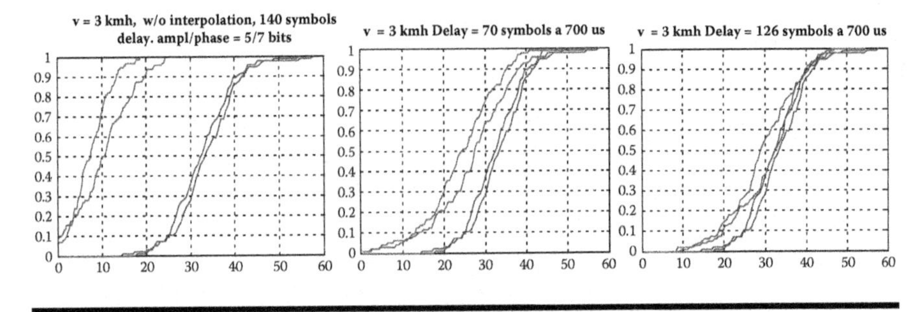

Figure 7.24 CDF of achieved SIR for COOPA HARQ concept for UEs 1 and 2. Left: without interpolation, FB delay = 10 ms; middle: with interpolation for worst-case FB delay of 5 ms; right: with interpolation for FB delay = 9 ms.

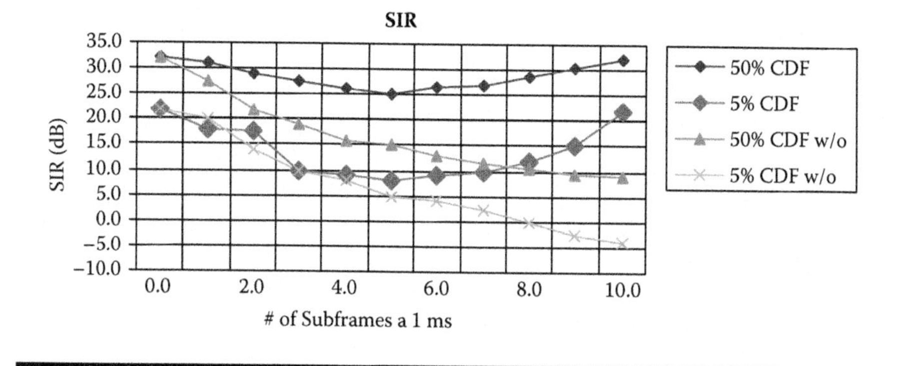

Figure 7.25 Achievable SIR for COOPA HARQ with ( ) and without (w/o) interpolation for different FB delays at 50 and 5% CDF values.

the signal can be scaled down by 1/SCd. Alternatively, Tx(t_3) is modified to Tx(t_3) $= \Delta\mathbf{W} \times \mathbf{H}_{\mathbf{delay}}\mathbf{d} + \mathbf{W} \times \mathbf{H}_{\mathbf{delay}}\mathbf{d}$, increasing the combining gain at the receiver.

COOPA HARQ can be easily combined with partial IRC at the UEs for suppression of residual IF, thus partly reducing the SDF at the UE side to other cell interference, but at the same time stabilizes the overall system further. COOPA HARQ leads to a robust system design but, as usual for HARQ systems, at the cost of an additional retransmission, thus increasing the required Tx resources. Optimum usage of these resources is currently being studied in more detail.

7.5 Conclusions

The motivation for cooperation, theoretical analysis, and simulation results have been presented in this chapter, giving a clear view of the high potential of cooperative antenna systems that promise performance gains in the order of several hundred percent or even more. At the same time, practical and robust implementations for COOPA systems are challenging, although feasible if properly designed. Specifically, active interference management that combines different techniques like intra-eNB cooperation with antenna tilting, IRC, and interference coordination seems to be promising because the sizes of so-called cooperation areas can be kept small and practical.

Different implementation strategies have been presented and some important issues for the core element of any cooperation area have been addressed for the example of intersector cooperation. The newly proposed COOPA HARQ concept allows reducing feedback overhead and, at the same time, solves the critical issue of feedback delay. In the meantime, many research groups and companies are confident that cooperative antenna systems are realizable in real systems. Therefore, further discussions will not be about the question of whether cooperation should be standardized but, rather, about what type of cooperation should be standardized. Due to the multitude of implementation options, many interesting discussions can be expected in this area.

References

1. Matthias, R. 2008. Future network infrastructure for a world connected. NGNM conference at CeBIT, Hannover, Germany, March 6, 2008.
2. ITU homepage: http://www.itu.int/net/home/index.aspx
3. Schulz, E., and J. Kahtava. 2007. Nokia Siemens networks and Nokia's view on IMT-advanced minimum requirements. FuTURE Forum: Seminar on the requirement and development of IMT-advanced technology and system, Beijing, September 25, 2007.
4. Hideaki Takagi, H., and B. H. Walke. 2008. *Spectrum requirement planning in wireless communications: Model and methodology for IMT-Advanced.* Chichester, England: John Wiley & Sons.
5. Nokia Siemens Networks GmbH & Co KG. December 2007. LTE trial will increase mobile broadband speed 10 times. Press release.
6. Karakayali, M. K., G. J. Foschini, R. A. Valenzuela, and R. D. Yates. 2006. On the maximum common rate achievable in a coordinated network. *ICC06,* March 3, 2006.
7. Marsch, P., and G. Fettweis. 2008. Rate region of the multicell multiple access channel under backhaul and latency constraints. *Proceedings of the IEEE Wireless Communications and Networking Conference (WCNC '08),* Las Vegas, NV, March 30–April 3, 2008.
8. Homepage of next generation mobile network (NGNM) alliance: http://www.ngmn.de/
9. Homepage of company NOMOR: http://www.nomor.de/home
10. Roy, B., C. Roy, and M. Einhaus. 2006. A case study on frequency reuse in OFDMA systems using hierarchical radio resource management. *Proceedings of Asian-Pacific Microwave Conference,* p. 4, Yokohama, Japan, December 2006.
11. Holma, H., and A. Toskala. 2007. *WCDMA for UMTS,* 4th ed. Chichester, England: John Wiley & Sons.
12. Homepage of third generation partnership program (3GPP): http://www.3gpp.org/
13. Karakayali, M. K., G. J. Foschini, R. A. Valenzuela, and R. D. Yates. 2006. On the maximum common rate achievable in a coordinated network. *ICC06,* June 2006.
14. Documents and results of 3GPP IMT advanced workshop in Shenzen, China, beginning 2008: http://www.3gpp.org/ftp/workshop/2008-04-07_RAN_IMT_Advanced/
15. Official IEEE802.16j page for MMR systems: http://wirelessman.org/relay/
16. 3rd Generation Partnership Project. Technical Specification Group radio access network; evolved universal terrestrial radio access (E-UTRA); physical channels and modulation: 3GPP TS 36.211/multiplexing and channel coding: 3GPP TS 36.212/physical layer procedures: 3GPP TS 36.213/V8.2.0 (2008-03) (release 8).

17. Thiele, L., M. Schellmann, W. Zirwas, and V. Jungnickel. 2007. Capacity scaling of multiuser MIMO with limited feedback in a multicell environment. *41st Asilomar Conference on Signals, Systems and Computers,* Monterey, CA, IEEE, November 2007. Invited paper.
18. Zirwas, W., E. Schulz, M. Schubert, W. Mennerich, V. Jungnickel, and L. Thiele. 2007. Cooperative antenna concepts for interference mitigation. *13th European Wireless Conference (2007),* Paris, France, on CD-ROM.
19. Jungnickel, V., T. Haustein, W. Zirwas, J. Eichinger, E. Schulz, et al. 2007. Demonstration of virtual MIMO in the uplink. Institution of Engineering and Technology Seminar on Smart Antennas and Cooperative Communication, London, October 2007.
20. Zirwas, W., E. Schulz, J. H. Kim, V. Jungnickel, and M. Schubert. 2006. Distributed organization of cooperative antenna systems. *12th European Wireless Conference (2006),* Athens, Greece, on CD-ROM.
21. Jungnickel, V., A. Forck, T. Haustein, C. Juchems, and W. Zirwas. 2006. Gigabit mobile communications using real-time MIMO-OFDM signal processing. In *MIMO system technology for wireless communications,* ed. G. Tsoulos. Boca Raton, FL: Taylor & Francis.
22. Döttling, M. et al. 2005. IST-2003-507581 WINNER, deliverable D2.7 v. 1.1, assessment of advanced beam-forming and MIMO technologies, 28.02.2005; see deliverable list on http://www.ist-winner.org/
23. Weber, T., M. Meurer, and W. Zirwas. 2004. Low complexity energy efficient joint transmission for OFDM multiuser downlinks. *Proceedings of IEEE PIMRC 2004,* Barcelona, Spain.
24. Jaeckel, S., and V. Jungnickel. 2006. Multi-cell outdoor MIMO measurements. *Proceedings of VDE Congress,* October 23–25, 2006, Aachen, Denmark.
25. Official OBSAI site: http://www.obsai.org/
26. Official CPRI site: http://www.cpri.info/
27. Internetworking technology handbook. Cisco: http://www.cisco.com/en/US/docs/internetworking/technology/handbook/IP-Multi.html
28. Jungnickel, V., S. Jaeckel, L. Thiele, U. Krueger, A. Brylka, and C. von Helmolt. 2006. Capacity measurements in a multicell MIMO system. *Proceedings of IEEE Global Telecommunications Conference (GLOBECOM 2006),* San Francisco (on CD-ROM).
29. Weber, T., M. Meurer, and W. Zirwas. 2005. Improved channel estimation exploiting long term channel properties. *ICT '05,* May 2005.
30. Roemer, F., M. Haardt, and G. Del Galdo. 2006. Higher order SVD based subspace estimation to improve multi-dimensional parameter estimation algorithm. *Signals, Systems and Computers* Oct.–Nov.:961–965.
31. Thiele, L., M. Schellmann, S. Schiffermüller, V. Jungnickel, and W. Zirwas. 2008. Multi-cell channel estimation using virtual pilots. *IEEE 67th Vehicular Technology Conference VTC2008–Spring,* Singapore, May 2008.
32. Schellmann, M., V. Jungnickel, A. Sezgdin, and E. Costa. 2006. Rate-maximized switching between spatial transmission modes. *Assilomar 2006.*
33. Schiffermüller, S., and V. Jungnickel. 2006. Practical channel interpolation for OFDMA. *Proceedings of IEEE Global Telecommunications Conference (GLOBECOM 2006),* San Francisco (on CD-ROM).

34. Zirwas, W., J. H. Kim, and M. Haardt. 2007. Efficient feedback via subspace based channel quantization for downlink transmission in distributed cooperative antenna systems. *WSA2007,* Wien, Germany.
35. Haustein, T., C. von Helmolt, E. Jorswieck, V. Jungnickel, and V. Pohl. 2002. Performance of MIMO systems with channel inversion. *Proceedings of IEEE 55th Vehicular Technology VTC Spring,* May 6–9, 2002, Atlantic City, NJ, 1: 35–39.
36. Deepshikha, D., and F. Adachi. 2005. Throughput comparison of turbo-coded HARQ in OFDM, MC-CDMA and DS-CDMA with frequency-domain equalization. *IEEE Transactions on Communications* E88-B (2): 664–677.
37. Homepage for BMB+F founded project ScaleNet: http://www.scalenet.de/index.htm
38. Homepage of BMB+F founded project 3GET: http://www.3get.de/
39. Ahlswede, N., S.-Y. Cai, R. Li, and R. W. Yeung. 2000. Network information flow. *IEEE Transactions in Information Theory,* 46(4) 1204–1216.
40. Boche, H., and M. Schubert. 2008. A superlinearly and globally convergent algorithm for power control and resource allocation with general interference functions. *IEEE/ACM Transactions on Networking* 16(2) 384–395.
41. Boche, H., and M. Schubert. 2008. Concave and convex interference functions—General characterizations and applications. *IEEE Transactions on Signal Processing.* Accepted for publication. 56(10) 4951–4965.
42. Boyd, S., and L. Vandenberghe. 2004. *Convex optimization.* Cambridge: Cambridge University Press.
43. Cover, T., and A. E. Gamal. 1979. Capacity theorems for the relay channel. *IEEE Transactions in Information Theory* 25 (5): 572–584.
44. Foschini, G. J., and M. J. Gans. 1999. On limits of wireless communications in a fading environment when using multiple antennas. *Signal Processing* 6: 311–335.
45. Fragouli, C., J-Y. Le Boudec, and J. Widmer. 2006. Network coding: An instant primer. *SIGCOMM Computer Communications Review* 36 (1): 63–68.
46. Huang, C., and R. Yates. 1998. Rate of convergence for minimum power assignment algorithms in cellular radio systems. *Baltzer/ACM Wireless Networks* 4: 223–231.
47. Kelly, F., A. Maulloo, and D. Tan. 1998. Rate control for communication networks: Shadow prices, proportional fairness and stability. *Journal of Operations Research Society* 49 (3): 237–252.
48. Koskie, S., and Z. Gajic. 2006. SIR-based power control algorithms for wireless networks: An overview. *Dynamics of Continuous Discrete and Impulsive Systems* 13: 187–220.
49. Leung, K. K., C. W. Sung, W. S. Wong, and T. M. Lok. 2004. Convergence theorem for a general class of power-control algorithms. *IEEE Transactions on Communications* 52 (9): 1566–1574.
50. Monzingo, R. A., and T. W. Miller. 1980. *Introduction to adaptive arrays.* New York: John Wiley & Sons.
51. Schubert, M., and H. Boche. 2005/2006. QoS-based resource allocation and transceiver optimization. *Foundations and Trends in Communications and Information Theory* 2 (6): 383–529.
52. Telatar, E. 1999. Capacity of multiple-antenna Gaussian channels. *European Transactions on Telecommunications* 10: 585–595.
53. Viswanath, P., D. N. C. Tse, and R. Laroia. 2002. Opportunistic beam-forming using dumb antennas. *IEEE Transactions on Information Theory* 48: 1277–1294.

54. Wolniansky, P., G. Foschini, G. Golden, and R. Valenzuela. 1998. V-BLAST: An architecture for realizing very high data rates over the rich-scattering wireless channel. *Proceedings of URSI International Symposium on Signals, Systems, and Electronics,* New York, pp. 295–300.
55. Yates, R. D. 1995. A framework for uplink power control in cellular radio systems. *IEEE Journal of Selected Areas in Communications* 13 (7): 1341–1348.
56. Yeung, R. W., N. Cai, S.-Y. R. Li, and Z. Zhang. 2005. Theory of network coding. Submitted.
57. Yeung, R.W., S.-Y.R. Li, N. Cal., and Z. Zhang. 2005. Network coding theory part I/II: Single/multiple source. *Foundations and Trends in Communications and Information Theory* 2 (4–5): 241–381.
58. Zander, J., and S.-L. Kim. 2001. *Radio resource management for wireless networks.* Boston: Artech House.
59. Jungnickel, V., M. Schellmann, A. Forck, H. Gäbler, S. Wahls, T. Haustein, W. Zirwas, J. Eichinger, E. Schulz, C. Juchems, et al. 2007. Demonstration of virtual MIMO in the uplink. IET invited paper, October 22, 2007, London.
60. Weis, M., G. Del Galdo, and M. Haardt. 2007. A correlation tensor-based model for time variant frequency selective MIMO channels. *Proceedings of International ITG/IEEE Workshop on Smart Antennas (WSA '07),* Vienna, Austria, February 2007.
61. Kim, J. H., W. Zirwas, and M. Haardt. 2008. Adaptive codebook: Ultra-efficient feedback reduction method for cooperative antenna systems. New Orleans: Globecom 2008.

Chapter 8

Multihop Extensions to Cellular Networks—the Benefit of Relaying for LTE

Rainer Schoenen

Contents

8.1 Introduction

In technologies developed a few years ago, the use of relays has hardly been studied, not to mention any standardization taking place. Long term evolution (LTE) documents so far do not contain many hints about the use of relays [1,6,7,8]. A relay can be built for a kind of stealth mode so that it appears exactly like a user terminal (UT) toward the wired base station (BS) and behaves like a BS toward the UTs associated with it [12]. However, radio resources are a scarce good, so it is much better to integrate multihop extensions right from the beginning into the protocol [2,3].

8.1.1 Motivation for Multihop

There will always be a demand for high data rates in large areas. Any public radio communication technology aims at providing high radio capacity and large coverage. Unfortunately, the properties of radio wave propagation [35] limit the ideals to certain bounds. What the conventional cellular architecture offers does not match the demand for several reasons.

- *The limited power problem.* The transmit power of any sender is limited by regulations so that there is no harm to humans in the neighborhood. This is an EIRP limit, so antenna gains do not give an improvement for the sending side. With limited given transmit power level, when using higher transmission rates we get lower energy per bit, and the Shannon bound dictates that there will be more bit errors. Whether we use forward error correction (FEC) or automatic repeat request (ARQ), there is no way to increase the effective net bit rate with a single antenna.

- *Radio propagation.* Above ultrahigh frequencies (e.g., 2 GHz), there is a much higher vulnerability to bad non-line-of-sight conditions, but this is a common scenario in densely populated urban areas due to buildings, cars, and other obstructions. In effect, the path loss is higher between BS and mobile.
- *Nonconstant rate offer.* The maximum data rate offered by a BS depends on the distance of the mobile from it. The closer the UT is to the base station, the higher the received SINR (signal-to-interference-plus-noise ratio) value will be. A high SINR allows the highest modulation and coding scheme (PhyMode), which offers the highest data rate. But, with increasing distance, weaker PhyModes must be used. At its extreme near the cell border, the maximum possible offered data rate is one order of magnitude lower. If we consider many UTs distributed equally over the area, the problem looks worse because, if some terminals operate at the lowest PhyMode, they require a 10-times bigger share of the spectrum than the same number of terminals operating at the highest PhyMode. This means the average cell capacity is more than proportionally determined by the maximum possible rate at the outer regions. Figure 8.1 gives an idea of this problem.

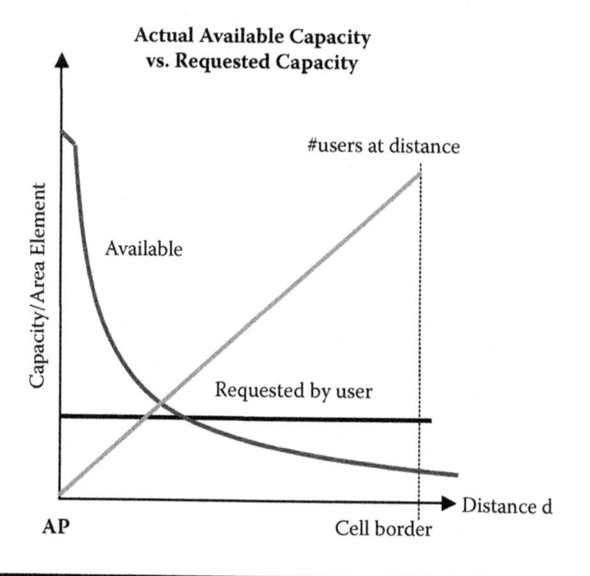

Figure 8.1 The single-hop problem: with a constant user density, the number of users increases with the distance *d*, so the cell capacity offered per area element differs from the capacity requested by the users.

One brute-force solution to this problem is to place more BSs per area, but this comes with much higher deployment costs because every base station needs its own access to the fiber backbone network. Future base stations cannot reuse the simple wired access that the global system for mobile communications (GSM)[5] over UMTS (universal mobile telecommunications system) BSs had due to their high access rate beyond 100 Mb/s. Investing these costs is economically feasible only if the number of subscribers increases at the same rate. This is unlikely because the penetration of mobile phones is already very high in modern societies. The same number of subscribers will demand more and more transmission rates, which makes the aggregate data rate of a cell the bottleneck in future wireless systems. Assuming that customers are not willing to pay more for the increasing demand, increasing the density of base stations does not seem economically reasonable.

This discussion leads to the conclusion that more fundamental enhancements are necessary to meet the future requirements. Overcoming the Shannon bound with multiple antennas will raise the data rate in certain circumstances. In addition, the wireless network architecture itself can be modified. A radio cell can then contain a number of relay nodes (RNs), also called fixed relay stations (FRSs), that do not need any fiber access. This proposal has existed for quite a long time [16,30].

Both "multihop" and "relaying" in the following discussion refer to layer-2 store-and-forward relays [32] rather than to amplify-and-forward relays or cooperative relays [12]. Relays like these allow planning of radio cells and enable deployment schemes for either coverage or capacity extension of a classical one-hop cell. In this method, the BS coordinates the partitioning of radio resources within the relay-enhanced cell (REC). In contrast to that, ad hoc and peer-to-peer multihop networks [31] have to work without any infrastructure; because they do not have a planned structure, they are not well suited for commercial infrastructure-based cellular networks. The goal of the described multihop-augmented, infrastructure-based networks is the almost-ubiquitous provision of coverage and throughput with very high data rates for any UT within the cell. Figures 8.2 and 8.3 show the two

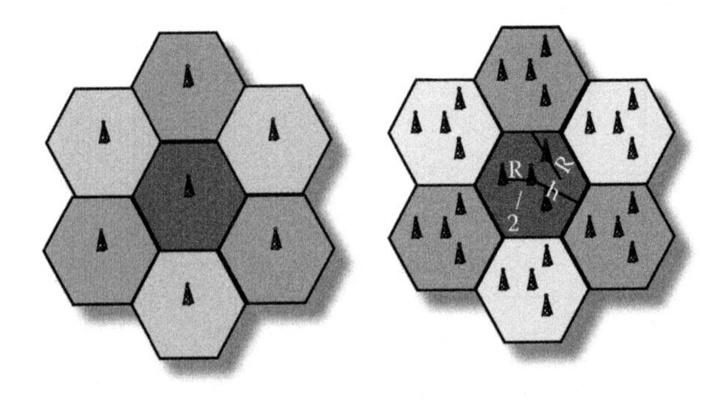

Figure 8.2 Left: single-hop cellular hexagonal layout (frequency reuse 3). Right: multihop geometry to increase the capacity of the cell.

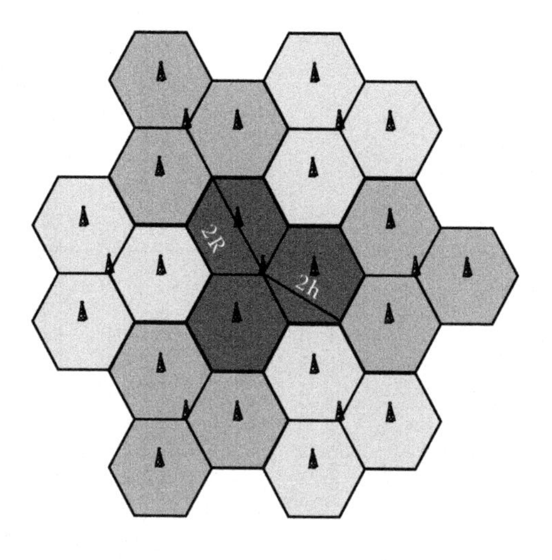

Figure 8.3 Multihop cellular geometry to increase the coverage area of a cell. With three relays, the covered area is increased +200%.

ways from a conventional cellular layout to multihop-augmented cells for both goals. A BS is located in the center of the cell and RNs are within the cell (to increase the capacity; Figure 8.2) or a normal cell radius apart (to increase the coverage of a single cell; Figure 8.3).

The following are some facts about cellular multihop networking:

- Relays are low-cost, low-transmit power devices (compared to BSs).
- They are not connected to a wired backhaul.
- Data received from the BS to the UT and vice versa are stored and forwarded later on the next transmit opportunity.
- Relays can reduce the propagation losses between BS and UT.
- Networks using fixed infrastructure relays compared to ad hoc networks do not need complicated distributed routing algorithms. The topology is a tree.
- Simultaneous transmissions of BS and RN, or only among RN, are possible with spatial multiplexing, so additional capacity gains are achieved.

8.2 Fundamentals

8.2.1 Protocol Stack

In the ISO–OSI reference model terminology, the physical layer, or layer 1, is responsible for the transport of bit streams over the channel, modulation and channel coding, equalization, synchronization, orthogonal frequency division

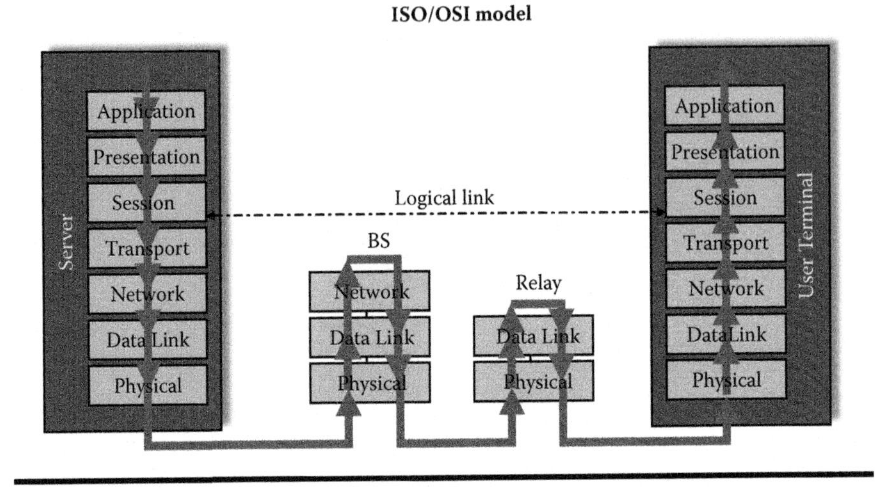

Figure 8.4 The ISO–OSI view on relays. On the user plane, they behave like a bridge by using the lowermost two layers.

multiplexing (OFDM) symbol creation, and decoding. This layer is common in all station types—BS, RN, or UT. They may only differ in points like transmit power level, RF (radio frequency) filter quality, and antenna gains. Thus, for a link between communication peers, the link between BS and RN is just like that from BS to UT, as is the link between RN and UT. It is the higher layer that gives the link a meaning. Whether it is a broadcast or unicast transmission or a data or control channel, uplink or downlink is something decided by the medium access control (MAC) functionality of the data link layer (DLL), or layer 2. This is where the radio system becomes a character and distinguishes LTE from WiMAX [4,9] (worldwide interoperability for microwave access), WiFi, or other OFDM-based systems.

A wireless relay is simply a device that receives incoming bits up to layer 2 and forwards them down again on layer 2 of the sending side, so this decode-and-forward relay acts like a repeater, bridge, or router known in the classical wired world. Figure 8.4 shows this layer model. In these systems, received protocol data units (PDUs) are fully error corrected, automatically repeated (ARQ), stored and scheduled for transmission, and, if necessary, even segmented and reassembled prior to their next-hop transmission.

Conventional cellular systems use one-hop bidirectional (duplex) communication between BS and UTs. UTs are immediately associated (linked) to the BS, which coordinates resources needed for the downlink (DL) and uplink (UL) transmission according to the traffic demand in both directions. Simply stated, for every certain rate to be supported, a proportional radio resource has to be allocated. Second-generation systems can only handle constant-rate traffic, while 3G systems have more flexibility in resource allocation. But they still suffer from assumptions that came from the circuit-switched, voice-only view on the customer. Only the next-generation systems will make no difference between data and voice communication. They even sacrifice the advantages of classical digital voice transmission for a more contemporary voice over Internet protocol (VoIP) convergence. Multihop systems are transparent for the UT. The RN acts like a BS toward the UT on the second (last) hop. Additional resources are required only for the first hop, where the RN acts like a UT toward the BS. This resource coordination was not foreseen in older systems—one reason why the deployment of smart relay concepts did not come at that time.

8.2.2 Generic Multimode Protocol Stack

Once the protocol stack for a relay is clear, it is simply the exchange of the uplink or the downlink side to construct a heterogeneous relay. For example, LTE could be used on the first hop; WiMAX or some backward-compatible GSM or UMTS could be used on the second hop. Each system requires different layers, especially on layer 2. Within a layer, we could think of these systems as a "mode" and simply supply an implementation for each mode. This is most useful for UTs that can adapt to the best (e.g., least expensive or best performing) radio interface at the current location. See Berlemann et al. [11] for the ideas behind it.

8.2.3 Link-Level Models and Performance

This chapter deals with performance results on the MAC layer. They are very much linked with the raw performance on the PHY layer. Therefore, link-level performance results are needed at first. Here a way is shown that is abstract enough not to require simulations anymore and is suitable for studies on higher layers.

For determining the required link-level results, we build upon the mutual information (MI) method [14]. This works by applying a formula from SINR to MI, and then from MI to BER (bit error ratio) and PER (packet error ratio) to get the packet error probability. For the first step, the performance data of modulation schemes typically come from link-level simulations. MI has the meaning of the number of effective bits that can be transported at a certain SINR level. It is always below the Shannon bound:

$$\mathrm{MI}_{\mathrm{shannon}}(\mathrm{SINR}) = \log_2\left(1 + 10^{\{\mathrm{SINR}/10\,\mathrm{dB}\}}\right)$$

In reality, each modulation grade comes with its own MI level, depending on SINR. For further analysis, we use an analytic expression developed by fitting the link-level result data with a suitable function [22]. Figure 8.5 shows the resulting graph. In low-SINR regions, MI is limited by the Shannon bound. In high-SINR regions, it is saturated and limited by the number of bits that the modulation scheme

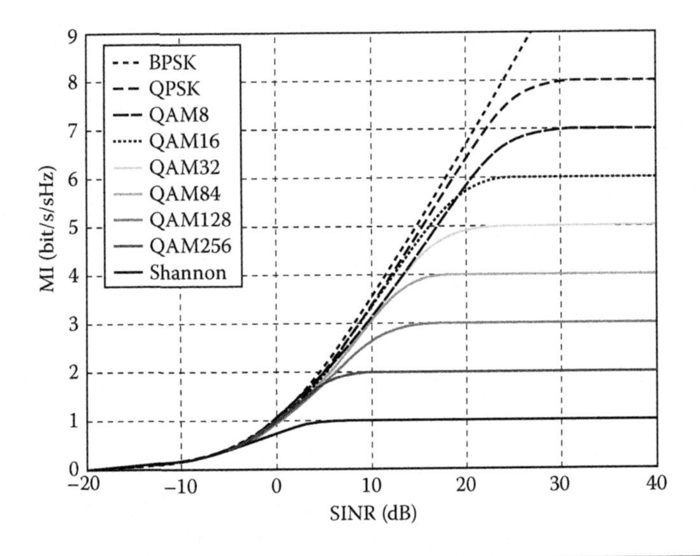

Figure 8.5 Mutual information as a function of the SINR. The modulation schemes perform close to the Shannon bound. This graph does not yet contain the error correction statistics.

supports (m). The region in between is influenced by both effects and handled by this formula:

$$MI(SINR, m) = 1 : ([s \times MI_{shannon}(SINR)]^{-w} + m^{-w})^{1/w}$$

using the following abbreviations

$$s = s(m) = 0.95 - 0.08 \times (\text{m } \textit{modulo } 2)$$

$$w = w(m) = 2\ m + 1$$

where m is the modulation index—that is, the number of bits per symbol (1 = QPSK (quadrature phase shift keying)…8 = QAM (quadrature amplitude modulation) 256). The scale factor $s(m)$ reveals the remarkable fact that square-shaped modulation constellations (m = 2, 4, 6, 8) perform slightly better than the other I/Q-asymmetric constellations.

The MI value has the unit of bits per second per hertz, so we can derive the data rate by multiplying by the bandwidth of the subchannel. The net bit rate of the PHY layer is less than this value because of channel coding. The net PHY throughput is obtained by multiplying by the coding rate. For LTE, coders have 1/3, 1/2, 2/3, and 5/6 [15]. The preceding approximation formula has been validated with tabular simulation results and differs from the simulation-measured data by at most 0.2 dB.

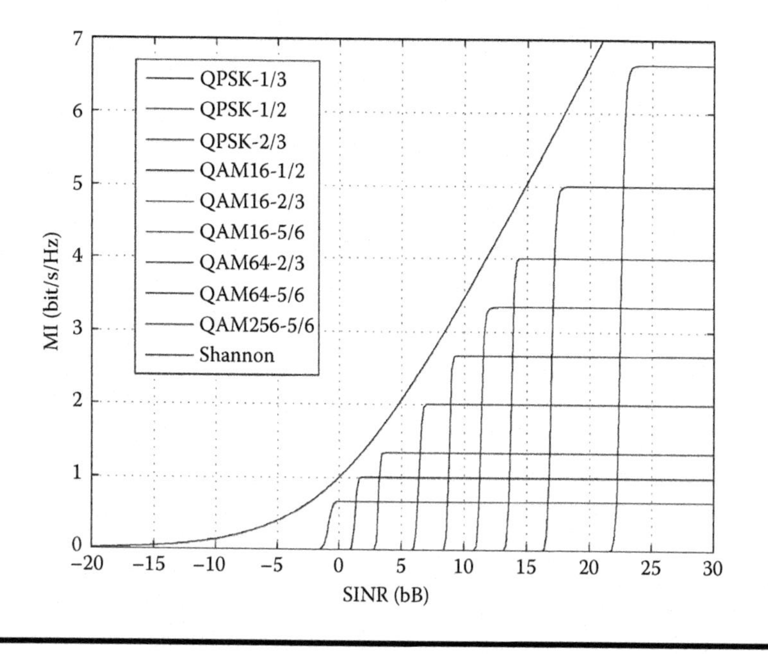

Figure 8.6 Mutual information as a function of the SINR. Here, an error correction coding using turbo codes of rates 1/3 to 5/6 is used.

On the MAC layer, bit trains become packets. The CRC unit detects packets with errors. The probability of a packet error is the PER. It depends on the type of channel coder used, the coding rate, and the packet length. This mapping must be taken from tabular results of simulations. With ARQ, these dropped packets are retransmitted. Retransmissions always come with an overhead because capacity is wasted. The resulting net rate is given by

$$r_{\text{aboveARQ}} = r_{\text{belowARQ}} \times (1 - \text{PER})$$

Taking these effects into account, the resulting $r_{\text{MAC}} = f(\text{SINR})$ can be derived.

Figure 8.6 shows this rate function for the 3G-LTE PhyModes. The mode QAM 256-5/6 was added because it can be used for the quasi-static BS-to-RN links to improve the relaying performance. The optimum switching points can be determined from the last result; Table 8.1 shows the suitable PhyModes and their best SINR intervals.

8.2.4 *Ad Hoc versus Fixed Relays*

Ad hoc networks have a different approach to multihop operation. As in IEEE 802.11 WiFi systems in ad hoc mode, user terminals can take over the role as access points. With extensions, all terminals can self-organize themselves as a mesh

Table 8.1 LTE PhyModes and Their Required $SINR_{min}$

PhyMode	*Modulation*	*Code Rate*	*$SINR_{min}$ (in Decibels)*
1	QPSK	1/3	0.9
2	QPSK	1/2	2.1
3	QPSK	2/3	3.8
4	QAM 16	1/2	7.7
5	QAM 16	2/3	9.8
6	QAM 16	5/6	12.6
7	QAM 64	2/3	15.0
8	QAM 64	5/6	18.2

network [34]. Due to the lack of centralized coordination and planning, this is not comparable (e.g., in terms of performance and reliability) to cellular systems, on which we focus in this chapter. All cellular systems require a centralized control of resources to avoid interference due to uncoordinated transmissions. With its frame-based transmission in TDD mode, WiMAX (IEEE 802.16) can also support a multihop operation (with RN) due to its centralized control [18]. This mode is preferred over the 802.16 mesh concept, which is not compatible. Mesh networks are not further covered in this chapter.

8.2.5 Theory of Relay-Enhanced Operation

"Relaying" means that any traffic destined from the fixed network toward the UT goes through the BS and one or more RNs [27,29]. We assume one relay (two hops) now, but the principles are inductive for more relays in a row [20,33]. The traffic is packetized in PDUs transmitted by the BS to the RN using dedicated downlink resource blocks (a fraction of the time and bandwidth). In the BS, the RN is treated like a UT. For the next hop, the PDUs require resources (sending opportunities) between the RN and UT in the downlink. These resources are assigned by the BS, which is typically relay aware in modern multihop designs. Thus, the BS has reserved DL resources that it does not use itself but, rather, lets the RN operate on it. The RN has full freedom of use for these resources. It decides like a BS which resource to use for which of its UTs. The size of the resources for the second hop may differ from that of the first hop, if the SINR situation is different, and different PhyModes are used. This fact has to be taken into account when reserving resources for hop two. UTs receive PDUs from an RN just like they would from a BS. They do not see a difference and do not need to be relay aware for that purpose.

For the uplink, a UT announces any available traffic to its master station, which is the RN. The RN assigns UL resources to UT for the PDU transmission while it also announces new uplink traffic toward the BS. PDUs received in the uplink of hop two are stored in the RN until the UL resources are available for hop one.

Therefore, the relaying problem simplifies to the control of resources, which is a task of any central, base-station-controlled cellular system, anyway.

Typical scenarios for cellular multihop operation are shown in Figures 8.2 and 8.3. In the coverage extension scenario, the supported area (regarding sufficient SINR) around relay nodes is of the same size as the single-hop area around the base station. The inner hexagon is typically served by BS within its range, and the outer regions are served by one of the three RNs.

For capacity extension, the whole hexagonal area can be served by the BS, but in regions closer to the cell border and near the RN, the RN offers a much higher data rate to user terminals than the BS would. Being served by a RN means that an intracell handover has been performed in which the decision has been made that the UT is better supported by the RN than by the BS. "Better" means not only that the UT receives a higher SINR at its current position, but also that the total number of resources needed for hop one (BS–RN) and hop two (RN–UT) is less than what would be required if it were a single-hop transmission between BS and UT. That is, relaying is the most efficient way to handle the UT traffic, both locally for the UT and globally for the spectral efficiency of the cell. Figures in Section 8.5 show the typical SINR conditions and the resulting supported rate around BS and RNs.

Another scenario typically used is the city or Manhattan scenario, where base stations placed on road crossings provide coverage only for those two streets with line-of-sight (LOS) contact to the UTs. Other areas are shadowed so that a transmission is not possible. Any relay placed on street corners immediately helps in serving another street [12,19].

Although relays are possible for all kinds of transmission schemes, OFDM has evolved as an efficient multiplex scheme of typically 1,024 small orthogonal subcarriers within the system bandwidth. Small subcarriers mean long symbols in time, so the problem of intersymbol interference (ISI) is relieved. The big advantage of OFDM is that each subcarrier can be modulated differently, so a robust BPSK (binary phase shift keying) can be used on frequencies where the channel is currently bad due to fading, while, for example, QAM 64 can be used on more stable subcarriers with higher SINR values. Channel coding can also be adjusted in wide ranges to adapt to the subchannel conditions. This is called dynamic adaptive modulation and coding (AMC) [26]. Channel adaptive behavior is an advanced topic. Together with adaptive power allocation (APA), it is far from comprehensive now.

Orthogonal frequency division multiple access (OFDMA) means that within one OFDM symbol, several different UT receivers can be addressed. For example, 256 subcarriers are used for each of four UTs that receive the symbol. Dynamic subcarrier assignment (DSA) selects the best resources for each UT based on channel state information (CSI), which is known due to channel quality indication (CQI). Because the distance and path loss may be very different among the UTs, it is likely that the PhyModes used for each UT are very different within the full OFDM symbol.

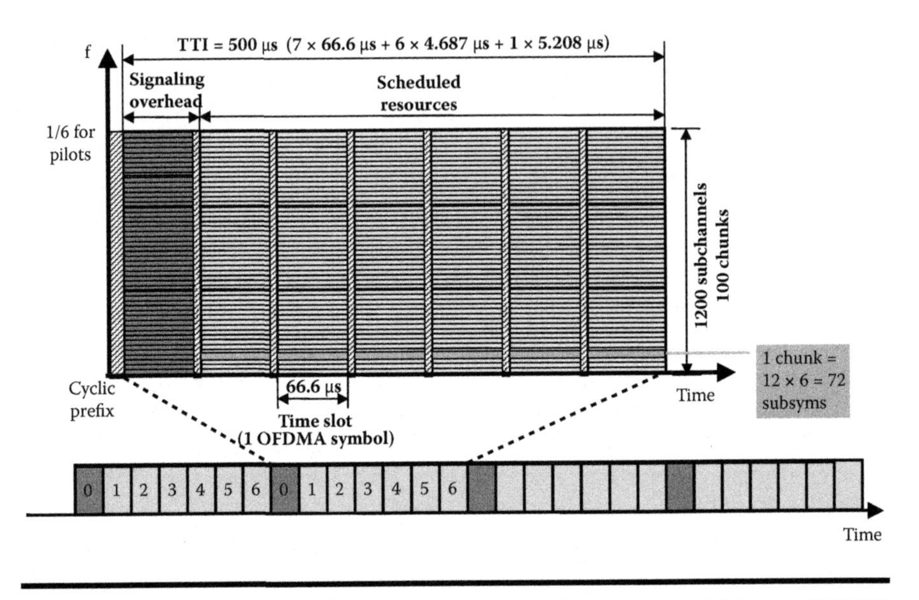

Figure 8.7 The principal layout of the orthogonal resources within one LTE TTI frame [23]. This frame contains seven OFDM symbols in 500 μs. The alternative is 14 symbols in 1 ms.

Radio resources are the valuable goods of which nobody can get enough. Traffic and tariffs are typically proportional to the used or allocated resources of a user. The Shannon bound limits the number of bits that can be transmitted in a given resource block of frequency (bandwidth) and time (transmission duration), given a certain SINR situation. In OFDMA, the granularity of these resources is typically a chunk (resource block)—that is, a group of subcarriers in frequency (f) direction and a number of OFDM symbols in time (t) direction. Figure 8.7 shows the frame format proposed for 3GPP-LTE [8]. In a real system, the pilot symbols are distributed over the resources. For simplicity, they are shown concentrated in symbol 0 in the figure. Resources shown in gray are (just arithmetically) reserved for signaling purposes (i.e., for announcing the outcome of the resource allocation decision). A chunk of size-12 subcarriers times six symbols can carry up to 360 b in PhyMode 64 QAM-5/6. Six symbols is the equivalent net size of a resource block. Smaller resources with finer granularity could fit smaller traffic better, but the overhead for segmentation, signaling, and control increases and leads to inefficiency.

An obvious additional resource dimension is space. Resources used at one location (cell) can be reused at another location (see reuse distance 3 in Figure 8.2). Also, within one cell, there can be a spatial reuse by using beam forming, multipath transmission, and spatial multiplexing by the use of distributed antenna elements. Spatial multiplexing is beneficial in multihop situations where relays belonging to one cell are separated enough from each other and can use the same $f + t$ resources in parallel [12]. The remaining orthogonality dimension is the code. CDMA (code division multiple access) systems excessively utilize this dimension. However, together with OFDMA, there is currently no useful case of combination.

8.2.6 Cross-Layer Issues

The support of sending PDUs with different PhyModes requires an interaction between layer 2 and layer 1 (e.g., an "interface for adaptive modulation and coding"). Adaptive power allocation per subchannel also requires a cross-layer interface. In the other direction (upstack), measurement reports about the received signal strength on certain subchannels must be sent in regular intervals from the PHY layer so that CQI messages can be prepared in higher-layer control plane units. These issues are on a link basis (i.e., there is no need to handle relays in any special way). The UT side of a relay (RNUT) acts like a UT and sends CQI information back to its base station. The BS side of a relay (RNBS) receives CQI messages from its UTs and uses this information for its local downlink scheduling.

8.3 Reducing Infrastructure Deployment Costs

Multihop networking claims to reduce deployment costs because it is assumed that RNs are much cheaper than BSs. Regarding CAPEX and OPEX, there is no need for any fixed line access to RNs, but BSs require expensive fiber lines. It is hard to get real expenditure numbers from operators. A simple analysis can help here. Let C_{BS} be the cost of a BS and C_{RN} that of an RN. All costs include CAPEX and OPEX over a period of, say, 8 years. For a coverage extension scenario with three RNs, we know that the covered multihop area is $A_{MH} = 3 \times A_{SH}$. Thus, the costs for serving this area with multihop are $C_{MH} = 1 \times C_{BS} + 3 \times C_{RN}$ compared to $C_{SH} = 3 \times C_{BS}$. Therefore, it is an advantage of multihop if $C_{MH} < C_{SH}$—that is,

$$C_{BS} + 3 \times C_{RN} < 3 \times C_{BS} \text{ or } C_{RN} < 2/3\ C_{BS}$$

It is not hard to meet this cost requirement. There is more on this topic in Timus [28].

8.4 Data Link Layer Algorithms for Multihop

The data link layer functionality is what makes a radio system characteristic [5]. Typically, each radio system has its own timing; that is, frames look different—some are frequency division duplex (FDD) and some are time division duplex [10]. (TDD), the number of resources in frequency and time direction is different, and so on. Also, the functional units inside layer 2 differ a lot. In this section, the detailed units in layer 2 are shown and discussed with respect to their importance for relaying.

8.4.1 Frame Timing (Radio Frame, TTI Frame)

The BS always sets up the master frame (i.e., generates the timing schedule for the next period). Beginning with synchronization pilots, a broadcast channel (BCH), and a random/contention access channel, a number of regular frames follow. The BS also reserves some of the frames for the second hop, to be used in the responsibility of the relay. With stealth relaying, the BS does not even distinguish between relays and ordinary UTs because the relay reserves the radio resources needed for its second hop by pretending to have a huge uplink traffic demand. This was a method for older non-relay-aware systems. However, a central, relay-aware BS-controlled resource allocation is preferable in all cases for future systems. The preamble is typically also sent by relays because UTs are often not able to receive the BS preamble in the coverage scenario. However, with transparent relays, only the BS sends the preamble.

Resources for the second hop that are located within the resources granted to the RN in stealth mode make up the frame-in-frame concept [16]. Modern multihop systems incorporate interleaved multihop frames, where frames for the first and second hops alternate in time, all controlled by the BS. In Figure 8.7, the downlink frame format for one TTI frame is shown. Some people define a TTI frame to be 1 ms long, but the principal properties remain the same. The smallest granularity resource unit is a chunk, and 100 of them fit into this basic frame. Symbols used for signaling and synchronization are out of focus here (shown in gray) and the only impact is the overhead (resources not available for data throughput). Downlink and uplink transmissions happen simultaneously in the same frame raster. There are three ways to integrate relaying (i.e., the frames used for the second hop—or beyond):

- *Time domain relaying.* Resources for hop one and hop two are separated in time (sequentially).
- *Frequency domain relaying.* Resources for hop one and hop two are separated in frequency (neighbor band).
- *OFDMA domain relaying.* Resources for hop one and hop two are separated in frequency (subchannels).

Frequency domain relaying is trivial. It simply means we need another center frequency for the second hop. Then the BS does not need to reserve extra (idle) frames. It treats RNs as UTs, and the second hop acts as if it were a standalone cell. The result in performance is the same as for a wireless feeder. The main drawback is the higher usage of spectrum (i.e., a reduced spectral efficiency of the system).

With time domain relaying, for a normal single-hop transmission (upper terminal in Figure 8.8), the BS uses dedicated resources on the downlink channel, which it can allocate for itself, knowing the traffic demand. Transmissions toward the RN happen in the same frame, just as with any other terminal. The traffic is known as well because the downlink scheduler knows the number of packets in its queues.

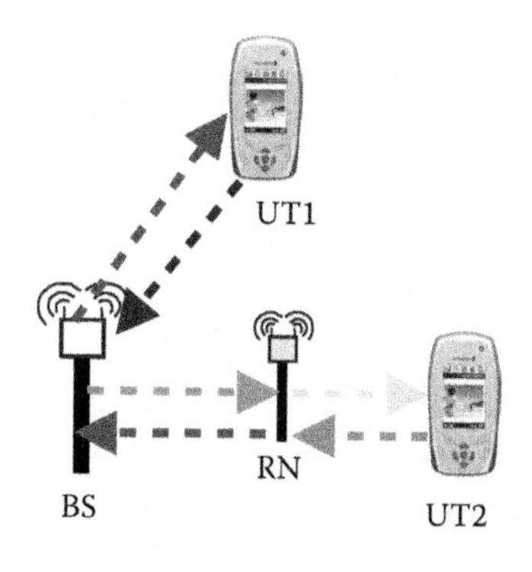

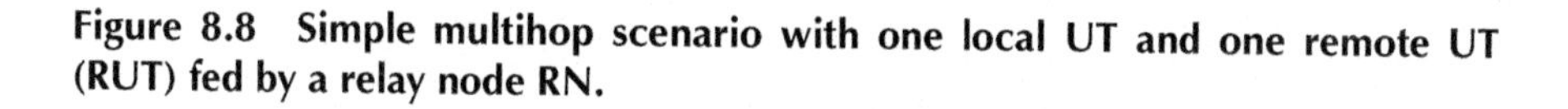

Figure 8.8 Simple multihop scenario with one local UT and one remote UT (RUT) fed by a relay node RN.

Power control can be used to assign a different transmit power to both blocks; the station closer to the BS does not need the full power compared, for example, to a UT at the cell border. Because of the fast Fourier transform (FFT) operation for OFDM, the full DL bandwidth is in use by the RF back end of the BS.

For the second-hop DL transmissions, the RF sender of the RN is active. Because RN and BS are hard to synchronize (especially for the required OFDM orthogonality), all the power of the other sender must be treated as side-band interference. Therefore, the transmission must happen in a separate frame to avoid interference and the BS must reserve the complete resources of this frame for the second hop. Thinking about guard band distances could be another approach; this is difficult, however, because of the huge variations of signal and interference power within the area covered by both senders.

The frame schedules at BS, RN, UT1, and UT2 for a small scenario are shown in Figure 8.9. In the downlink direction, either the BS sends to UTs and RNs or the RN uses this time slot. The uplink is simply used in parallel at the same time, but this is not necessarily required. For duplex and simplex (half-duplex) FDD UTs, the schedule is different. The right-hand side of Figure 8.10 shows that the send and receive phases must not happen simultaneously for simplex terminals. Without the need for a duplex filter, simplex allows for cheaper radio hardware.

OFDMA offers some additional benefits for relaying. The resources can be subdivided in a finer granularity than what would be possible with OFDM only. Figure 8.9 shows that first-hop transmissions are all treated the same way. They just occupy the required resources for their traffic. There is no waste due to completely assigned but incompletely used frames. In the uplink, also, several UTs share the full bandwidth, each of them transmitting on a subset of subchannels. The BS or RN coordinates the orthogonal noninterfering use of these subchannels by the UTs. OFDMA subdivision can also take place in the downlink so that BS and RNs send on distinct subchannels.

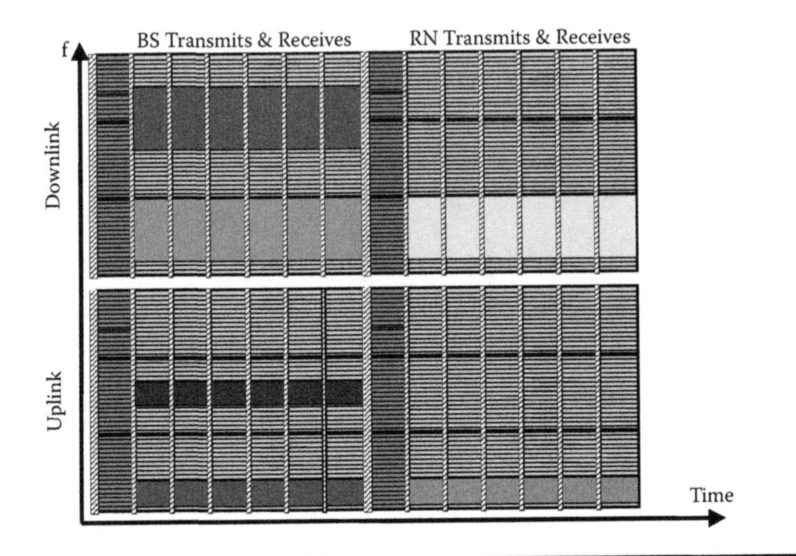

Figure 8.9 Resource usage in uplink and downlink. Shades correspond to the ones used in the previous figure. The RN toggles its task phase between UT and BS modes.

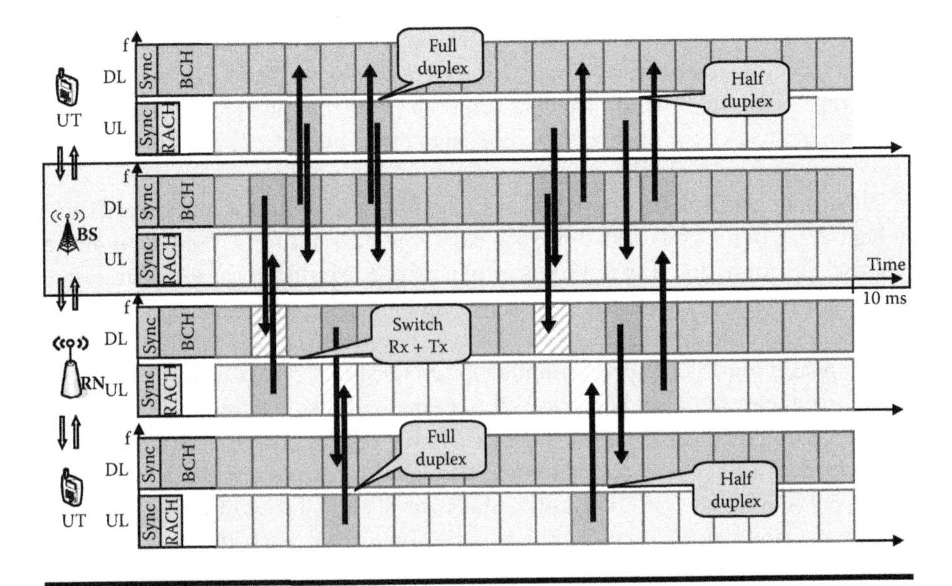

Figure 8.10 Radio frame format for the LTE FDD operation. Frame at BS: BS sends to downlink and receives in uplink resources. Top and bottom frames: UTs send in uplink and receive in downlink resources. Frame 3: RN has transceivers that can send and receive in both UL and DL.

8.4.2 Scheduling

In layer 2, the resource scheduler is one of the most important functional units. Its purpose is the assignment of PDUs to free radio resources. The PDUs waiting in the send queues are ordered by connection identifiers, and the scheduler can read out their current occupancy. A part of the scheduler's functionality has nothing to do with wireless resources—for example, managing the queues and overflow situations, handling quality of service (QoS) classes, ensuring fairness for data traffic, and prioritizing real-time critical traffic. The wireless part is the solution to the problem—which resources are used best to support a particular UT (or RN because this is treated just like a UT). The scheduler takes care that, within a frame and inside a cell, there is no interference (i.e., no two senders are allowed to send at the same time on the same resource). There is one scheduler for the DL and one for the UL parts of the transmission. Both work independently.

For the DL, the queue occupancy is known from the queues nearby; however, for the UL, there must be a signaling from the UT (or RN) to the BS telling about the number of bits and PDUs waiting inside the client stations. For QoS support, this information must be available per QoS class or priority. Relays simply work similarly. They receive resource requests from their UTs, schedule their DL and UL resources autonomously, and communicate their decision just like a BS toward its UT, within a resource map. The only difference from a BS is that the available resources on which the RN can decide make up a subset of all available resources on the link, and this subset has been assigned by the BS earlier. This resource partitioning is necessary so that, within a cell, all resources are distinct and interference is mitigated. The following functions are located in a wireless resource scheduler, but only their relevance with regard to relaying is emphasized here.

8.4.2.1 Dynamic Subcarrier Assignment (DSA)

DSA is a free decision of the scheduler responsible for the (single-hop) link. The set of available resources is given by the next-higher hierarchy leader, up to the BS. But the local decision is independent and can be based on CQI measurements, for example.

8.4.2.2 Adaptive Modulation and Coding (AMC)

The PhyMode to use within each resource is also a link-local decision. The scheduler knows an SINR estimation of the link and decides which PhyMode to take. This can be different for each subchannel if, for example, the channel is frequency dependent due to fading. Figure 8.11 shows such a frequency- and time-dependent channel. Due to changing PhyModes and probably different PhyModes on hop one and hop two of a multihop connection, a PDU frequently takes more or fewer resources on hop two than on hop one. This is the reason for buffers required to

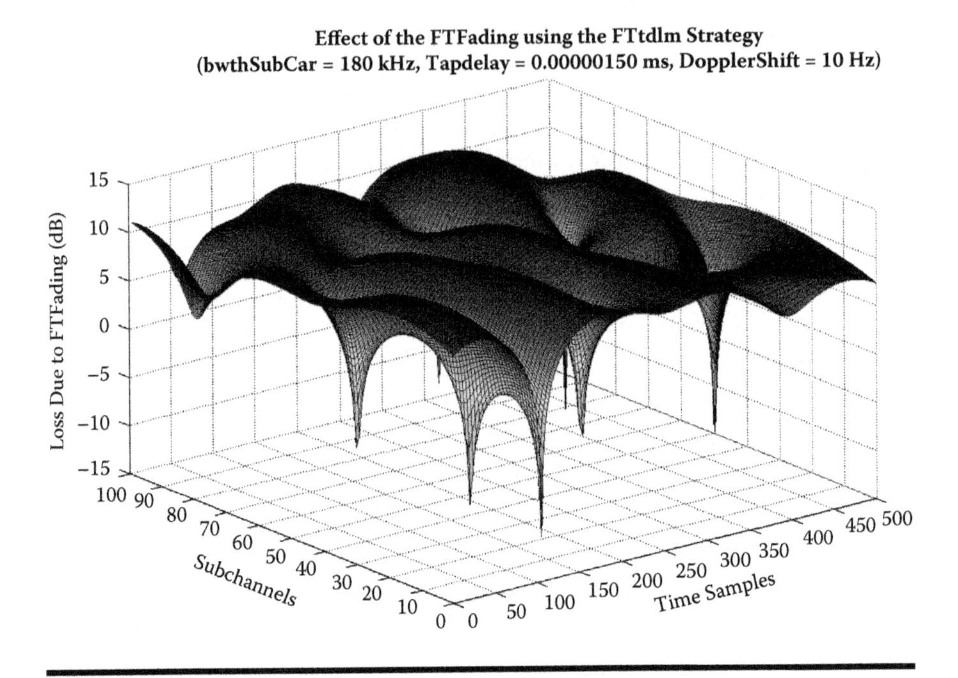

Figure 8.11 Typical frequency- and time-dependent behavior of a broadband OFDM channel.

store PDUs in flight and a proper resource partitioning so that an adequate resource set is always available on both hops [36].

8.4.2.3 Adaptive Power Allocation (APA)

Within certain bounds, the transmit power can be adjusted, even per OFDM subchannel. Transmissions to UTs nearby need less power than to UTs in the distance, so this decision is made locally.

8.4.3 Closed-Loop Control

Figure 8.12 shows that the methods for adaptivity are nothing less than functional blocks within a control loop. The value to be controlled (optimized) is the throughput, w.r.t. the least possible transmit power possible. This can be transformed into controlling the received SINR value to be about 20 dB, which is just enough to support the highest PhyMode. This control happens per subchannel because each one may be subject to fading (Figure 8.11). The bound of the total available RF power (maxPower) gives a global limiter block, which proportionally affects each

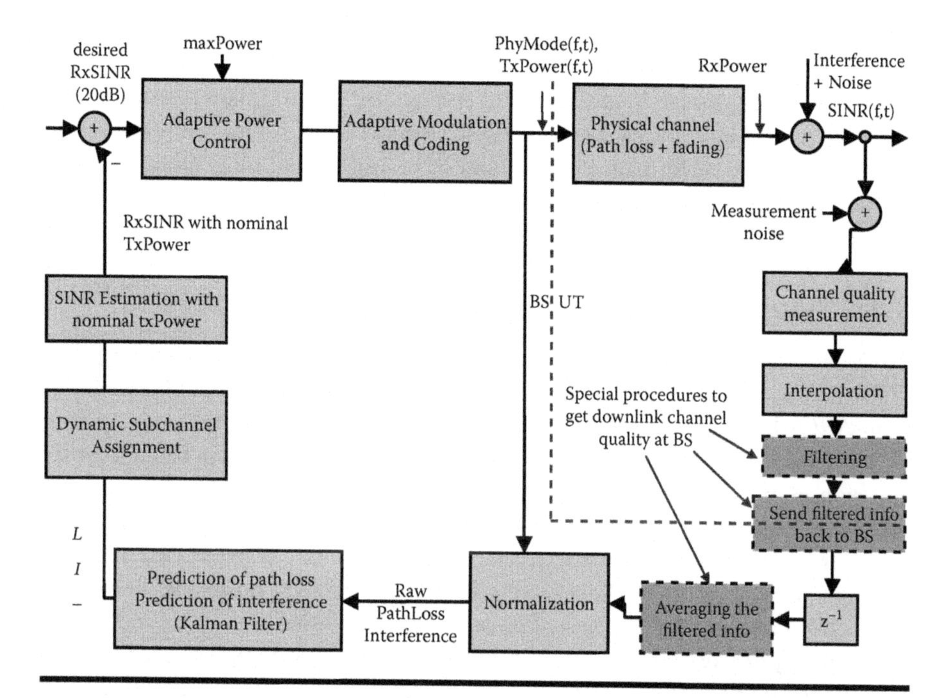

Figure 8.12 Closed-loop control theoretic view on the link adaptation. The scheduler relies on channel measurements, which allow dynamic subcarrier assignment, adaptive power control, and adaptive modulation and coding. In this control circuit, the desired SINR of approximately 20 dB at the receiver is the target value. In border regions of the cell where the path loss is too high, limiters within the power control block reduce the actual SINR. The control is vectorized. Each subchannel is one element.

subchannel. Interference, noise, and time-dependent path loss are disturbing values for which the control loop aims to compensate.

In the lower part of the block diagram is the feedback loop. The BS (or RN) receives CQI measurements in regular intervals, which arrive delayed compared to the current state (block z^{-1}). The received signal power is then normalized by the known transmit power so that, after normalization, the path loss and interference measurements are available. A prediction unit processes the data and estimates the path loss for the short-term future on all subchannels. DSA is now done based on the path loss vector, assigning the best subchannels to each user (multidimensional control). On the chosen subchannels, the SINR at the UT is estimated, assuming the use of a nominal transmit power level. At the output of the comparator, it is now known if there was too much power (reduce it) or too little power (reduce the PhyMode in AMC). The reaction time of the control system is at least one round-trip time or two TTI frame lengths, depending on the frequency of CQI updates.

8.4.4 Forward Error Correction (FEC)

The PhyMode consists of modulation and coding. This coding allows for correcting some bit errors before further processing so that the PDU is received without any packet errors. The performance is incorporated in link-level result graphs, which are also shown in this chapter. Because this is a one-way, single-link mechanism, there is no relay-specific issue with that.

8.4.5 Automatic Repeat Request (ARQ)

ARQ is the most important function to ensure error-free reception above layer 2. All PDUs that still have bit errors after FEC are detected by an additional CRC checksum. ARQ then sends negative acknowledgments (NACKs) to the sender asking for the transmission of this PDU again. In a multihop setup, this can be done either hop by hop or end to end. From simulative studies and real-world experiments, it is known that a hop-by-hop solution is robust and performs best. End-to-end ARQ suffers from the problem that there is not only one link in between, but also two (or three) links, as well as buffers in the relay in between. Any buffer overflow looks like a link failure and triggers a retransmission, which leads to more traffic and even more buffer loss. This situation is both severe and common because the links of hop one and hop two are typically not the same in terms of capacity; thus, any unmatched link capacity leads to transient buffer overflows from time to time.

8.4.6 Hybrid ARQ (HARQ)

HARQ is a technique to combine FEC with ARQ already in the physical layer. In case of a transmission error, the old received bits are not discarded but, instead, combined with a new, retransmitted PDU, and the result is more reliable than a normal retransmission would be. Because this is a link-local mechanism, there is no effect on relaying.

8.4.7 Radio Resource Management (RRM)

The resource schedulers decide upon the resources given to them. In a long-term view, these resources vary as new UTs come into the cell and some are handed off. Each time, the coarse resources must be redistributed. The RRM unit knows the number of UTs associated with the BS and each RN and its traffic contract, so it can make a proper decision [36]. This unit must be relay aware because it must send this partitioning decision to the RNs.

8.4.8 Signaling

There are several information elements that need to be exchanged between the BS and UTs and RNs and UTs. For these purposes, dedicated control channels exist. Without association, the BCH or random access channels (RACHs) must be used. Later, there are dedicated channels for which resources were provided. On each link (hop one or hop two), such channels must exist, but the information exchange is local (i.e., only meant for this hop). An exception is the arrival of a new UT at an RN, where the second hop operates just like BS-to-UT. However, the information about changed configuration and the additional terminal is notified from the RN to the BS so that routing and resource reservation can be adapted accordingly.

8.4.9 Handover Decision

Handover (HO) is the disassociation of a UT from any BS (or RN) and association to another BS or RN. In a multihop setup, the signaling for HO goes over all hops until the BS is informed and acknowledges. A mobility management entity (MME)/system architecture evolution (SAE) gateway in the network is responsible for switching the routing path if the BS is changed [8]. Here, with a focus on relaying; the new case is an intracell handover between two relays or between the BS and RN. Figure 8.13 shows the message sequence chart (MSC) for it. The BS is always in control of associations and directs its RNs to release (disassociate) or bind (associate) a UT to it.

8.4.10 Channel Quality Indication (CQI)

In a fading channel environment, the broad OFDM channel is frequency selective. The resource scheduler must know the channel state so that it can use this information for the DSA and AMC mentioned earlier. Therefore, the UT does channel measurements (e.g., by analyzing the receive signal strength of pilot signals). The measurements are packed into a compact format and sent back to the BS (or RN) for their further use on the DL. For the uplink, the BS (or RN) itself performs the measurements, uses the knowledge for scheduling, and tells its UL decision (PhyMode and power) to the UT (see Figure 8.14 for the MSC). The CQI handler in the BS keeps all information of DL and UL channel quality (CQ). In regular intervals, the UT sends DL-CQI measurements and uplink resource requests (UL-RREQs). Resources are granted in the DL-MAP and UL-MAP, together with commands concerning which power and PhyMode to use on the uplink.

8.4.11 Routing

Because fixed relay infrastructures do not have an ad hoc or mesh component, the interconnection graph is a tree, so routing is not an issue and is easily performed in the BS. Tables in the BS know the location of a destination MAC address and the RN that is responsible for it.

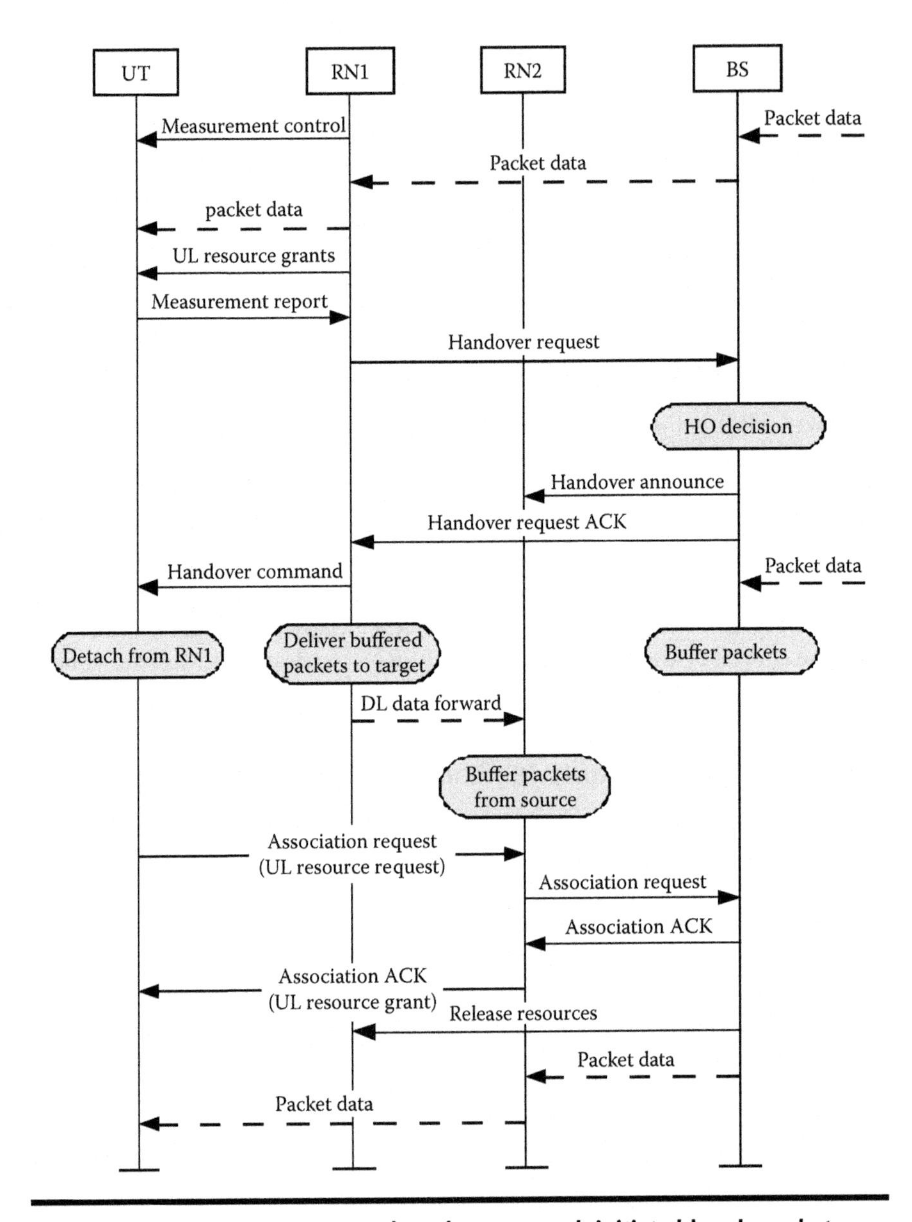

Figure 8.13 Message sequence chart for a network-initiated handover between relays RN1 and RN2.

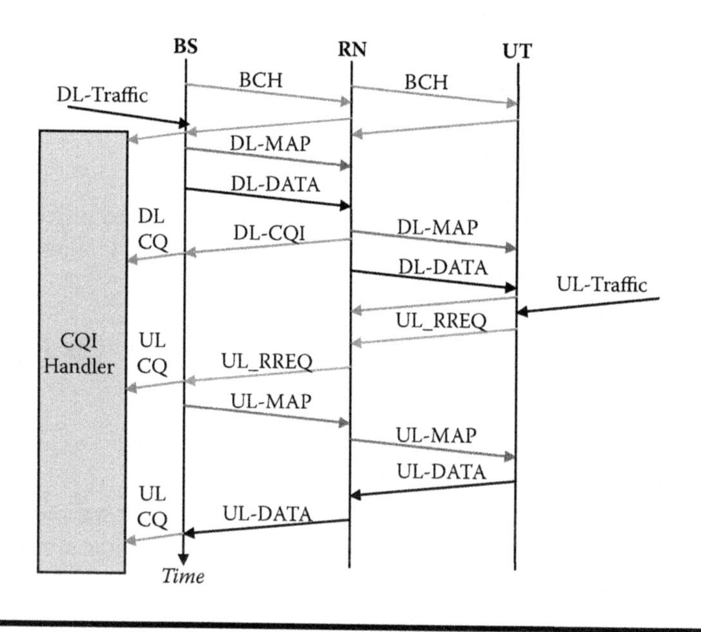

Figure 8.14 Message sequence chart for the CQI over a multihop link.

8.4.12 Intercell Interference Mitigation

Radio cells have a frequency repetition factor (also called cluster order) that specifies the density of cells operating on the same frequency band. For LTE, it is often assumed that there is no frequency reuse distance, so all cells have the same frequency band. In any case, all cells using the same frequency band interfere with each other. This is especially harmful at the cell border, where the UT might have the same distance to any BS1 and BS2 nearby. It is the purpose of an interference mitigation (IM) functionality in the RRM unit to coordinate neighbor BSs. For the best performance, the IM unit must know the position of neighbor BSs and RNs, as well as its own RNs. It can then decide to share resources in parallel, if the distance between interferers is high enough and the transmit power is reduced according to the proximity of the UT, or to assign a resource exclusively to a UT at the cell border, for example.

8.4.13 Full- and Half-Duplex FDD

It is complex to coordinate a radio system with both full- and half-duplex FDD terminals. It looks even trickier to have this working with relays. But it is just a matter of frame organization. Figure 8.10 shows a way to support half-duplex UTs. More details about half-duplex and full-duplex coexistence—especially the timing problem of resource request and grant phases, the scheduling of resources for several frames in advance, and the avoidance of wasted resources if only a half-duplex is present—are available. The work of Otyakmaz, Schoenen, and Walke [36] is recommended for more information.

8.5 Performance Evaluation

The benefit of multihop deployment is now studied in a number of scenarios [13]. The common part is the use of the following parameters, which allow calculation of the throughput from link level to MAC layer:

- transmit power: 40 W_{peak} at the BS (37 dBm), 34 dBm at the RN;
- bandwidth: b = 18 MHz net (20 MHz system);
- center frequency: 2.5 GHz;
- path loss I: non-line-of-sight propagation;
- path loss II: slow and fast fading effects;
- interference: neighbor cell BSs interfere (100% load, cluster order 7);
- noise: accounted for but not serious in interference-limited systems;
- SINR: the first performance measure below PHY layer;
- MI: mutual information determined from SINR and modulation;
- BER: bit error ratio, the PHY performance result;
- PER: packet error ratio, the result after channel decoding;
- delay: determined by PER (ARQ retransmissions) and roundtrip times;
- throughput: determined by bandwidth, PhyMode (modulation and code rate), and ARQ overhead;
- second-hop throughput: reduced by resources required on first hop; and
- third-hop throughput: reduced by resources required on first and second hops.

First, the sequential multihop scenario is studied. Second, the coverage and capacity extension scenarios (shown in Figures 8.2 and 8.3) are analyzed. Third, a realistic deployment study in Jersey is presented. Whenever the DL SINR results are plotted over the cell area, they show the SINR of the best station (BS, RN), not the maximum SINR. The important difference is that the *best station* is determined by the highest rate any of the stations can offer. The rate/throughput results contain the maximum achievable rate at a certain position within the cell, taking the required first hop resources into account also. Therefore, the second-hop maximum rate near the relay cannot be as much as near the BS. The relay is chosen as the serving station (association) if this is an advantage in fewer resources used, which is here the same as the maximum rate. In both scenarios, in huge areas the relay offers an advantage over the single-hop case, so there is more than just high SINR around RNs.

On the layer two, there is overhead due to framing, signaling, and ARQ retransmissions. The latter depends on PER, which can be taken into account when assuming selective repeat ARQ by the following equation:

$$r_{\text{aboveARQ}} = r_{\text{belowARQ}}\,(1 - \text{PER})$$

In total, we obtain a MAC overhead of MAC/PHY = 132.3%. The PHY overhead of PHY/RAW = 107.1% comes from OFDM cyclic prefix duration. Under multihop operation, individual resources are needed on every hop. The constant packet length requires a different resource share depending on the used PhyMode, which determines the maximum rate $r_{i,\max}$ usable on each hop. Therefore, we can get the maximum rate on the second hop to be

$$r_2 = ({r_{1,\max}}^{-1} + {r_{2,\max}}^{-1})^{-1}$$

and, on the third hop,

$$r_3 = ({r_{1,\max}}^{-1} + {r_{2,\max}}^{-1} + {\mathrm{r}_{3,\max}}^{-1})^{-1}$$

For every location (x,y), we can now determine the best rate out of $\{r_1, r_2, r_3\}$, which was used to generate the result figures. One of the three rates is maximum, and the index i of the maximum r_i determines the "best server" (i.e., it shows the station with which the UT at that location should be associated).

The system capacity is determined by assuming an equal traffic load for each user terminal and a homogeneous user density over the area [17]. This means that a UT far outside that has a low PhyMode requires more share of the resources than a UT close to the BS. The following equation for the capacity C considers this:

$$\frac{1}{C} = \iint\limits_{area} \frac{1}{Capacity}(x, y)dxdy$$

8.5.1 Sequential Multihop Scenario

For the first impression on relay performance, assume one BS1 at a center position, the next (interfering) BS2 at a distance of 9,000 m, and two relays, RN1 and RN2, positioned at 3,000 and 6,000 m, respectively. Let there be non-LOS (NLOS) propagation from the BS and RN to UTs and LOS between the BS and RN1. Figure 8.15 shows the SINR condition at any receiving UT on the line between BS1 and BS2. The BSs were placed such that the SINR is strong enough to carry at least the lowest PhyMode within the range of BS1, RN1, and RN2. The relays themselves receive enough SINR, so let the first hop carry the highest PhyMode. Figure 8.16 shows the effective rate that can be achieved around the BS and around RN1 and RN2. According to the preceding formula, the maximum rate around RN1 and RN2 is, of course, reduced because of the resources used for the (first and second) hops. Nevertheless, there is a high rate around RN1 and RN2, even up to a distance of 7,500 m. Intuitively, this shows that the capacity (high rate) can be increased and the coverage is extended (7,500 instead of 2,000 m). The next scenarios show this in a cellular context, which is the typical geometry used for radio network planning. A Manhattan scenario [19] is not studied here.

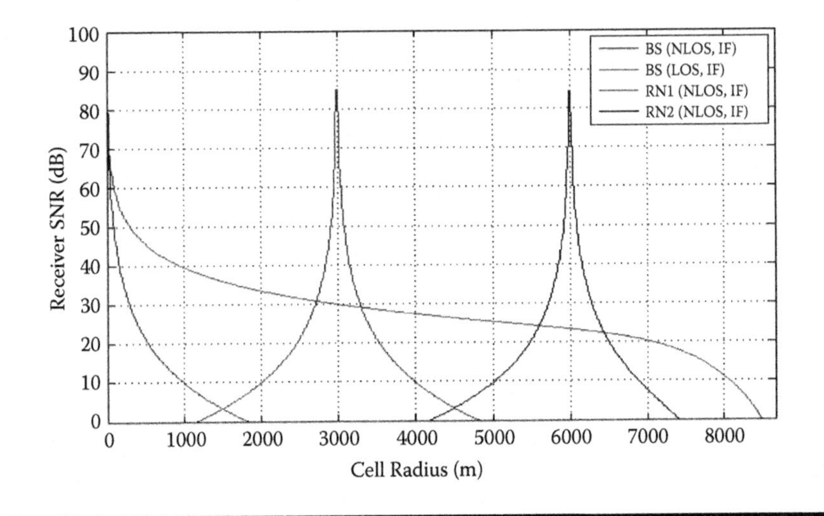

Figure 8.15 SINR at certain positions apart from the BS. Around the relays RN1 and RN2, the signal level is high enough to support a good PHY mode.

8.5.2 *Ideal Cellular Scenario*

A coverage increase layout means the cell geometry is extended, for example, with three relays (as shown in Figure 8.3) so that the covered area is three times the original area (+200%). Another geometry is shown in Figure 8.17; six relays

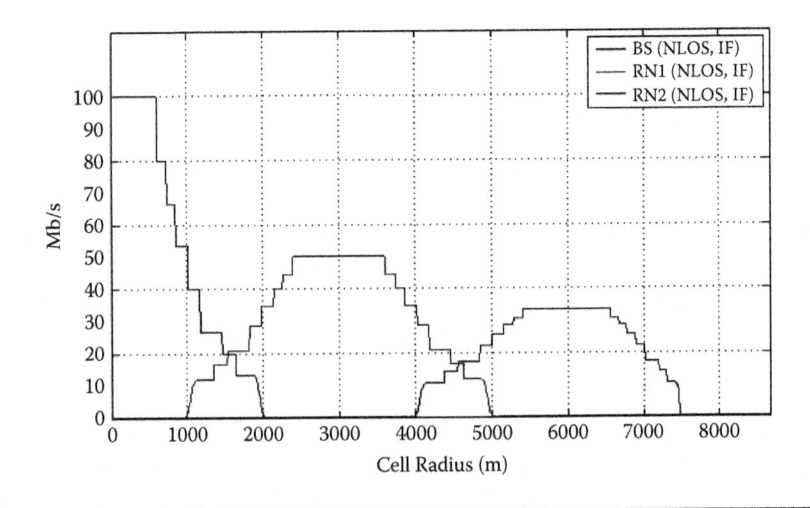

Figure 8.16 Supported rate around BS, RN1, and RN2. The first hop resources are taken into account; this is the reason for the reduced maximum rate by the RNs. Nevertheless, they are better beyond 1,800 m.

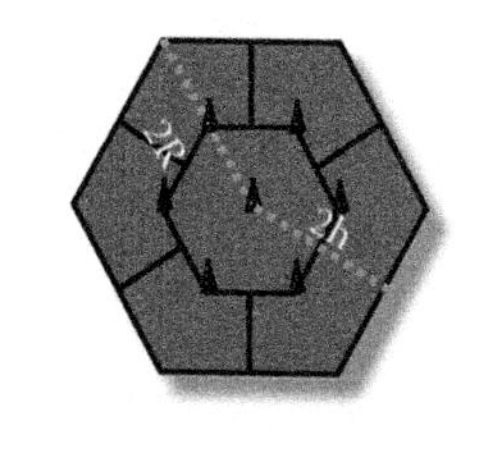

Figure 8.17 Multihop cellular geometry with six relays.

provide fourfold coverage (+300%) compared to a single cell. In Figure 8.18, the area served by the BS (the best serving station is indicated by shade) is small compared to the coverage achieved by the relays. The SINR condition at the possible receiver positions is shown in decibels in Figure 8.19. Figure 8.20 shows that the maximum rate (numbers in megabits per second) around the RNs is only half of the BS rate, but in areas the BS would never cover. Thus, although the inner cell still has capacity to offer (though not in saturated traffic load conditions), this can be used to extend the coverage very economically.

A capacity increase layout is when relays are placed within the normal radius of the single-hop cell (Figure 8.2). In this case, the BS is able to provide coverage for all UTs in the cell, even without relays. But there are areas near the cell border where an RN provides much better service in terms of SINR (Figure 8.22) and especially regarding the offered capacity for the UT (Figure 8.23). Thus, the cell border is no longer an area of the worst PhyMode, the highest (transmit)

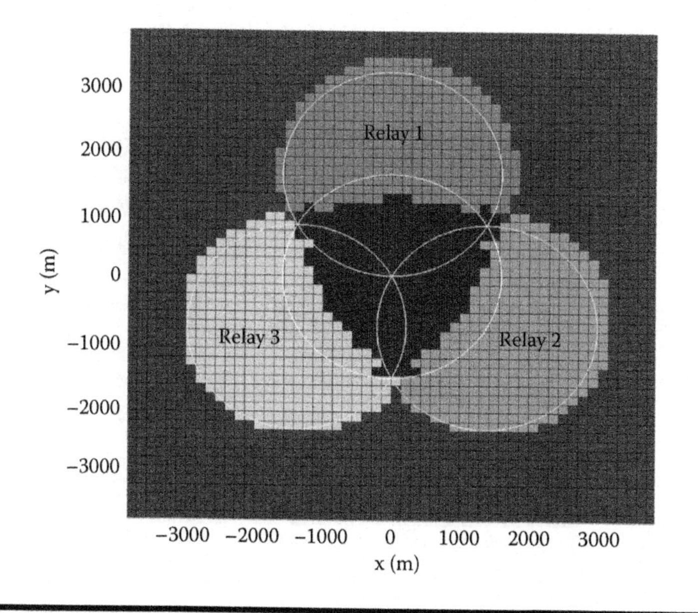

Figure 8.18 Indication of which element of (BS, RN1, RN2, RN3) serves best (i.e., gives the highest data rate to a UT) over the cell area in a coverage-increasing scenario.

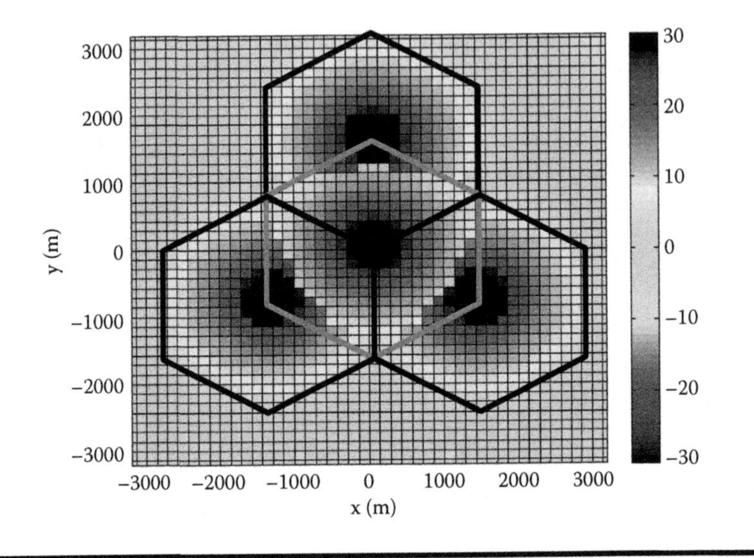

Figure 8.19 SINR over the cell area. The different colors indicate the signal strength of the best serving element (numbers in decibels).

power consumption, and the worst performance. This scenario makes sense for urban areas where the capacity demand per area at the border is the same as near the BS, so we approximate the "constant rate offer" to match the demand (see Figure 8.1).

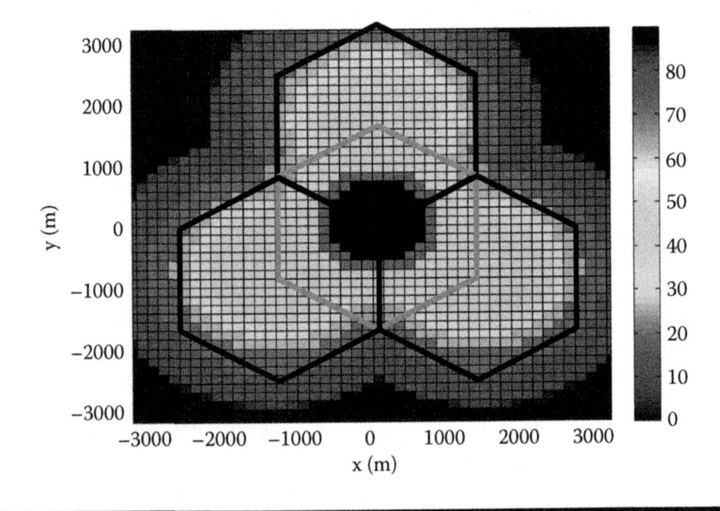

Figure 8.20 Maximum data rate achievable over the area in a coverage-increasing scenario. Due to the first hop resources, there is a lower maximum level in the relay-supported areas.

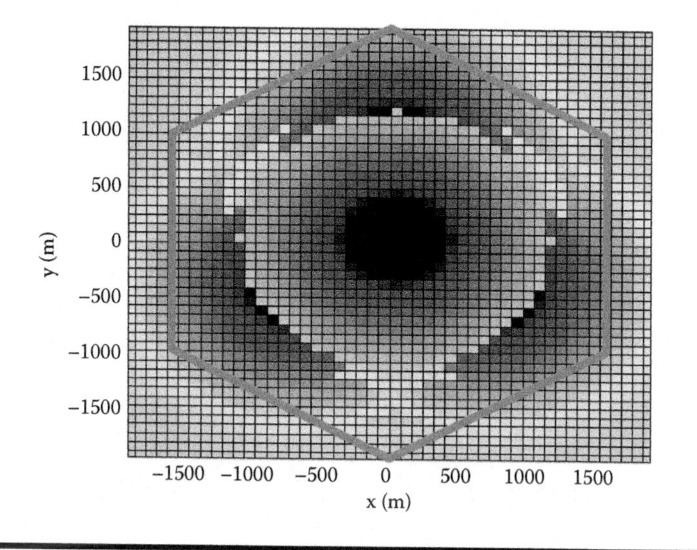

Figure 8.21 SINR in a multihop cell to increase the capacity. Here, only the SINR of the best server is shown. This produces the sharp edges at half of the radius. Beyond that, the relays are chosen as the better server.

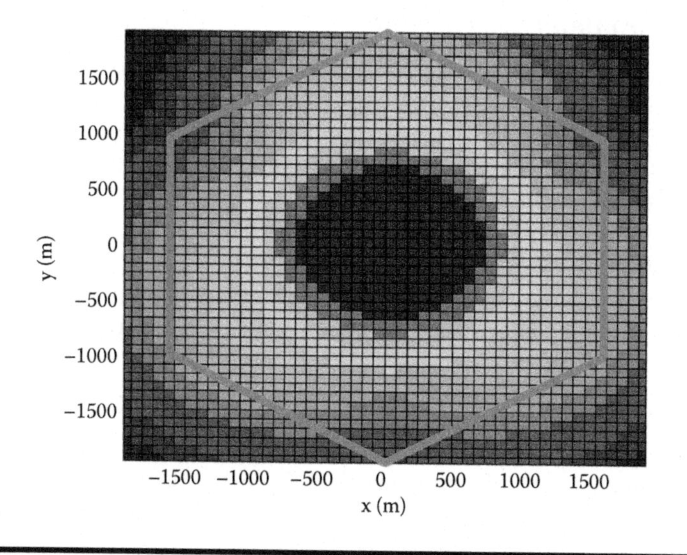

Figure 8.22 Maximum data rate achievable over the area in a capacity-increasing scenario.

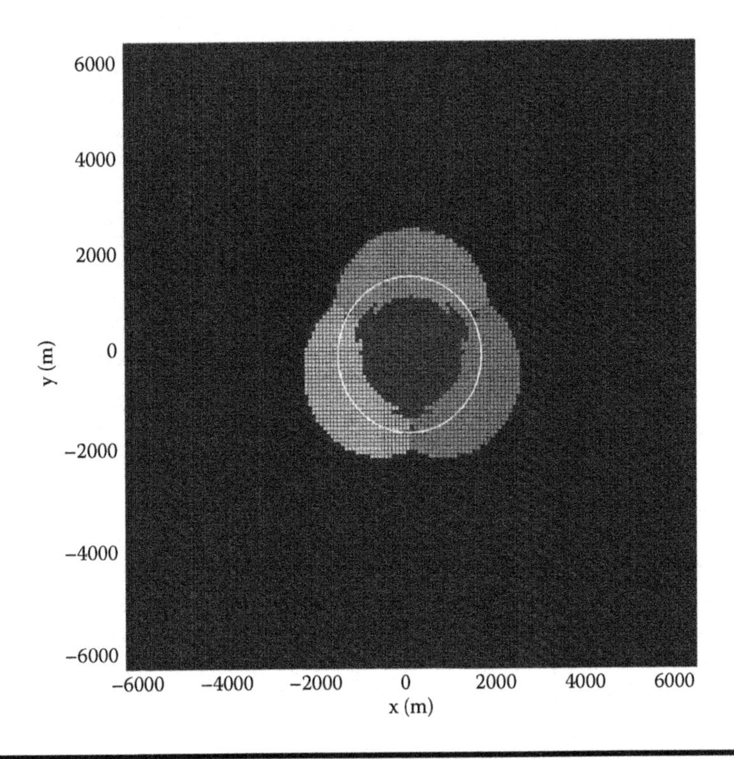

Figure 8.23 Best server chosen in a capacity-increasing scenario.

8.5.3 Parameter Discussion

Some parameters can be tuned [24]; their default values include:

- cluster order: $C = 7$;
- spatial reuse: time division multiplexing (TDM) or spatial division multiplexing (SDM);
- relay antenna gain: $G = 14$ dB or $G = 0$ dB;
- sectors per cell: $S = 1$;
- coverage-optimized (CO) versus throughput-optimized (TO) scenario; and
- directional antenna (DA; $A = 4/3\ \pi$) versus omnidirectional antenna (OA; $A = 2\ \pi$).

The 240° directional antenna of RNs is adjusted to radiate only into the outer hexagons. This is the area that is fully served by RNs as the best server. Figure 8.23 shows that an outer ring of approximately 30% within the hexagon is best served by RNs. We obtain the performance results in Table 8.2 for these two scenarios under "throughput TDM DA G14" and "coverage TDM DA G14." Switching on SDM means all RNs may transmit to and receive from their UTs in parallel and their interference is accounted for. The results are provided in Figure 8.24 for the

Table 8.2 Downlink Capacity and Spectral Efficiency of the Discussed Scenarios

Scenario	*DL Capacity (b/s)*	*DL Spectral Efficiency (b/s/Hz)*
Throughput TDM DA G14	50,973,029	2.8318349554
Throughput SDM DA G14	58,642,110	3.2578950216
Throughput TDM DA G0	48,737,548	2.7076415712
Throughput TDM OA G14	50,730,651	2.8183695086
Throughput SDM OA G14	53,799,503	2.9888612873
Coverage TDM DA G14	34,087,801	1.8937667754
Coverage SDM DA G14	64,801,466	3.6000814652
Coverage TDM DA G0	21,385,809	1.1881005356
Coverage TDM OA G14	34,937,225	1.9409569593
Coverage SDM OA G14	65,309,985	3.6283325128

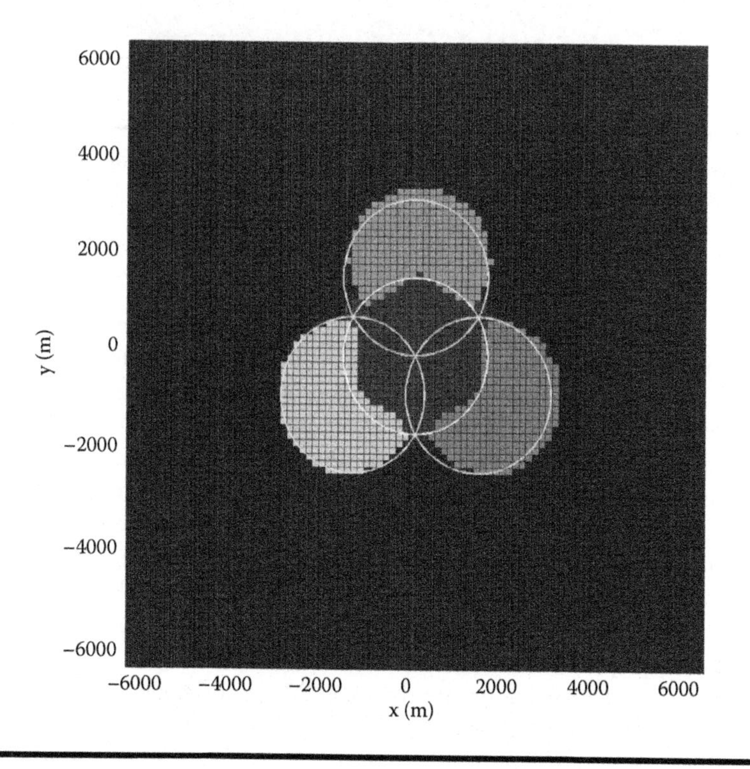

Figure 8.24 Best server chosen in a coverage-increasing scenario with sectorized relay antennas.

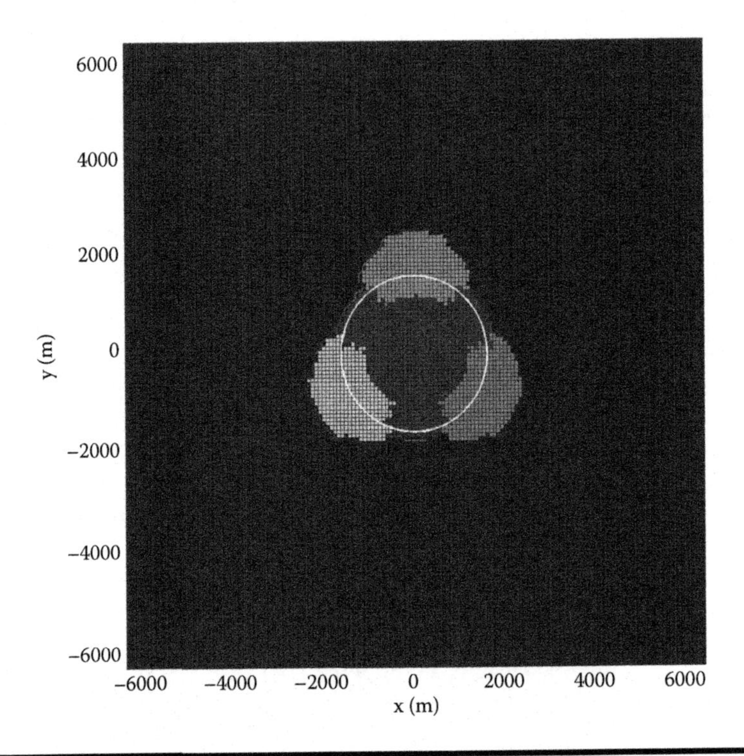

Figure 8.25 Best server chosen in a capacity-increasing scenario when spatial multiplexing is switched on (i.e., all relays may send in parallel). This causes intracell interference, reducing the effective relay-supported area.

coverage scenario and in Figure 8.25 for the throughput-enhancement scenario. Especially in the boundary area between RNs, we observe that no relay can cover this in a useful way because of the strong interference between those RNs. The preceding results were using directional $A = 4/3\ \pi$ antennas in the RNs, which point to the outside of the cell. This is especially useful to reduce SDM intracell interference. In contrast, Figure 8.18 shows TDM with omnidirectional antennas. Compared to the directional antennas, we observe a bigger area covered by the RNs at the edge of the inner hexagon. Results without antenna gain (G0) between the BS and RN are not shown graphically, but they are shown numerically in Table 8.2.

The main points about the parameters are that an antenna gain (G14) for the stationary first-hop link should be possible; directional RN antennas trade off less area near the center against more area between RNs, but the spectral efficiency does not change much; and using the SDM resource reuse among RNs, we gain a lot of capacity, especially in the coverage scenarios.

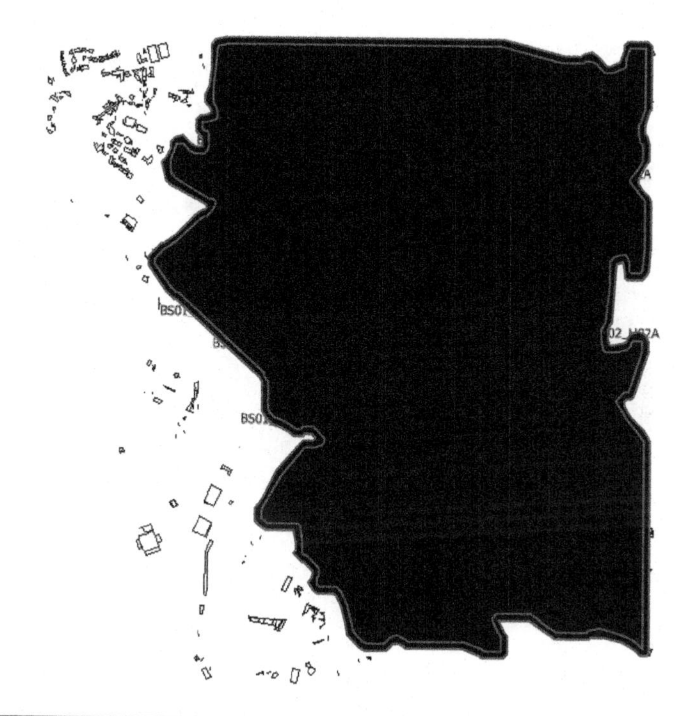

Figure 8.26 Map and bounding area of the Jersey scenario. The BS and RN are marked red.

8.5.4 Realistic Single-Cell Scenario

Now we come to a more realistic scenario for the use of fixed relay nodes. The city of Jersey was taken, with its geographical and building structure, and realistic path loss maps were simulated by ray tracing and used in this analysis [25]. Compared to the previous abstract and regular cellular scenarios, this provides a good proof of concept. Jersey has been chosen for this scenario because the area of approximately 4.439 km^2 is a typical cell size. By using a radio network planning tool, base stations and all relay nodes are placed at the best possible locations (see Figure 8.26). The relays are either fed by the BS (set RN_{H1}) or by a RN out of RN_{H1} (set RN_{H2}). Thus, there are one-hop, two-hop, and three-hop links between the BS and UTs. The topology (building placement) was known in advance, and ray tracing tools have been used to obtain the received signal power $P_{R,i}$ at each location from each possible transmitter site. The physical parameters and calculation steps have been given in the previous section.

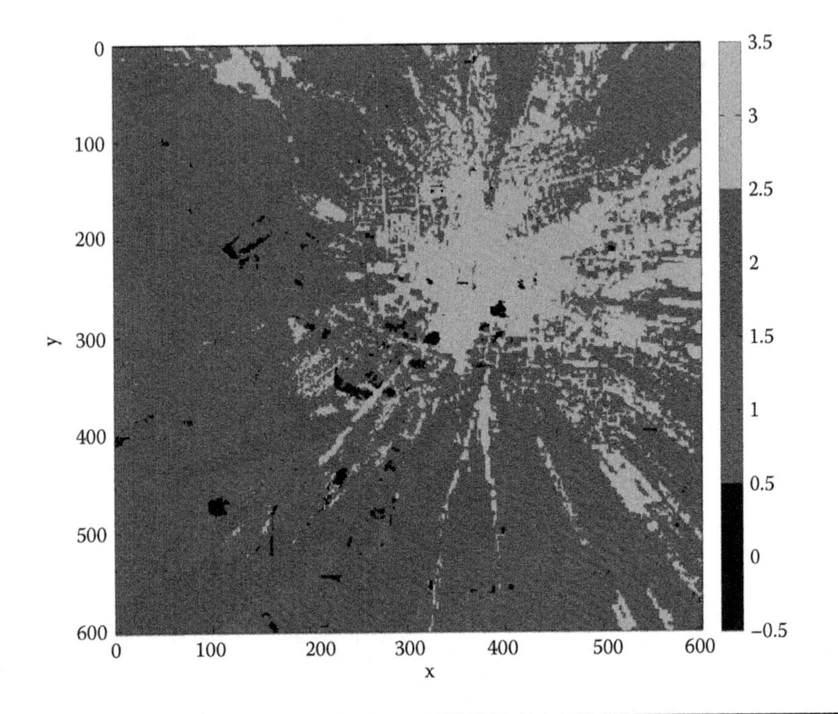

Figure 8.27 Best server and coverage in the Jersey scenario. 0 means that uncovered (SINR too low), 1 means that three-hop, 2 means that two-hop, and 3 means that single-hop transmission is better at the current position.

In the end, we obtain a coverage map that shows the serving station with which a terminal is best associated (best server) for each point on the map. This can be the BS, if this is the rate-optimal association, but it can also be one of the relays in RN_{H1} or RN_{H2}. The maximum achievable rate at a certain point is then determined. From these matrix (two-dimensional) results, scalar performance measures are calculated to show that the overall coverage and capacity are increased compared to using only one BS. Therefore, we compare the scenarios with (1) one BS only, (2) the BS plus a ring of four RNs in RN_{H1}, and (3) the BS plus a ring of four RNs in RN_{H1} plus a second ring of the nine RNs in RN_{H2}.

The performance results have been obtained over three different areas. First, the full area was used, including less densely populated parts ("square"). Second, a circular area with a radius of 800 m around the BS was defined, which gives a realistic cell size ("circle"). Third, the area within a polygon (Figure 8.26) was studied, which is exactly the urban populated area.

The analysis has been carried out to generate the two-dimensional data in Figure 8.28 that show the best server (BS = 3; $RN_{H1} = 2$; $RN_{H2} = 1$, NC = 0) and also indicate the coverage or area not covered (NC). Figure 8.27 shows the SINR of the best server and Figure 8.29 shows the resulting maximum data rate. The

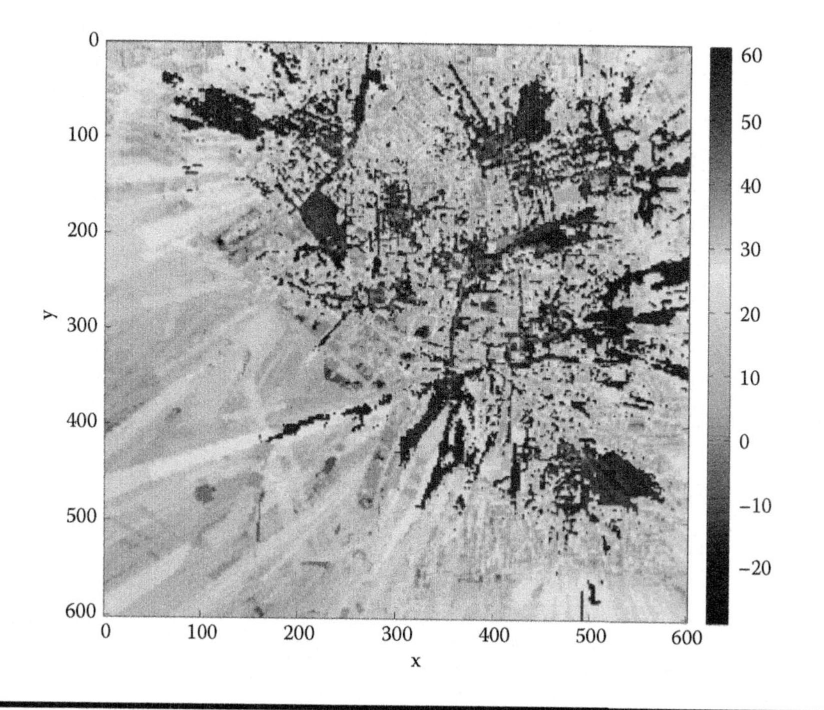

Figure 8.28 SINR distribution using all relays in the Jersey scenario.

best server has been determined by this maximum data rate that accounted for the overhead on hops one and two.

Scalar results were only counted in the polygon area defined in Figure 8.26. The coverage (in percent of the area) of the scenarios differing by the number of relays involved is determined by counting all locations with $SINR > SINR_{min}$. For LTE, $SINR_{min} = 0.9$ dB holds.

Figure 8.30 shows that, in a multihop scenario, increasing amounts of the coverage area of the BS are taken over by RNs. The capacity C in bits per second can be used to calculate the spectral efficiency $e = C/b$ using the system bandwidth b. The performance metrics are shown in Table 8.3 for the full area and in Table 8.4 for the densely populated polygon. According to this, two tiers of relays, compared to the BS-only scenario, increase the coverage by a factor of 2.08 overall and 1.53 within the city, as well as the capacity and spectral efficiency by a factor of approximately 1.33 overall and within the city.

Figure 8.31 shows the partition of the area served by the BS, H1, or H2 dependent on the (circular) radius of the cell. We observe that a first ring of relays comes in early, and a second ring is worth using beginning from 350 m. Again, these numbers show the benefit of using relays. It is not the radio power (which is, of course, more here) but, rather, the expensive fiber access to the BS that is the limitation in early rollout phases.

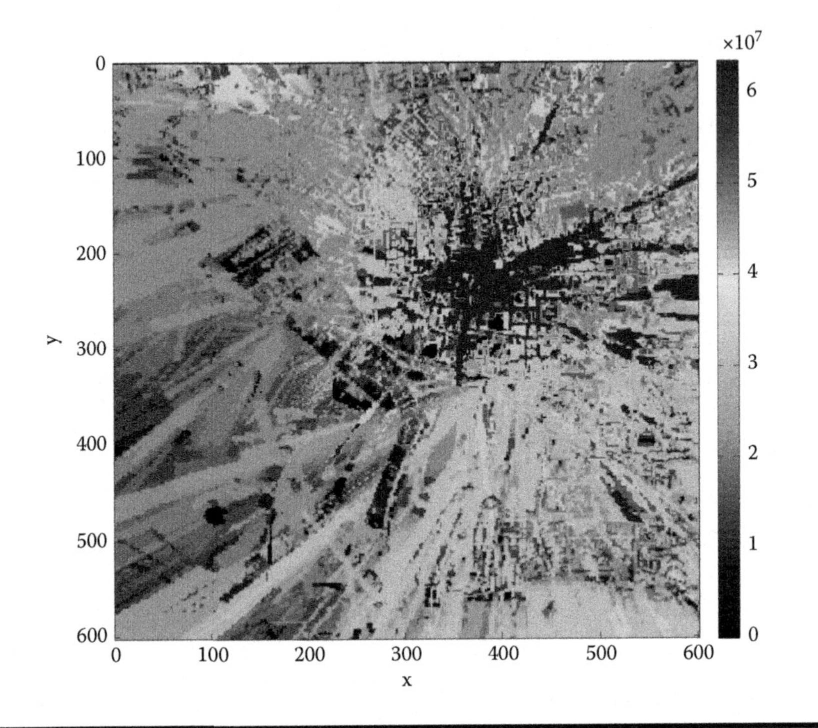

Figure 8.29 Maximum data rate distribution using all relays in the Jersey scenario.

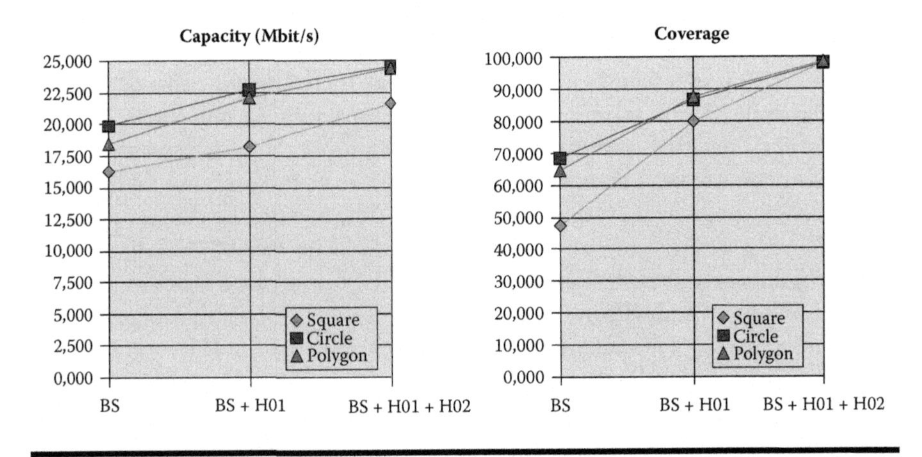

Figure 8.30 The increase of capacity and coverage using relays in the Jersey scenario.

Table 8.3 Scalar Results for the Relay Scenario within the Square Area

Scenario	*Coverage (%)*	*Capacity (Mb/s)*	*Spectral Efficiency (b/s/Hz)*
BS only	47.257	16.223	0.901
BS + H1	79.891	18.217	1.012
BS + H1 + H2	98.519	21.583	1.199

Table 8.4 Scalar Results for the Relay Scenario within the Polygon Area

Scenario	*Coverage (%)*	*Capacity (Mb/s)*	*Spectral Efficiency (b/s/Hz)*
BS only	64.816	18.519	1.029
BS + H1	87.536	22.091	1.227
BS + H1 + H2	98.925	24.375	1.354

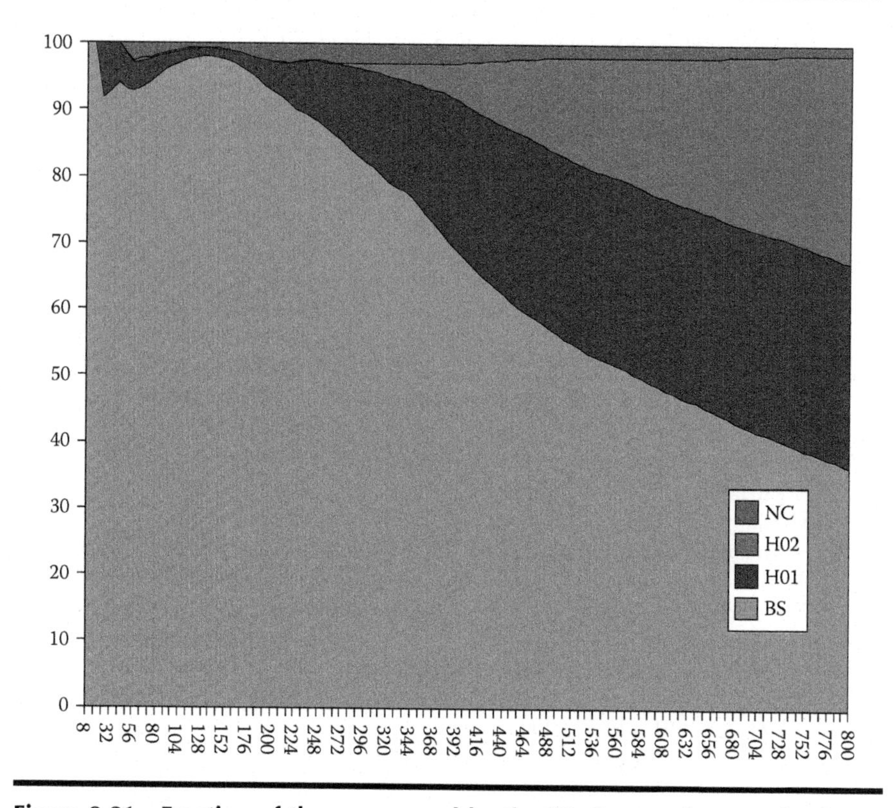

Figure 8.31 Fraction of the area served by the BS, the two-hop, or the three-hop links. This is shown as a function of the circular radius of the accounted area around the base station. Even close to the base station, relays help to improve in 8% of the area. Within an 800-m circle, relays cover 60% of the area.

References

1. LTE introduction at http://www.3gpp.org/Highlights/LTE/LTE.htm
2. IST-4-027756. Wireless world initiative new radio II annex.
3. The WINNER forum at http://www.ist-winner.org/
4. The WiMAX forum at http://www.wimaxforum.org/
5. ETSI. GSM recommendations 04.05, 1993. Data link layer—General aspects.
6. 3GPP. 2001. TS 25 211. Physical channels and mapping of transport channels onto physical channels (FDD).
7. 3GPP. 2001. TS 25 211. Physical channels and mapping of transport channels onto physical channels (TDD).
8. 3GPP. 2007. TS 36.300. Technical specification group radio access network E-UTRAN V8.0.0.
9. IEEE. 2004. IEEE Std. 802.16-2004. IEEE standard for local and metropolitan area networks, part 16: Air interface for fixed broadband wireless access systems.
10. Chan, P. W. C. et al. 2006. The evolution path of 4G networks: FDD or TDD? *IEEE Communications Magazine* 44 (12): 42–50.
11. Berlemann, L., R. Pabst, M. Schinnenburg, and B. Walke. 2005. A flexible protocol stack for multi-mode convergence in a relay-based wireless network architecture. *Proceedings of IEEE Personal Indoor and Mobile Radio Conference 2005 (PIMRC05),* September 2005.
12. Pabst, R. et al. 2004. Relay-based deployment concepts for wireless and mobile broadband radio. *IEEE Communications Magazine* 42 (9): 80–89.
13. WINNER II Deliverable D6.13.7. *Test Scenarios and Calibration Cases,* Issue 2, December 2006. Available: http://www.ist-winner.org
14. Brueninghaus, K., D. Astely, et al. 2005. Link performance models for system level simulations of broadband radio access systems. *Proceedings of the 17th Annual IEEE International Symposium on Personal, Indoor and Mobile Radio Communications,* September 2005.
15. Ekstrom H., A. Furuskar, J. Karlsson, et al. 2006. Technical solutions for the 3G long term evolution. *IEEE Communications Magazine,* March 2006, 38–45.
16. Esseling, N., H. Vandra, and B. Walke. 2000. A forwarding concept for HiperLAN/2. *Proceedings of European Wireless 2000,* September 2000.
17. Hoymann, C., M. Dittrich, and S. Göbbels. 2007. Dimensioning cellular WiMAX: Multihop networks. *Proceedings of European Wireless 2007.*
18. Hoymann, C., K. Klagges, and M. Schinnenburg. 2006. Multihop communication in relay enhanced IEEE 802.16 networks. *Proceedings of the 17th Annual IEEE International Symposium on Personal, Indoor and Mobile Radio Communications,* September 2006.
19. Irnich, T., D. Schultz, R. Pabst, and P. Wienert. 2003. Capacity of a multi-hop relaying infrastructure for broadband radio coverage of urban scenarios. *Proceedings of VTC Fall,* Orlando, FL, 2003.
20. Lott, M., M. Weckerle, W. Zirwas, et al. 2003. Hierarchical cellular multihop networks. *Proceedings of EPMCC 2003,* April 2003.
21. Otyakmaz, A., R. Schoenen, and B. Walke. 2008. Parallel operation of half- and full-duplex FDD in future multi-hop mobile radio networks. *Proceedings of the European Wireless Conference,* June 2008.

22. Schoenen, R., and B. Walke. 2007. On PHY and MAC performance of 3G-LTE in a multihop cellular environment. *IEEE WiCom 2007,* September 2007.
23. Schoenen, R., J. Eichinger, and B. Walke. 2007. On the OFDMA FDD mode in 3G-LTE. *Proceedings of the 12th International OFDM Workshop (InOWo '07),* August 2007.
24. Schoenen, R., R. Halfmann, and B. Walke. 2008. MAC performance of a 3GPP-LTE multihop cellular network. *Proceedings of the IEEE ICC '08,* May 2008.
25. Schoenen, R., W. Zirwas, and B. Walke. 2008. Raising coverage and capacity using fixed relays in a realistic scenario. *Proceedings of the European Wireless Conference,* June 2008.
26. Song, G., and Y. Li. 2005. Utility-based resource allocation and scheduling in OFDM-based wireless broadband networks. *IEEE Communications Magazine* December, 127–134.
27. Sreng, V., H. Yanikomeroglu, and D. Falconer. 2003. Relayer selection strategies in peer-to-peer relaying in cellular radio networks. *Proceedings of the VTC '03.*
28. Timus, B. 2006. Deployment cost efficiency in broadband delivery with fixed wireless relays. Dissertation, KTH Stockholm.
29. Walke, B., R. Pabst, and D. Schultz. 2003. A mobile broadband system based on fixed wireless routers. *Proceedings of ICCT 2003 International Conference on Communication Technology,* April 2003.
30. Walke, B., and R. Briechle. 1985. A local cellular radio network for digital voice and data transmission at 60 GHz. *Proceedings of Cellular & Mobile Communications,* November 1985.
31. Weiss, E., O. Klein, G. Hiertz, and B. Walke. 2006. Capacity and interference aware ad hoc routing in multi-hop networks. *Proceedings of 12th European Wireless Conference 2006,* April 2006.
32. Yanikomeroglu, H. 2002. Fixed and mobile relaying technologies for cellular networks. *Second Workshop on Applications and Services in Wireless Networks (ASWN '02),* July 2002.
33. Zirwas, W., T. Giebel, N. Esseling, E. Schulz, and J. Eichinger. 2002. Broadband multihop networks with reduced protocol overhead. *Proceedings of European Wireless 2002,* February 2002.
34. Zhao, R., B. Walke, and G. Hiertz. An efficient IEEE 802.11 ESS mesh network supporting quality of service. *IEEE Journal on Selected Areas in Communications* November, 2006. 24(11) 2005–2017.
35. Yacoub, M. D. et al. 1993. *Foundations of mobile radio engineering.* Boca Raton, FL: CRC Press.
36. Schoenen, R. and, A. Otyakmaz. 2008. Concurrent operation of half-and full-duplex terminals in future multihop FDO-based cellular networks. *Proceedings of the 4th International Conference on Wireless Communications* (WiCom 2008). October 2008.

Chapter 9

User Plane Protocol Design for LTE System with Decode-Forward Type of Relay

Jijun Luo

Contents

9.1 Introduction

In the 3rd Generation Partnership Project (3GPP) WCDMA (wideband code division multiple access)/HSPA (high-speed packet access) standards, the radio link control sublayer [1] is defined on top of the medium access control (MAC) sublayer and under other higher sublayers such as RRC (radio resource control) and PDCP (packet data convergence protocol) [2]. Its main function is to guarantee reliable data transmission by means of segmentation, ARQ (automatic retransmission request), in-sequence delivery, etc. The radio link control (RLC) layer has three functional modes: the TM (transparent mode), UM (unacknowledge mode), and AM (acknowledge mode), enabling services with different speed and reliability for the upper layer. In this chapter, we mainly study the AM mode of RLC for the unicast traffic on the downlink user plane. The AM mode incorporates the ARQ functionality, so it is by far the most important operation mode for reliable data transfer.

9.2 RLC Layer Overview

In the traditional WCDMA context, the responsibility of providing ARQ lies solely on the RLC layer because the MAC layer assumes no retransmission functionality. Moreover, the RLC layer and MAC layer are separated by the Iub interface [2] between the radio network controller (RNC) and the base station (BS). Such separation determines that the RLC layer transmission is much slower than the MAC layer, which resides on the BS and communicates directly to the air interface.

The RLC layer functions (in AM mode) mainly include the following:

- Segmentation, concatenation, and reassembly: RLC service data units (SDUs) that are received from the upper layer are segmented or concatenated to RLC protocol data units (PDUs) of a certain size. Each PDU is assigned a sequence number (SN). On the receiver side, the PDUs are reassembled to the same RLC SDUs and delivered to the upper layer.
- Error correction by ARQ: the RLC AM mode basically realizes the selective-repeat ARQ protocol. The ARQ is realized by the mechanisms of PDU SN, polling, and status report. Polling is sent by the RLC sender by marking a polling bit in the PDU, which orders the RLC receiver to send a new status report indicating which RLC PDUs have been received and which have not. The status report is triggered by the polling or self-triggered at the receiver. By examining the status report, the sender RLC retransmits the lost PDUs.
- In-sequence delivery of RLC SDUs: this can be configured for the RLC layer to deliver SDUs in its original order. Various studies (e.g., Teyeb et al. [3]) have suggested that in-sequence delivery could benefit the telephone communication power (TCP) throughput. Hence, it is normally configured as the default for the data transmission over the RLC AM mode.
- Flow control: this provides a mechanism for the RLC receiver to control the sending rate of the RLC sender. While the RLC receiver maintains a fixed-length reception window, it orders the sender to change the transmission window size if necessary.

- Duplicate detection: this ensures that the RLC SDUs are delivered only once to the upper layer.
- Ciphering function: this provides protection of privacy.

Because our study is on the RLC layer performance issue, the focus is necessarily on the RLC ARQ function. The standard [1] defines various triggers of sending the polling at the sender and sending the status report by the receiver. The request of polling includes the following:

- Timer-based polling: the RLC sender sends polling periodically as decided by a timer.
- Poll timer: this starts whenever a polling message is sent and stops when a status report is received. This primarily retransmits a poll if it is lost in the link.
- Last PDU in buffer: a poll should be sent with the last PDU in the RLC sender's transmission buffer.
- Last PDU in retransmission buffer: the RLC sender always puts the unacknowledged PDUs into the retransmission buffer to wait for possible retransmission. The last PDU in the retransmission buffer should be sent with a poll, if it is scheduled for retransmission.
- Poll every N PDU: this triggers the poll for every *N* PDU sent.
- Poll every N SDU: this triggers the poll for every *N* SDU sent.
- Window-based polling: this sets the polling bit when a certain percentage of the transmission window is sent.

Also, a number of triggers for the receiver RLC to send the status report are listed as follows:

- The "polling bit" in a received AMD (acknowledged mode data) PDU is set to "1."
- "Missing PDU indicator" is configured and a missing AMD PDU is detected. The missing PDU indicator is optional; this tells the RLC receiver to send a status report whenever a hole in the received PDU sequence is detected.
- The "timer-based STATUS transfer" is configured and the timer Timer_Status_Periodic has expired.
- A status report can be triggered by the MAC, following the MAC-hs (the MAC entity for high-speed downlink packet access [HSDPA]) reset operation.

On the other hand, the status report can be prohibited even if any of the preceding situations holds true, except if the status report is triggered by the MAC-hs reset. This is controlled by the timer Timer_Status_Prohibit, which prohibits the status report transfer if it is started. Upon its expiry, a status report is sent if it is triggered and prohibited before. Thus, practically, the status prohibit timer delays the status report and also reduces the number because only one status report would be sent upon timer expiry, even if multiple reports have been triggered. The status prohibit timer prevents too many status reports from being sent to congest the uplink. *It can be easily figured out that the status prohibit timer should be set slightly higher than one RTT (round-trip time).*

As can be seen, the RLC protocol itself defines many ways to configure the ARQ operation. The goal is always to have timely discovery of lost PDUs and to have timely retransmission of them. Other issues have to be considered also, such as reducing the number of unnecessary status reports and reducing the number of duplicate retransmissions. It can be foreseen that the previously listed methods to trigger polling and status report should be selectively configured, in order to reduce negative interactions between each other. Some comprehensive studies of RLC configuration and performance for the WCDMA data channel (DCH) can be found in references 4 and 5; however, no comprehensive study has been found for downlink transmission with shared channel as in the case of HSDPA or the future orthogonal frequency division multiple access (OFDMA)-based system.

9.3 Impact of Broadband RAN

The RLC was designed at the time WCDMA was introduced; the transmission speed of the air interface was only 384 kb/s (for one user). Also, the separation of RLC and MAC by the Iub interface makes the RLC ARQ operation rather slow. The RTT easily amounts to several hundreds of milliseconds. Although the introduction of HSDPA makes the air interface speed much higher (in the order of 1 Mb/s), the Iub interface still limits the RLC layer performance. What has changed significantly from WCDMA to HSDPA is that the RLC is no longer the sole mechanism for data loss recovery. Because the MAC layer HARQ (hybrid ARQ) recovers most of the errors on the air interface, the RLC layer only has to handle the residual errors after the HARQ operation.

As the 3GPP sets its target for the next generation of broadband radio access network (RAN)—commonly referred to as long term evolution (LTE) [6]—the RLC layer also has to evolve. Two significant changes of the RLC layer are quite clear. First, due to architectural changes [6], the RLC layer now resides also on the BS. Second, due to the much faster air interface, the RLC layer should also strive to make error recovery much faster than before. As indicated in reference 7, this is partly enabled by closer interaction between the MAC HARQ and the RLC ARQ. The draft standard mentions three optional ways of interaction:

- If the HARQ transmitter detects a failed delivery of a PDU—for example, due to a maximum retransmission limit being reached—the relevant transmitting ARQ entities are notified and potential retransmissions can be initiated.

- If the HARQ receiver is able to detect a NACK (negative acknowledge) to ACK (acknowledge) error, it is FFS (for further study) if the relevant transmitting ARQ entities are notified via explicit signaling.
- If the HARQ receiver is able to detect transmission failure, the receiving ARQ entities could be notified.

Among the three possible strategies, the first one is rather straightforward. Because the RLC and MAC reside in the BS, a transmission failure can be reported by the MAC to the RLC to trigger faster retransmission without waiting for the status report. The latter two strategies mostly try to solve the problem of NACK ® ACK error with the HARQ because this cannot be recovered by the first strategy. However, for the moment, let us leave the NACK ® ACK error out of the picture. In our simulation model (introduced in Section 9.6), we incorporate the first HARQ/ARQ interaction listed before and we will neglect the NACK ® ACK error.

9.4 Discussion of RLC Configurations

The RLC operation in the future broadband RAN must be able to recover the HARQ transmission failures efficiently. Within the same framework as introduced in Section 9.2, we now analyze the appropriate RLC configurations. First, let us look at the RLC receiver; we surely have to send the status report when receiving the poll. Also, it should be better to enable the missing PDU indicator. No polling mechanism itself could instantly discover a transmission failure at the MAC layer, and, although the missing PDU should be recovered as soon as possible, the autonomous status report by detecting "holes" in the received sequence generally makes the error recovery faster. Given the presence of the missing PDU indication, the status prohibit timer becomes unnecessary unless excessive polling comes from the RLC sender. As to the timer-based status transfer, previous study [5] has suggested a setting of slightly higher than one RLC RTT. The RTT here is the time between issuing a status report and receiving the retransmission from the sender.

However, such an RTT-based timer has an inherent weakness because, in the broadband RAN, the RTT may vary a lot more than in the WCDMA case. Because WCDMA assumes a dedicated code channel with determined bit rate, the RTT variation is quite small. In the shared physical channel structure assumed by HSDPA and the future OFDMA-based system, the RTT becomes highly variable, depending on the scheduling and the current traffic load. The negative side of such timer-based status reporting is the difficulty of accurately estimating the RTT and figuring out appropriate timer settings. In addition, the timer-based status transfer issues a status report regardless of the current PDU error rate; thus, an unnecessary status report would be sent. This not only wastes uplink resources but also triggers spurious retransmissions from the sender.

On the other hand, let us look at the triggers of polling at the sender side. First, polling should always be sent with the last PDU in the RLC transmission buffer or the retransmission buffer, in order to protect the last PDU from being lost. Also, the poll timer should be enabled in order to detect possible loss of the polling PDU and triggering retransmission. Because the RLC sender could often forward data instantly to the MAC (if allowed by the sending window), a timer should be set to determine the last PDU in the buffer. The value of the timer should be generally very small, in order not to add extra delay to the PDU. In our simulation, we set it to four times the physical layer TTI (transmission time interval).

The other polling triggers—that is, timer-based polling, poll every *N* SDU (or PDU), and windows-based polling—are generally periodic methods. Given the sender-side HARQ/ARQ interaction and the missing PDU indicator at the receiver side, errors would be unheeded only if a HARQ NACK ® ACK error happened to the retransmission PDUs because the "missing PDU indicator" would not respond to a "hole" in the PDU sequence that has been reported. Our suggestion is to select at least one of the periodic polling mechanisms to detect such kinds of errors. If the other two kinds of HARQ/ARQ interaction could recover such errors, the period timers would become unnecessary. However, as has been said, the HARQ/ARQ interaction and NACK ® ACK error constitute a complicated issue that is out of our consideration here. Because we assume no NACK ® ACK error, *we simply select only the window-based polling, which sets the polling bit whenever 50 or 100% of the RLC sender window is sent.* In reality, special polling/status report triggers might be needed for special events such as handover; however, those are out of the boundaries of our discussion here.

It should be noted that RLC cannot be 100% reliable; the protocol itself defines a way to discard SDUs if any PDU fails to be delivered successfully after MaxDAT [1] times of transmission or after a predefined time. This operation is surely the last resort because the data recovery would then depend only on the higher layer, which usually takes longer. Given the low HARQ residual error rate and the HARQ/ARQ interaction, the probability of SDU discard operation should be extremely low. *In our simulation, we do not impose a limit on the maximum time or the MaxDAT for RLC transmission.*

Furthermore, due to the HARQ operation at the downlink, the MAC layer tends to disorder the PDUs. Because the in-sequence delivery function of RLC is configured and the missing PDU indicator triggers a status report whenever disordering is detected with the received PDU sequence, reordering functionality should be implemented beforehand. Reordering of PDUs could take place in the receiver MAC or the RLC. If it is placed in the MAC, the MAC PDU header has to carry an additional sequence number to support ordering. In addition, not all RLC PDUs need in-sequence delivery; reordering at the MAC delays the delivery of all PDUs, regardless of the ordering requirement. Thus, the reordering function should be placed in the RLC; only if the reordering fails at the expiry of a timer would the missing PDU be reported to trigger a status report.

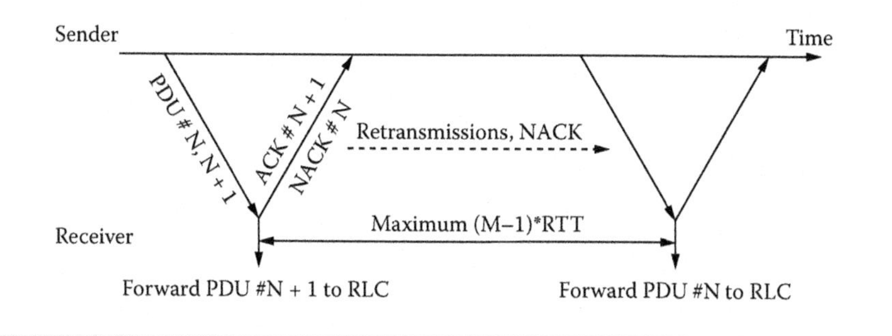

Figure 9.1 Worst-case HARQ reordering delay.

Figure 9.1 shows how to set the reordering timer. Because the RLC sender always send PDUs to the MAC in their original order, the worst-case delay situation happens when PDUs N and $N + 1$ are sent in the same TTI. PDU N successfully arrives at the receiver, but PDU $N + 1$ takes the maximum number of transmissions (M) to get to the receiver. The times in which the receiver RLC gets those two PDUs differ by $(M - 1) \times$ RTT, where RTT is the HARQ round-trip time, if we assume HARQ retransmissions are always scheduled instantly. Thus, the RLC reordering timer should be set to $(M - 1) \times$ RTT because a smaller value may trigger a status report prematurely, ignoring the unfinished HARQ retransmissions.

To summarize, given the sender side ARQ/HARQ interaction in the new Beyond 3G system architecture, we suggest the following principles for RLC layer configuration:

- The receiver side should configure the "missing PDU indicator" to report errors as quickly as possible. To prevent immature status reports due to PDU disordering by HARQ operation, a reordering timer should be configured with $(M - 1) \times$ RTT and started whenever a "hole" in the received PDU sequence is observed.
- The sender side should enable polling for "the last PDU in the buffer" and "the last PDU in the retransmission buffer," as well as for the "poll timer," to prevent the loss of polling PDU.
- Due to the possibility of HARQ NACK ® ACK errors that happen to the retransmitted PDU, other triggers of polling/status report are still needed (if no indication from the MAC exists to handle such errors) but could be reduced to a minimum extent. The reason is that the probability of RLC retransmission equals that of the HARQ residual PDU error rate, which is about 10^{-3} [8]. The probability of NACK ® ACK is normally about 10^{-3}, so the probability of a HARQ NACK ® ACK error happening to an RLC retransmission PDU is only about 10^{-6}. Thus, the periodic triggers of polling/status reports are largely unnecessary and could result in spurious retransmissions. A window-based polling could be used because it results in very few additional status reports.

9.5 Cellular Network with Relay

For the future broadband RAN, ubiquitous high data rate transmission is a demanding requirement. Because the RAN is still likely to inherit the cellular structure, inevitably the transmission quality at the cell edge area would be inferior to that at the cell center. Relay stations may be a good candidate to enhance the cell edge data rate and to provide homogeneous quality of service (QoS) experience for the users. We investigate the relay-enhanced cell (REC) with the aim of enhancing the data rate. As shown in Figure 9.2, a three-sector site is assumed in which each sector still takes the hexagonal form. For each sector, one relay node (RN) is placed at the direction of the antenna bore sight. The UT (user terminal) could directly connect to the BS or be connected via the RN. No diversity from the BS and the RN is assumed for a single UT.

It is expected that the RN should not only act as an analog amplifier, as it has traditionally, but also follow the "decode and forward" paradigm. The RN has a full-fledged MAC layer with the HARQ functionality. Also, it has to take the responsibility of scheduling and resource allocation for those UTs connected to it. Figure 9.3 shows how the downlink transmission is organized in the OFDMA-based REC for every two TTIs. In the first TTI, the BS transmits to the RN and the UTs directly connected to it. In the second TTI, the resources for RN transmission and BS transmission are separated, in order to eliminate interference between the two. We assume a fixed allocation of resources for the RN to UT transmission; the detailed parameter settings are listed in Table 9.1.

Our focus is to investigate the role of the RLC layer on the RN. Basically, two RLC deployment modes can be chosen by the REC, as shown in Figure 9.4. In the upper case, the RLC layer is also implemented on the RN, so each hop can

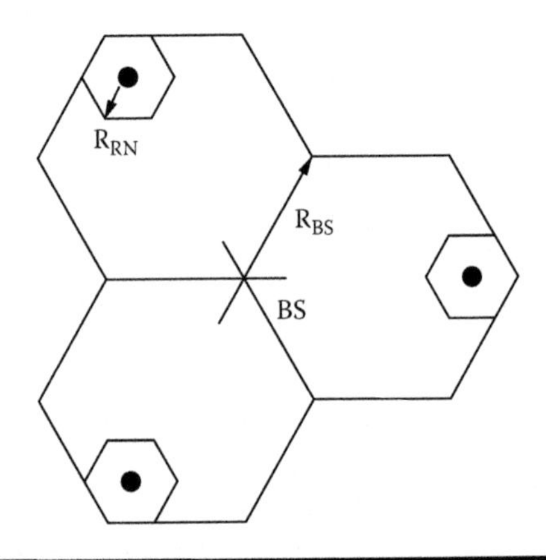

Figure 9.2 Relay-enhanced cell deployment.

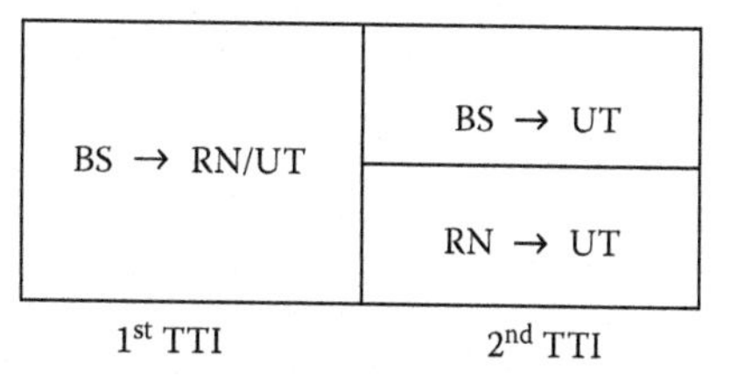

Figure 9.3 Downlink transmission with relay.

Table 9.1 Physical Layer Parameters

Inter-BS distance	1 km
BS–RN distance	543 m
Carrier frequency (DL)	2.5 GHz
Bandwidth	2 × 20 MHz
BS antenna gain	14 dBi
BS–RN antenna array gain	6 dBi
RN antenna configuration	Single antenna, omnidirectional
Receiver noise figure	7 dB
BS max. transmit power per sector	46 dBm
RN max. transmit power	37 dBm
TTI length	1 ms
Chunk size	12 OFDM symbols
No. of chunks for RN downlink	16
No. of chunks for BS–RN downlink	16
User speed	30 m/s
RN–UT channel	Non–line of sight
RLC PDU size	40 b
HARQ round-trip time	4 TTI
RLC status report feedback delay	Fixed to 5 ms/hop

have a separate RLC connection that handles hop-wise ARQ; this is referred to as per-hop RLC. In the lower case, no RLC layer is implemented on the RN; a single RLC connection handles ARQ over the two hops, which is referred to as two-hop RLC.

9.6 Simulation Model and Results

We are interested only in the protocol behavior; thus, in the simulation, only UTs connecting to the RN are modeled. The OFDMA-based system takes the parameters listed in Table 9.1 and operates in the FDD mode. Also, we assume that the

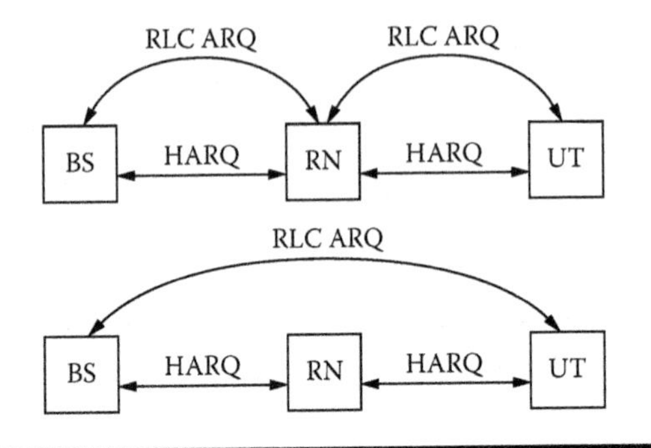

Figure 9.4 RLC operation with REC.

same number of resources is allocated to the BS–RN link and the RN–UT link. This means that in every first TTI such as shown in Figure 9.3, a fixed number of resources is dedicated to the BS–RN transmission. In reality, this is not necessary because the actual number of resources dedicated to the BS–RN link depends on the scheduling and the traffic load. We use a Pareto distribution-based on–off traffic model to simulate the data traffic, of which the parameters are listed in Table 9.2.

As in the REC, the relay area is not big and is supposed to serve a limited number of users; we always assume 16 users are served by the RN, which keeps the multiuser diversity. In order to simulate different traffic load scenarios, we assign different numbers of traffic flows to each user; each flow is independent of the other and follows the Pareto on–off traffic model. With the REC dimension and the pointed antenna beam from the BS to the RN (6-dB additional gain), the BS–RN link has a very stable and high signal-to-noise ratio (SNR) and the transmission on it has no errors. For the RN to UT link, we assume that chase combining HARQ could have, at most, three times of transmissions for one PDU. This would result in a HARQ residual PER of about 1.8%. Because we simulate only one REC cell and

Table 9.2 Traffic Generator Parameters

Period	*Shape (α)*	*Location (β)*	*Mean*
ON	1.05	1,024 bytes	14,026 bytes
OFF	1.4	1 second	2.4939 seconds
Pareto distribution	$F(X<x)=1-\left(\frac{\beta}{x}\right)^{\alpha}, x>\beta$		
Packet length	200 bytes		
Peal rate	150 kb/s		
Mean rate	31.5 kb/s		

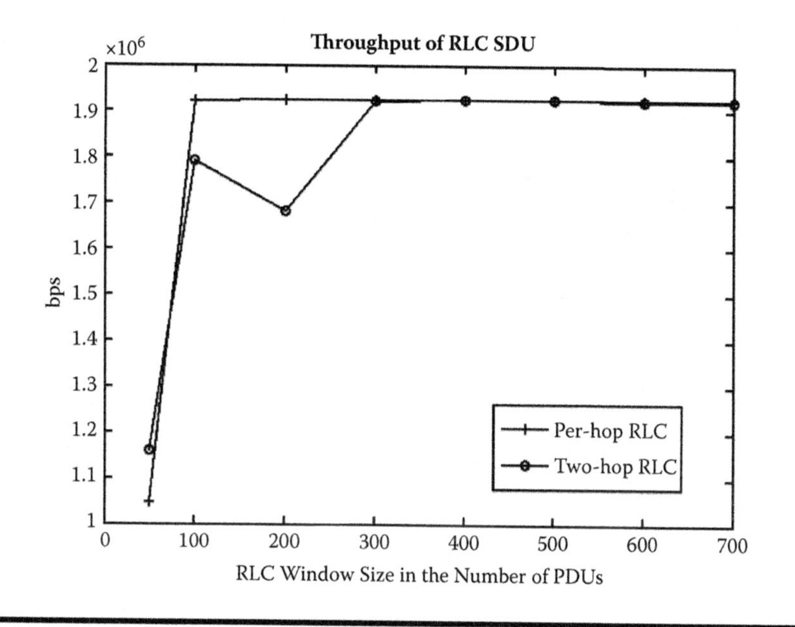

Figure 9.5 Throughput of RLC operation with REC.

leave interference out of consideration, the HARQ residual PER with a maximum of four times of transmission would be very small (about 0.1%). A residual PER that is too small makes RLC layer retransmission a rare event, and, hence, we make only a maximum of three transmissions on the HARQ.

In Figures 9.5 and 9.6, the throughput and CDF of RLC SDU (IP packet) delays are shown for both the two-hop RLC and per-hop RLC cases, with different RLC transmission window sizes and unlimited RLC receiving window sizes. If the SDU is not successfully delivered to the UT, its delay is the simulation time minus the packet generation time. A big problem with single RLC over two hops is that the tuning of RLC transmission window size is quite difficult. In single-hop transmission, if, without the constraint of the receiver RLC window, the RLC transmission window is adapted to the link capacity, overflow or underflow at the MAC layer can be avoided. In the REC, where the two hops could have distinctive link capacity, the two-hop RLC scheme could only adapt the transmission window size to one of the two links.

The RLC standards with a universal mobile telecommunications system (UMTS) [1] allow the receiver to adjust the window size of the transmitter. However, in the two-hop RLC transmission case, the UT is ignorant about whether the RN buffer undergoes overflow or underflow due to inappropriate sender window size; hence, it cannot tell the sender the suitable window size. Because the first hop normally transmits more quickly than the second hop, setting a big transmission window size

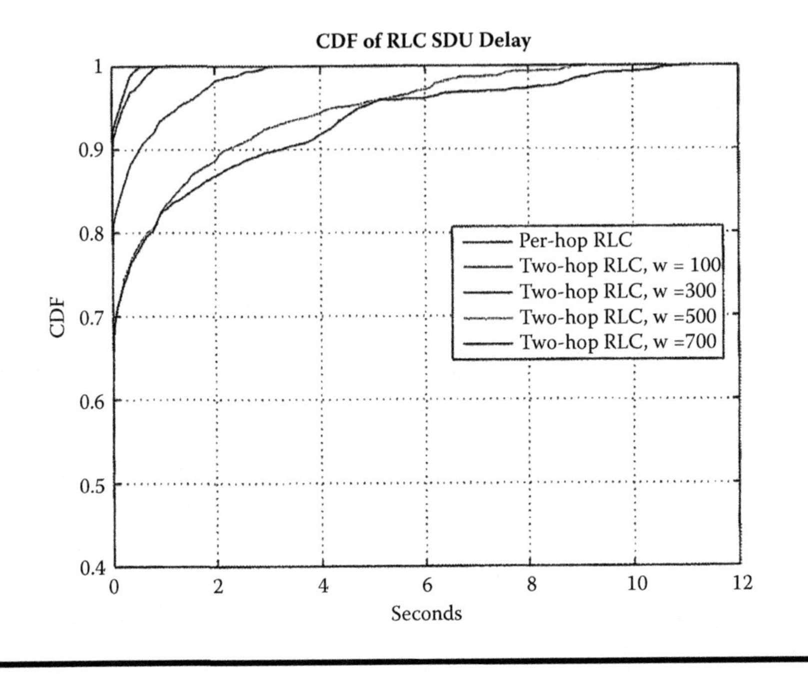

Figure 9.6 Delay of RLC operation with REC.

would make data queue up at the RN; setting a small window size might underutilize the capacity provided by the first hop and might result in a window stall problem. Also, the threat of overflow on the RN buffer—having too much data queuing at the RN—slows down the retransmissions.

Although the BS always gives RLC retransmission higher priority, the retransmitted PDU nevertheless must queue behind the already transmitted data that are waiting on the RN buffer. This makes retransmission rather slow and, even worse, the data queuing before the retransmitted PDU might contain the polling, which triggers a status report asking again for the retransmission of the same PDU. Duplicate retransmissions would then be issued by the BS because the first retransmission is still on the way.

9.7 Conclusion and Discussion

In this chapter, we examined the legacy UMTS RLC protocols and discussed their evolution for the future broadband network. We investigated different operating schemes of RLC for the multihop REC. The conclusion is that the two-hop RLC scheme performs poorly compared to the per-hop RLC scheme, due to the asymmetric capacity on the two hops. We also concluded that, even with the more efficient per-hop RLC scheme, a flow control mechanism on the BS–RN hop has to be developed. However, the topic of flow control, as well as the impact of the HARQ/ARQ cross-layer interaction, is left for our future study.

References

1. 3GPP. 2006. TS 25.322 V7.2.0. Radio link control (RLC) protocol specification (2006-09).
2. Holma, H., and A. Toskala. *WCDMA for UMTS: Radio access for third generation mobile communications,* 3rd ed. New York: John Wiley & Sons, 2000.
3. Teyeb, O., M. Boussif, T. Sorensen, J. Wigard, and P. Mogensen. 2005. The impact of RLC delivery sequence on FTP performance in UMTS. *WPMC 05,* Aalborg, Denmark, September 2005.
4. Zhang, Q., and H.-J. Su. 2002. Performance of UMTS radio link control. *IEEE International Conference on Communications* (ICC 2002).
5. Li, J., D. Y. Montuno, J. Wang, and Y. Q. Zhao. 2004. Performance evaluation of the RLC protocol in 3G UMTS. *Wireless Networks and Emerging Technologies (WNET).*
6. 3GPP. 2006. TR 25.913 V7.3.0 (2006-03). Requirements for evolved UTRA (E-UTRA) and evolved UTRAN (E-UTRAN) (release 7), 16.
7. 3GPP. 2007. TS 36.300 V8.1.0 (2007-06). Evolved universal terrestrial radio access (E-UTRA) and evolved universal terrestrial radio access network (E-UTRAN). Overall description (release 8).
8. 3GPP. TR 25.848 V4.0.0. Physical layer aspects of UTRA high-speed downlink packet access.

Chapter 10

Radio Access Network VoIP Optimization and Performance on 3GPP HSPA/LTE*

Markku Kuusela, Tao Chen, Petteri Lundén, Haiming Wang, Tero Henttonen, Jussi Ojala, and Esa Malkamäki

Contents

* This chapter previously appeared in *VoIP Handbook: Applications, Technologies, Reliability, and Security,* published by Taylor & Francis, 2008.

10.1 Introduction

Circuit switched (CS) voice used to be the only way to provide voice service in the cellular networks, but during the past few years there has been growing interest to use cellular networks for real-time (RT) packet-switched (PS) services such as Voice over Internet Protocol (VoIP) to provide voice service without circuit switched service. The main motivation for the operators to use VoIP instead of CS voice are the savings that could be achieved when the CS-related part of the network would not be needed anymore. It is also expected that VoIP can bring better capacity than CS voice due to more efficient utilization of resources. Supporting VoIP in any radio access technology faces certain challenges due to VoIP traffic characteristics (strict delay requirements, small packet sizes), which make the efficient exploitation of radio interface capacity difficult due to control channel constraints. The solutions to these challenges vary for different technologies.

The introduction of 3G networks with integrated IP infrastructure [1] included in WCDMA Release 99 made it possible to run VoIP over cellular networks with reasonable quality, although with lower spectral efficiency than the circuit switched voice [2]. 3GPP Releases 5 and 6 have brought High Speed Packet Access (HSPA) [3] to WCDMA downlink (DL) and uplink (UL). HSPA consists of High-Speed Uplink Packet Access (HSUPA) [4,5] in the UL direction and High Speed Downlink Packet Access (HSDPA) [6] in the DL direction and was originally designed to carry high-bit rate delay tolerant non-real time (NRT) services like web browsing. Even though HSPA was not originally designed to support RT services, with careful design of the system, the RT services can be efficiently transported over HSPA, and a number of new features have been introduced to 3GPP Releases 6 and 7 to improve the efficiency of low-bit rate delay sensitive applications like VoIP. It has also been shown that VoIP can provide better capacity on HSPA than CS voice [7].

Long-term evolution (LTE) of 3GPP [8] work (targeted to 3GPP Release 8) has been defining a new packet-only wideband radio-access technology with a flat architecture, aiming to develop a framework for a high-data-rate, low-latency, and packet-optimized radio access technology called E-UTRAN. As E-UTRAN is purely a PS radio access technology, it does not support CS voice, which stresses the importance of efficient VoIP traffic support in E-UTRAN.

This chapter provides an overview of the challenges faced in implementing VoIP service over PS cellular networks, in particular 3GPP HSPA and LTE. Generally accepted solutions by 3GPP leading to an efficient overall VoIP concept are outlined and the performance impact of various aspects of the concept is addressed. The chapter concludes with a system simulation-based performance analysis of the VoIP service, including a comparison between HSPA and LTE.

10.2 System Description

10.2.1 HSPA

High-Speed Packet Access, a 3GPP standardized evolution of Wideband Code Division Multiple Access (WCDMA), has become a huge success as the world's leading third-generation mobile standard. HSPA consists of HSUPA and HSDPA

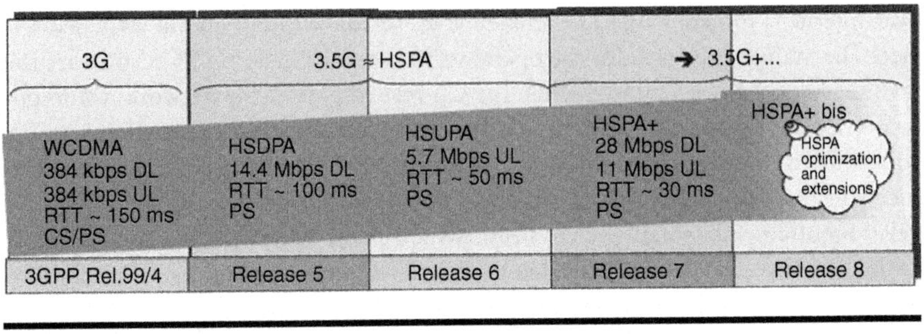

Figure 10.1 WCDMA/HSPA key performance indicators.

which enhance WCDMA in the uplink and the downlink separately. Several key techniques such as HARQ, fast base station-controlled scheduling, and the shorter frame size were introduced into HSPA targeting the higher data rates with more efficient spectrum usage and lower transmission delay (see Figure 10.1).

In HSDPA, a high-speed downlink shared channel (HS-DSCH) is associated with a 2-ms frame size, in which the users share at most 15 fixed spreading factor 16 (SF 16) high-speed physical downlink shared channels (HS-PDSCHs) by code multiplexing. In this way, radio resources can be utilized in a more efficient manner than earlier dedicated resource allocation in WCDMA. Furthermore, HARQ in HSPA can provide physical layer (L1) retransmissions and soft combining, which reduce the amount of higher-layer ARQ transmissions and frame selections (i.e., hard combining). This also improves the efficiency of the Iub interface between the base station and radio network controller (RNC) significantly.

Rather than the conventional dedicated transport channel (DCH), the enhanced DCH (E-DCH) mapping on the enhanced dedicated physical data channel (E-DPDCH) is used in HSUPA to further improve the uplink data rate up to 5.76 Mbps. The 2-ms TTI frame length introduced into HSUPA can further reduce the transmission delay in the air inter-face: the minimum Round Trip Time (RTT) is decreased from about 150 ms in WCDMA to 50 ms in HSPA. Besides, base station controlled fast scheduling can select the user in a good channel condition for the transmission reaching the higher throughput due to the multiuser diversity gain. This is impossible in the WCDMA system, where the reaction of the RNC-controlled scheduling is too slow to follow the fast-changing channel conditions. This is because of the transmission delay between the UE and the RNC via the base station.

10.2.2 LTE

To ensure competitiveness of 3GPP radio access technologies beyond HSDPA and HSUPA, a LTE of the 3GPP radio access was initiated at the end of 2004. As a result, the Evolved Universal Terrestrial Radio Access Network (E-UTRAN) is now being specified as a part of the 3GPP Release 8. Important aspects of E-UTRAN

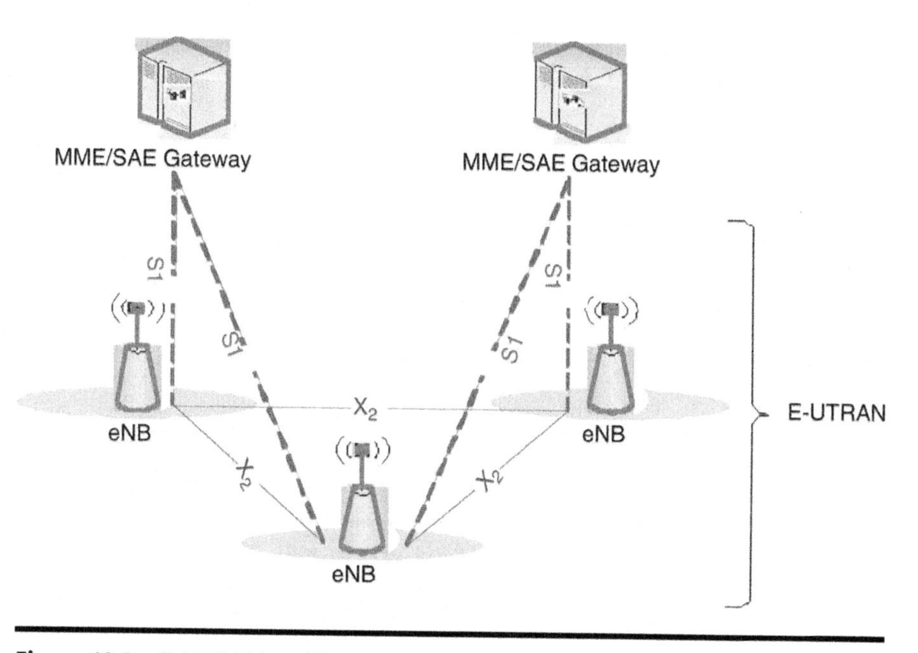

Figure 10.2 E-UTRAN architecture [8].

include reduced latency (below 30 ms), higher user data rates (DL and UL peak data rates up to 100 Mbps and 50 Mbps, respectively), improved system capacity and coverage, as well as reduced CAPEX and OPEX for the operators [9].

In order to fulfill those requirements, 3GPP agreed on a simplified radio architecture. All user plane functionalities for the radio access were grouped under one entity, the evolved Node B (eNB), instead of being spread over several network elements as traditionally in GERAN (BTS/BSC) and UTRAN (Node B/RNC) [8]. Figure 10.2 depicts the resulting radio architecture where:

- The E-UTRAN consists of eNBs, providing the E-UTRA user plane (PDCP, RLC, MAC) and control plane protocol terminations (RRC) toward the UE and hosting all radio functions such as Radio Resource Management (RRM), dynamic allocation of resources to UEs in both uplink and downlink (scheduling), IP header compression and encryption of user data stream, and measurement and measurement reporting configuration for mobility and scheduling.
- The eNBs may be interconnected by means of the X2 interface.
- The eNBs are also connected by means of the S1 interface to the Evolved Packet Core (EPC), which resides in the MME/SAE gateway.

Figure 10.3 shows the E-UTRA frame structure, which consists of 10 subframes, each containing 14 symbol blocks. In the E-UTRAN system, one

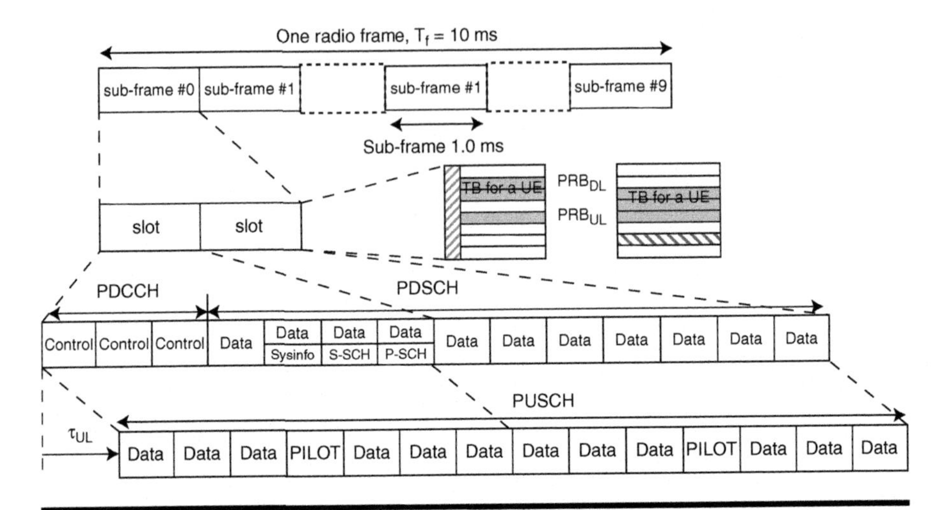

Figure 10.3 Frame structure in E-UTRAN system [44]. (*Source:* M. Rinne, K. Pajukoski et al., "Evaluation of recent advances of the Evolved 3G (E-UTRA) for the VoIP and best effort traffic scenarios," IEEE SPAWC, June 2007.)

sub-frame (1 TTI) of length 1.0 ms is regarded as the minimum time allocation unit. In the frequency domain, a minimum allocation unit is a Physical Resource Block (PRB), which consists of 12 subcarriers (each subcarrier is 15 kHz). In DL, a OFDMA scheme is designed to allow signal generation by a 2048-point FFT. This enables scaling of system bandwidth to the needs of an operator. At least the bandwidths of {6, 12, 25, 50, 100} PRBs in the range of 1.4 MHz to 20 MHz are available, yielding a bandwidth efficiency of about 0.9. In DL, user data is carried by the downlink shared channel (DL-SCH), which is mapped to the physical downlink shared channel (PDSCH). The UL technology is based on Single-Carrier FDMA (SC-FDMA), which provides excellent performance facilitated by the intra-cell orthogonality in the frequency domain comparable to that of an OFDMA multicarrier transmission, while still preserving a single carrier waveform and hence the benefits of a lower Peak-to-Average Power Ratio (PAPR) than OFDMA. In the UL direction, user data is carried by the uplink shared channel (UL-SCH), which is mapped to the physical uplink shared channel (PUSCH).

With all the radio protocol layers and scheduling located in the eNB, an efficient allocation of resources to UEs with minimum latency and protocol overhead becomes possible. This is especially important for RT services such as VoIP, for which high capacity requirements were set: at least 200 users per cell should be supported for spectrum allocations up to 5 MHz, and at least 400 users for higher spectrum allocations [9].

10.3 Properties and Requirements for VoIP

This section illustrates the characteristics of VoIP traffic by describing the properties and functionality of the used voice codec in HSPA/LTE (Section 10.3.1) and by presenting the requirements and the used quality criteria for VoIP traffic in Sections 10.3.2 and 10.3.3, respectively.

10.3.1 Adaptive Multirate (AMR)

Adaptive Multirate (AMR) is an audio data compression scheme optimized for speech coding. AMR was adopted as the standard speech codec by 3GPP in October 1998 and is now widely used. The AMR speech codec consists of a multirate speech codec, a source controlled rate scheme including a voice activity detector, a comfort noise generation system, and an error concealment mechanism to combat the effects of transmission errors and lost packets.

The multirate speech codec is a single integrated speech codec with eight source rates from 4.75 kbps to 12.2 kbps and a low rate background noise encoding mode. For AMR, the sampling rate is 8 kHz. The usage of AMR requires an optimized link adaptation that selects the best codec mode to meet the local radio channel and capacity requirements. If the radio conditions are bad, source coding is reduced and channel coding is increased.

In addition to an AMR audio codec, an extension of it is an AMR-wideband (AMR-WB) audio codec. The sampling rate for AMR-WB is 16 kHz, which is double the sampling rate of AMR. Therefore, AMR is often abbreviated as AMR-NB (narrowband). AMR-WB is supported if 16-kbps sampling for the audio is used in the UE. Similarly, like AMR-NB, AMR-WB operates with various bit rates. For AMR-WB the bit rates range from 6.60 kbps to 23.85 kbps. It is emphasized that both AMR-NB and AMR-WB are used in HSPA and LTE radio systems. A more detailed description of AMR-NB and AMR-WB can be found in [10].

General functionality of the AMR codec is illustrated in Figure 10.4. During voice activity periods there is one voice frame generated every 20 ms (number of

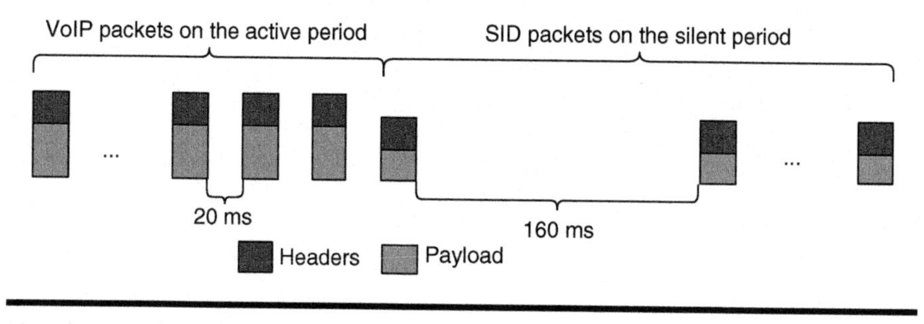

Figure 10.4 Illustration of VoIP traffic.

bits/frame depending on the AMR codec mode), whereas during the silent periods Silence Descriptor (SID) frames are generated once every 160 ms.

10.3.2 Delay Requirements

Voice over Internet Protocol (VoIP) is a conversational class service, and the packet delay should be strictly maintained under reasonable limits. The maximum acceptable mouth-to-ear delay for voice is in the order of 250 ms, as illustrated in Figure 10.5. Assuming that the delay for the core network is approximately 100 ms, the tolerable delay for MAC buffering/scheduling and detection should be strictly below 150 ms. Hence, assuming that both end users are (E-)UTRAN users, the tolerable one-way delay for MAC buffering and scheduling should be under 80 ms, which is the used air interface delay for HSPA. To improve voice quality further on LTE, the tolerable air interface delay was reduced to 50 ms for LTE [11]. For LTE FDD, the average HARQ RTT equals 8 ms, implying that at most 6 HARQ retransmissions are allowed for a VoIP packet when air-interface delay of 50 ms is used. For HSPA the HARQ RTT would be 12 ms in downlink and 16 ms in uplink with 2 ms TTI, which means at most 6 HARQ retransmissions in downlink and 4 HARQ retransmissions in uplink with 80 ms HSPA delay budget. Similarly, at most 1 HARQ retransmission could be allowed in the uplink, corresponding to 40 ms HARQ RTT with 10 ms TTI.

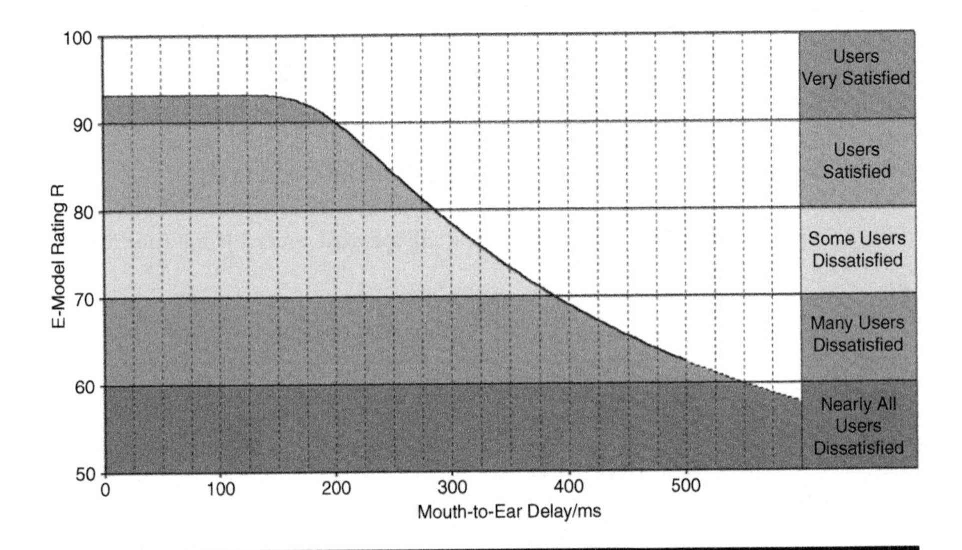

Figure 10.5 E-model rating as a function of mouth-to-ear delay (ms) [42]. (*Source*: ITU-T Recommendation G. 114. "One way transmission time." With permission))

10.3.3 Quality Criteria

Considering the nature of radio communication, it is not practical to aim for 100% reception of all the VoIP packets in time. Instead, a certain degree of missing packets can be tolerated without notably affecting the QoS perceived by the users. For VoIP traffic, the system capacity is measured as the maximum number of users who could be supported without exceeding 5% outage. During the standardization phases of HSPA and LTE in 3GPP, slightly different outage criteria were used when evaluating VoIP system performance. Used outage criteria are defined according to the system as follows: For HSPA, a user is in outage if more than 5% of the packets are not received within the delay budget when monitored over a 10-second window. For LTE [11], an outage is counted if more than 2% of the packets are not received within the delay budget when monitored over the whole call. It is emphasized that even though the used outage criteria differ slightly between HSPA and LTE, the performance difference originating from different outage criteria is rather small, being inside the error margins of the simulations.

In the UL direction there exists an alternative method to measure VoIP performance, which is used in the context of this chapter for HSPA. That is, UL VoIP capacity can be measured as the allowed number of users in a cell with an average noise rise of X dB measured at the base station. The optimal value for X may depend on the used system, and for HSPA the used value was 6 dB.

10.4 Radio Interface Optimization for VoIP

The most important mechanisms to optimize air interface for VoIP traffic in (E-)UTRAN are presented in this section.

10.4.1 Robust Header Compression (ROHC)

In streaming applications such as VoIP, the overhead from IP, UDP, and RTP headers is 46 bytes for IPv4 and 66 bytes for IPv6. For VoIP traffic, this corresponds to about 144–206% header of overhead, assuming that the most typical VoIP codec (AMR 12.2 kbps) is used with 31 bytes of payload. Such a large overhead is excessive in wireless systems due to the limited bandwidth and would lead to serious degradations in the spectrum efficiency.

In order to overcome this, a robust header compression (ROHC) technique [13] was adopted as a nonmandatory feature to 3GPP specifications from Release 4 onwards—hence, the ROHC feature is supported in HSPA and LTE radio systems as well. With ROHC, the IP/UDP/RTP header is compressed down to a few bytes, with the minimum size being 3 bytes, which is also the most commonly used header size. Hence, with ROHC the header overhead may be compressed down to 25%, implying that the utilization of ROHC leads to significant improvements in the VoIP air interface capacity.

10.4.2 Multiuser Channel Sharing

Due to the characteristics of the VoIP traffic (strict delay constraints, regular arrival of small size packets), multiuser channel sharing plays an important role when optimizing the radio channel interface for VoIP traffic. With multiuser channel sharing, multiple users can be multiplexed within the same TTI, and this will lead to significant improvements in VoIP capacity. To stress the importance of multiuser channel sharing as a building block to optimize radio interface for VoIP traffic, it is concluded that VoIP capacity is directly proportional to the average number of scheduled users per TTI. In HSPA, multiuser channel sharing is done by separating users multiplexed into the same TTI in the Orthogonal Variable Spreading Factor (OVSF) code domain, whereas in LTE user separation on PDSCH is done in the frequency domain.

An important issue related to multiuser channel sharing is how the transmission resources are divided between simultaneously scheduled users. Conventional methods use an even-share method, where each multiplexed user is given the same amount of resources. The main shortcoming of this method with VoIP traffic is that it wastes resources: The size of the resource allocation has to be done according to the needs of the weak users to guarantee sufficient QoS at the cell border as well. If users in good channel conditions are assigned too many transmission resources, part of the resources are wasted, as the delay requirements do not allow to buffering of enough packets to take full benefit of the allocated transmission resources. At the same time, users in unfavorable channel conditions might not have enough transmission resources to transmit even one VoIP packet at the target block error rate (BLER).

A more sophisticated method to assign transmission resources for the scheduled users is to determine the size of the resource assignment dynamically based on the channel quality and on the supportable payload of a user. With a dynamic approach, each user is allocated the amount of transmission resources that is necessary to reach the target BLER. This can be estimated by the link adaptation unit based on the reported channel quality indicator (CQI) and the selected modulation and coding scheme (MCS). With the dynamic approach, the size of the resource allocation matches more accurately to the requirements of the user. Thus the same total amount of transmission resources can be used more efficiently, which increases the resource utilization efficiency. Moreover, weak users can be boosted with a higher share of transmission resources, which also improves the DL coverage of the VoIP service.

The impact of the used resource allocation algorithm to the VoIP system performance on HSDPA was studied in [14]. The results indicated that the dynamic resource allocation on top of the VoIP-optimized scheduling provides substantial (up to 20%) capacity gain over the even-share approach.

10.4.3 Utilization of VoIP-Specific Packet Scheduling (PS) Algorithm

Perhaps the most essential building block for the efficient support of VoIP traffic in HSPA/LTE is the packet scheduling (PS) algorithm. This algorithm at (e)NodeB is responsible for user selection and the resource allocation for the selected users. Considering VoIP traffic characteristics, the PS buffer status, and individual users' channel conditions when making scheduling decisions and resource allocation for the selected users, VoIP capacity can be boosted up to 115% [14] when compared to the blind Round Robin (RR) scheduling method without any VoIP specific user prioritization. In this section, a brief description of the proposed VoIP optimized PS algorithm is given (details on the algorithm and corresponding performance analysis on HSDPA and LTE are given in [7,14–16]). Moreover, it is assumed that the proposed algorithm schedules each transmission of data by L1 control signaling. Henceforth, this kind of PS method is referred to as fully dynamic PS.

10.4.3.1 Design Criteria of the Packet Scheduling Algorithm for VoIP

The following issues should be considered when designing a PS algorithm for VoIP traffic:

- *Strict delay requirements of VoIP service* —As described in Section 10.3, the tolerable air interface delay for VoIP traffic (including packet scheduling and buffering) should be strictly within 80 ms. When this delay is exceeded, the quality of the voice perceived by the user starts to deteriorate rapidly. Due to the stringent delay limit, time domain scheduling gains for VoIP traffic are very limited, as the buffered VoIP packet(s) should be scheduled within the given delay budget in order to sustain good speech quality. For users whose channel is in a deep fade, the delay limitation implicitly means that there is not enough time for PS to wait until the conditions become more favorable. Instead, the data should be scheduled early enough to allow enough time for the needed retransmissions. This implies that weak users, that is, users relying on HARQ retransmissions, should be prioritized in the user selection.
- *Low user data rate of VoIP service* —The size of the VoIP packet is typically very small; for an example, for an AMR 12.2-kbps codec, the size of the most common ROHC compressed packet is 40 bytes (every 20 ms). Together with strict delay requirements of VoIP traffic, this leads to frequent transmission of small payloads, as it is not possible to wait long enough to send more than a few VoIP packets at a time. Therefore, when high capacities are targeted, it

is necessary for several users' transmissions to share the DSCH resources in each TTI, and each user needs to be scheduled fairly often.

- *Managing the control overhead*—Due to the low user data rates, the number of scheduled users per TTI can become quite large with multiuser channel sharing. When each of these users is scheduled by L1 control signaling, the control channel overhead can become the bottleneck of VoIP system performance. In that case the dynamic packet scheduler is unable to fully exploit the PDSCH air interface capacity, and some of the PDSCH resources remain unused while the control channel capacity is already exhausted. One attractive technique with a fully dynamic scheduler to improve VoIP capacity within control channel limitations is to bundle multiple VoIP packets together at L1 in one transmission for a user. Packet bundling is CQI based, which means that PS decides whether to utilize it or not for a user based on the user's channel conditions. Packet bundling is described in more detail in Section 10.6.2.1. An alternative way of addressing the control overhead problem is semipersistent scheduling, which, instead of scheduling each transmission individually, relies on long-term allocations. Semi-persistent scheduling is discussed in more detail in Section 10.6.2.2.

10.4.3.2 Proposed Fully Dynamic Packet Scheduling Algorithm for VoIP

The following packet scheduling algorithm aims to address the particulars of the VoIP service listed in Section 10.4.3.1. The proposed user selection procedure relies on the concept of the scheduling candidate set (SCS) [14,16], which classifies the schedulable users into three groups. These three groups in the SCS are dynamically updated in each TTI. The groups are further ordered in the following priority:

1. Users with pending retransmissions in their HARQ manager
2. Users whose head-of-line (HOL) packet delay is close enough to the delay budget, where HOL delay is the delay of the oldest unsent packet in the (e)NodeB buffer
3. Remaining users (other than the retransmission users and HOL users).

The priority metric used for ordering users within each group is a relative CQI metric. The relative CQI metric for each user is given by *CQI_inst*/*CQI_avg*, where *CQI_inst* is the instantaneous CQI calculated over the full bandwidth and CQI_avg represents the average wideband CQI value in the past. The relative CQI metric reflects the user's instantaneous channel condition compared to the average and therefore can indicate a favorable moment for scheduling the user. Of course, the time domain scheduling gains are in the end somewhat limited because of the tight delay budget.

Table 10.1 Packet Scheduling and Resource Allocation Algorithm Alternatives

Scheduler	*Description*
RR	RR scheduling with even-share resource allocation
PF	Proportional fair scheduling with even-share resource allocation
PF + SCS	SCS-based VoIP optimized scheduling algorithm using PF metric to order the users in the SCS. Even-share resource allocation
PF | SCS | RA	As PF | SCS, but with dynamic resource allocation

As a further improvement, compared with sending one packet per TTI, bundling multiple packets together in one transmission can reduce related control channel signaling overhead and thus improve air interface efficiency (see Section 10.6.2.1 for further details). Therefore, in the user selection within each group, priority is given to users who can support bundling. Further, in Table 10.1 the abbreviations of considered scheduler alternatives are listed and explained.

Table 10.2 illustrates the relative HSDPA capacity improvements that can be obtained with a packet scheduler specifically designed for VoIP service over the blind RR scheduling method without any VoIP specific user prioritization. The table shows capacities for an 80-ms delay budget assumption. The selected set of scheduler alternatives compares the enhancements incrementally.

These numbers verify that packet scheduling and allocation of transmission resources play a key role in achieving good VoIP capacity. There is a large difference between the best and the worst scheduler option. PF gives the gain over RR, as it is, on average, able to schedule users at better channel conditions. Introducing SCS is an important improve-ment, as it considers the delay requirements of VoIP traffic by prioritizing urgent transmissions and retransmissions. Moreover, SCS makes

Table 10.2 Relative Capacity Gain with Different Scheduler Improvements Compared to Round Robin Scheduler with 80 ms Delay Budget

Scheduler	*Capacity gain (%)*
RR	0
Proportional fair (PF)	20
PF | SCS	42
Dynamic resource allocation (PF + SCS + RA)	80

Source: P. Lundén and M. Kuusela, "Enhancing performance of VoIP over HSDPA," VTC 2007 Spring.

the packet bundling more likely, thus improving the overall efficiency. The capacity gain of dynamic resource allocation (PF + SCS + RA) over even-share resource allocation is approximately 20% to 30% for the PF + SCS scheduler.

10.4.3.3 RNC Controlled Nonscheduled Transmission in the Uplink for VoIP over HSUPA

There are two scheduling schemes defined for HSUPA: Node B controlled scheduled transmission and RNC controlled nonscheduled transmission. The former scheme does not guarantee a minimum bit rate since the scheduling algorithm decides whether to allocate any power to the user or not. The latter scheme (nonscheduled transmission) defines a minimum data rate at which the UE can transmit without prior request. When nonscheduled transmission is configured by the serving RNC (SRNC), the UE is allowed to send E-DCH data at any time, up to a configured bitrate, without receiving any scheduling command from Node B. Thus, signaling overhead and scheduling delay are minimized. Therefore, RNC controlled nonscheduled transmission is the most suitable choice for VoIP traffic.

10.4.4 Advanced Receivers

By using advanced receiver techniques at UE, the performance of VoIP can be significantly improved. The following advanced receiver techniques were introduced to HSDPA terminals: 2-antenna Rake receiver with Maximum Ratio Combining (MRC) and 1-antenna LMMSE time domain chip level equalizer were adopted by 3GPP Release 6 onwards, and 2-antenna LMMSE chip level equalizer was adopted by 3GPP Release 7. When comparing the performance improvements achieved with 3GPP Release 6 advanced receiver techniques, it was concluded that the biggest improvement is achieved with plain receive diversity. By having an additional receive antenna at the terminal, a 3 dB array gain is added to the realized signal-to-noise ratio (SNR) for all terminals, due to which the VoIP DL capacity is significantly boosted. Further, the utilization of an LMMSE chip equalizer improves VoIP system performance as well by reducing the intracell interference of the users in the cell, but the achieved performance improvement is much smaller compared to utilization of receive diversity, as the equalizer is not able to improve the performance of the cell edge users, that is, the users that most probably are in the outage, as the performance of those users is mainly limited by the intercell interference. On the basis of system level simulations, the utilization of receive diversity provides approximately 80% gains in HSDPA capacity compared to the case with 1-antenna RAKE receivers at the terminals. For a 1-antenna LMMSE chip level equalizer, the corresponding gain is in the order of 20–30%.

In LTE, the baseline receiver technique is a 2-antenna receiver with either MRC combining or Interference Rejection Combining (IRC), where the diversity antennas are used to reject the co-channel interference by estimating the interference covariance matrix and selecting combining weights accordingly to maximize SNR. According to [17], IRC combining provides approximately 0.5 dB gain in the received SNR over MRC combining, and this would correspond to 10–15% gains in VoIP LTE DL capacity.

10.5 Practical Constraints in VoIP Concept Design and Solutions

When optimizing air interface for VoIP traffic, several practical constraints are to be considered in VoIP concept design on HSPA and LTE radio systems. This section provides descriptions for the most important practical limitations in radio interface optimization. Existing solutions (if any) to avoid these limitations are covered in Section 10.6.

10.5.1 Control Channel Limitations

With VoIP traffic the number of supported users may increase drastically. To support VoIP transmission of DL and UL transport channels, there is a need for certain associated DL and UL control signaling. Therefore, the overhead from associated control signaling in UL and DL may become a bottleneck for the VoIP system performance. Sections 10.5.1.1 and 10.5.1.2 describe the main contributors for control channel overhead for HSPA and LTE radio systems. Methods to reduce control channel overhead are described in Sections 10.6.1 and 10.6.2.

10.5.1.1 HSPA

For VoIP traffic, the packet size is considerably small, corresponding to a low data rate transmission on the data channel, for example, 32 kbps with 10 ms TTI and 160 kbps with 2 ms TTI in HSUPA. In other words, it implies that the overhead from control channels would consume relatively more power for the transmission of VoIP traffic than the other packet data traffic transmitted with the high data rate. On the other hand, the control channels cannot be subject to HARQ retransmissions; that is, the information carried by the control channels cannot be soft combined. Thus, more retransmissions may lead to the higher overhead from the control channels associated with data transmission. It makes sense to study the effect of the control channel overhead to performance and investigate how the overhead could be reduced.

In HSPA, there are mainly three types of dedicated physical control channels in the UL and one in the DL: the UL Dedicated Physical Control Channel (UL DPCCH), the E-DCH Dedicated Physical Control Channel (UL E-DPCCH), the UL Dedicated Physical Control Channel associated with HS-DSCH transmission (UL HS-DPCCH), and the DL Shared Control Channel for HS-DSCH

(HS-SCCH). Their constraints on the VoIP performance are addressed in the following sections.

DPCCH overhead in the uplink transmission

The uplink DPCCH in 3GPP Release 6, that is, HSUPA, is used for carrying physical layer control information which mainly consists of known pilot bits to support channel estimation for coherent detection and transmit power-control (TPC) commands to be used for the DL transmission. When the high capacity (in terms of users in a cell) is targeted for VoIP on HSDPA and HSUPA, the interference caused by continuously transmitted UL DPCCHs becomes the limiting factor for capacity.

To illustrate this, let us calculate the UL capacity by using the following UL load formula [1]:

$$NR_{dB} = -10\log 10\left(1 - \frac{\rho}{W/R} N \cdot v \cdot (1+i)\right), \qquad (10.1)$$

where ρ is the E_b/N_0 target after antenna combining, W is the chip rate, R is the E-DPDCH bit rate, N is the number of users, v is the equivalent activity factor, i is the other cell to own cell interference ratio, and NR_{dB} is the noise rise in decibels. When the UL load formula (10.1) is used, the overhead from DPCCH, E-DPCCH, and the retransmissions are included in the equivalent activity factor v.

Further, it is also possible to divide this formula into components as shown in Equation 10.2 and then have separate requirements for E_c/N_0 (i.e., the signal to noise target per chip after the antenna combining) and activity factors for each channel. Here E-DCH is assumed to include both E-DPCCH and E-DPDCH, which would always be transmitted together, thus having the same activity factor:

$$NR_{dB} = -10\log 10\left(1 - \left[\left(\frac{E_c}{N_0}\right)_{E-DCH} v_{E-DCH} + \left(\frac{E_c}{N_0}\right)_{DPCCH} v_{DPCCH} + \left(\frac{E_c}{N_0}\right)_{HS-DPCCH} v_{HS-DPCCH}\right] N \cdot (1+i)\right), \qquad (10.2)$$

For DPCCH gating (see Section 10.6.1.1), it is pessimistically assumed that E-DPDCH and HS-DPCCH transmissions never overlap and thus the DPCCH activity was obtained as the sum of activities on E-DPDCH and HS-DPCCH.

Thus, the noise rise (NR) contributed by inactive UEs (i.e., UEs with inactive packet transmission) can be calculated as shown in Figure 10.6, assuming that i is

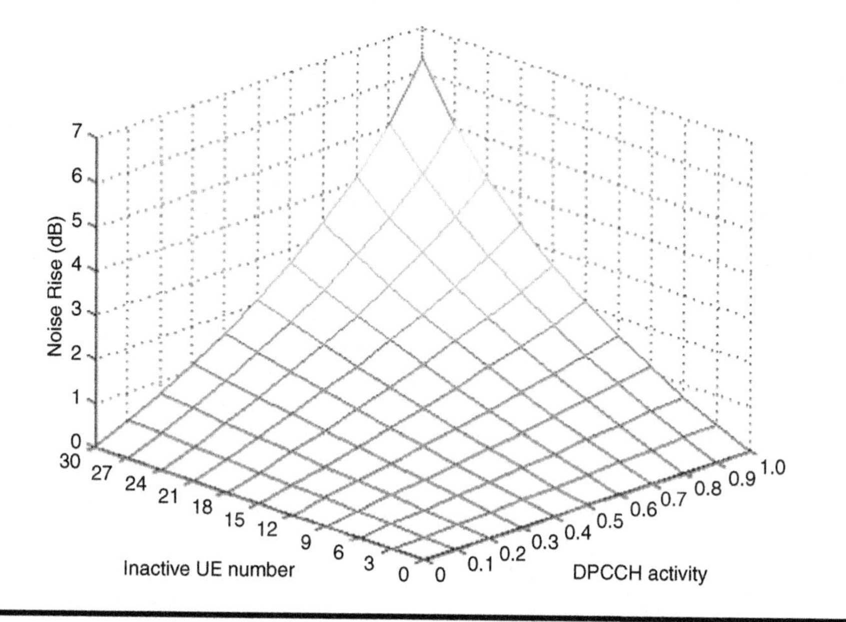

Figure 10.6 Noise rise as a function of the inactive UE number and the DPCCH activity factor. (*Source:* Tao Chen et al., "Uplink DPCCH gating of inactive UEs in continuous packet connectivity mode for HSUPA," IEEE WCNC, March 2007. With permission.)

0.65 for the typical 3-sector macro cell scenario and the required DPCCH E_c/N_0 is about −18 dB for the inactive UE [18]. Obviously, up to 30 inactive users with the continuous DPCCH transmission in this case can eat up all reserved radio resources, that is, a 6 dB noise rise target. However, the less the DPCCH activity or the less inactive the UEs, the less interference is experienced in UL.

With VoIP over E-DCH, the UL transmission of voice is not continuous as the CS voice over Dedicated Channel (DCH). One voice frame is transmitted every 20 ms, or two voice frames every 40 ms if bundling of two voice packets is allowed. With VoIP, the continuous transmission of the UL control channel DPCCH would consume the scarce radio resources and become inefficient.

Overhead from HS-DPCCH carrying CQI report

Except for the DPCCH channel, there are certain other control channels such as HS-DPCCH which may consume the UL radio resource. HS-DPCCH, carrying CQI information for DL channel sensitive schedulers and adaptive modulation scheme selections, may be transmitted periodically according to a CQI reporting period controlled by RRC. Correspondingly, the effect of the CQI reporting period can be studied according to the earlier-stated equations as well. Reducing CQI activity (i.e., increasing the CQI reporting period) from once per 10 ms to once per 20 ms increases the VoIP capacity by about 10% and going down to one CQI transmission per 80 ms increases the VoIP capacity by 20% (from CQI once per 10 ms). On the other hand, the larger transmission interval of CQI may worsen the downlink performance due to less reliability of CQI. Therefore, the CQI reporting period should be leveraged for the UL and DL performance.

HS-SCCH and E-DPCCH overhead associated with the data transmission

Other control channel constraints for VoIP over HSPA would be the overhead from the VoIP packet transmission associated control channels such as HS-SCCHs in the DL and E-DPCCHs in the UL. The HS-SCCH is a DL physical channel used to carry DL signaling related to HS-DSCH transmission, whereas the E-DPCCH is a physical channel used to transmit control information associated with the E-DCH. When a large number of users are served at a time, the control overhead (i.e., HS-SCCH overhead and E-DPCCH overhead) increases and the respective share of resources for the VoIP packet transmission of the user is reduced. Otherwise, more HARQ retransmissions would also increase this overhead. Thus, there would be a trade-off between the allowed retransmission number and the control overhead.

10.5.1.2 LTE

To support the transmission of DL and UL transport channels, certain associated UL and DL control signaling is needed. This is often referred to as L1/L2 control signaling, as the control information originates from both the physical layer (L1) and the MAC (L2). The DL L1/L2 control signaling, transmitted by using the Physical Downlink Control Channel (PDCCH), includes both PDSCH scheduling related information for each scheduled terminal and PUSCH related scheduling messages (aka scheduling grants). PDSCH scheduling information includes information about the used resource assignment and MCS, and it is necessary to the scheduled terminal to receive, demodulate, and decode PDSCH properly. PDCCH overhead limitations to VoIP DL capacity are discussed in the "PDCCH overhead" subsection.

Like LTE DL, LTE UL needs for certain associated L1/L2 control signaling to support the transmission of DL and UL transport channels. The L1/L2 control signaling is carried by the Physical Uplink Control Channel (PUCCH), and the carried L1/L2 control signaling includes HARQ acknowledgments for scheduled and received PDSCH resource blocks in DL, CQI information indicating the channel quality estimated by the terminal, and UL scheduling requests indicating that the terminal needs UL resources for PUSCH transmissions. The biggest part of UL L1/L2 control signaling originates from CQI signaling, which is required by the channel aware scheduler at eNodeB to exploit the time and frequency domain scheduling gains of PDSCH. With VoIP traffic the CQI signaling overhead may start to limit VoIP UL capacity, as the number of supported users in E-UTRA is large. The CQI overhead issue is handled in the "CQI overhead" subsection.

PDCCH overhead

As in HSPA, the baseline scheduler for VoIP traffic is the fully dynamic packet scheduling method, where each packet is transmitted by L1 control signaling. As described in Section 10.4.2, multiuser channel sharing plays an important role when optimizing air interface capacity for VoIP traffic. As PDCCH should be transmitted to each scheduled terminal, PDCCH overhead may become a limiting factor for VoIP capacity. Due to this reason, there are not enough control channel resources to schedule all physical resource blocks (PRBs), and hence part of PDSCH capacity is wasted. For example, at 5 MHz bandwidth with an AMR 12.2 kbps codec, the PDSCH utilization rate for fully dynamic PS is only 35%.

One way to avoid control channel limitations with fully dynamic PS, is that several of a user's VoIP packets can be bundled at L1 for one transmission, as described in Section 10.6.2.1. The main benefit from bundling is that more users could be fitted to the network with the same control channel overhead, as good users are scheduled less often. This will lead to significant capacity improvements in the control channel limited situation. For example, at 5 MHz bandwidth with an AMR 12.2 kbps codec, VoIP DL capacity is improved by 75–80% and the PDSCH utilization rate is increased from 35% to 70% when packet bundling (up to two packets bundled together) is used.

An alternative method to avoid control channel limitations for VoIP performance is to utilize the semi-persistent packet scheduling method [8,19–21]. This approach has been heavily studied in 3GPP. With this method, savings in control channel overhead are realized as the initial transmissions of VoIP packets are scheduled without L1 control signaling by using persistently allocated time/frequency resources. The semi-persistent PS method works very well in control channel limited circumstances, which can be seen as an increased PDSCH utilization rate. As an example, by using semi-persistent PS, the utilization rate of PDSCH is approximately 90% at 5 MHz bandwidth with an AMR 12.2 kbps codec. Semi-persistent resource allocation is described in more detail in Section 10.6.2.2.

CQI overhead

As EUTRA is based on the OFDMA scheme, in LTE system, the channel-dependent scheduling can be conducted in both time and frequency domains. To achieve maximal frequency domain scheduling gains in DL, each UE has to measure CQI in the frequency domain for all PRBs across the whole bandwidth and report the information to eNodeB for link adapta-tion and scheduling decisions. In practice, the feasible CQI feedback resolution in the frequency domain is limited by the UL control overhead related with CQI feedback, which may become prohibitively large, especially with VoIP traffic, where the number of supported users may be large. Therefore, CQI feedback should be reduced both in time and in the frequency domain to keep the UL control overhead at a reasonable level. The quantity of the reduction of CQI feedback in the time domain depends on the velocity of a user; for example, at 3 km/h the feasible CQI update rate for VoIP traffic is 5–10 ms. For this traffic, the impact of the CQI update rate is not that crucial, as due to the delay limitations of the traffic, the time domain scheduling gains are rather limited.

When reducing CQI feedback in the frequency domain, the challenge is how much it could be reduced while maintaining the benefits of frequency domain scheduling. A simple method to reduce CQI feedback in the frequency domain is to relax CQI measurement granularity in the frequency domain and calculate one CQI feedback over N consecutive PRBs. Additionally, frequency selective CQI feedback schemes have been adopted by 3GPP, such as the best-M average scheme, where the UE reports only the average CQI for the M CQI blocks having the highest CQI value and indicates the position of those M blocks within the band-width (wideband CQI for the whole band-width is reported as well in this scheme). Detailed descriptions of the various CQI reporting options available in LTE can be found in [22]. The impact of CQI feedback reduction on the performance of fully dynamic PS with VoIP traffic is analyzed in [23].

Due to the nature of the VoIP traffic, the realized time and frequency domain scheduling gains are lower than, for example, the best effort (BE) traffic. This implies that VoIP traffic is rather robust against reduced CQI feedback. This is also verified by the results presented in [23]—according to the results, 84% reduction in CQI overhead implied only 7% loss in capacity compared to the case, where full CQI (one CQI per PRB) was used. On the other hand, this compressed CQI overhead still corresponds to a 4 kbps UL channel bit rate, which might be too much considering the limited capacity of PUCCH.

As the number of supported users in LTE with VoIP traffic is high, it may necessitate the usage of wideband CQI in order to keep the overhead from CQI feedback at a reasonable level. This would mean the lowest possible UL signaling overhead from CQI feedback at the cost of reduced capacity, as all frequency domain scheduling gains will be lost. In order to keep the capacity reduction at a minimum, the impact of lost frequency domain scheduling gains to the performance should be compensated with more efficient utilization of frequency diversity. Means to achieve this are described in Section 10.6.2.3.

10.5.2 Co-Channel Interference and Noise Limitations for LTE UL

When the frequency re-use factor is set to one (1) for LTE, co-channel interference is generated in the network, degrading the UL performance of LTE significantly. In order to mitigate the impact of co-channel interference in LTE UL, interference aware scheduling [24,26], described in Section 10.6.2.5, could be used.

Further, in extreme noise limited scenarios such as macro-cell scenario case 3 [27], LTE UL may suffer from UL coverage problems. This is caused by poor UE power utilization from the small 1 ms TTI duration. To overcome this, a TTI bundling technique [28,29] was adopted as a part of LTE 3GPP standard. The TTI bundling technique is described in Section 10.6.2.6.

10.6 VoIP Solutions to Avoid System Related Limitations

10.6.1 Solutions in HSPA

10.6.1.1 Uplink Gating Associated with Packet Bundling

To address the practical constraints for VoIP over HSPA, several solutions have been proposed to improve VoIP performance.

In 3GPP releases up to Release 6, the DPCCH channel is transmitted continuously regardless of whether there is actual user data to be transmitted or not, thereby highly loading the cell (see Section 10.5.1.1). The idea of gating, introduced in 3GPP Release 7 and presented in [30], is to stop the transmission of DPCCH when there is no data to be sent on E-DCH and no L1 feedback signaling on HS-DPCCH (see Figure 10.7). This would reduce interference to other users and increase the UL capacity. Besides, it can increase UE standby time efficiently because of less power consumption.

DPCCH gating might pose a difficulty for the network to distinguish an inactive period from a lost connection. Hence it is desirable that DPCCH transmission is not totally silent but transmitted periodically following a predefined pattern. Another issue to consider is transmission power. After a gating period without tracking the received SIR on DPCCH, the channel response variations could degrade the transmission and increase the interference between the UEs. Thus, in the 3GPP specifications for DPCCH gating, the optional power control preambles are defined to be sent prior to the data channel reactivation. As a further improvement,

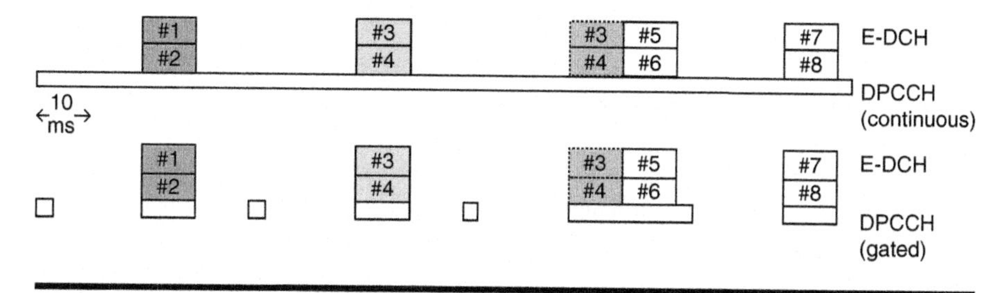

Figure 10.7 Illustration of VoIP transmission on E-DCH assuming 10-ms TTI and bundling of two VoIP packets into one TTI with continuous and gated DPCCH transmission.

compared with sending one packet per TTI, bundling multiple packets together in one transmission can reduce related control channel signaling overhead and thus improve air interface efficiency. More details about packet bundling can be found in Section 10.6.2.1. Packet bundling is typically utilized in the downlink direction only, due to the UE power shortage. However, packet bundling is applicable in the UL direction as well if 10 ms TTI is used [12].

10.6.1.2 HS-SCCH-Less Operation

3GPP Release 7 also allows HS-SCCH-less operation in the DL to minimize the interference from HS-SCCH. The high VoIP capacity typically requires 4–6 code multiplexed users on HS-DSCH in the DL of Release 6. Each code multiplexed user requires an HS-SCCH channel. HS-SCCH-less operation can be applied for VoIP when the packet interarrival time is constant. In HS-SCCH-less operation it is possible to send HS-PDSCH without associated HS-SCCH by using predefined transmission parameters (e.g., modulation and coding scheme, channelization code set, transport block size) configured by higher layers and signaled to UE via RRC signaling. HS-SCCH-less operation implies that UE has to blindly decode the HS-PDSCH that has been received without associated HS-SCCH. The interference from HS-SCCH transmissions can be avoided with HS-SCCH-less operation, leading to higher VoIP capacity.

The dynamic HS-DSCH power allocation on HS-PDSCH and HS-SCCH channels in the case of HS-SCCH-less operation for VoIP service enables supporting more than four code-multiplexed users on HS-DSCH due to the more efficient use of the scarce power resources. This improves performance, especially with the short 80 ms delay requirement, as buffering of VoIP packets in MAC-hs is becoming very restricted, leading to the higher power requirement.

10.6.2 Solutions in LTE

10.6.2.1 Packet Bundling

The size of the VoIP packet is typically very small; for example, for the AMR 12.2 kbps codec, the size of the most common ROHC compressed packet is 40 bytes. This, together with strict delay requirements of VoIP traffic, necessitates the use of PDSCH multiuser sharing when high capacities are targeted. As each user is scheduled by L1 control signaling, with multiuser channel sharing the control channel overhead becomes the limiting factor for VoIP system performance and the dynamic packet scheduler is not able to fully exploit PDSCH air interface capacity. One attractive technique with the baseline scheduler to improve VoIP capacity within control channel limitations is to bundle multiple VoIP packets (of one user) together and transmit the bundled packets within the same TTI to a scheduled user. Packet bundling is CQI-based, meaning that it is used only for such users whose channel conditions are favorable enough to support bundling. With packet bundling, good users are scheduled less frequently, implying that transmission resources can be more efficiently used and L1/ L2 control overhead can be reduced. Hence packet bundling can be seen as an attractive method in DL to improve VoIP capacity for fully dynamic PS while keeping L1/L2 control overhead at a reasonable level. From the perspective of voice quality, it is important to keep the probability of losing consecutive packets as small as possible. This can be achieved by making the link adaptation for the TTIs carrying bundled packets in a more conservative manner, leading to a reduced packet error rate for the first transmission.

In UL, packet bundling could be used as well, but it is not an attractive technique because of the limited UE transmission power and non-CQI-based scheduling.

The impact of packet bundling on the PDSCH utilization rate in macro cell case 1 [27] with a AMR 12.2 kbps codec at 5 MHz bandwidth is presented in Figure 10.8, which contains a cumulative distribution function for scheduled PRBs per OFDM subframe. Here, it is assumed that the statistics are plotted with the load corresponding to a 5% outage point–this assumption is valid for the other statistics presented in this chapter as well. In Case 1, approximately 70% of the users use bundling (see Figure 10.9), so approximately 70–80% more users can be fitted to the network without exceeding an outage, as is shown in Section 10.7. Due to the higher load compared to the nonbundling case, the average PDSCH utilization rate is approximately doubled with bundling.

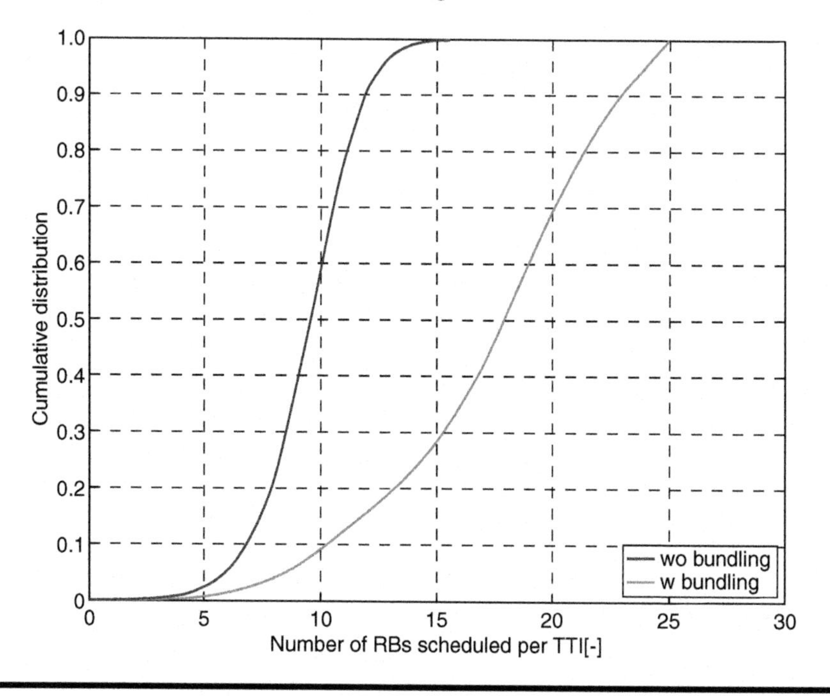

Figure 10.8 Cumulative distribution function for scheduled PRBs per TTI.

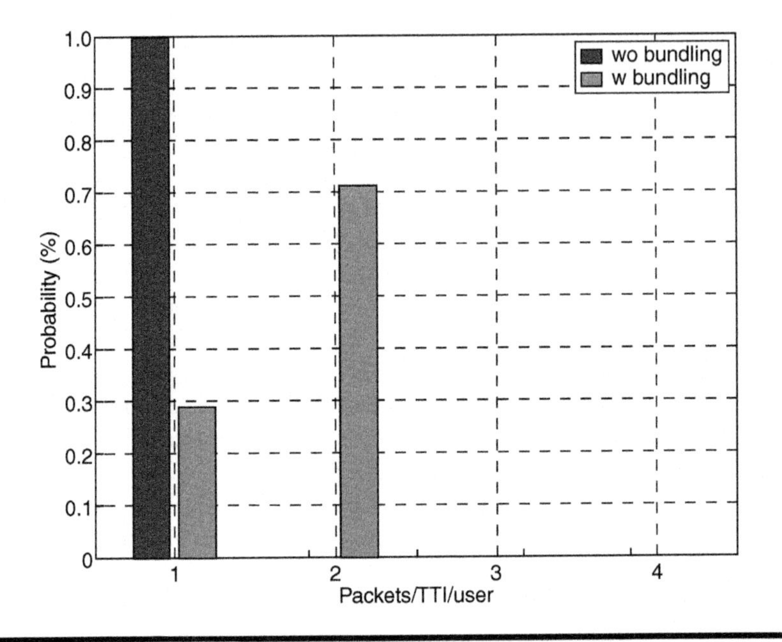

Figure 10.9 Probability distribution function for bundled packets TTI/user.

10.6.2.2 Semi-Persistent Resource Allocation

In order to avoid control channel limitations for VoIP traffic in E-UTRAN, a concept of semi-persistent packet scheduling (PS) [8,19] was adopted in 3GPP for LTE. Semi-persistent PS can be seen as a combination of dynamic and persistent scheduling methods, where initial transmissions of VoIP traffic are scheduled without assigned L1/L2 control information by using persistently allocated time and frequency resources, whereas the possible HARQ retransmissions and SID transmissions are scheduled dynamically. For VoIP traffic the used BLER target in DL is of the order 10–20%, and this together with SID transmissions implies that only 20–30% of the transmissions require L1/L2 control signaling in DL. As initial transmissions are scheduled by using persistently allocated time and frequency resources, savings in control channel overhead are achieved with the cost of reduced time and frequency domain scheduling gains.

The semi-persistent resource allocation method adopted in 3GPP LTE is talk spurt based persistent allocation, and in the DL direction the method works as follows. At the beginning of a talk spurt, a persistent resource allocation is done for the user, and this dedicated time and frequency resource is used to transmit initial transmissions of VoIP packets. At the end of the talk spurt, persistent resource allocation is released. Thus, the released resource can be allocated to some other VoIP user, which enables efficient usage of PDSCH bandwidth.

DL

UL

←20 ms→

CQI: NAK: ACK: DL allocation: DL VoIP packet: Retransmission: Reserved resource:

Figure 10.10 Talk spurt based persistent allocation in downlink.

The persistent allocation at the beginning of the talk spurt is signaled by using L1/L2 signaling. RRC signaling is used to signal some semi-static parameters such as the periodicity of the allocation. A slow, talk spurt based link adaptation is possible since the modulation scheme as well as the amount of resources are signaled with L1/L2 signaling.

The release of the persistent allocation at the end of the talk spurt is done via explicit signaling from the eNodeB with L1/L2 signaling. It is still open whether implicit release (without signaling) of DL semi-persistent resources is supported, too. DL operation for the talk spurt based persistent allocation is illustrated in Figure 10.10.

In the UL direction, the UE should send a scheduling request (SR) to the eNodeB at the beginning of a talk spurt to get the radio resource allocated for it. Similarly, as in DL, the allocated radio resource is used to carry initial transmissions of VoIP packets without any scheduling related L1/L2 signaling in DL. SID frames can be allocated dynamically (thus requiring SR for each SID). The required signaling in the UL direction is similar to the DL. The persistent UL allocation at the beginning of the talk spurt is signaled by using L1/L2 signaling.

When the talk spurt ends, the resource is released either explicitly with release signaling or implicitly by noticing that no more data is coming. Thus, the released resource can be allocated to some other VoIP user.

Figure 10.11 shows the required UL and DL signaling as well as the transmission resources. Here we assume that SID frames are scheduled dynamically.

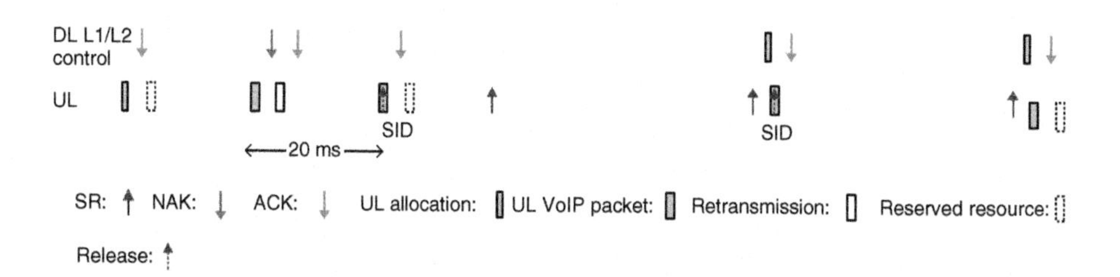

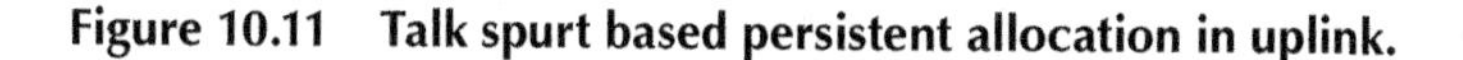

Figure 10.11 Talk spurt based persistent allocation in uplink.

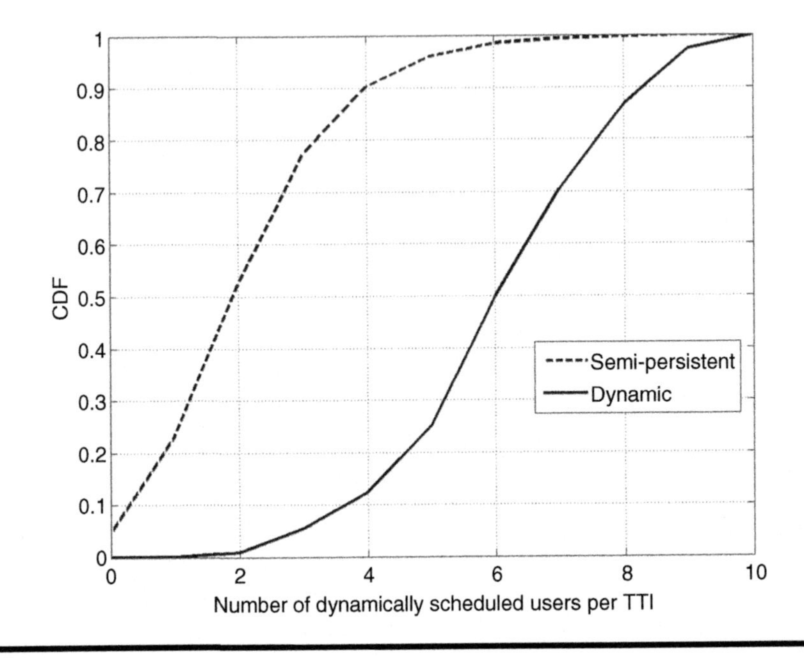

Figure 10.12 Cumulative distribution function for the dynamically scheduled users per TTI (Out of the maximum 10 users).

The impact of the semi-persistent resource allocation method on DL control overhead is illustrated in Figure 10.12, which contains the cumulative distribution function (CDF) of the dynamically scheduled users per TTI. As is evident from the figure, with the semi-persistent resource allocation method, DL control overhead is approximately 40% of the control channel overhead of fully dynamic PS. In Figure 10.13 the CDF for the number of scheduled PRBs per OFDM subframe is presented. Performance of fully dynamic PS is control channel limited; that is, there are not enough control channel resources to utilize the total transmission bandwidth: with fully dynamic PS, only 70% of the total transmission bandwidth (5 MHz ~ 25 PRBs) is used on average. Here we assume that packets are used for the dynamic scheduler. Semi-persistent PS does not suffer from control channel limitation, but its performance is data limited: on average only 10% of the total transmission bandwidth is unused.

10.6.2.3 Means to Overcome CQI Imperfections

CQI feedback is utilized by the channel aware scheduler at eNodeB when selecting the scheduled users and allocating resources for them. Additionally, CQI reports are used by the link adaptation algorithm when selecting the used modulation and coding schemes for the scheduled users. CQI feedback is subject to reporting delays

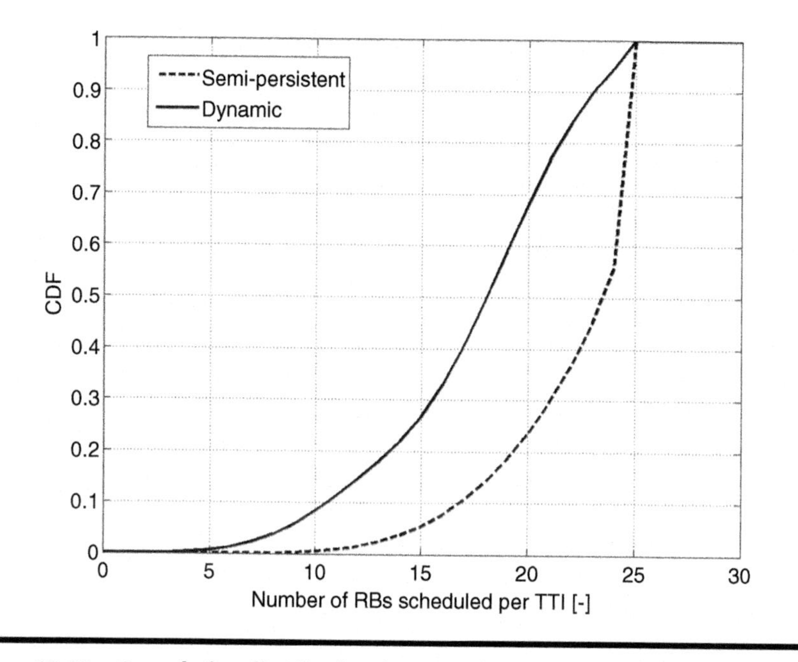

Figure 10.13 Cumulative distribution function for scheduled PRBs per TTI.

and UE measurement/estimation imperfections, and these are compensated by the Outer Loop Link Adaptation (OLLA) algorithm, which uses HARQ ACK/NAK feedback from the past to maintain the average BLER at the target level. OLLA has been shown to provide a good mitigation mechanism for the LA errors caused by the imperfect CQI feedback information [31–33].

As described earlier, wideband CQI may be required for VoIP traffic in order to keep the UL control overhead at a reasonable level. This implies that frequency domain packet scheduling gains are lost: the eNodeB packet scheduling can no longer track the frequency selective fading and will have to rely on a transmission scheme that offers maximal frequency diversity. Or similarly, frequency diversity transmission should be favored if, due to the increased velocity, frequency dependent CQI information is unreliable (velocity >25 km/h). Frequency diversity transmission can be achieved by one of the following methods [34]:

(1) Using localized transmission where a user is scheduled on multiple PRBs that are scattered over the full system bandwidth to offer maximal frequency diversity; or
(2) Using distributed transmission where a number of PRBs (scattered over the full system bandwidth) are shared by a set of users.

Distributed transmission refers to the case where one PRB of 12 contiguous subcarriers is shared between multiple users on a subcarrier resolution. The distributed transmission feature was adopted by 3GPP Release 8 with the restriction that

Table 10.3 Impact of Wideband CQI to Capacity

Used Packet Scheduling Method	*Dynamic*	*Semipersistent*
Relative loss in capacity	14	7

a user scheduled with distributed transmission shall always be scheduled on groups of Nd × 12 subcarriers, where Nd is limited to two (2).

Method (1) offers a high degree of frequency diversity for cases where there is sufficient data for a user to be scheduled on multiple PRBs, that is, at least 3–4 PRBs. Unfortunately, with VoIP traffic the data amount for one user typically requires at most 2 PRBs, and therefore it is difficult to achieve efficient frequency diversity with localized transmission. Frequency diversity offered by method (1) can be further improved if HARQ retransmissions are scheduled in different locations in frequency than previous transmission(s).

VoIP LTE DL performance of semi-persistent PS with wideband CQI was analyzed in [34] by comparing localized transmission with method (1) (with HARQ retransmissions scheduled so that frequency diversity is maximized) and distributed transmission with Nd = 2. It was shown that distributed transmission provided only 2% higher capacity than localized transmission with the enhanced method (1).

In Table 10.3 relative losses in capacity due to utilization of wideband CQI instead of narrowband CQI are given for fully dynamic and semi-persistent PS methods. Here it is assumed that localized transmission with method (1), assuming diversity enhancement over HARQ retransmissions, is used. As can be seen from the results, when the frequency domain scheduling losses are compensated with efficient utilization of frequency diversity, degradation in performance due to usage of wideband CQI stays tolerable. Without frequency diversity enhancements, corresponding losses would be higher. Moreover, the relative losses in performance realized by the semi-persistent PS are only half of the corresponding losses for fully dynamic PS, as only HARQ retransmissions and SID transmissions benefit from FD scheduling gains.

10.6.2.4 Adaptive HARQ in UL

For trade-off purposes, the currently agreed-on proposal in 3GPP is to use a synchronous nonadaptive HARQ protocol as much as possible for LTE UL. It means adaptive HARQ will be enabled for use if fully nonadaptive HARQ is impossible to be used. Such a decision is due to the drawbacks of fully nonadaptive HARQ. The nature of the nonadaptive solution is that retransmissions occur at a predefined (normally fixed) time after the previous (re)transmission using the same resources. The benefit of synchronous nonadaptive HARQ is that the UL control signaling can be minimized since only a NAK needs to be signaled back to start a HARQ retransmission. However, due to orthogonality and single carrier requirements in

LTE UL, there are some obvious problems for nonadaptive HARQ retransmissions combined with semi-persistent allocation:

(a) Resource fragmentation: When several UEs are scheduled in one TTI and some users require retransmissions whereas others do not, the UL resources can be fragmented and the scheduling of new users becomes more difficult or the required resources cannot be scheduled to a user due to single carrier requirement [36].

(b) Low resource efficiency: For semi-persistent allocation, part of the resources are allocated persistently and part are scheduled dynamically. A nonadaptive HARQ sets unnecessary restrictions and leads to poor resource utilization. Since with synchronous nonadaptive HARQ the possible retransmission would overlap with the persistent allocation, a persistent allocation in a given time-frequency resource implies that nothing can be scheduled on the same frequency resource at a HARQ RTT earlier (or a few HARQ RTTs earlier).

(c) Separate ACK/NAK channel in DL: With synchronous nonadaptive HARQ, a separate ACK/NAK has to be sent strongly coded (repetition) and with high power in order to guarantee low error probability.

Synchronous adaptive HARQ can then naturally be used to solve the problems: when resource fragmentation happens, an adaptive HARQ allows moving of the retransmissions to the edges of the band, thus avoiding the fragmentation; when the collision happens between initial persistent transmissions and retransmissions, by allowing adaptive HARQ, the possible retransmission can be sent on different frequency resources, thus avoiding the problem; adaptive HARQ implies that all the retransmissions are scheduled. This has the advantage that no separate ACK/NAK channel is needed in the DL. Retransmissions are requested with UL allocation (implicit NAK), and a UL allocation for a new transmission implies that retransmissions are not continued. Though the ACK/NAK channel still exists in terms of 3GPP agreement, it is overridden by UL allocation signaling.

10.6.2.5 Interference Aware Packet Scheduling for LTE UL

As proposed in [24,26], an interference aware packet scheduling method to alleviate the co-channel interference for larger cells (e.g., macro cell Case 3 [27]) could be used. Briefly, the method is as follows: First, users in the same sector are sorted on the basis of DL path loss measurement. Second, UEs closer to BS are allocated into middle frequencies and UEs at the cell edge are allocated into edge frequencies. The SNR target is first set to cell edge users, and the SNR target of users closer to BS may be gradually increased. In other words, the basic idea in this method is that users with similar path loss use the same frequency in all sectors/cells. Hence, with this method the interference level is comparable to the useful signal level, but,

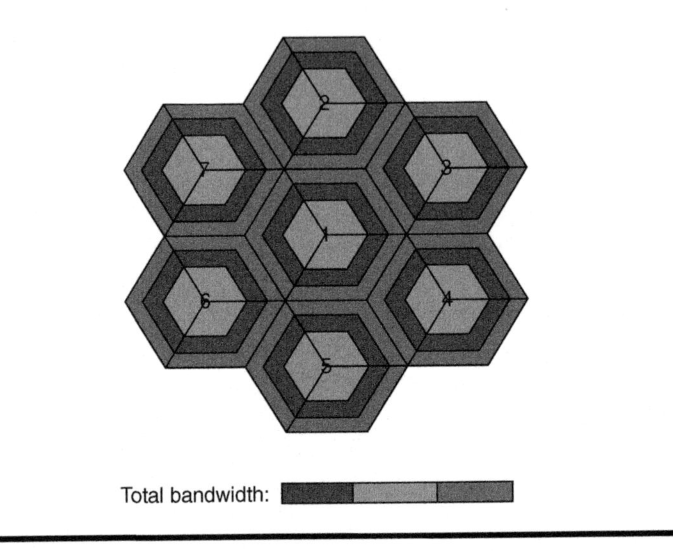

Figure 10.14 IC scheme for large cell.

as can be seen from Figure 10.14, the cell edge users may interfere with each other. Therefore, for small cells (e.g., macro cell Case 1 [27]), different re-use patterns to minimize intercell interference can be used. This is depicted in Figure 10.15. Similarly, as for the interference control scheme for larger cells, UEs are sorted according to DL path-loss measurements. Then, the required transmission power levels of UEs are fi gured out according to the SNR target. The SNR target is the same for all UEs in the network; that is, for re-use a factor three power sequence [0 −4 −4] is used. The key point here is that users with similar path loss use different frequencies in different sectors within one cell. Hence, the cell edge users in

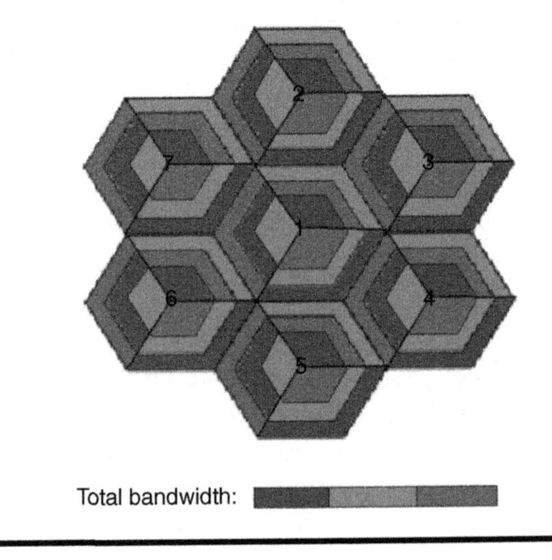

Figure 10.15 IC scheme for small cell.

different sectors use different sub-bands, implying that they do not interfere with each other. On the other hand, the interference among the three sectors within one cell seems to be a bit larger.

Of course, some kind of combined scheme by utilizing the advantages of the two IC schemes can be designed to further enhance the performance as well.

10.6.2.6 TTI Bundling for LTE UL

A natural solution to improve the coverage in a noise limited scenario would be to use RLC layer segmentation of VoIP packets. The drawback of this method is mainly the significantly increased overhead. To overcome this problem, the TTI bundling method is adopted in 3GPP [28,29], in which a few consecutive TTIs are bundled together and a single transport block is first coded and then transmitted by using this bundled set of consecutive TTIs. With TTI bundling only one L1/L2 grant is needed to schedule the transmission and only one HARQ feedback signal is sent from the eNodeB for the bundled subframes. Moreover, the same hybrid ARQ process number is used in each of the bundled subframes. The operation is like an autonomous retransmission by the UE in consecutive HARQ processes without waiting for ACK/NAK feedback. The redundancy version on each autonomous retransmission in consecutive subframes changes in a predetermined manner. There is a possibility that the code rate might be larger than one for the first subframe, but decoding becomes possible at the eNodeB as soon as the code rate becomes less than or equal to one. TTI bundling is a technology that improves the data rate, reducing overheads and leading to a packet-optimized transmission.

10.7 Network Performance of VoIP

10.7.1 HSPA

10.7.1.1 Simulation Methodology and Assumptions

The quasi-static system level simulators for HSUPA and HSDPA, where all necessary RRM algorithms as well as their interactions are modeled, are used to investigate the performance of VoIP on HSPA. These tools include a detailed simulation of the users within multiple cells. The fast fading is explicitly modeled for each user according to the ITU Vehicular-A profile. Regarding the methodology, this kind of quasistatic simulator is based on descriptions in [3]. A wraparound multi-cell layout modeling several layers of interference is utilized in this study.

The main parameters used in the system simulation are summarized in Table 10.4, where UE moving speeds other than 3 kmph are mostly applied to investigate the effect of the mobility on the performance of the DL.

Table 10.4 Simulation Parameters Settings

Parameter	*Value*
System Configuration	
Inter-site distance	2.8 km
Cell configuration	ITU Veh-A, macrocell
Voice call mean length	60 seconds
UE speed	3 (30, 50, 90) kmph
Outage observation window length	10 seconds
Cell outage threshold	5%
Residual FER	1%
HSUPA Specific Configuration	
Receiver	2 antenna RAKE with MRC combining
Frame size	2 ms/10 ms
Channels	E-DCH/DPCCH
Number of HARQ channels	8 (2ms TTI)/4 (10ms TTI)
Max number of L1 transmissions	4 (2ms TTI)/2 (10ms TTI)
Uplink delay budget	80 ms
Scheduling algorithm	Non-scheduled
HSDPA Specific Configuration	
Receiver	1 antenna RAKE with MRC combining
Frame size	2 ms
Channels	HS-DSCH/CPICH/HS-DPCCH
Number of HARQ channels	6
Max number of L1 transmissions	4
Max code-multiplexing users	4
Number of codes	10
Total BS transmission power	20 W
HS-DSCH power	10 W
Downlink delay budget	80 ms

10.7.1.2 Simulation Results and Analysis

VoIP performance in HSUPA

The UL capacities for VoIP over HSUPA are presented to evaluate the performance of some advanced solutions as described in Section 10.6.1.1. For the 2 ms TTI without UL packet bundling, Figure 10.16 shows UE power distribution for 60 UEs/cell for gating DPCCH and continuous DPCCH. It may be noted that there can be huge power savings by applying DPCCH gating. The exact power savings level would depend

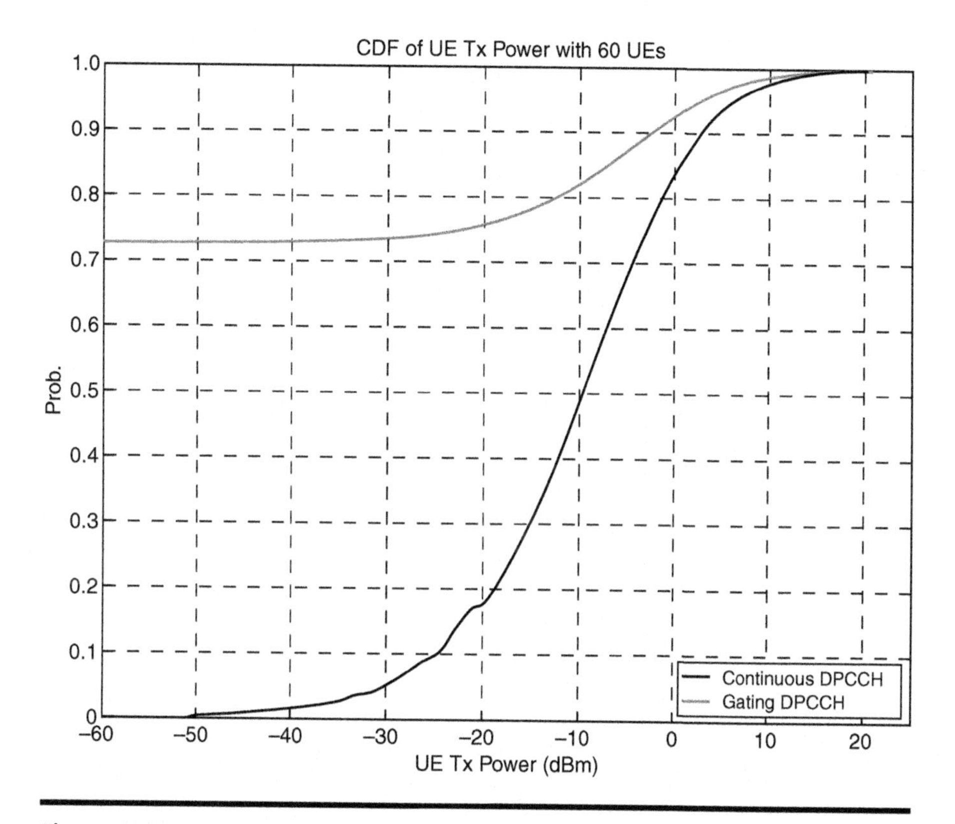

Figure 10.16 UE power CDF distributions with 60 UEs/cell for gating DPCCH and continuous DPCCH.

on the activity of DPCCH, which is affected by the transmission of E-DCH and HS-DPCCH. It also implies that the frequent CQI reporting carried on HS-DPCCH would reduce the power savings gain because of having fewer opportunities to use DPCCH gating. Similarly, if the DL data transmission over HS-DSCH is quite active, the transmission of ACK/NAK feedback information carried over HS-DPCCH in the UL would be quite frequent, which will also constrain the gating gain.

Table 10.5 summarizes the VoIP capacity results obtained with a quasi-static system simulator for both 2 ms and 10 ms TTI in the case of 3 kmph UE moving speed. The capacity is calculated with two different criteria presented in Section 10.3.3: the number of VoIP users per sector not exceeding the 5% FER target measured over 10 s or the number of VoIP users allowed per sector with an average noise rise of 6 dB. Here the delay budget of 80 ms is assumed for HSUPA. This would not be the bottleneck at the performance since HARQ retransmissions will be completed within this delay budget when the maximum allowed HARQ retransmission number is 1 for 10 ms TTI and 3 for 2 ms TTI. The results show that VoIP capacity over HSUPA is comparable to or to a great extent better than CS

Table 10.5 Summary of the UL VoIP Capacity Results without and with UL DPCCH Gating

		Capacity Criteria: 5% FER Over 10 s			*Capacity Criteria: Noise Rise 6 dB*		
TTI	*Average Number of Transmissions*	*Continuous DPCCH (Users)*	*Gated DPCCH (Users)*	*Gating Gain (%)*	*Continuous DPCCH (Users)*	*Gated DPCCH (Users)*	*Gating Gain (%)*
2 ms	~3	82	123	50	75	106	41
10 ms*	1.25	65	115	77	61	93	52
	1.67	80	120	50	73	103	41

* 10ms TTI results using bundling of 2 VoIP frames in a single TTI.

Source: 3GPP TR 25.903 V1.0.0 (2006–05). "Continuous connectivity for packet data users."

voice over DCH (about 65 voice users per sector) [3] with continuous DPCCH, and gating gives 40–50% further gain. Otherwise, fewer transmissions can increase the DPCCH gating gain in the capacity from 50% to 77% with 10 ms TTI because of the increasing opportunities for DPCCH gating from fewer transmissions on E-DCH. (For more results and simulation assumptions, see, e.g., [3,12,30,37].)

VoIP performance in HSDPA

VoIP system performance in HSDPA achieved with the best mode scheduling alternative ("PF + SCS + RA" described in Section 10.4.3.2) is shown in Figure 10.17 for both Release 6 and Release 7. Release 6 results are achieved with a 1-antenna LMMSE chip equalizer, whereas Release 7 results are achieved with a 2-antenna RAKE with MRC combining.

Corresponding UL capacities for Release 6 and 7 are also presented in the figure for comparison. As the results show, receive diversity provides superior performance compared to the 1-antenna LMMSE chip equalizer. Furthermore, results indicate that HSPA capacity is uplink limited.

Effect of the mobility on the performance

In the real world, the UE is always moving at a varying speed. Therefore, it makes sense to understand the effect of UE velocity and handover delay in practice. VoIP

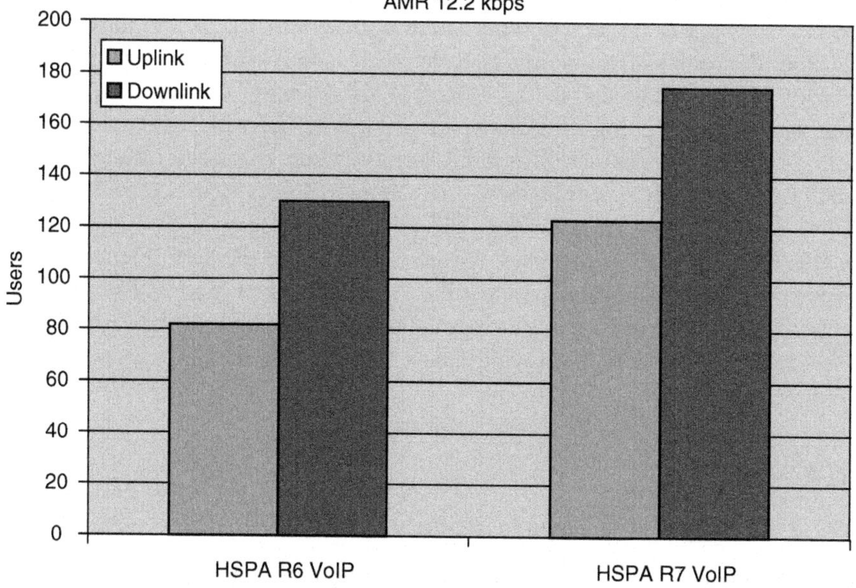

Figure 10.17 VoIP HSPA capacity comparison between downlink and uplink.

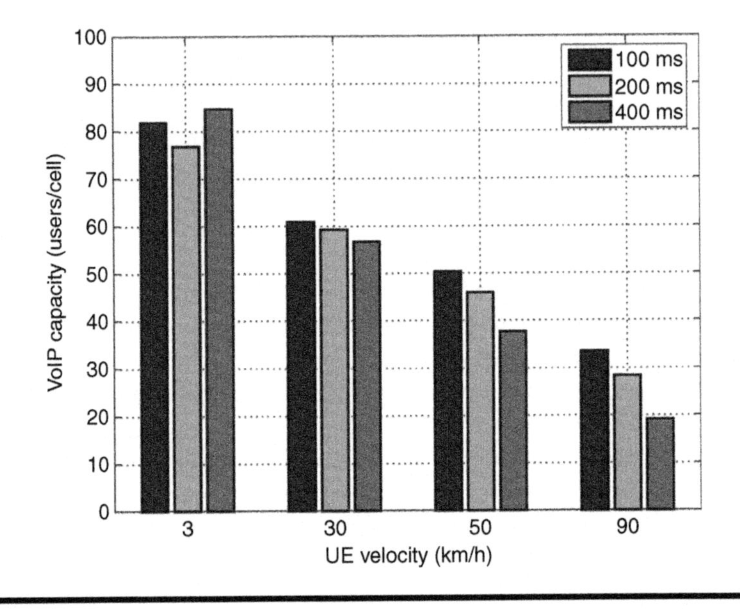

Figure 10.18 Average VoIP cell capacity with different UE velocities and handover delays. (*Source*: P. Lundén, J. Äijänen, K. Aho, and T. Ristaniemi, "Performance of VoIP over HSDPA in mobility scenarios," VTC 2008 Spring. With permission.)

cell capacities are presented in Figure 10.18. The figure shows the achieved capacities for 100 ms, 200 ms, and 400 ms handover delays.

From the results it can be seen that the VoIP capacity is highly sensitive to UE velocity. Higher velocities lead to lowered VoIP capacity: the capacity drops by roughly 20% when going from 3 km/h to 30 km/h, and with higher velocity the capacity drops even further. This is because the radio channel is changing more rapidly and scheduling is thus more challenging. Moreover, the number of handovers is increased significantly when UE velocity increases.

The results show that the handover delay is critical to the performance of VoIP service over HSDPA, especially when the UE velocity gets higher. This is caused by the fact that with high UE velocities the performance is already greatly affected by the radio channel conditions, and thus it is required to have fast response to the handover procedures.

10.7.2 LTE

10.7.2.1 Simulation Methodology and Assumptions

The performance evaluation for LTE is conducted with a quasi-static system level simulator [11]. The simulated network is composed of 19 cell sites with 3 sectors each. The scheduling algorithm and other radio resource management (RRM)

Table 10.6 Summary of Main Simulation Parameters

Description	*Settings*
Number of cells	19 Sites with 3 cells
Carrier center frequency	2.0 GHz
Simulation scenarios	Macro case 1 and macro case 3
System bandwidth	5 MHz
Path loss	According to [27], minimum coupling loss (MCL) 70 dB
Standard of log-normal shadowing	8 dB
Velocity	3 km/h
Channel	SCM-C with correlations for spatial channels
Link to system model	DL EESM (realistic ChEst), UL AVI (realistic ChEst)
PDCCH modeling	Realistic

functions are explicitly modeled in 3 center cells, and statistics are also gathered from these 3 center cells. The remaining cells generate intercell interference with full load. The propagation characteristics of the physical link as a distance dependent path loss, shadowing and frequency selective fast fading are modeled in the system simulator, whereas the link performance is fed to the system simulator through an Exponential Effective SNR Mapping (EESM) interface [25] or, alternatively, through an Actual Value Interface (AVI). Furthermore, for LTE the Spatial Channel Model [38] SCM-C shall be employed. The main simulation parameters are based on [17] and are also listed in Table 10.6. UEs are uniformly distributed in 3 center cells and assigned with VoIP traffic. The VoIP traffic is modeled as in [11] and the most important characteristics are summarized in Table 10.7. Moreover,

Table 10.7 Main Traffic Related Assumptions

Description	*Settings*
Codec	AMR 12.2 kb/s, AMR 7.95 kb/s
Header compression	ROHC
Payload including all overhead	40 bytes for AMR 12.2 kbps, 28 bytes for AMR 7.95 kbps
SID payload including all overhead	15 bytes
Voice activity	50%
Talk spurt	Negative exponential distribution, mean 2.0 s
One-way delay (ms)	50

Table 10.8 Main DL Related Simulation Parameters

Description	*Settings*
Multiple access	OFDMA
eNode-B Tx power	43 dBm (5 MHz)
Max C/I limit in the receiver	22 dB
Transmission scheme	2 × 2 Space-Time Transmit Diversity (STTD)
Receivers	2-antenna with MRC combining
CQI settings	Zero mean i.i.d. Gaussian error with 1 dB std, 5-ms measurement window, 2ms reporting delay
DL L1/L2 control channel overhead	4 OFDM symbols per TTI including pilot overhead
MCS set	As in [17]
HARQ	8 HARQ SAW channels. Asynchronous, adaptive HARQ with chase combining. Maximum 6 retransmissions
Packet scheduler	VoIP optimized PS, semi-persistent PS
Packet bundling	Can be used for fully dynamic PS (up to 2 packets could be bundled per TTI)
PDCCH assumption	10 CCEs reserved for DL traffic scheduling per TTI (max 10 users could be scheduled dynamically per TTI)

the main DL related parameters and UL related parameters are summarized in Table 10.8 and Table 10.9, respectively.

The performance is evaluated in terms of VoIP capacity. VoIP capacity is defined according to the quality criteria given in Section 10.3.3; that is, capacity is given as the maximum number of users that can be supported without exceeding 5% outage. The user is in outage if more than 2% of the packets are lost (i.e., lost or erroneous) during the whole call.

10.7.2.2 Simulation Results and Analysis

VoIP performance in LTE

VoIP capacity numbers for LTE are summarized in Table 10.10 for both DL and UL directions. Let us analyze the DL results first.

As described in Section 10.5.1.2, the performance of the fully dynamic PS without packet bundling is limited by the available PDCCH resources, due to which savings in VoIP packet size are not mapped directly to capacity gains. This can be verified from the results showing almost identical performance for AMR 7.95 and AMR 12.2. With packet bundling, control channel limitations can be partly avoided and, hence, the VoIP capacity with fully dynamic PS can be boosted up

Table 10.9 Main UL Related Simulation Assumptions

Description	*Settings*
Multiple access	SC-FDMA
UE transmission power	Max 24 dBm
Power dynamics of UEs at eNodeB	17 dB
Transmission scheme	1 × 2
Receivers	2-antenna with MRC combining
UL control overhead	2 PRBs per TTI
MCS set	As in [17]
HARQ	8 HARQ SAW channel. Synchronous, adaptive HARQ with incremental redundancy (IR). Maximum 6 retransmissions
Packet scheduler	Dynamic PS, semi-persistent PS
PDCCH assumption	10 CCEs reserved for UL per TTI (max 10 users could be scheduled dynamically per TTI)

to 90%. With bundling, the performance of fully dynamic PS starts to suffer from data limitation as well, due to which small gains in capacity (8%) can be achieved when AMR 7.95 is used instead of AMR 12.2. On the other hand, the performance of semi-persistent PS is data limited instead and, hence, savings in VoIP packet size are nicely mapped to gains in capacity—approximately 35% higher capacity is obtained if AMR 7.95 is used instead of AMR 12.2. When comparing the performances of fully dynamic PS and semi-persistent PS with each other, it is concluded that if packet bundling is not allowed, then the performance of fully dynamic PS is badly control channel limited and, hence, semi-persistent PS is able to have 50–100% higher capacities than fully dynamic PS. As the performance of

Table 10.10 VoIP Capacity in LTE

VoIP Codec	*AMR 7.95*	*AMR 12.2*
Downlink Capacity		
Dynamic scheduler, without bundling	210	210
Dynamic scheduler, with bundling	400	370
Semi-persistent scheduler	430	320
Uplink Capacity		
Dynamic scheduler	230	210
Semi-persistent scheduler	320	240

semi-persistent PS suffers from data limitation, gains in capacity over fully dynamic PS are smaller for the AMR 12.2, which has a higher data rate and hence a bigger VoIP packet size. When packet bundling is used with AMR 12.2, the control channel limitation for the performance of fully dynamic PS is not that significant compared to the corresponding case with AMR 7.95, due to a lower number of supported users at 5% outage. Therefore, fully dynamic PS is able to have 15% gains in capacity over semi-persistent PS if AMR 12.2 is used with packet bundling. When bundling is used with AMR 7.95, control channel limitation again starts to dominate the performance of fully dynamic PS due to the high number of supported users at 5% outage, and hence semi-persistent PS achieves 8% higher capacities than fully dynamic PS. However, in practice it will be most likely necessary to use wideband CQI with VoIP traffic to keep the CQI related UL signaling load reasonable. This will reduce the performance of the dynamic PS more than that of the semi-persistent PS (as the dynamic PS relies more on the frequency selective CQI information). Therefore, the semi-persistent PS can be seen as an attractive option for supporting a high number of simultaneous VoIP users.

When analyzing VoIP results for LTE UL, it is emphasized that similar to DL, the semi-persistent scheduling suffers much less from control channel limitations, whereas the performance of dynamic scheduling greatly relies on the number of control channels per TTI, as illustrated in Figure 10.19, which contains the cumulative distribution

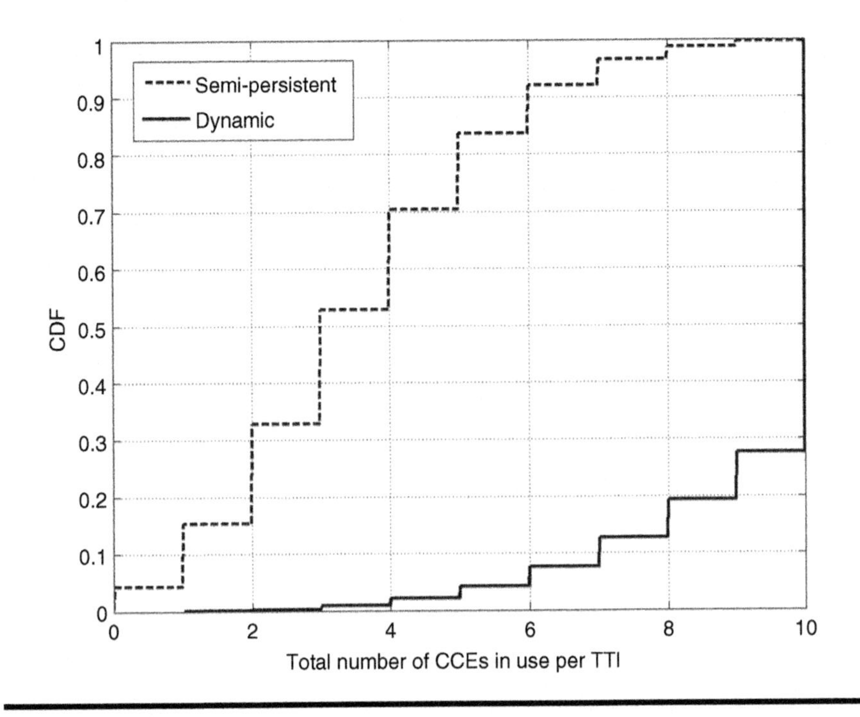

Figure 10.19 Distribution of used CCEs per TTI.

function of the used control channel elements (CCEs) per TTI. As control channel limitation becomes a bottleneck for VoIP UL system performance with fully dynamic PS, especially if a very high number of users are accessed into the system, that is, as is the case with AMR 7.95. Due to the control channel limitations, fully dynamic PS can support only 230 users per sector in Case1 with AMR 7.95, whereas with semi-persistent scheduling, control channel limitations can be avoided and 320 users with a gain of 40% can be supported. However, with AMR 12.2, the impact of control channel limitations is much smaller for the fully dynamic PS due to bigger VoIP payload sizes and hence a smaller number of supported users at 5% outage. Therefore, the capacity gain of semi-persistent PS over fully dynamic PS is only about 14%.

Finally, when comparing the performances of DL and UL together, it is concluded that the VoIP performance in LTE is UL limited, similar to HSPA. This is mainly due to the coverage limitation in UL, which is caused by the relatively low maximum transmit power of the UEs. The extra DL capacity is not wasted but can be used, for example, for supporting additional best effort traffic (web browsing, etc.), which is typically asymmetric and geared more toward DL.

VoIP performance with mobility in LTE

Figure 10.20 shows the average behavior of different UE mobility scenarios in LTE. The simulations were run with dynamic scheduling, but it is assumed that similar behavior would result with semi-persistent scheduling.

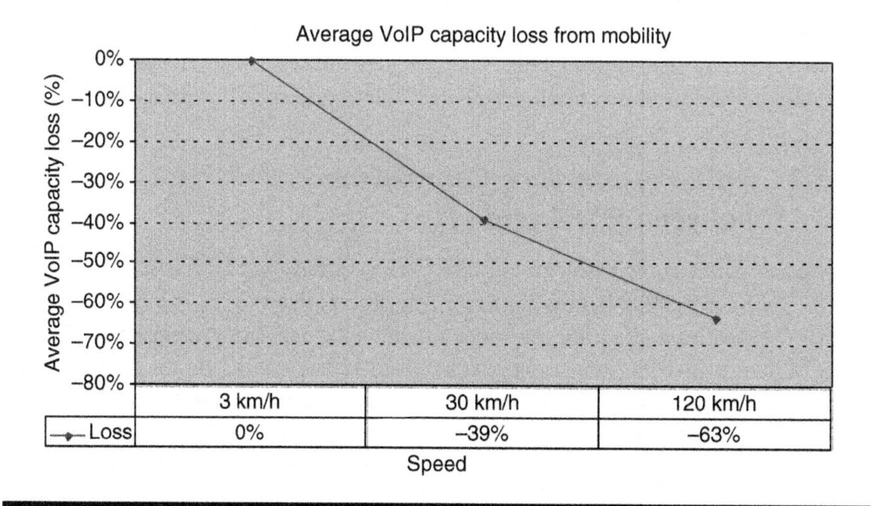

	3 km/h	30 km/h	120 km/h
Loss	0%	−39%	−63%

Figure 10.20 Average VoIP cell capacity with different UE velocities and handover delays. (*Source:* T. Henttonen, K. Aschan, J. Puttonen, N. Kolehmainen, P. Kela, M. Moisio, and J. Ojala, "Performance of VoIP with mobility in UTRA long term evolution," IEEE VTC Spring, Singapore, May 2008.)

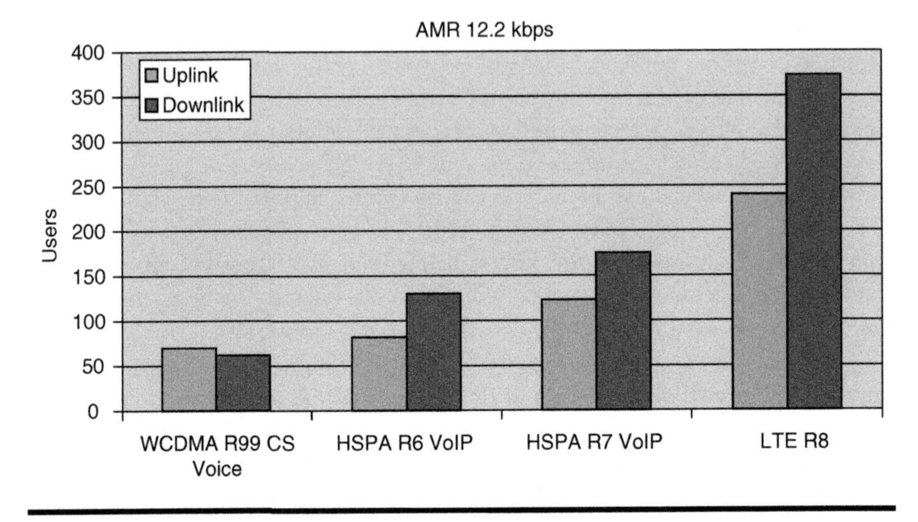

Figure 10.21 Voice capacity with WCDMA/HSPA/LTE.

Mobility clearly degrades VoIP capacity. With high speed, the loss (compared to the case with the pedestrian speed of 3 km/h) is about 63%, so the capacity is roughly only one-third of the nonmobility case. The loss in capacity is caused by several factors: First, the CQI measurements become less accurate as the speed increases, which causes the scheduler to lose most of the efficiency of frequency-dependent scheduling. Link adaptation also performs badly as it is tied to the CQI reports, leading to an increase in retransmissions. And the faster the users move, the more handovers are done, with all the resulting effects (L1 packets flushed when a handover is done, delays due to handover execution delay) also affecting the VoIP capacity [43].

10.7.3 VoIP Performance Comparison between HSPA and LTE

The voice capacity comparison among WCDMA, HSPA, and LTE is summarized in Figure 10.21 for 5 MHz bandwidth. From the comparison, it is obvious that HSPA Release 7 provides a clear improvement to the voice capacity compared to the CS voice. When comparing voice capacities between HSPA and LTE, it is observed that LTE achieves more than 100% higher capacity than HSPA Release 7. Furthermore, it can be noted that both VoIP capacity is UL limited in both HSPA and LTE.

10.8 Summary and Conclusion

The potential for simplifications in the core network and the cost benefits that an "All IP network" can provide are clear. However, the VoIP solution has to work well to enable a stand-alone cellular system with all the commonly used services.

The network has to be able to guarantee the same service quality and higher voice capacity with VoIP than has been required from CS-based voice systems. The principles of capacity and coverage gaining mechanisms, explained in Section 10.4, are: limiting the IP-header overhead, multiuser channel sharing, VoIP specific packet schedulers, and advanced receivers. ROHC, a solution for the first mechanism, enables the system to perform more efficiently by reducing the overhead and has been widely studied and recognized as an essential optimization for VoIP. The other gaining mechanisms enhance the capacity and the coverage by helping to guarantee the quality also at the cell edge.

Control channel overhead and the co-channel interference, as explained in Section 10.5, are severe obstacles to the VoIP performance. To enable the whole potential offered by the different gaining mechanisms in the networks with the practical limitations and further gain in capacity and coverage, several methods were explained in Section 10.6. These solutions (packet bundling, UL gating, HS-SCCH-less operation, semi-persistent allocation, distributed allocation, adaptive HARQ, TTI bundling) with VoIP optimization provide an optimistic sight for the VoIP success. The simulation results of Section 10.7 show that especially the control channel overhead reduction solutions adopted by 3GPP are attractive, as they provide very good performance.

References

1. H. Holma and A. Toskala, "WCDMA for UMTS-HSPA evolution and LTE," 4th ed., John Wiley, 2007.
2. R. Cuny and A. Lakaniemi, "VoIP in 3G networks: An end-to-end quality of service analysis," IEEE, VTC Spring, vol. 2, April 2003, pp. 930–934.
3. H. Holma and A. Toskala, "HSDPA/HSUPA for UMTS," John Wiley, 2006.
4. 3GPP TS 25.309, "FDD enhanced UL; Overall description; Stage 2."
5. 3GPP TR 25.896, "Feasibility study for enhanced UL (FDD)."
6. 3GPP TS 25.308, "HSDPA; Overall description; Stage 2."
7. H. Holma, M. Kuusela, E. Malkamäki, K. Ranta-aho, and T. Chen, "VoIP over HSPA with 3GPP, Release 7," PIMRC 2006, Helsinki.
8. 3GPP TS 36.300, "Evolved universal terrestrial radio access (E-UTRA) and evolved universal terrestrial radio access network (E-UTRAN); Overall description; Stage 2 (Release 8)."
9. 3GPP TR 25.913 V7.1.0, "Requirements for evolved UTRA (E-UTRA) and evolved UTRAN (E-UTRAN)."
10. 3GPP TS 26.071, v7.0.1, "Mandatory speech CODEC speech processing functions; AMR speech CODEC; General description (Release 7)."
11. 3GPP R1-070674 "LTE physical layer framework for performance verification," Orange, China Mobile, KPN, NTT DoCoMo, Sprint, T-Mobile, Vodafone, Telecom Italia.
12. O. Fresan, T. Chen, K. Ratna-aho, and T. Ristaniemi, "Dynamic packet bundling for VOIP transmission over relapses; 7 HSUPA with 10 ms TT1 length." ISWCS. 2007.

13. IETF RFC 3095: "RObust Header Compression (ROHC): Framework and four profiles: RTP, UDP, ESP, and uncompressed."
14. P. Lundén and M. Kuusela, "Enhancing performance of VoIP over HSDPA," VTC 2007 Spring.
15. Y. Fan, M. Kuusela, P. Lunden, and M. Valkama, "Downlink VoIP support for evolved UTRA," *IEEE Wireless Communication and Networking Conference (WCNC) 2008*, Las Vegas, United States.
16. B. Wang, K. I. Pedersen, T. E. Kolding, and P. Mogensen, "Performance of VoIP on HSDPA," VTC 2005 Spring, Stockholm.
17. "Next generation mobile networks (NGMN) radio access performance evaluation methodology," *A White Paper by the NGMN Alliance*, January 2008.
18. T. Chen, E. Malkamäki, and T. Ristaniemi, "Uplink DPCCH gating of inactive UEs in continuous packet connectivity mode for HSUPA," IEEE WCNC, March 2007.
19. 3GPP TS 36.321, "Evolved Universal Terrestrial Radio Access (E-UTRA); Medium Access Control (MAC) protocol specification (Release 8)."
20. 3GPP R2-070475, "Downlink scheduling for VoIP," Nokia, 2007.
21. 3GPP R2-070476, "Uplink scheduling for VoIP," Nokia, 2007.
22. 3GPP TS 36.213 V8.3.0 (2008-05), TSG-RAN; Evolved Universal Terrestrial Radio Access (E-UTRA); Physical layer procedures (Release 8).
23. Y. Fan, P. Lundén, M. Kuusela, and M. Valkama, "Performance of VoIP on EUTRA DL with limited channel feedback," IEEE ISWCS 2008, Reykjavik, Iceland.
24. 3GPP R1-050813, "UL interference control considerations," Nokia.
25. K. Bruninghaus et al., "Link performance models for system level simulations of broadband radio access systems," *Proceedings of IEEE International Symposium on Personal, Indoor and Mobile Radio Conference (PIMRC-2005)*, Berlin, September 2005.
26. 3GPP R1-060298, "UL inter-cell interference mitigation and text proposal," Nokia.
27. 3GPP TR 25.814 v7.0.0: "Physical layer aspect for evolved universal terrestrial radio access (UTRA)."
28. 3GPP R2-074061, "HARQ operation in case of UL power limitation," Ericsson.
29. 3GPP R2-074359, "On the need for VoIP coverage enhancement for E-UTRAL UL," Alcatel-Lucent.
30. 3GPP TR 25.903 V1.0.0 (2006–05), "Continuous connectivity for packet data users."
31. A. Pokhariyal et al., "HARQ aware frequency domain packet scheduler with different degrees of fairness for the UTRAN long term evolution," *IEEE Proceedings of the Vehicular Technology Conference*, May 2007.
32. K. I. Pedersen et al., "Frequency domain scheduling for OFDMA with limited and noisy channel feedback," *IEEE Proceedings of the Vehicular Technology Conference*, October 2007.
33. I. Z. Kovács et al., "Effects of non-ideal channel feedback on dual-stream MIMO OFDMA system performance," *IEEE Proceedings of the Vehicular Technology Conference*, October 2007.
34. 3GPP R1-074884, "On the impact of LTE DL distributed transmission," Nokia, Nokia-Siemens-Networks.
35. 3GPP TS 36.201, "TSG-RAN; Evolved universal terrestrial radio access (E-UTRA); LTE physical layer—General description, Release 8."

36. 3GPP R2-071841, "Resource fragmentation for LTE UL," Ericsson, 2007.
37. O. Fresan, T. Chen, E. Malkamaki, and T. Ristaniemi, "DPCCH gating gain for voice over IP on HSUPA," IEEE WCNC, March 2007.
38. 3GPP TR 25.996, "Spatial channel model for MIMO simulations," v.6.1.0, September 2003.
39. 3GPP TS 36.211, "TSG-RAN; Evolved universal terrestrial radio access (E-UTRA); LTE physical layer—Physical channels and modulation, Release 8."
40. P. Lunden, J. Äijänen, K. Aho, and T. Ristaniemi, "Performance of VoIP over HSDPA in mobility scenarios," *Proceedings of the 67th IEEE Vehicular Technology Conference*, Singapore, May 2008.
41. A performance summary for the evolved 3G (E-UTRA); VoIP and best effort traffic scenarios.
42. ITU-T Recommendation G.114, "One way transmission time."
43. T. Henttonen, K. Aschan, J. Puttonen, N. Kolehmainen, P. Kela, M. Moisio, and J. Ojala, "Performance of VoIP with mobility in UTRA long term evolution," IEEE VTC Spring, Singapore, May 2008.
44. M. Rinne, K. Pajukoski et al., "Evaluation of recent advances of the Evolved 3G (E-UTRA) for the VoIP and best effort traffic scenarios," IEEE SPAWC, June 2007.

Chapter 11

Early Real-Time Experiments and Field Trial Measurements with 3GPP-LTE Air Interface Implemented on Reconfigurable Hardware Platform

A. Forck, T. Haustein, V. Jungnickel,
V. Venkatkumar, S. Wahls, T. Wirth, and E. Schulz

Contents

11.1. Description of a Scalable MIMO-OFDM Experimental System for Real-Time Measurements

The experimental test bed that allowed the first-ever multiple input, multiple output orthogonal frequency division multiple access (MIMO-OFDM) transmission according to the new 3rd Generation Partnership Project long term evolution (3GPP-LTE) air interface was planned at the end of 2005. This was during the early stages of the LTE standardization. Since that time, the standardization body has agreed to a few minor changes that have no impact on the validity and relevance of the measurement results obtained with the LTE demonstration system described in this chapter.

Table 11.1 gives an overview of the system parameters of the experimental system and, for comparison, the current parameter values of the standard. The main change has been in the definition of a physical resources block (RBs) from 25 subcarriers over 7 OFDM symbols to 12 subcarriers and 14 OFDM symbols, which in spirit has exchanged time against frequency granularity. At moderate speed of approximately 5–10 km/h and for moderate numbers of users, the reported results maintain full validity.

11.1.1 Functional Description of the Implemented LTE Downlink and Uplink

Figures 11.1 and 11.2 show the functional description of the downlink and uplink transmission chains. The uplink has a discrete Fourier transform (DFT) and inverse discrete Fourier transform (IDFT) in addition to the OFDM transmission chain, while the downlink also incorporates a sync channel at time transmission interval (TTI) 0.

11.1.2 Channel Estimation in Downlink and Uplink

11.1.2.1 Complexity of Optimal Interpolation

High-quality channel estimation for 3G LTE is a challenge. Pilots are available only on a sparse grid in the time-frequency domain. On intermediate subcarriers and time slots, channel interpolation has to be used. Because the same set of pilots

Table 11.1 Transmission Parameters for the Measurement Results

System Parameter	*Experimental System*	*Current Status of LTE Standard*
Downlink/uplink, full duplex	2.53/2.68 GHz	Various bands including UMTS extension band
Antennas base station/terminal	2 TRx/2 TRx	2 TRx/1 Tx, 2 Rx
Sampling rate	30.72 MHz	
Bandwidth used	Up to 18 MHz	Up to 18 MHz
Downlink symbol period/ uplink long (short) block	71.4/70.1 (37.5) ns	71.4/70.1 (37.5) ns (LBs only)
Cyclic prefix downlink/uplink	4.6875/4.13 ns	4.6875/4.13 ns
Total number of subcarriers	2.048	Up to 2.048
Number of used subcarriers	Up to 1.200	Up to 1.200
Transmission scheme downlink	OFDMA BLAST	OFDMA BLAST
Modulation in downlink	4, 16, 64 QAM	4, 16, 64 QAM
Transmission scheme uplink	SC-FDMA CDD	SC-FDMA
Modulation in uplink	2, 4, 16, 64 QAM	4, 16, 64 QAM
Radio frame duration	10 ms	10 ms
TTI	0.5 ms	1.0 ms
Symbols/TTI in downlink	7	14
Symbols/TTI in uplink	6 Long, 2 short	12 Long data, 2 long pilots
Resource block size	25 Carriers in 1 TTI	12 Carriers in 1 TTI
Channel coding	Convolutional	Turbo Coding
Code rates	1/2, 3/4	Various, including 1/2, 3/4

is broadcast to all users in the downlink, the optimal solution would be to use all pilots for interpolation. However, we show this to be impractical with current digital signal processors (DSPs).

For example, in order to interpolate the LTE channel in 20-MHz bandwidth, one needs 1,200 channel coefficients for each pair of transmit and receive antennas from, say, each fourth subcarrier being a pilot. For the optimal Wiener filter, this leads to computation of an interpolation matrix of size 1200×300. This process employs a 300×300 matrix inversion. Next, the interpolation matrix is multiplied with a vector containing received pilots of size of 300×1. The complexity of matrix inversion is given as $300^3 = 2.7 \times 10^7$. Fortunately, these matrices can be precomputed for certain signal-to-noise ratio (SNR) values (see reference 16). Therefore, the real constraint lies in the complexity of matrix-vector multiplication $1200 \times 300 = 3.6 \times 10^5$, which needs to be realized in real time. If a TTI with 0.5-ms duration is considered as a coherence interval and four times this complexity is needed in a 2×2 MIMO system, it requires roughly 4×10^9 real-valued instructions per second.

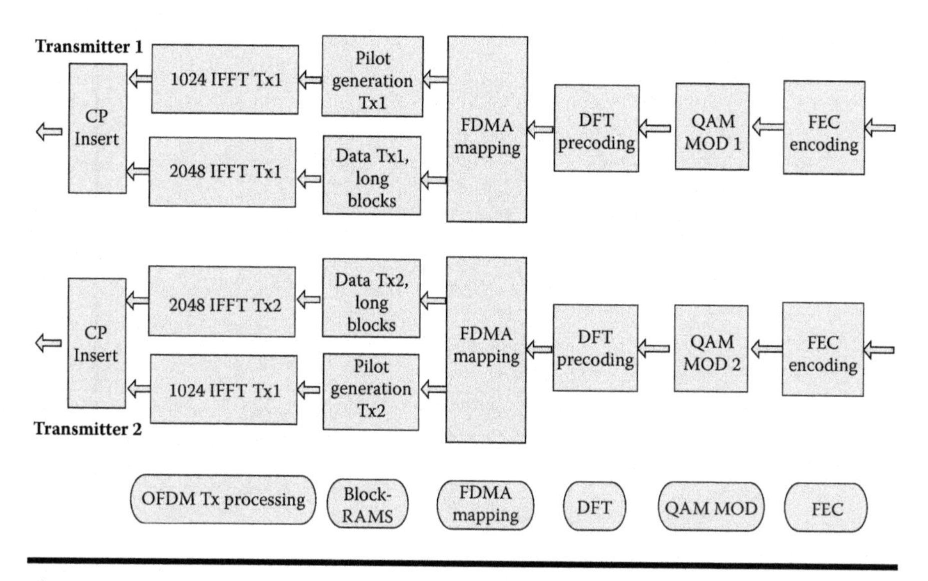

Figure 11.1 Functional blocks of the downlink base band chain. Top: transmitter at BS; bottom: receiver at one UE.

This exceeds the processing power of our DSPs and, therefore, optimal interpolation cannot be performed. An alternative simplified algorithm is considered.

11.1.2.2 Localized Interpolation Algorithm

The localized interpolation method has been discussed recently to simplify the channel interpolation process in a broadband system. It relies on a subset of pilots in the local area around the desired resource unit. Decision-aided interpolation [16] reportedly provides the best performance at high SINR (signal-to-interference-plus-noise ratio). But cellular radio systems generally operate in low SINR. For low SINRs, linear techniques have been shown to perform better. Haustein et al. [12] show that the loss in interpolation gain is only a factor of 2 to reduce the computational burden by a factor of 20. A prerequisite for this is an SINR estimation [12], which makes the filter robust.

A localized interpolation method has been implemented in Haustein et al. [12] on a Texas Instruments (TI) 6713 DSP operating at 300-MHz clock by applying local interpolation separately in each resource unit. This is done for each pair of transmit and receive antennas. Implementation is simplified by exploiting the fact that interpolation matrices are independent of frequency when the local pilot grids are identical. This has been taken into account when defining the pilot patterns described later. The precalculated filter matrices are stored according to the estimated signal-to-noise ratio (SNR). Runtime for interpolation over the entire 20-MHz LTE band is 95 μs for a 2×2 MIMO configuration (i.e., well within the coherence time).

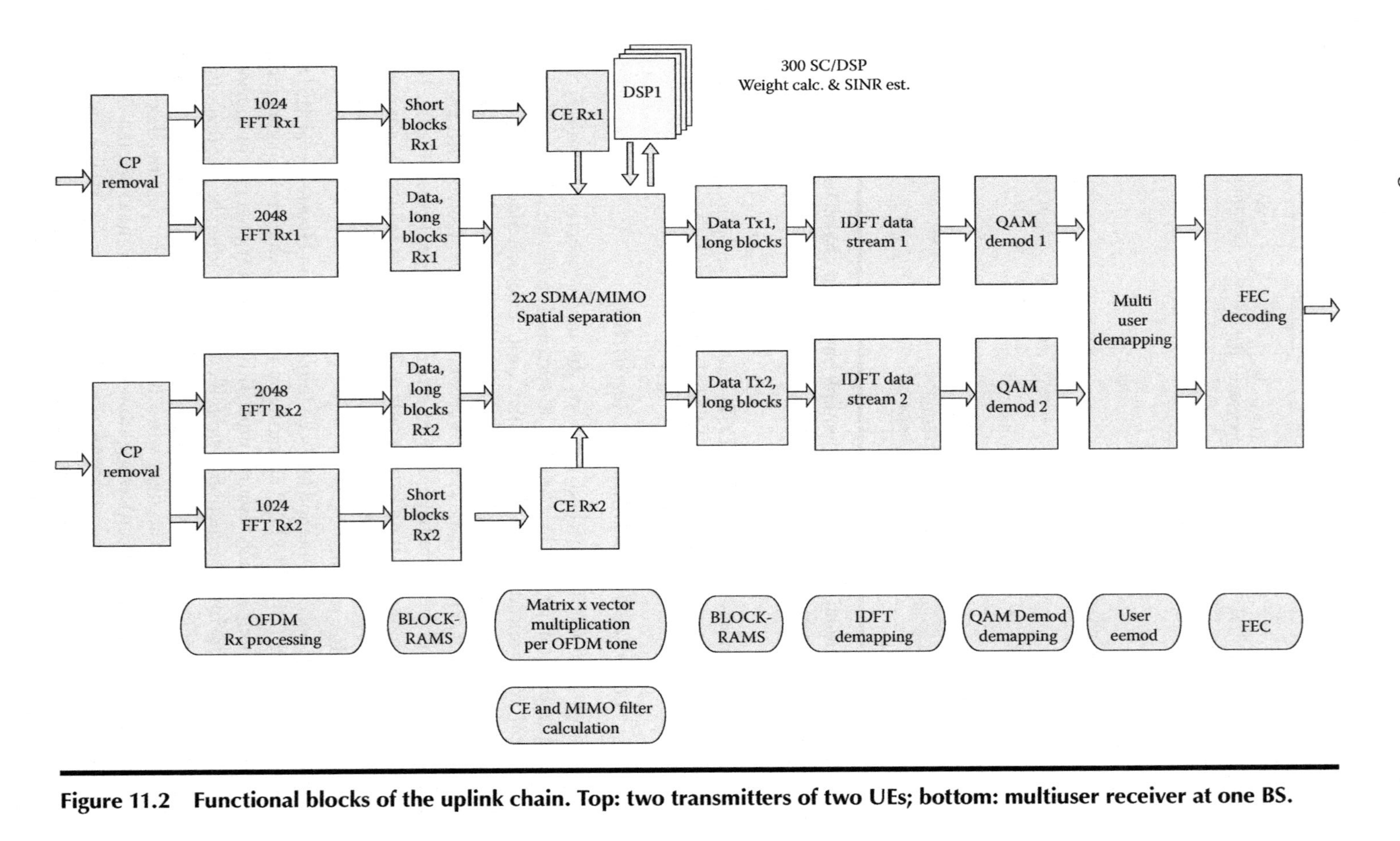

Figure 11.2 Functional blocks of the uplink chain. Top: two transmitters of two UEs; bottom: multiuser receiver at one BS.

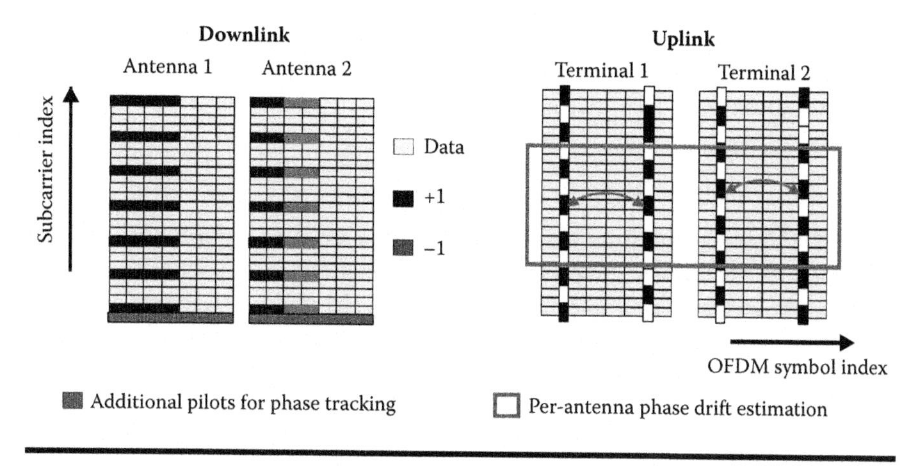

Figure 11.3 Local downlink (left) and uplink (right) pilot grids.

11.1.2.2.1 Downlink Pilots

Our local downlink pilot grid is shown in Figure 11.3 (left) for a single resource unit on both antennas. A code-multiplexed approach is used along the time axes and each fourth subcarrier is used. By this spacing, pilots get placed at the edges of resource units; this aids localized interpolation. Antennas are identified by sequences with a length of four along the time axes taken from an orthogonal set. Additional scrambling is applied along the frequency axes. Correlation over multiple OFDM symbols along the time axes [2] is applied prior to the interpolation along the frequency axes. This is again carried out individually for each pair of transmit and receive antennas.

11.1.2.2.2 Frequency and Time Synchronization in the OFDM Downlink

A coarse frequency estimation is performed at the user equipment (UE) before the fast Fourier transform (FFT) based on a special preamble transmitted in TTI 0. The remaining center frequency offset between the base station (BS) and the UE is measured in the frequency domain after FFT, exploiting the dedicated phase tracking pilots from the channel estimation and maximum ratio combiner (MRC) detection. Using different loop feedback, the remaining carrier frequency offset (CFO) can be reduced down to a few tens of hertz in a static environment at SNR above 10 dB.

The OFDM frame synchronization is based on another special preamble using a Schmidl–Cox correlator before FFT at the UE. For the current single-user/close-range scenario, no timing advance estimation or compensation was required to be implemented. For the sake of robustness, the synchronization preambles are transmitted with cyclic delay diversity (CCD) in order to sum up the power of the two transmit antennas without suffering from static beam patterns caused by coherent transmission from two antennas.

11.1.2.2.3 Uplink Pilots

In an FDMA-based multiuser scenario, each RB can be allocated to a different user in the uplink. Therefore, the pilot structure was defined locally within each RB. For the case of spatial multiplexing in the uplink, the multiple transmit antennas have to use orthogonal pilot patterns to be distinguishable. Due to the fact that one pilot in the short block (SB) represents every second subcarrier in the long block (LB), the choice of 25 SCs per RB is unfavorable (meanwhile, standardization has decided on 12 subcarriers per RB, which corresponds to exactly 6 subcarriers in the SB). Considering this, we combined two RBs of 25 subcarriers each to a double RB, which is considered in the resource allocation scheduler.

The uplink resource unit comprises six LBs (i.e., OFDM symbols with 2,048 subcarriers) and two SBs (i.e., OFDM symbols with 1,024 subcarriers). Pilots are situated in SBs (see Figure 11.3, right). Each terminal is identified by an orthogonal sequence along the frequency axes. Time and frequency interpolation are separately performed. In the time domain interpolation, possible amplitude differences inside a resource unit have been ignored. The mean phase drift $\Delta\phi$ is evaluated for each terminal inside the rectangle indicated in. Pilots available on identical subcarriers in the two SBs are used for estimation of $\Delta\phi$. To reduce noise, raw estimates H are averaged as

$$H = \frac{1}{2}(H^{SB1} + H^{SB2} \cdot e^{-j\Delta\phi}) \tag{11.2}$$

Frequency domain interpolation in uplink exploits the fact that the coefficient measured at a given pilot carrier i in an SB corresponds to the carrier with index $2 \times i$ in an LB. The odd number of 25 carriers causes overlap at the edges of resource units (see Figure 11.3, right). We have performed local interpolation over two adjacent resource units, accordingly. While the channel estimates are used to compute weight matrices for multiuser detection, the phase values are passed to a phase tracking unit placed after the multiuser detector operating at frequency $\Delta\phi/(4.5 \times T_{LB})$, where T_{LB} is the LB duration.

11.1.3 Spatial Signal Separation Using a MIMO-MMSE Equalizer

The spatial separation of the data signals transmitted from two transmit antennas is performed inside the FPGA in the frequency domain after FFT per subcarrier, based on MIMO processing; the minimum mean square error (MMSE) [3] or MRC

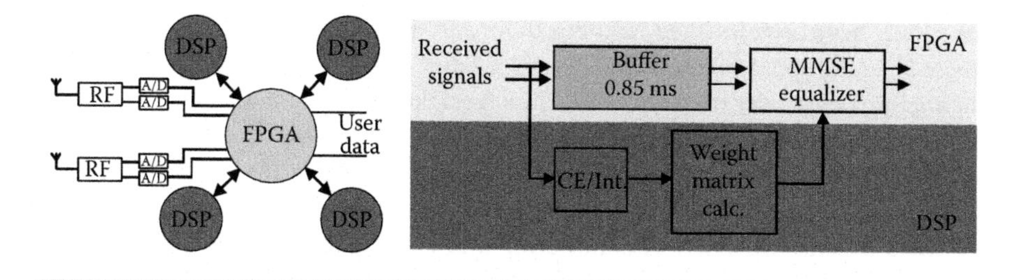

Figure 11.4 Partitioning of signal processing. Processing is implemented in a Virtex 2Pro/100 FPGA except for interpolation and MMSE weight matrix computation.

filter weights per subcarrier are calculated in the DSP every 500 μs. Furthermore, the SNR [2] per chunk is estimated and made available to the Viterbi decoder for soft decoding. Due to the stringent timing requirements of calculating 1,200 2 × 2 MIMO matrices every 500 μs, we decided on a DSP star architecture (as seen in Figure 11.4) using four floating point DSPs TI 6713, each processing 300 SCs.

11.1.3.1 Latency Issues

In order to separate the two spatially multiplexed data streams in the downlink or to separate the two user signals in the uplink in a simple and robust manner, we have implemented the linear MMSE filter [13]. In a preceding 1-Gb/s trial [2] with a 3 × 5 antenna configuration and 64 subcarriers, one TI 6713 DSP was used to obtain all weight matrices in 2 ms. For LTE, the number of subcarriers is increased by a factor of 32. In turn, the number of antennas is approximately halved. The complexity of weight calculation scales linearly with the number of carriers but cubically with the number of antennas at one side of the link. The complexity per coherence interval is roughly multiplied by a factor of four compared to Jungnickel et al. [2].

Due to hardware constraints, signal processing has been divided among four TI 6713 DSPs (Figure 11.4) synchronously triggered by the 0.5-ms TTI clock period. Each DSP is connected over an individual bus with the FPGA via the external memory interface and is responsible for 300 carriers.

Customized routines for multiplication of small matrices provided by TI have been used in the MMSE equalizer weight calculation. A complex-valued calculation is always used. For the matrix inversion step in the MMSE weight calculation, a closed-form expression of the inverse 2 × 2 matrix is used instead of Gauss–Jordan. This is less complex for two transmit and two receive antennas. A noise variance estimate is needed for the MMSE calculation. The estimated values from the localized channel interpolation are used for this purpose.

Reading coarse channel estimates into the DSP, performing channel interpolation and MMSE weight matrix calculation, and writing weight matrices back into the FPGA have been finished in about 0.3 ms in downlink and 0.45 ms in uplink, where the additional phase estimation is needed. In this way, the 20-MHz LTE channel variation over frequency and time is tracked in real time.

The stream of received signals is buffered after FFT for 12 OFDM symbol durations before entering the MIMO-MMSE equalizer (see Figure 11.4). In this way, weight matrices arrive time aligned with received signals. The buffer consumes 90-kbyte on-chip memory. This buffering allows a higher mobility, but time for buffering (≈0.85 ms) must be regarded when latency is evaluated.

The MIMO-MMSE equalizer implementation is implemented in the FPGA (see *VTS Magazine*). It uses a pipelined matrix-vector multiplication where multiple on-chip multipliers are operated in parallel and weight matrices are exchanged from one subcarrier to the next.

11.1.3.2 Frequency Offset Issues

Despite the frequency advance introduced at the terminal, there is an inevitable frequency offset for each terminal due to the residual estimation error for the downlink CFO. It can be visualized directly after the MMSE equalizer in the frequency domain. In our test case, both terminals map 16-QAM (quadrature amplitude modulation) data signals directly onto the inner OFDM transmitter (i.e., without DFT precoding). While top and bottom traces show I1 and Q2 signals of the first and second terminals (subcarrier index on horizontal axes), respectively, the full 16-QAM constellations are typically rotated by less than 5° for each terminal at the receiver. Because the lag in the fine synchronization loop at the terminal is 0.5 ms, the residual CFO is smaller than 3 Hz under static laboratory conditions when frequency advance at both terminals is activated.

The residual phase drift is a random process that must be corrected instantaneously at the base station, for each terminal individually. As mentioned previously, the residual offset is evaluated instantaneously as a part of the uplink channel estimation procedure. The offset values for all resource units are delivered from the DSPs to the FPGA as a vector, where they are corrected out of the user signals after the separation in the frequency domain (i.e., immediately after the MMSE detector).

11.1.4 DFT Precoding

In order to reduce the peak-to-average-power-ratio (PAPR) (especially for cell-edge users), 3GPP-LTE agreed on DFT precoding at the UE and a matching IDFT at the BS. The resulting signal envelope in the time domain shows a significantly reduced PAPR, especially for low-order modulation levels (e.g., binary phase shift keying [BPSK] and quadrature phase shift keying [QPSK]) compared to standard OFDM. For the demo system, we focused on localized subcarrier allocation [4]; nevertheless, distributed subcarrier allocation similar to interleaved frequency division multiple access (IFDMA) can also be supported with slight changes in the

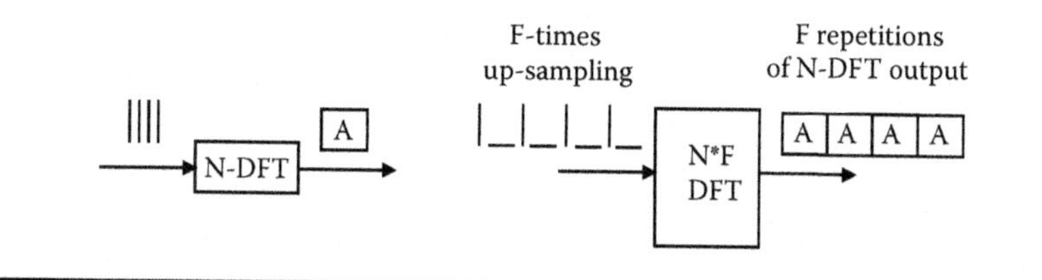

Figure 11.5 A DFT with variable size can be realized by up-sampling using a fixed-size DFT.

subcarrier mapping module that, in the localized mode, maps the DFT precoded symbols to adjacent subcarriers in the frequency domain.

DFT-precoded OFDM is a particular kind of single-carrier transmission with cyclic prefix (CP) that has been introduced in LTE for two major reasons. First, single-carrier transmission reduces the power fluctuations of the output waveforms at the transmitter and allows a better usage of the power amplifier. Second, the DFT precoding spreads the information across the assigned set of subcarriers and enables the use of multipath diversity.

Currently available DFT cores for FPGAs have a fixed number of points [XX].* In order to realize a variable bandwidth, one must be able to change the number of points at run time. A variable DFT with a selected number of points has been realized with variable up-sampling in front of a custom-made 1.200-point DFT (see Figure 11.5). In order to realize DFTs of size $N = 1.200/F$, where F is an integer, the input signal is F-times up-sampled with F-1 zeros inserted between input values. This gives F periodic blocks after the 1.200-point DFT, where each block is identical to the desired N-point DFT output. One of these blocks is further processed, mapped to the desired resources, and fed into the IFFT of the inner OFDM transmitter. Note that this solution lacks flexibility and is more complex than a variable DFT. On the other hand, at the time of writing this book, this is the only functional FPGA-based DFT implementation for which even multiuser multiplexing is feasible [16].

11.1.4.1 Power Scaling Issues

Care must be taken with the scaling after the DFT before the signal is fed into the inner OFDM transmitter. At first, the output must be scaled depending on bandwidth used. Signals are fed with the same mean amplitude per carrier into the IFFT. Furthermore, signals must be decimated properly after the 23-b DFT output before passing them into the 12-b IFFT input. Because 1,200 is not a power

* The fixed-size 1.200-point DFT soft core needs as many resources in the FPGA as four 2.048-point FFT cores.

of two, proper decimation is performed by dividing the DFT output by a number representing the occupied bandwidth. Taking the 1.200-point DFT, we have used divisors 300, 150, 75, and 19 to realize variable bandwidths of 20, 10, 5, and 1.25 MHz, respectively. From the resulting 23-b output signal, the sign and least significant 11 b are further used (see Figure 11.5). The power fluctuations of DFT-precoded waveforms are significant despite DFT precoding because raised cosine filtering is not included in LTE.

11.1.5 Uplink System Integration

The complete physical layer for the joint detection of two users with a base station using two antennas in the uplink is summarized in Figure 11.6. At the terminal side, it comprises the DFT precoding, mapping to resource units, IFFT, multiplexing with pilot signals, adding of the CP, and the frequency advance, based on the fine synchronization in the downlink. At the BS side, after removal of the CP and FFT, pilots are separated, the channel is interpolated, and the residual phase drift is obtained. The equalizer weight matrices are computed for each subcarrier and the users are separated in the frequency domain using the linear MMSE detector. Finally, because of the residual fine frequency estimation error in the downlink, the instantaneous phase offset is individually corrected for each terminal before the signals are fed into the IDFT.

The entire physical layer of one LTE transceiver is implemented in two *FFP basic* motherboards, each containing two Virtex2Pro/100 FPGAs. At the receiver, in both uplink and downlink directions, four TI 6713 DSPs are used for channel interpolation and equalizer weight matrix calculations. FPGAs operate at a 30.72-MHz clock. The functional mapping on hardware is sketched in Figure 11.7 at the terminal side. Framing and forward error encoding (FEC) in the uplink direction and decoding and demapping for the downlink direction are hosted in the left FPGA on board 2. The right FPGA on board 2 hosts the transmit processing for the uplink (see top of Figure 11.6) as well as the interfaces to the radio frame (RF) unit. The left FPGA on board 1 contains the MIMO-OFDM MMSE equalizer for the downlink. Coarse and fine synchronizations in downlink are implemented in the right FPGA on board 1.

In total, four DSPs are mounted in a sandwich-like fashion on board 1 as well. There are external interfaces to the RF via low-voltage differential signaling (LVDS), to 1 Gb/s Ethernet for real-time data applications, and to the local monitoring terminal (LMT) where received constellations, channel estimates, and other information are displayed. Terminal RF units operate in full-duplex frequency division duplex (FDD) mode at 2.53 and 2.68 GHz. The spectrum on the air is shown in Figure 11.8. Note that the second transmitter, which is available in our experimental terminals, has been fed with the same signal, and cyclic delay diversity (CDD) has been applied. At the base station, commercial RF units and duplex filters matched to the new 3G spectrum have been used. Terminal and base station

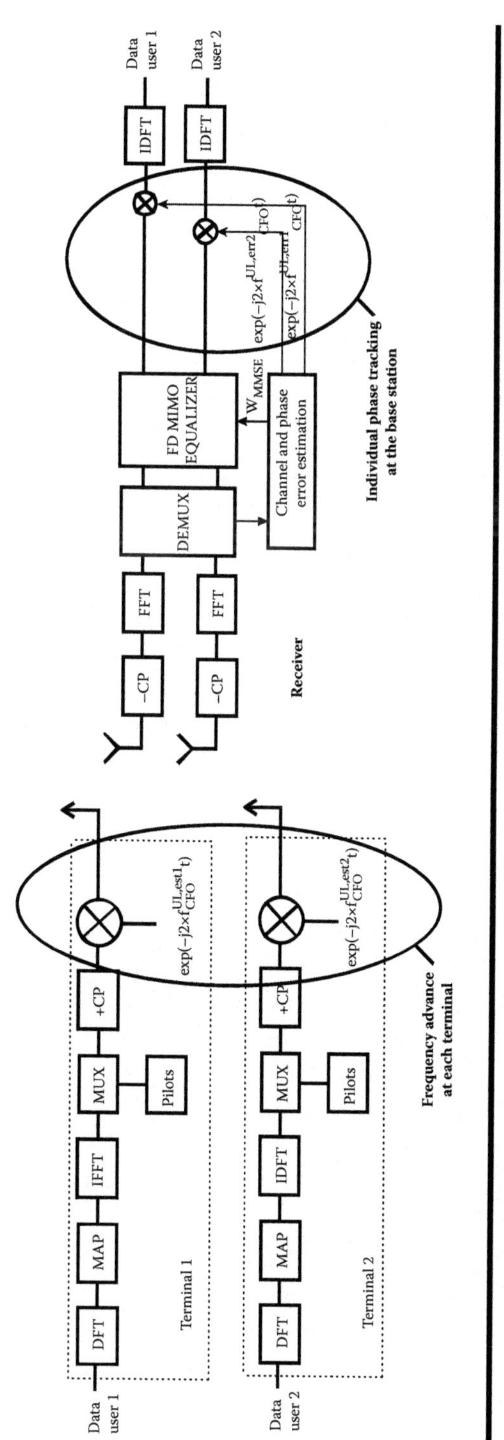

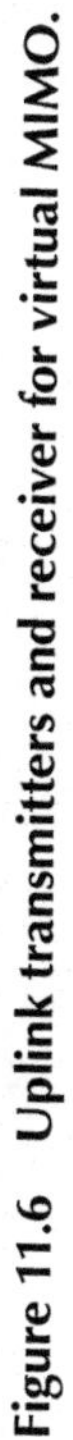

Figure 11.6 Uplink transmitters and receiver for virtual MIMO.

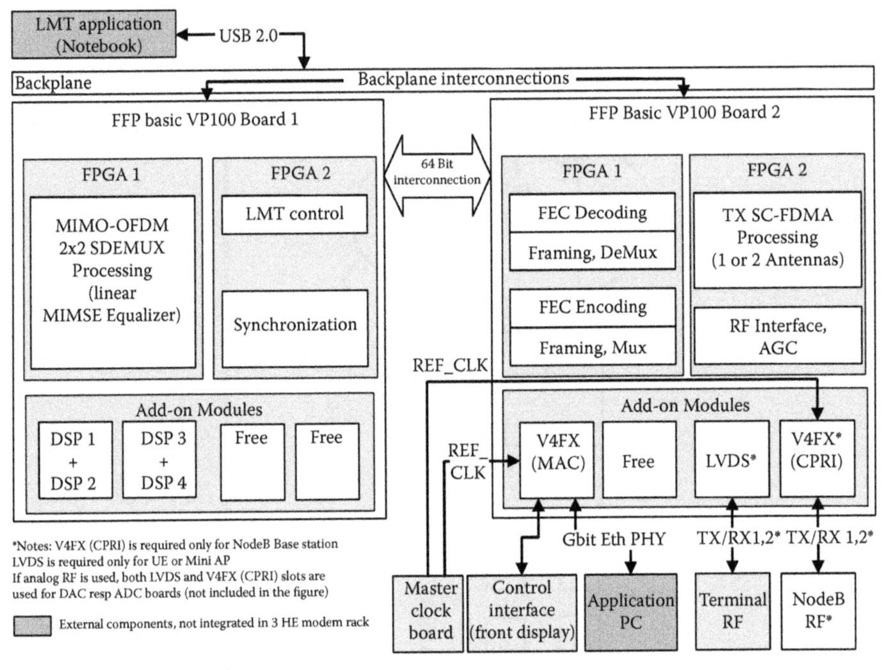

Figure 11.7 Functional mapping at the terminal side.

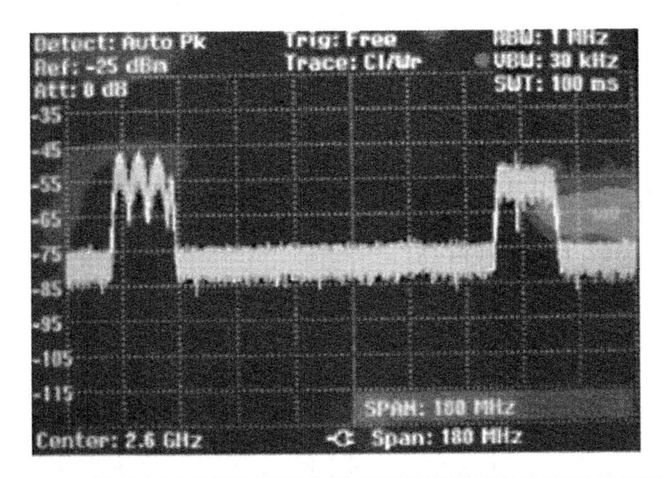

Figure 11.8 Uplink showing a CCD spectrum (left) and downlink (right) plain OFDM spectrum on air.

RF front ends are both digitally coupled via parallel cable using LVDS and serial 1.2 Gb/s common public radio interface (CPRI), respectively.

11.2 Measurements of Multiantenna Gains Using a 3GPP-LTE Air Interface in Typical Outdoor and Indoor Scenarios

Multiple antenna systems have been considered straightaway for open-loop wireless standards such as IEEE 802.11n because the gains in theoretic results by Telatar [1] are easier to realize in wireless LAN systems. However, in cellular systems, there are intricacies to realize the information theoretic gains. Fundamental issues such as cell size, delay spread, and lack of richness in spatial structure often dampen the theoretic MIMO gains. To overcome these problems, closed-loop concepts of feedback and adaptation as well as optimal antenna placements are often needed, even in an interference-free channel. In this section, we discuss the important ingredients of our cellular MIMO-OFDM system design. The main focus of this chapter is discussion of multiple antenna gains in the downlink for a single-user case/single-cell scenario.

Measurement results with our MIMO configuration test bed have shown throughput exceeding 100 Mb/s, a significant increase over the existing single-antenna system. These MIMO-OFDM measurement results were performed in an office and in a suburban outdoor scenario. We argue that frequency-dependent link adaptation in combination with transmit mode selection is a vital feature in our system design.

In principle, we show that this parallelized link adaptation provides robust gains in a cellular broadband system. The robustness is seen in both indoor and outdoor measurement results. The key feature of our work is implementation of multiple-antenna concepts in a broadband wireless system.

11.2.1 Channel-Aware Link Adaptation

A complete schematic of frequency-dependent link adaptation is shown in Figure 11.9. Two key features of this link adaptation are described next.

11.2.1.1 Adaptive Modulation

One important advantage of OFDM compared to CDMA (code division multiple access) techniques is that it exploits frequency-dependent resource allocation. Resources are used independently, with specific modulation and coding schemes (MCSs) for a specific user.

The broadband system is decomposed into parallel channels. The effective SINR after MIMO detection at receiver is calculated. This value acts as input for

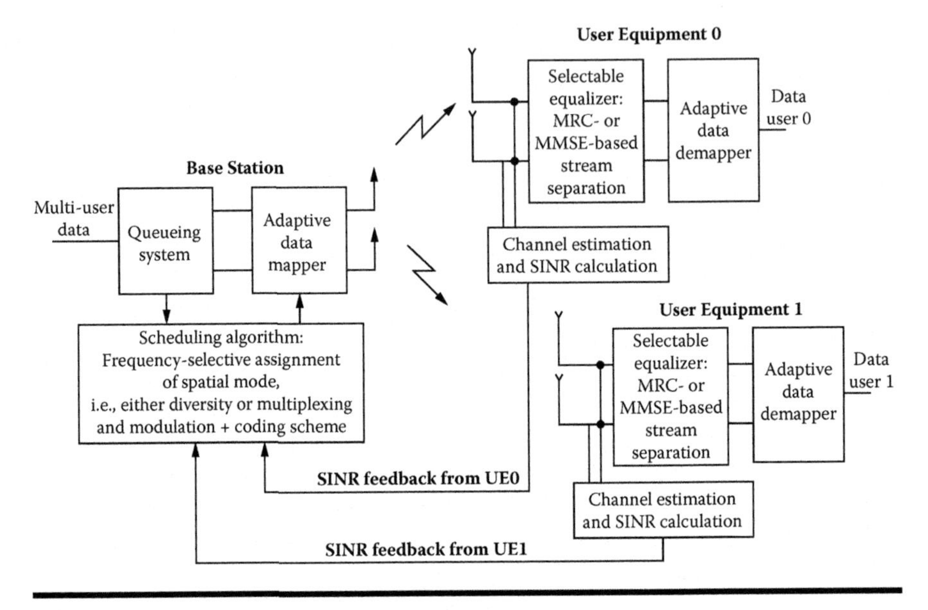

Figure 11.9 Principle of frequency-dependent link adaptation.

an adaptive bit-loading algorithm. The calculation of the effective SINR includes the estimated channel coefficients, the automatic gain control values, the estimated effective noise power, and the spatial detector used at the receive side. The extension to spatial precoding techniques like CDD or unitary precoding as proposed in 3GPP-LTE [8] is fairly straightforward. The dedicated uplink control channel is protected with rate coding, a CRC, low-order modulation BPSK/QPSK, DFT precoding, and spatial diversity reception using CDD at the UE and MRC at the BS. The uplink bandwidth is limited to a 5-MHz bandwidth. This low uplink bandwidth has been selected in order to achieve a higher spectral power density. The setup allows us to compensate for the imbalance in power budget between uplink and downlink. The UE feedback rate is set to 9 kb/s. By this very basic link adaptation loop, we achieve full flexibility in active antenna numbers at each end of the link.

11.2.1.2 Transmit Mode Selection

The measurement prototype supports antenna subset selection. The number of active antennas at each side of the link is varied by disconnecting one BS antenna or one UE antenna, allowing for four basic antenna setups for the downlink transmission: single input, single output (SISO); multiple input, single output (MISO); single input, multiple output (SIMO); and MIMO. A key feature of our implementation is that transmit mode selection and link adaptation are performed jointly.

In essence, the receiver performs single-stream or dual-stream mode selection. It also calculates the MCS based on achievable SINR for all transmit modes. The best combination of transmit mode and MCS that achieves the highest data rate is reported to the base station. This is performed for every resource block. However, a practical implementation issue occurs in transmit mode selection. The transmission power loaded per stream is always fixed. This means that the radiated power from MIMO dual-stream transmission is twice that of single-stream transmission. This constraint has implications in our transmit mode selection probabilities in MIMO configuration.

11.2.2 Detection Algorithms

In the downlink with K transmitters, we assume an MMSE filter at the receiver for the spatial separation of the multiplexed streams. An MRC is applied if only a single stream is transmitted. The receiver processing is performed in the frequency domain per subcarrier. For the experiments, a constant power profile over the allocated subcarriers was implemented and no beam forming was used.

11.2.2.1 MIMO-OFDM Downlink

The parameter set for the PHY implementation was chosen according to working assumptions around November 2005. A detailed description of these assumptions is found in Table 11.1.

11.2.3 Measurement Scenarios

For the experiments, we set up a basic configuration of one BS and one UE in an interference-free environment (no other BS or UEs are active during the experiments). Note that this is a single-cell, single-user setup, where we cannot benefit from multiuser diversity. On the other hand, we also do not suffer from intercell interference. The setup allows us to study the gains from the implemented multiantenna techniques as in an isolated hot-spot scenario.

11.2.3.1 XPD Antennas

It is common knowledge that MIMO performance degrades with the lack of spatial decorrelation. This is of concern in outdoor scenarios, where line of sight conditions exist. However, cross-polarized discrimination (XPD) antennas can be used for polarization multiplexing [4,5,7] to overcome this problem. A detailed analysis of XPD antennas and copolarized antennas is offered in the next section.

Indoor. The cross-polarized BS antenna (±45°) is positioned in the corner of a typical office room of size 20 m × 8 m × 3 m and is looking into the center of the room. The office is equipped with standard office furniture and has doors at three sides of the room; the third side consists of a glass front. The UE is moved on a

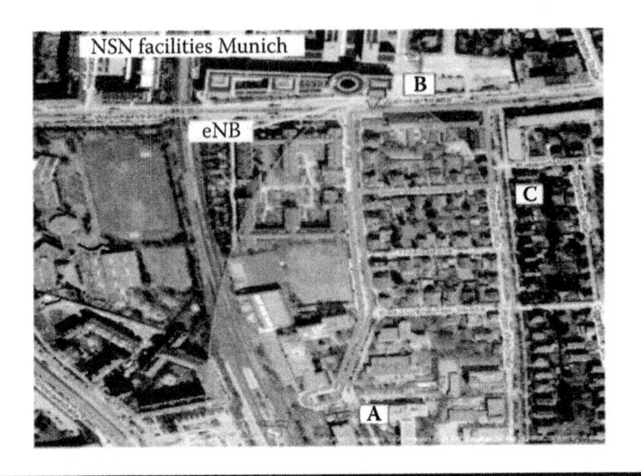

Figure 11.10 Picture depicting the measurement track.

small trolley through the room along a predefined line with constant velocity. The start is from a door 5 m away from the BS and the end is outside the room on the floor about 35 m from the BS.

Outdoor. The XPD base station antenna (±45°) is positioned outside the window of the office facing away from the glass front of the building. The covered area is depicted in Figure 11.10. The UE is installed inside a measurement van and the antennas are mounted 30 cm above the roof of the car. The two UE antenna polarizations are positioned vertically and horizontally. The same predefined route was followed in every experiment, with different antenna configurations at a constant velocity of about 3–5 km/h. The route consisted of areas with very low, medium, and very high path loss covering a dynamic range between –92 and –40 dBm at the receive (Rx) antennas of the UE.

11.2.4 Indoor Results

11.2.4.1 Singular Value Distribution of the Measured Channel

First experiments were conducted in a typical indoor office environment in order to verify the performance of channel estimation and MIMO detection for a single-user scenario. We focused primarily on the downlink, using 2×2 MIMO and link adaptation; the uplink was operated in a robust 1×2 SIMO scheme with MRC at the receiver.

In order to validate the system under operation, a specially prepared LMT was writing measured data sets, such as channel state information, SINR, and AGC values, and achieved a bit error rate (BER) with and without coding and bit-loading information from the link adaptation into log files. From the log files, we extracted plots over time and frequency that will be discussed in the following sections. The

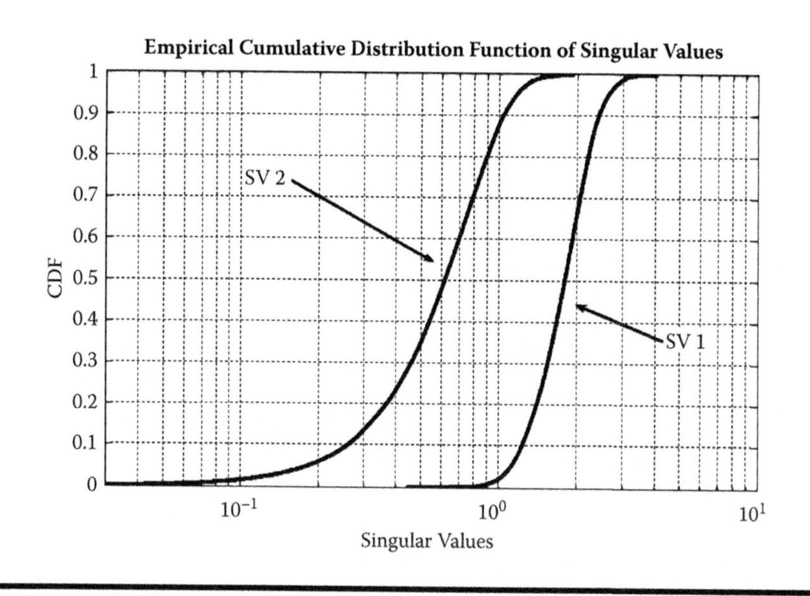

Figure 11.11 CDFs of the singular values during the experiments in the 2 × 2 MIMO downlink in 20-MHz bandwidth, channels normalized.

system performance of the RF and the base band was prechecked in a two-cable link supporting modulation schemes from BPSK to 64 QAM with an uncoded BER below 10^{-5}.

A first prediction about the statistical properties in the transmission scenario can be derived from the singular value (SV) distribution depicted in Figure 11.11. Note that all channels were normalized to allow comparison with simulations. Here, the long tail of SV2 to very small values indicates that spatial multiplexing will sometimes not be supported due to the resulting variation in the achievable SINR at the demodulator. This is a direct consequence of the lack of spatial diversity with the 2 × 2 symmetric antenna configurations. Exploiting link adaptation including adaptive MIMO mode and MCS selection should enable us to benefit significantly from the rich scattering multipath environment found in the office scenario. Very similar results were reported with 100-MHz bandwidth at 5-GHz carrier frequency [3] and omnidirectional antennas at both sides of the link.

11.2.4.2 SINR Distribution

A practical parameter to choose the optimum MCS level and the MIMO mode is the effective SINR seen at the QAM demodulator. In order to separate the path loss variances from the spatial structure of the channel, all measured channels are normalized before calculation of the effective SINR. Channel estimation is performed using broadcasted pilots at the receiver.

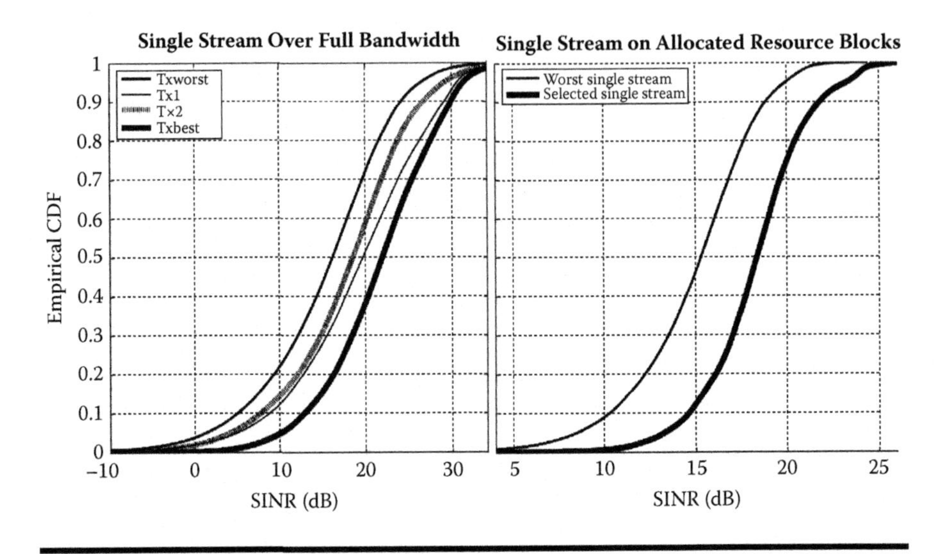

Figure 11.12 Left: SINR distribution for single-stream mode over the full bandwidth. Right: SINR distribution only for the allocated resource blocks.

Single-stream mode. A fixed transmission mode on all resource blocks would result in an effective SINR distribution as experienced in IEEE 802.11n transmission techniques. An example is a fixed two-receive-antenna MRC with transmission over the full bandwidth. Alternatively, one could also employ transmit antenna selection based on feedback from the UE. The other two curves represent this technique. It is straightforward that choosing the better single stream achieves effective SINR gain compared to a fixed choice of the transmit antenna. The gap between the worst single stream and the best single stream is 6 dB in Figure 11.12.

In the figure (right), SINR plots on those resources allocated single-stream mode are shown. There is ambiguity as to whether the selected single stream is, in fact, the best single stream. This is because the bit-loading algorithm has additional constraints, such as protocol data unit (PDU) size matching over the entire bandwidth. To verify this, we plot the third curve showing the effective SINR on the selected single stream. Observe that the selected single stream coincides with the best single stream choice. The gap between the worst single stream and the best single stream is 3 dB.

Dual-stream mode. Figure 11.13 compares the SINR for dual-stream transmission. The left-hand bunch of curves represents achieved SINRs with dual-stream operation on the entire bandwidth. This is obtained with a linear MMSE algorithm. The other two curves show plots for Tx1 and Tx2, which transmit simultaneously on a time-frequency resource. The effective SINR shows a longer tail toward the left. This happens because a second stream has to be separated at the receiver. The plot also shows a significant gap between the worst of two streams and the best of

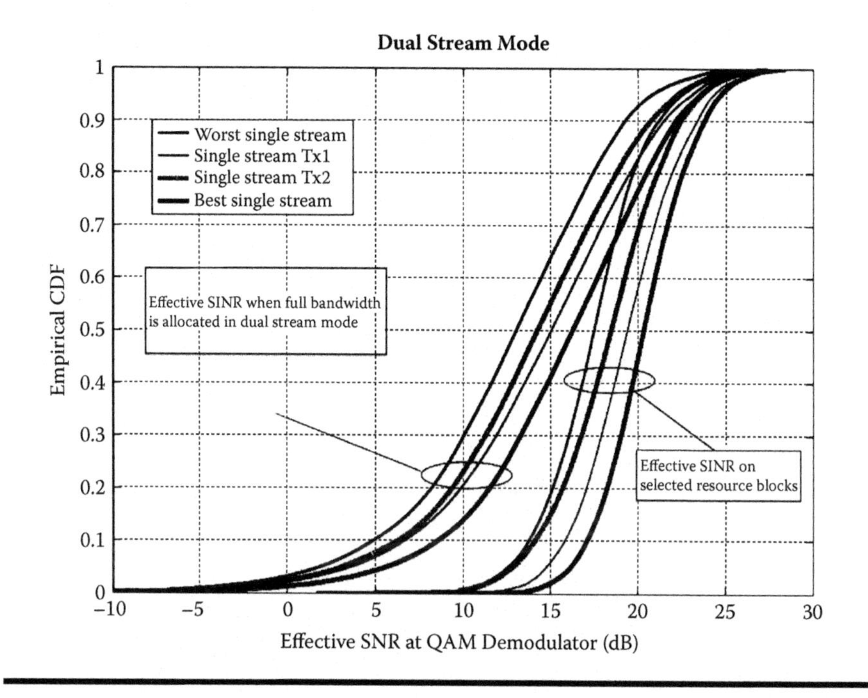

Figure 11.13 SINR distribution for the dual-stream mode. Left group of curves: dual mode over full bandwidth. Right: Dual mode only on the resources allocated during link adaptation. Thick curve: best of the two streams with antenna ordering.

two streams. This gap indicates that different MCS levels will be assigned to the streams. The right-hand group of curves represents the SINR achieved in dual-stream mode on RBs selected for dual-stream transmission. The curves validate that dual-stream transmission is chosen in MIMO configuration only when the SINR and channel structure are both suitable. Remember that resource blocks are assigned single-stream transmission if dual stream transmission is not optimal.

Remark. Dual-stream transmission by means of assigning each stream to a separate user can be exploited for spatial division medium access (SDMA) operations. The rightmost curve in Figure 11.13, representing the better of two streams, shows a significant SINR increase. In SDMA techniques, a UE is given only the better of the two streams and the second stream is allocated to another UE in the cell. It is well known that a medium to high SINR range allows for dual-stream operation, but at low SINR single-stream transmission is preferable. The probability of finding another user in the medium to high SNR range on a resource block in a cell is, in fact, quite high. Initial trials with only two users have shown that this SDMA mode is chosen quite often. Thus, we expect SDMA multiuser diversity to bring additional throughput gains to the system. The SDMA gain is in addition to the OFDMA multiuser scheduling gains.

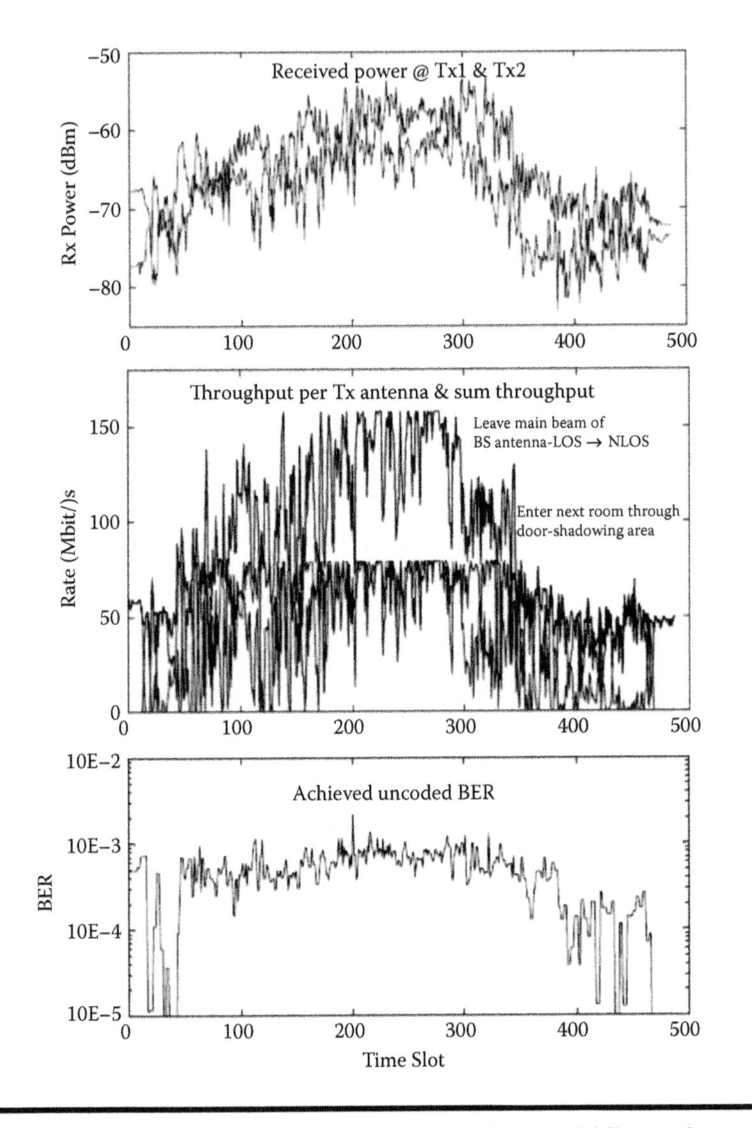

Figure 11.14 Plots of received power (top), throughput (middle), and BER (bottom) in an office scenario while the UE is moved from one room to another.

11.2.4.3 Measured Throughput with Link Adaptation

Figure 11.14 shows the received power at the UE antennas, the achieved throughput with link adaptation when moving the UE through the office, and the resulting uncoded BER. The link adaptation including MIMO mode selection and MCS selection (with fixed code rate) was adapted for every radio frame (10 ms). From the logged bit-loading pattern per transmit antenna, we calculate the achieved throughput within one TTI of 500 μs per transmit antenna from the MCS used in each RB (note that 144 symbols per RB carry data):

$$R = \sum_{k=1}^{48} Mod(RB_k)bits \cdot 144/500\mu s$$

The results in Figure 11.14 show the throughput for each transmit antenna, considering a factor of 19/20 because TTI-0 was excluded from data transmission. From the top plot in Figure 11.14, we observe that one Rx antenna "red" always receives about 5–0 dB less power than the second Rx antenna. This is easily explained by the fact that one antenna points more toward the BS and the other more to the wall. This situation is a natural situation for a multiantenna system with more or less arbitrary antenna positions in the room. In our indoor experiments, the received broadband power within the 20-MHz bandwidth for the two Rx antennas rarely differed by more than 10 dB due to the strong non-line-of-sight (NLOS) component. However, for a cellular outdoor scenario, this difference is expected to be significantly higher.

The measurement track consisted of the first 60% of sample points* measured within the main lobe of the BS antennas while the UE moved from about 25-m start distance between UE and BS to almost 5 m before the main lobe was left (LOS ® NLOS). Although the received power stays almost the same, the rate drops significantly because the channel rank decreases for many RBs in the 2 × 2 antenna configuration. At about time slot 350, we leave the office through a door and continue behind a wall of steel cupboards. The received power drops by about 10 dB when we leave the office. Due to the higher received power from transmit (Tx) antenna 1 (blue color), the scheduler starts to favor this antenna on all RBs for certain UE positions. This can be seen where the sum throughput for the "red" Tx antenna drops to zero. The measured average sum gross throughput for the single UE using the full bandwidth and spatial multiplexing when possible was above 120 Mb/s in the high-SNR indoor scenario, with a maximum rate of 157 Mb/s when all RBs on the two Tx antennas used 64 QAM. Depending on the chosen code rate (e.g., 3/4), a net throughput above 100 Mb/s could be realized, as targeted by 3GPP.

The bottom plot shows the uncoded BER when exploiting fast link adaptation. The target BER was set to 10^{-3} to ensure error-free transmission after forward error correction. reconfigurable hardware test bed. First experiments with a single user showed net peak rates above 100 Mb/s in very good channels and robust transmission when combined with basic link adaptation functionalities in the downlink.

* The sample points were taken every 300 ms with an average BER measurement over this time and a slow movement through the room. Therefore, the located bit-loading pattern does not vary much over the 300 ms.

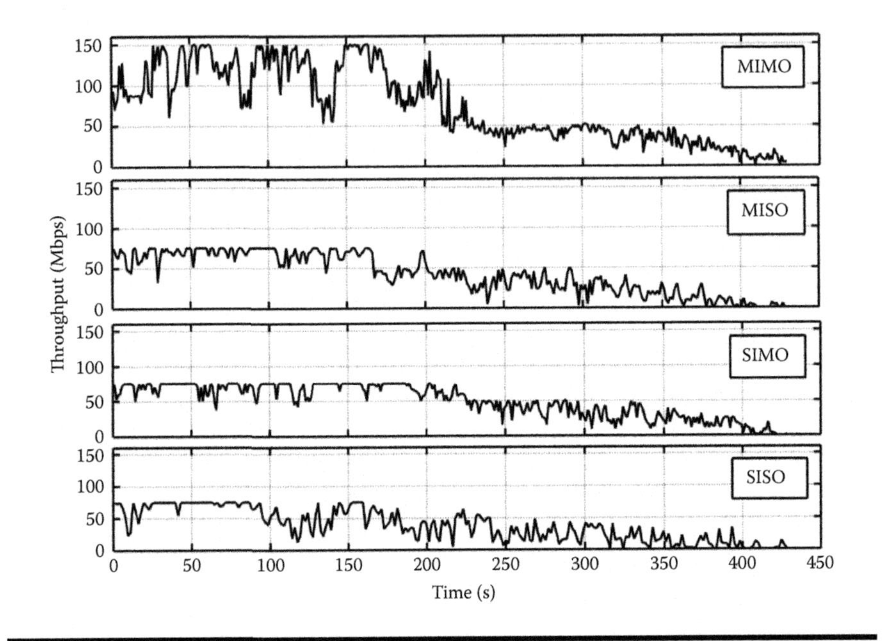

Figure 11.15 Measured sum throughput along the office with various Tx and Rx antenna combinations. Sum rate measured at UE.

11.2.4.4 Measured Rates with Different Antenna Configurations

In Figure 11.15, the measured data rates for the various Tx and Rx antenna configurations are plotted versus the position on the track through the office. For SISO (bottom plot), robustness seems to be the issue. The sum rate shows fluctuations because no spatial diversity can be exploited. Frequency diversity is also limited in the indoor scenario, with a coherence bandwidth of 5–10 MHz. For such indoor scenarios, the case for CDD of 10- to 20-tap delay to be applied is ideal. The CDD technique introduces additional frequency diversity into the channel. One would expect fewer rate variations after applying CDD.

Figure 11.15 (MISO plot) shows improvement in throughput and fewer rate fluctuations toward low values. Here, any spatial diversity between the antennas can be exploited. Moreover, if one of the antennas shows frequency selectivity, this additional gain is also realized. The blue and red lines represent the throughput contributions from Tx1 and Tx2. The third plot from the top depicts the rate in SIMO mode by applying an MRC diversity detector. The diversity gain in both MISO and SIMO is on the order of two. The SIMO mode has an additional 3-dB power gain over MISO mode. The top plot in Figure 11.15 shows the achieved rate with MIMO antenna configuration. First, we point out the difference between MIMO configuration and MIMO mode transmission. MIMO mode transmission

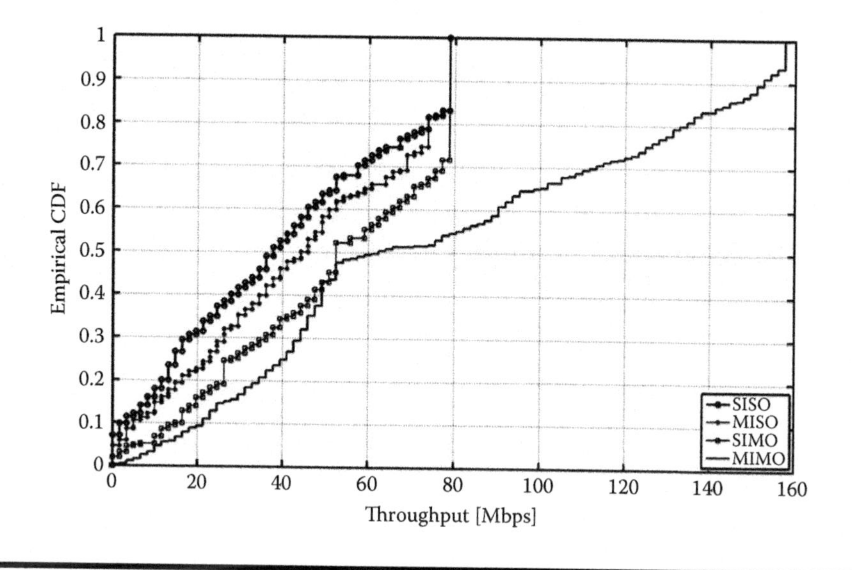

Figure 11.16 CDF of the measured throughput in the office scenario with single (SISO, MISO, SIMO) and dual (MIMO) streams.

is a subset of MIMO antenna configuration. In spirit, the advantage of MIMO antenna configuration is that it includes SIMO and MISO transmission modes by link adaptation. Therefore, the RBs are loaded in single- or dual-stream mode based on actual channel realization. Further, flexible mode selection is applied independently to each RB in a broadband OFDMA system. It is not straightforward whether fixed MIMO mode transmission alone would improve the performance in a cellular system because a random user could be experiencing a low SINR, in which case single-stream mode is capacity optimal. The maximum achieved rate is 157 Mb/s.

Figure 11.16 characterizes the throughput performance gains with cumulative distribution function (CDF) plots. We see clear diversity gains from antenna selection of MISO and the receive diversity of SIMO. Adaptive MIMO transmission outperforms the single-stream modes.

11.2.5 Outdoor Results

The antenna configuration measurements were repeated in the outdoor scenario. The target uncoded BER was set to 10–4, which allows for stable IP transmission after rate coding, even without a hybrid automatic repeat request (HARQ). Figure 11.17 (top plot) shows the measured sum throughput (black) and the transmit rate allocations to Tx1 (marked line) and Tx2 (unmarked line). The bottom picture shows the received power at the two Rx antennas covering a range of –50 to –90 dBm.

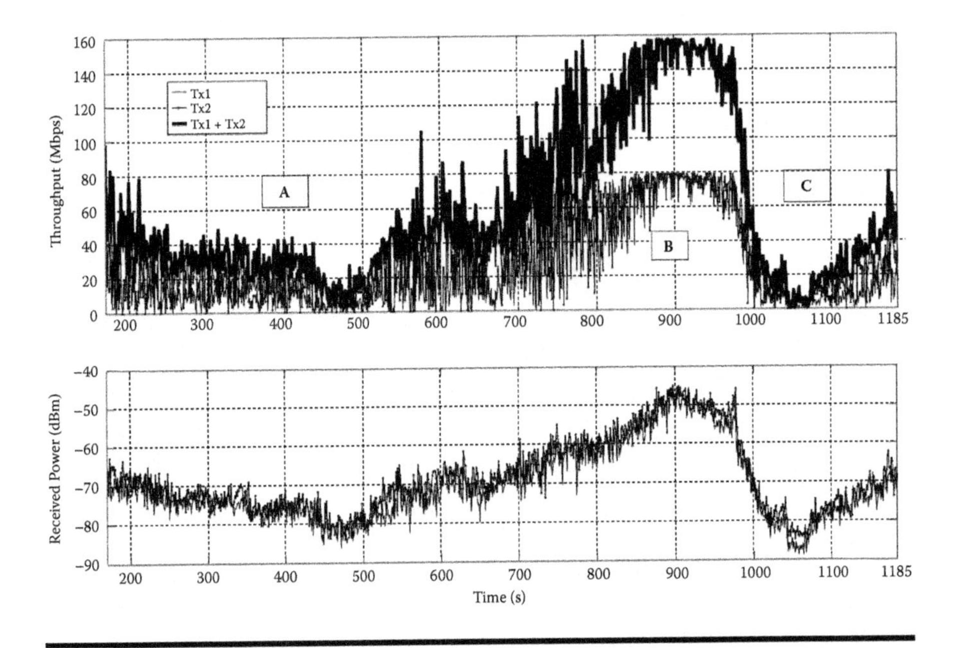

Figure 11.17 Top: measured throughput along the measurement track in megabits per second. Bottom: measured attenuation per Rx antenna in decibels of measured power (dBm). Positions A, B, and C correspond to letters in Figure 11.10.

Figure 11.18 provides a comparison of the achieved rate as a CDF. Similar to the indoor scenario, SIMO significantly improves the throughput outdoors. The dual-stream mode increases the throughput for received power above –70 dBm. This is for the single-cell/single-user scenario. Comparing Figures 11.18 and 11.16, we find that almost the same throughput is achieved in both indoor and outdoor scenarios. Moreover, this measurement result is seen with all antenna configurations. However, a naive understanding states that MIMO performance suffers performance loss in outdoor channels because of lack of scattering. We do not see this in our results. This benefit is obtained from XPD antennas and frequency-dependent link adaptation. In the outdoor scenario, the channel shows more frequency selectivity. Therefore, frequency-dependent resource allocation helps us to achieve good throughput performance. This effect is even seen in SISO results in Figures 11.16 and 11.18. Therefore, significant throughput improvement over the SIMO system is also achieved.

Figure 11.19 further explains the relationship of transmit mode selection to received transmit power. We clearly see that in low SINR (below –70 dBm), single-stream transmission is preferred. At higher SINR, the MIMO dual stream is selected.

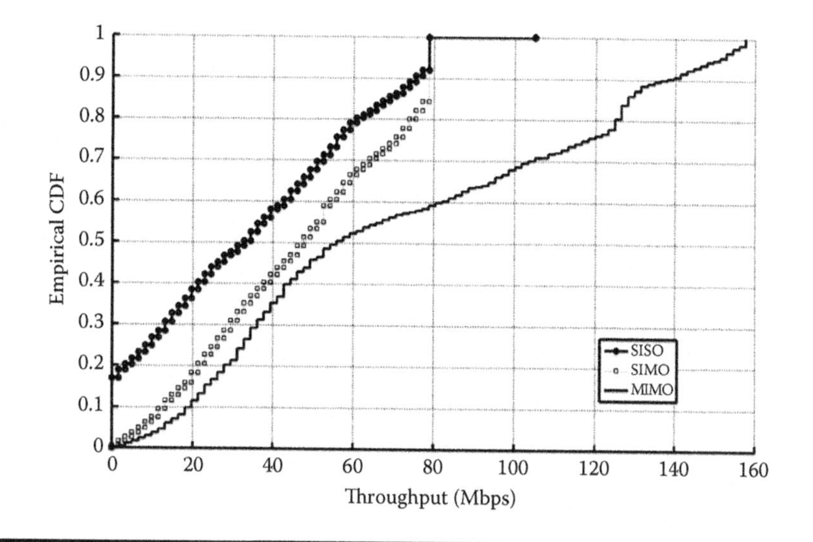

Figure 11.18 CDFs of the measured throughput in the outdoor scenario with single-stream (SISO, SIMO) and dual-stream (MIMO) modes.

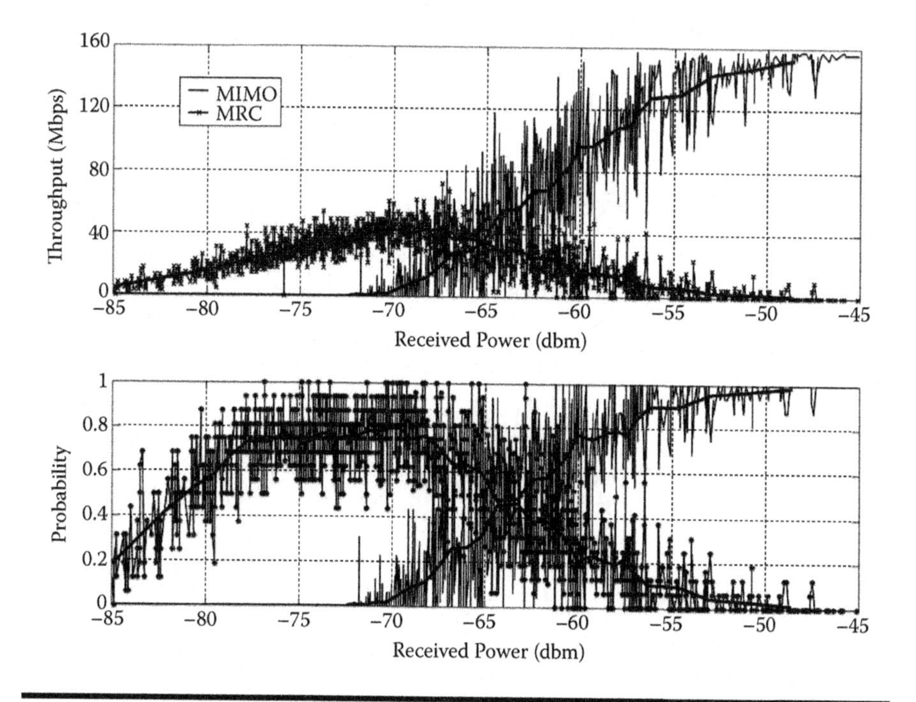

Figure 11.19 Top: achieved throughputs over average received power at the Rx antennas. Bottom: probability of single- and dual-stream selection versus averaged received power at Rx.

11.2.6 Discussion of Results

The measurements in indoor and outdoor scenarios showed the advantages of resource-based link adaptation. The RB-based transmit mode selection along with link adaptation achieved more than 100 Mb/s on 30% of the locations. Full coverage for dual-stream transmission up to 400 m between BS and UE when using multiple antennas was also observed. Our results also exhibited robustness in throughput in going from indoor to outdoor scenarios.

11.3 Real-Time Multiuser Downlink Measurements

We now present real-time broadband multiuser measurements in typical indoor and outdoor scenarios employing multiple antennas. This is a pioneering work of implementation of combined frequency-dependent fair and efficient resource assignment to multiple users along with spatial mode selection and adaptive modulation in real time on standard digital signal processing hardware. This represents a new medium access control (MAC) layer [12,13,15]. This MAC layer is steered over wireless feedback and control channels in the uplink in a closed-loop manner. The implementation uses parameters close to the forthcoming LTE of the 3G air interface (as in Table 11.1), thus illustrating the feasibility of the new functions in next-generation cellular radio systems. This section describes our real-time implementation and reports several test results. These results were taken from a small-cell measurement scenario.

11.3.1 Fair and Efficient Scheduling

The opportunism inherent in multiuser wireless communication with mobile users is that each user experiences a unique channel (see Figure 11.20). The channel is

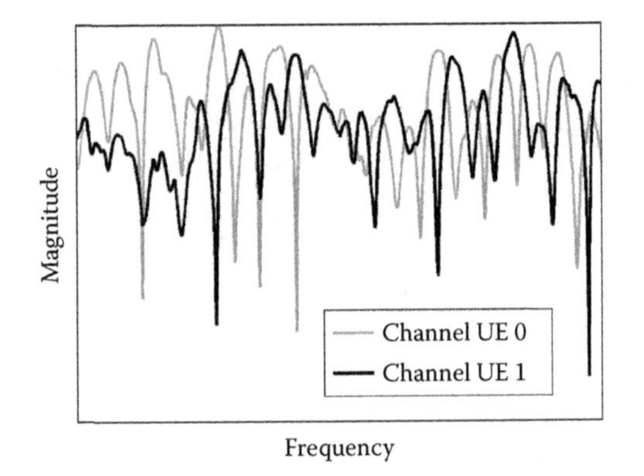

Figure 11.20 Examples of frequency responses within 20-MHz bandwidth.

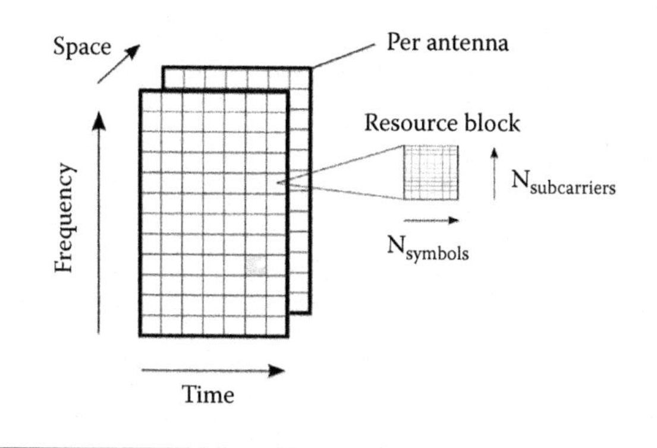

Figure 11.21 Resource allocation.

time variant and frequency selective. In the case of a multiuser environment, the system throughput benefits from this multiuser diversity. An additional requirement, however, is that in order to support terminals on their best resources, regular feedback links are required. To support each user on its best resources, OFDMA, in combination with MIMO, allows allocation of resource blocks flexible in time, frequency, and space domains (see Figure 11.21). There are clearly several possibilities to find a scheduling strategy that allocates resources in an efficient and fair manner over these domains.

The basic idea of multiuser MIMO-OFDM is that there is a fundamental trade-off between spatial multiplexing and spatial diversity. In transmit mode selection, we consider spatial multiplexing using multistream transmission using per-antenna rate control (PARC) and an MMSE detector for multistream transmission. Single-stream transmission with antenna selection and maximum ratio combining is used otherwise. For feedback, each terminal computes the effective SINR for all transmission modes prior to demodulation. A quantized version of the SINR is considered as a regular transmission packet and fed back over the physical uplink shared channel (PUSCH) to the base station. The resulting feedback rate is 9.6 kb/s.

Frequency-dependent multiuser scheduling is implemented on a longer time scale. In our implementation, SINR feedback is updated every 10 ms for every RF. Frequency-dependent scheduling allows us to exploit the resources in a broadband OFDMA system by allocating RBs to users with better spectral efficiency. The scheduling algorithm assigns resources in time, frequency, and space domains according to a scheduling policy. The result is a resource grant map for each radio frame. The grant map contains information about the targeted user, the MCS, and the spatial mode used on each RB. The implemented scheduling policy is an opportunistic proportional fair scheduler according to Viswanath, Tse, and Laroia [14], which additionally incorporates a fairness utility function for the scheduler input.

11.3.2 Multiuser Measurement

For the measurements, we deployed a multiuser MIMO scenario with one BS and two UEs. There was no interference from other BSs. Measurements were conducted in indoor office and outdoor scenarios. The indoor scenario consisted of a typical rectangular office room of size 20 m × 8 m × 3 m. The room contained typical office furniture (e.g., chairs, tables, and computers). The room had three doors at each side and a window front on the fourth side. The BS was located in the corner of the room. Both UEs were moved along predefined routes through the room, using a constant velocity of approximately 1 m/s. All antennas were cross-polarized.

The outdoor scenario was a typical suburban scenario. The scenario is characterized by NLOS and multipath propagation with houses of 6–9 m height in the surrounding area. The BS antenna covered a single 120° sector and was mounted at a window of a modern office building. For the BS antenna, a commercial cross-polarized outdoor antenna (±45°) with 11 dBi gain and built-in down tilt of 3° was used. The antenna height was 15 m above the ground and above rooftops of the surrounding buildings. The UE antennas were installed at 2-m height on a mobile trolley. A fixed route was taken on several runs with constant speeds of 1 m/s. The path loss on the measurement track ranged from −80 to −60 dB. The distance between BS antenna and UE was between 30 and 80 m.

The first results taken from the indoor scenario (Figure 11.22) show that the targeted BER of 2×10^{-2} can always be met by the closed-loop scheduling algorithm. The fluctuations toward lower BERs are caused by the discrete MCS. If the

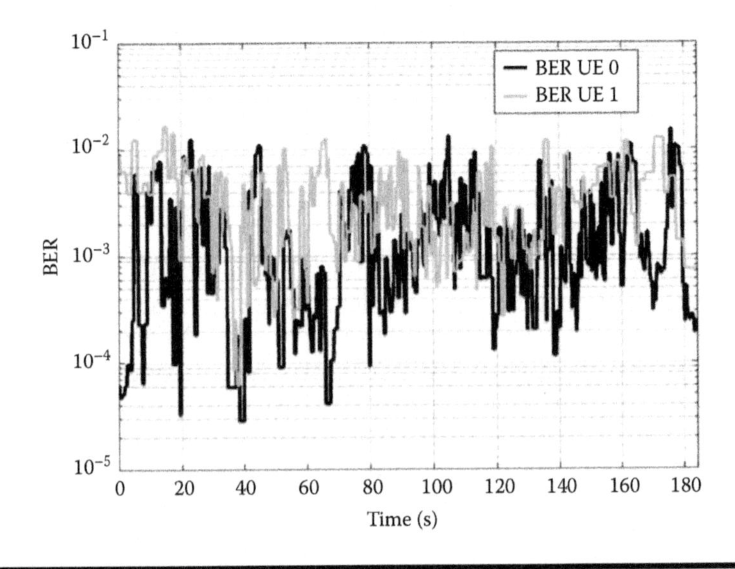

Figure 11.22 Measured BER versus time.

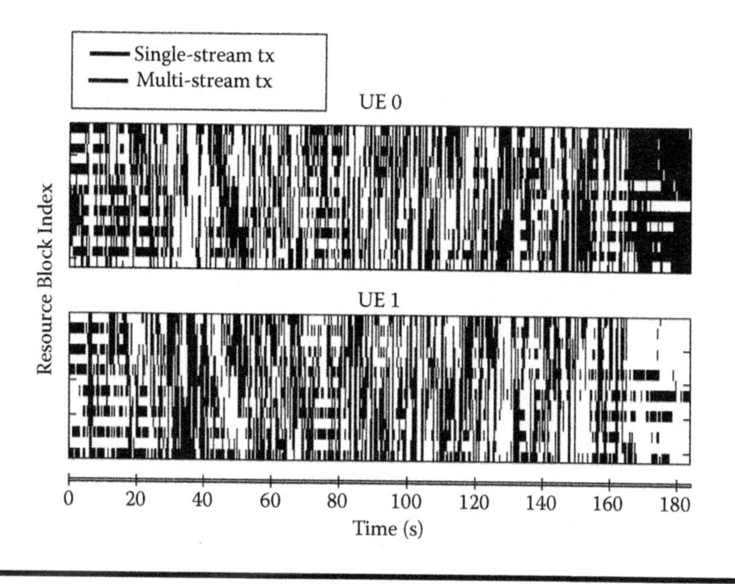

Figure 11.23 Resource block allocation, spatial modes.

BER requirement cannot be reached, the next MCS is applied with a reduced data load and hence fewer errors.

The spatial mode selection over time (Figure 11.23) shows the RB allocation for both UEs in the indoor scenario. The resource block index contains 48 RBs per antenna in the frequency domain. The color of each RB indicates the spatial mode assigned by the scheduler to the UE. Black indicates single stream, and gray indicates the dual-stream mode. White indicates no resources assigned to this UE. The allocation table shows that all modes are utilized for both UEs. If two users have the same priority at the same resource, the scheduler assigns the odd resources to UE 0 and the even resources to UE 1. This explains the striped structure in the allocation plot and is a very simple solution for the two-user case to ensure instantaneous fairness among the users regarding the number of allocated RBs.

11.3.3 Discussion of Results

To conclude, we show the throughput CDF for the outdoor measurements. Figure 11.24 shows the statistics for the individual user rates and the sum rates. The individual user rates ranged from approximately 60 to 70 Mb/s for the median with a total system throughput of 146 Mb/s.

We have presented a first real-time PHY and MAC implementation for frequency-selective multiuser MIMO using parameters close to the forthcoming 3G LTE standard. High aggregate data rates of more than 100 Mb/s can be realized with incredible flexibility for the resource allocation. We have tested the new MAC

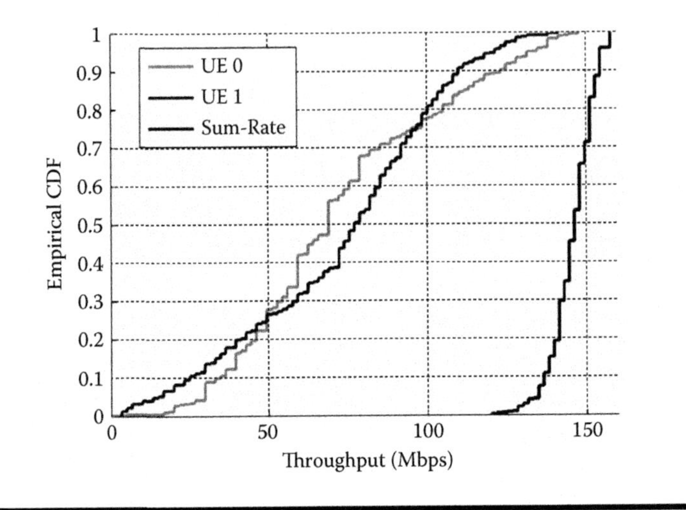

Figure 11.24 Throughput statistics with fair scheduling.

functions in indoor and outdoor measurements with two users moving at pedestrian speeds. The results confirm the high potential throughput in real propagation environments due to the combination of frequency-selective resource assignment to multiple users, spatial mode selection, and adaptive modulation. Frequency-selective multiuser MIMO is technically feasible for application in next-generation cellular radio, and it implies a significant improvement of the overall system performance.

11.4 Polarization Multiplexing for Multiuser Downlink

Antenna polarization techniques have been considered for many years in channel measurement studies. These studies were primarily based on enhancing the signal propagation efficiency. Diversity reception using polarized antennas was utilized as a form of antenna diversity in multiple antenna systems.

However, recently, multiple antennas have been observed to provide far more influential multiplexing gains [1] apart from diversity gains. For MIMO systems, signal propagation efficiency alone does not translate to higher realizable throughput. Channel scattering is essential for delivering the theoretical capacity growth. For instance, the capacity of spatially multiplexed MIMO systems is known to degrade in LOS channel conditions, as compared to NLOS conditions, because a dominant LOS path component reduces the channel scattering effects. In practice, LOS conditions are commonplace occurrences in cellular systems.

Antenna polarization can be used to compensate the lack of spatially multiplexing via polarization multiplexing [4–7]. Polarization multiplexing exploits the second strong singular value in the MIMO transfer matrix by using orthogonally polarized antennas, also known as cross-polarized antennas. The benefits of using XPD antennas in cellular MIMO-OFDM systems are twofold: Cross-polarized antennas yield a more compact antenna design and they enhance the throughput as compared to copolarized antennas.

In this section, we present the measurement results with XPD antennas for a MIMO-OFDMA system. We show a comparison of performance results between XPD and copolarized antennas. The results highlight the importance of XPD antennas in MIMO throughput enhancement in cellular systems. Further analysis is also made on the importance of these antennas for OFDMA scheduling, which is an important cross-layer aspect of LTE networks. It is seen that XPD antennas are a better choice for providing user fairness along with sum throughput maximization.

11.4.1 2 × 2 Cross-Polarized MIMO Channel

Antennas are distinguished as cross-polarized, copolarized, or dual polarized, depending on the plane of signal propagation. Various studies have been conducted to model an XPD MIMO channel closely. For instance, Erceg, Sampath, and Catreux-Erceg [5] presented a simple MIMO channel model that predicts the eigenvalue distribution and the capacity of single polarized MIMO channels for specific antenna configurations. Shafi and others [6] presented a simple three-dimensional cross-polarized model and extended the 3GPP channel model to incorporate the three-dimensional component. In Oestges, Erceg, and Paulraj [11], polarization and scattering matrices are used to propose a stochastic geometry-based scattering model for the multipolarized MIMO channel. Jiang, Thiele, and Jungnickel recently proposed a simple representation of the polarized MIMO channel [10]. In their work, the cross-polarized two-dimensional channel is represented as a modified copolarized two-dimensional channel according to the polarization mismatch.

We follow the representation in Jiang et al. [10] later to explain intuitively the measurement results with XPD antennas. Therefore, we represent

$$\mathbf{H}_{XPD} = \mathbf{H}_{2\times2} \bullet \begin{bmatrix} \cos\theta_p & -\sin\theta_p \\ \sin\theta_p & \cos\theta_p \end{bmatrix}, \tag{11.1}$$

where • means element-wise multiplication, θ_p is the polarization rotation angle, and $\mathbf{H}_{2\times2}$ is the standard 2 × 2 MIMO channel model using copolarized antennas given as

$$\mathbf{H}_{2\times2} = \begin{bmatrix} h_{11} & h_{12} \\ h_{21} & h_{22} \end{bmatrix}.$$

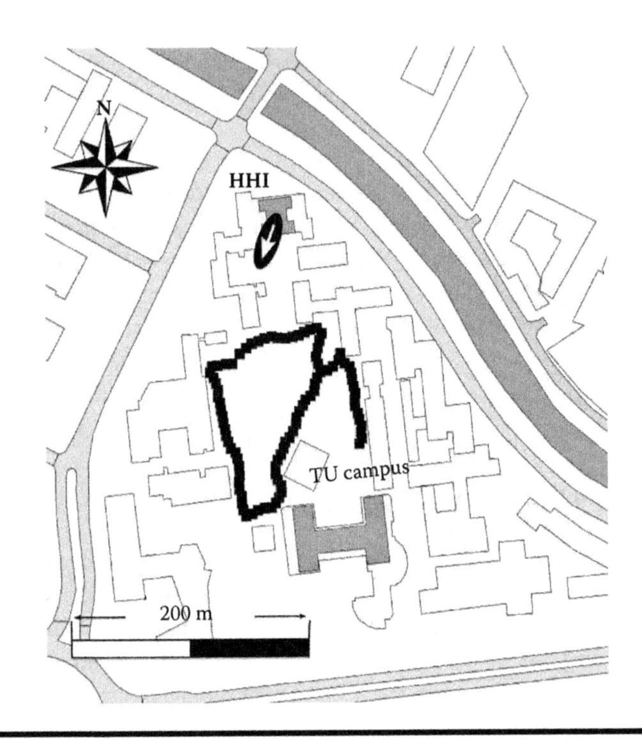

Figure 11.25 A depiction of measurement track.

11.4.2 Measurement Setup

A brief description of the measurement setup is as follows. For the downlink, two transmit antennas were used at the BS and two receive antennas at the UE. Two sets of measurements were performed: one with XPD antennas (±45°) at the BS and the other with copolarized antenna elements (each ±45°). Both the antenna measurements were taken without precoding. At the UE, XPD antennas were used in all measurements. The two copolarized antennas at the BS were placed on top of each other with an antenna spacing of 10λ. This was to ensure uncorrelated transmit signals. The same antennas were used in the cross-polarized measurement. The BS antennas had a down tilt of 12°. The UEs were separated by about 1 m and faced different directions. The measurement track is depicted in Figure 11.25. For brevity, we summarize the transmission parameters in Table 11.2.

11.4.3 Single-User Throughput Analysis

11.4.3.1 Channel Characteristics

The measured channel characteristics for validation of throughput results are shown in Figures 11.26 and 11.27. Figure 11.26 shows the channel path loss variation over

Table 11.2 Transmission Parameters for the Measurement Results

Parameter	*Value*
Transmit power	2.5 W, 34 dBm
Beam width	60°
Down tilt	12°
Bandwidth	20-MHz DL, 5-MHz UL
Antenna gain	11 dBi

time. The measured channel is normalized over time and frequency so that the average channel power is unity for all channel coefficients. This averaging is performed over all the frequency subcarriers and over both the receiver antennas. The channel path loss variation is the same for cross-polarized and copolarized antennas. Figure 11.27 shows the condition number distribution for XPD and copolarized antennas. Studies of throughput versus average received SNR are performed in Figures 11.28–11.31 for the following schemes:

- fixed dual-stream transmission with and without SVD precoding;
- adaptive transmission scheme with SVD precoding over a water-filling power allocation policy, which will automatically include single-stream adaptation; and
- unitary codebook-based precoding with dual-stream or single-stream adaptation.

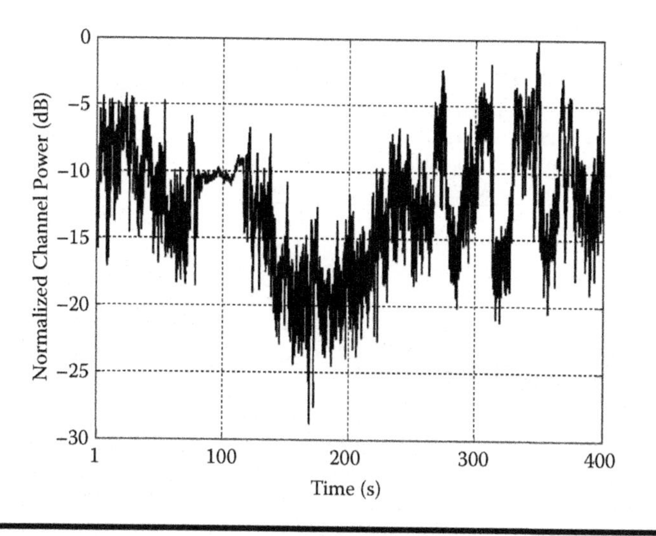

Figure 11.26 Channel variation in the measurement.

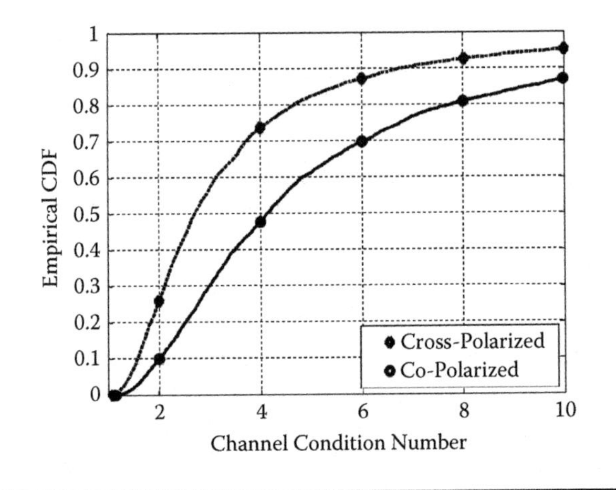

Figure 11.27 Comparison of channel condition numbers.

These analyses are done in order to characterize the single-user achievable throughput based on the measured channel. The average SNR is varied in simulations after normalizing the channel as previously described.

11.4.3.2 Channel Condition Number

A 2 × 2 copolarized MIMO channel is generally represented in a matrix form of $\mathbf{H}_{2\times2}$ as used in Equation (11.1). A key measure related to the throughput performance of the MIMO channel is the condition number of the channel matrix. This

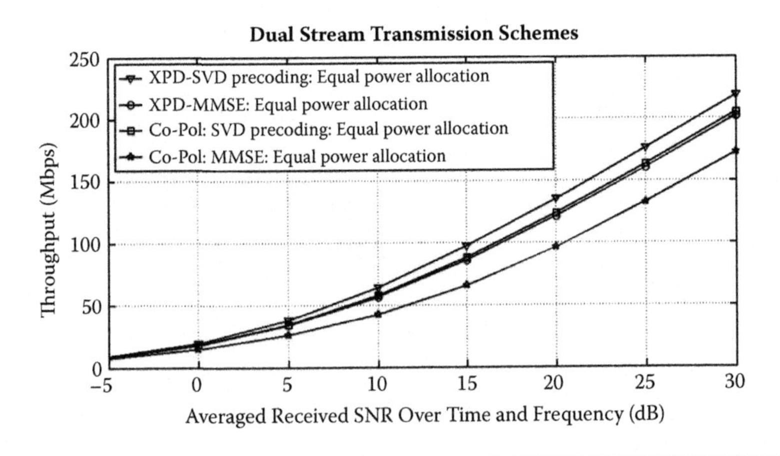

Figure 11.28 Performance gains with SVD precoding–dual stream–equal power allocation.

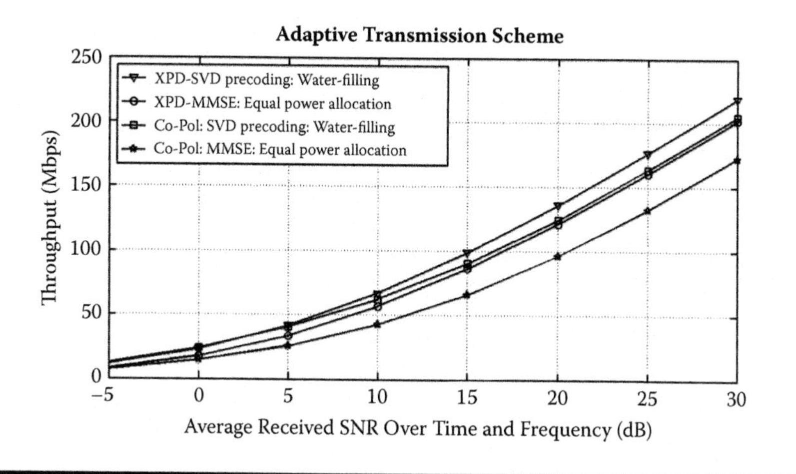

Figure 11.29 Water-filling power allocation gains.

is even more relevant if suboptimal decoding algorithms, such as a zero-forcing decorrelator or an MMSE filter, are used at the receiver because these algorithms commonly employ a channel inversion step, which boosts the noise power after signal processing. The condition number of the channel is an indicator of the resulting noise enhancement. A high condition number means a higher noise enhancement, and a lower condition number means a lower noise enhancement. Channel correlations because of LOS channel conditions tend to increase the condition number of the matrix.

Referring to Equation (11.1), the worst-case channel condition in an LOS copolarized channel occurs when all the channel coefficients are equal to a constant h,

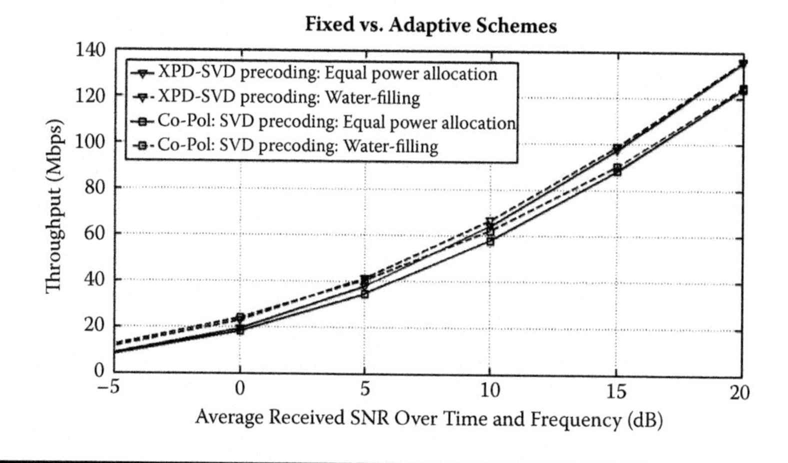

Figure 11.30 Water-filling power allocation upper bound.

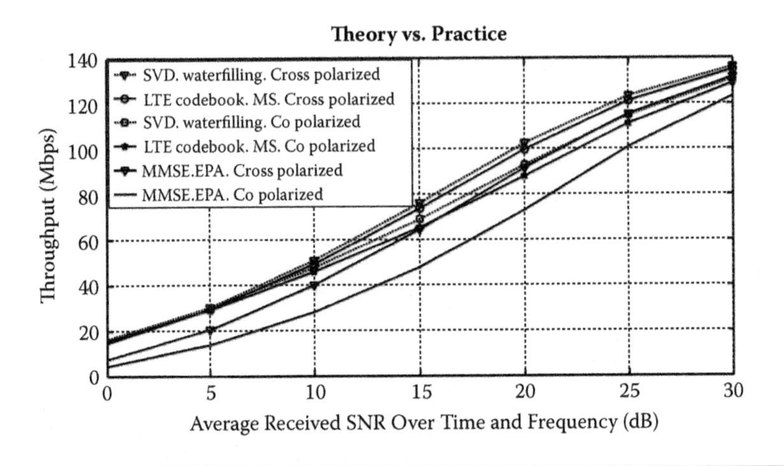

Figure 11.31 Codebook precoding throughput gains with bit loading.

the dominant path without any scattering. The condition number of this rank-one channel matrix equals infinity. However, the equivalent cross-polarized channel is a unitary matrix with entries

$$\mathbf{H}_{XPD} = h\begin{bmatrix} \cos\theta_p & -\sin\theta_p \\ \sin\theta_p & \cos\theta_p \end{bmatrix} \tag{11.1}$$

The condition number of the unitary matrix in Equation (11.1) is one. This transformation of condition is the key argument for usage of XPD antennas in cellular systems. In reality, factors such as azimuth displacement make the XPD channel matrix nonunitary but still considerably reduce the condition number of the channel.

Figure 11.27 shows a comparison of the measured condition numbers between XPD and copolarized antenna measurements. It is observed that, on median, the condition number of the channel reduces from 4 to approximately 2.5 with XPD antennas.

11.4.3.3 SVD Precoding

SVD precoding of the MIMO channel matrix is known to be a simple capacity-achieving strategy for a MIMO channel with fixed transmit power constraint. We show the results with this precoding technique in Figure 11.28 and compare them to dual-stream equal power allocation (EPA). In the result, we first observe that for MMSE EPA, XPD antennas always outperform copolarized antennas by approximately 20% at 30 dB. The gains with SVD precoding gains are also consistent at all

SNRs. With SVD precoding, XPD antennas are better than copolarized antennas by 10%.

11.4.3.4 Spatial Water Filling

SVD precoding, along with spatial water-filling power allocation, has been shown to be the capacity achieving strategy for a transmitter with sum transmit power constraint. Here, prior to the SVD operation, optimum power allocation for each stream is performed according to the water-filling solution [9]. This solution inherently includes adaptive single-stream or dual-stream mode selection. With SVD spatial water filling, we observe a crossover point at approximately 6 dB between XPD and copolarized antennas. Below this SNR, copolarized antennas are observed to be capacity optimal. This is motivated by the fact that low SNR is a power-limited region. It is therefore optimal to allocate the sum transmit power to the best eigenmode using single-stream transmission. In this context, we observe that the maximum eigenmode for copolarized antennas is greater than for cross-polarized antennas. This structure arises from dual properties: trace constraint on the channel power matrix $\mathbf{HH}^H$ and higher condition number (as shown in Figure 11.27) of copolarized antennas.

In Figure 11.29, we compare the performance of the SVD-water-filling strategy to that of MMSE equal power allocation. XPD antennas are a better choice for SNRs greater than 6 dB. To isolate the gains of water-filling power allocation from SVD precoding, throughput CDF of SVD with water filling is compared to that without water filling in Figure 11.30. It is seen that benefits of water filling are primarily at low SNRs. At high SNRs, the water-filling strategy converges to equal power allocation.

11.4.3.5 Precoding with Unitary Codebooks and Bit Loading

Next, we present the results with codebook-based precoding. This is done along with dual-stream or single-stream transmit mode selection. Mode selection can be viewed as a water-filling strategy with just three quantized power levels of 0, 0.5, and 1. In the codebook-based strategy, the terminal reports only the codebook index selected from the three codebooks, specified in IEEE 3GPP TS 36.211 V8.0.0 [8].

The codebook is selected by the terminal such that the achievable rate is maximized. To reduce feedback, the terminal reports only on the bit loading per stream. In computation, the terminal also takes into account single-stream or dual-stream mode selection.

A natural question to ask is whether the performance loss is significant with codebook-based precoding as compared to optimal precoding. We observe that it is not significant (Figure 11.31). This result is intuitively explained on the following basis. Channel inversion at the receiver is the main contributor to performance degradation for dual-stream transmission. This happens if the channel matrix is ill

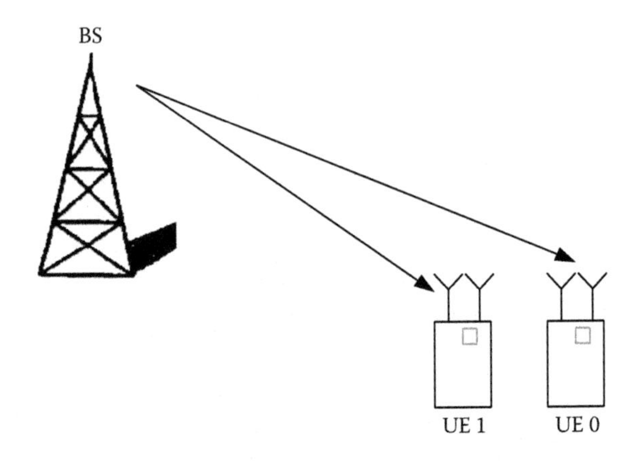

Figure 11.32 Downlink transmission scheduling for colocated terminals.

conditioned. However, this drawback is overcome by single-stream transmit mode selection. Therefore, the observation in Figure 11.31 is that codebook-based precoding is close to optimal when performed in conjunction with transmit mode selection.

11.4.4 Multiuser Scheduling

This section presents the results for OFDMA scheduling for two users. The throughput results are the real-time measured throughput from the measurement campaign. Two colocated user terminals are scheduled in the downlink, depicted in Figure 11.32. The implemented MAC architecture is the same as that described in the previous section.

The two terminals move along the same measurement track, shown previously in Figure 11.25. The two users are scheduled according to the proportional fair algorithm. Results are compared to the OFDM-TDMA (time division multiple access) scheduler, for which the sum throughput is given by

$$S_K(t) = \frac{\sum_{k=1}^{k=K} \mathbf{1}^T \mathbf{u_k}}{K} \tag{11.3}$$

where $\mathbf{u_k}$ is the supportable throughput for the kth user at time instance t.

The plots of the CDFs of individual user rates and sum rates are shown in Figure 11.33. It is seen that, on median, XPD antennas give a sum rate gain of approximately 16.5 Mb/s with OFDMA proportional fair scheduling (PFS). In the same plot, we also compare the performance of a round-robin TDMA scheduler. OFDMA scheduling appears to be a clearly better strategy. Surprisingly, multiuser scheduling gains are observed even when the users are colocated and in LOS

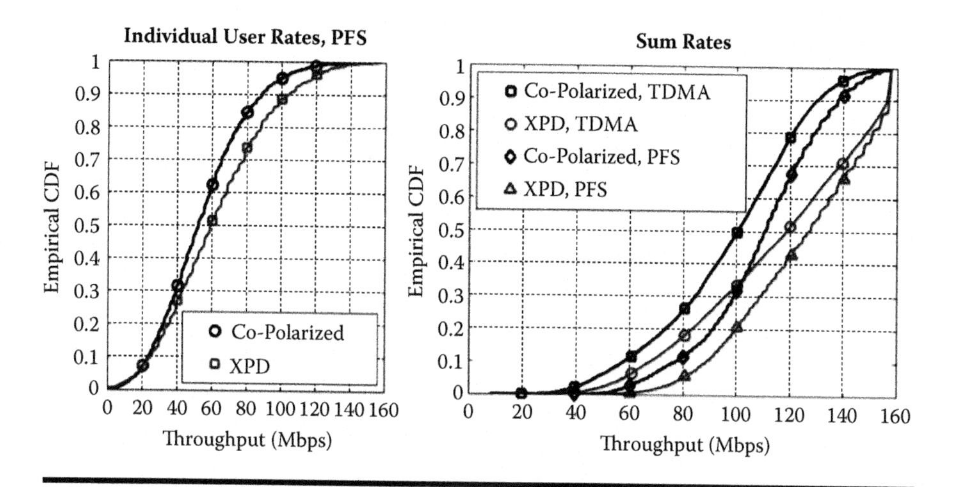

Figure 11.33 System performance. Left: individual user throughput; right: sum throughput.

channel conditions. TDMA shows a lower throughput because frequency resources are exclusively assigned to one UE per OFDM symbol. For instance, if a UE has a zero-supportable rate in a time slot, TDMA still allocates the spectrum to the UE. This makes inefficient use of the spectrum. In cases of OFDMA scheduling, resources are assigned opportunistically to users. This shows that a lot can be gained from multiuser diversity in LTE systems.

11.4.4.1 XPD Antennas for Multiuser Scheduling

To understand the results of Figure 11.33 more clearly, we decouple the gains of multiuser scheduling and XPD antennas in Figure 11.34. We see that the XPD antennas increase the probability of dual-stream selection from 76 to 92% with TDMA scheduling. With OFDMA scheduling, this probability of selection increases to 98%. Further, the probability of loading 64 QAM on the dual streams also improves markedly from 24 to 70% with XPD antennas.

11.4.4.2 Impact of More Users on System Performance

It is known that the MIMO channel-hardening effect reduces multiuser scheduling gains. This leads to two questions:

- Are XPD antennas in fact beneficial if the number of users in the system is large?
- How large can the number of users be before this effect occurs?

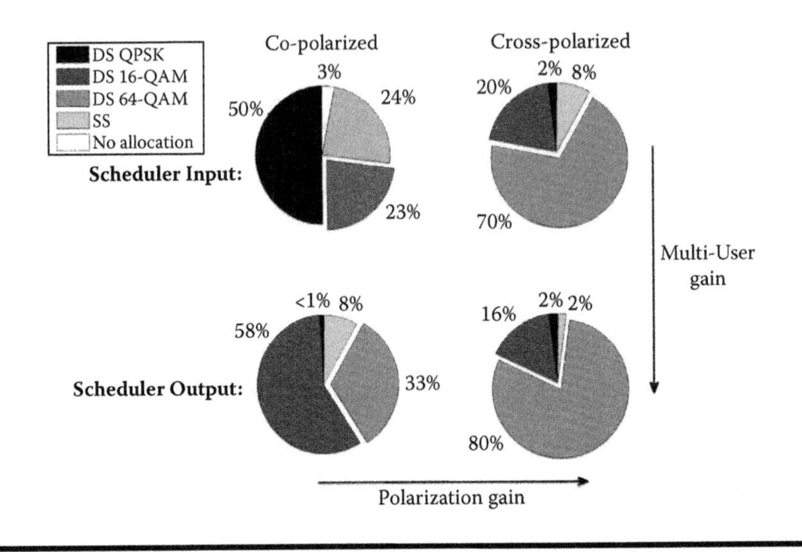

Figure 11.34 Gains of XPD antennas and multiuser scheduling gains.

For this further analysis, off-line simulations were conducted by increasing the number of users scheduled per OFDM time slot. The number of users is increased from 2 to 10. The users are taken to be distributed along the measurement track.

The CDFs for the sum rates for two users are shown in Figure 11.35. The two users are equidistantly placed on the measurement track and move along the same

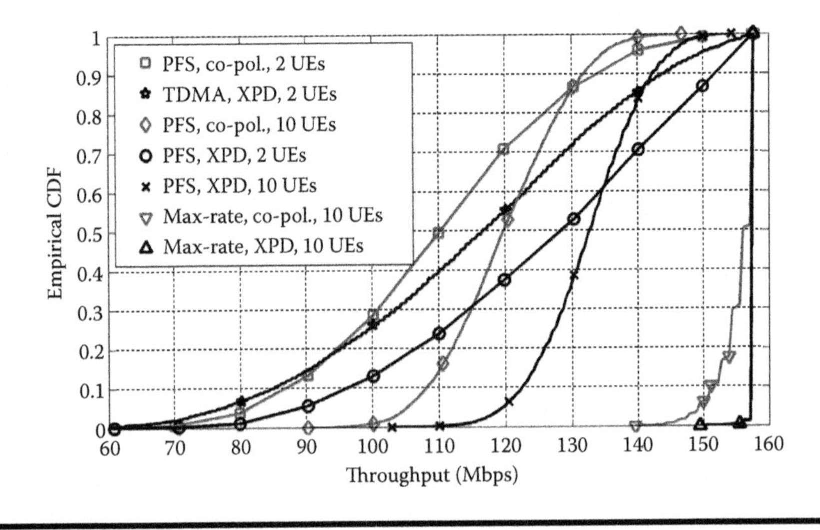

Figure 11.35 System sum throughput with multiple users.

path over time. XPD antennas are always seen to outperform copolarized antennas. This observation is consistent with that of colocated users, even though marginal gains are observed due to path loss diversity.

The plots are repeated for 10 users, who move along the same path and experience different path loss in each time slot. In OFDMA max-rate scheduling, the sum throughput using copolarized and XPD antennas converges with only 10 users, as in Figure 11.35. This shows that, even for 10 users in the system, the gains of XPD antennas saturate as a result of the channel-hardening effect.

One would wonder whether this result is still applicable if additional user fairness is incorporated. With a proportional fair scheduler, a performance gap between XPD antennas and copolarized antennas with 10 users is observed. The throughput difference is approximately 12 Mb/s. These gains are derived from the higher single-user throughput results in Figure 11.28. There it was observed that single-user supportable throughput is higher with XPD antennas for a mean SNR of more than 6 dB. The mean SNR in the measurement track is 27 dB. Therefore, single-user supportable throughput is clearly higher for XPD antennas. This higher supportable throughput is more favorable for the purpose of fairness scheduling.

For completeness, we show the individual user rate for a random user in the scenario in Figure 11.36 for one, two, and ten UEs. It is observed that the user rate with XPD antennas is always higher. This relates directly to the higher overall spectral efficiency of XPD antennas. The gap for 10 users between XPD and copolarized antennas is approximately 1.2 Mb/s for the median, a more or less even distribution of sum-throughput gains.

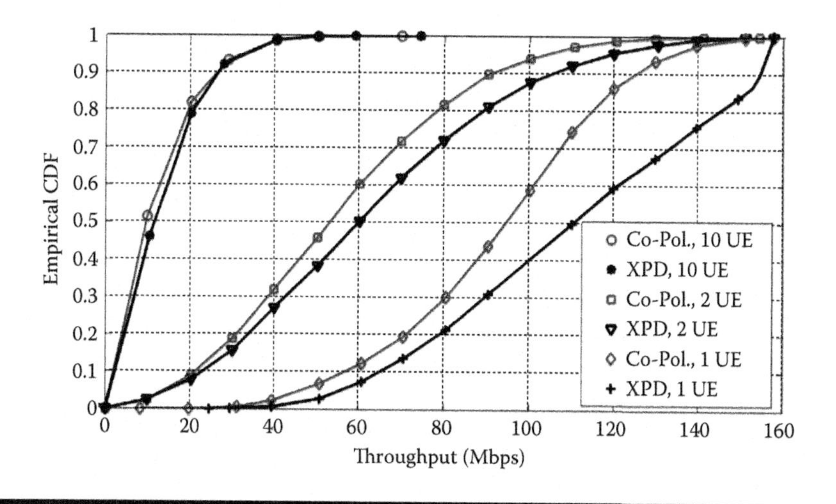

Figure 11.36 Individual user throughput.

11.4.5 Discussion of Results

Spatial multiplexing with multiple antennas is an important PHY layer aspect to deliver high throughput in OFDMA systems. Spatial multiplexing relies on channel scattering effects in wireless multipath environments. In practical cellular systems, such channel conditions may not commonly exist to support multiplexing. Multiplexing with polarized antennas is then shown to be a suitable solution. Our measurement results, which were performed in LOS conditions, demonstrate the throughput gains by using polarized antennas. Throughput analysis shows that orthogonally polarized antennas considerably outperform copolarized antennas at high SNRs. This spectral efficiency gain is shown to better influence multiuser scheduling. It is seen that orthogonally polarized antennas are more suitable for user fairness in OFDMA scheduling.

11.4.5.1 Summary of Field Trial Results

As shown in the previous sections, we could verify that the 3GPP LTE air interface, which is still under discussion as this book is edited, can be operated indoors as well as outdoors and achieve very high data rates. Robustness to channel conditions has also been shown for coverage areas up to 800 m with mobility. Furthermore, we could demonstrate the benefits of the new 3GPP-LTE air interface with a new cross-layer MAC architecture of link adaptation, transmit MIMO mode selection, and MCS level selection, all of which are frequency dependent. The plot in Figure 11.37 gives an overview of the coverage in a single-cell/single-sector scenario measured in downtown Berlin.

The base station was installed at a height of about 60 m on top of the HHI building and the XPD BS antenna pointed southwest toward the campus of the University of Technology in Berlin. The sector antenna had an 11-dBi antenna gain and a 60° horizontal angle; the transmit power was 2.5 W at 2.6 GHz and a 20-MHz bandwidth FDD. During the coverage measurements, two UEs were colocated inside one car and driven around the area with speeds during the measurements of 5–40 km/h. In order to demonstrate the capacity and robustness of the new high-rate air interface, high-rate video streaming to each individual user was showcased.

By exploiting adaptive MIMO mode selection (MIMO [2Tx, 2Rx], SIMO [1Tx, 2Rx]) and fast-loop MCS level adaptation (QPSK, 16 QAM, 64 QAM, 1/2, and 2/4 rate coding), a target BER of 10^{-3} was achieved robustly throughout the coverage region. The coverage area was more than 800 m from the base station.

A HARQ retransmission scheme was also used so that high-rate videos could be streamed reliably within the whole measurement scenario. In Figure 11.37, we see the sum rate measured with two colocated users around the campus area.

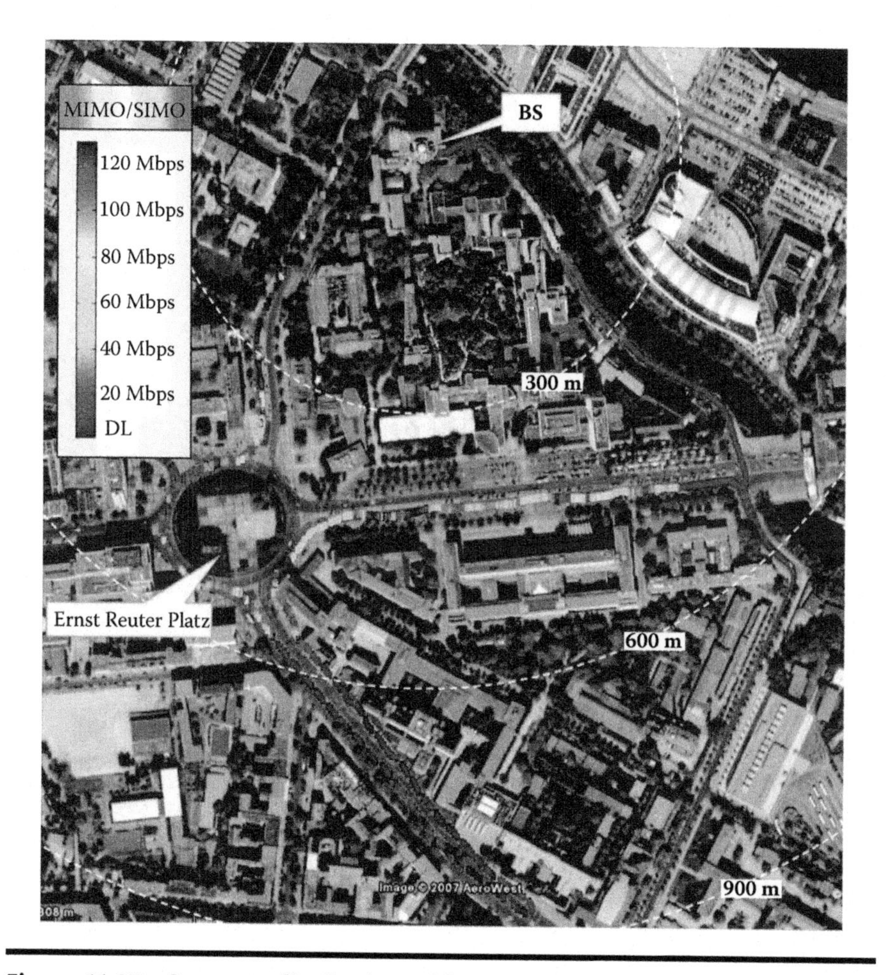

Figure 11.37 Sum rate distribution with two colocated users using a proportional fair scheduler.

The capability of achieving a PHY throughput of more than 60 Mb/s has a remarkable correspondence to the chances of dual-stream transmission. For these high rates, in order to exploit the spectrum more efficiently, one relies significantly on dual-stream transmission made possible by the high SNR available in the cell. In Figure 11.38, the area of dual-stream availability is depicted in black (red), and the areas of single stream only are depicted in white (blue). The area is considered as dual-stream capable when at least one resource block can be operated in dual-stream mode as the spectrally more efficient transmit mode compared to single-stream transmission.

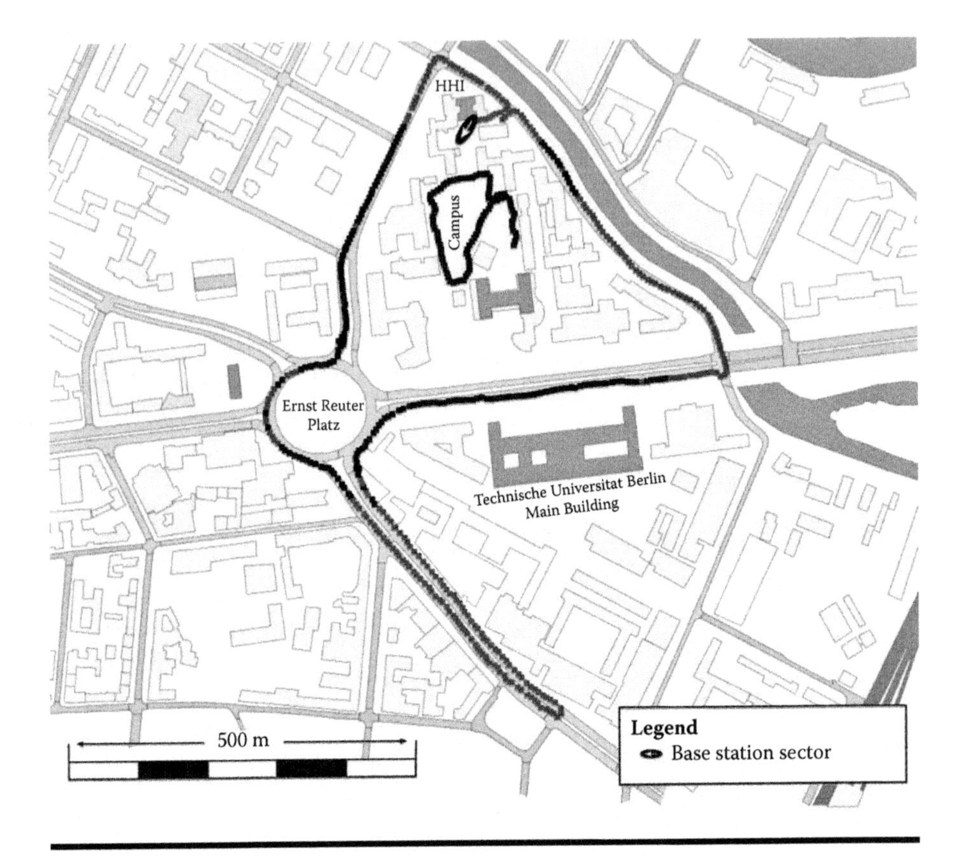

Figure 11.38 Distribution of single-stream and dual-stream opportunities in the measured scenario.

Due to the significant shadowing along some of the roads and the low transmit power at the base station, we see MIMO or dual-stream transmission to be available up to a distance of about 600 m. The high granularity of the link adaptation and the multiuser diversity exploited by the channel-aware proportional fair scheduler allowed for a smooth transition of the sum rate along the measurement track.

References

1. Telatar, E. 1999. Capacity of multiantenna Gaussian channels. *European Transactions on Telecommunications* 10 (6): 585–596.
2. Jungnickel, V., A. Forck, T. Haustein, S. Schiffermuller, C. von Helmolt, F. Luhn, M. Pollock, C. Juchems, M. Lampe, W. Zirwas, et al. 2005. 1 Gb/s MIMO-OFDM transmission experiments. *Proceedings of VTC 2005,* Dallas, TX, September 2005, 2:861–866.

3. Jungnickel, V., S. Jaeckel, L. Thiele, U. Krueger, A. Brylka, and C. von Helmolt. 2006. Capacity measurements in a multicell MIMO system. *Proceedings of IEEE Global Telecommunications Conference (GLOBECOM 2006),* San Francisco, CA, November 2006, pp. 1–6.
4. Bolcskei, H., R. Nabar, V. Erceg, D. Gesbert, and A. Paulraj. 2001. Performance of spatial multiplexing in the presence of polarization diversity. *Proceedings of IEEE International Conference on Acoustics, Speech, and Signal Processing (ICASSP '01)* 4: 2437–2440.
5. Erceg, V., H. Sampath, and S. Catreux-Erceg. 2006. Dual-polarization versus single-polarization MIMO channel measurement results and modeling. *IEEE Transactions on Wireless Communications* 5 (1): 28–33.
6. Shafi, M., M. Zhang, A. Moustakas, P. Smith, A. Molisch, F. Tufvesson, and S. Simon. 2006. Polarized MIMO channels in 3-D: Models, measurements and mutual information. *IEEE Journal on Selected Areas in Communications* 24 (3): 514–527.
7. Degen, C., and W. Keusgen. 2002. Performance of polarization multiplexing in mobile radio systems. *Electronics Letters* 38 (25): 1730–1732.
8. IEEE 3GPP TS 36.211 V8.0.0. September 2007. Physical channels and modulation (release 8).
9. A. Goldsmith, A., and P. Varaiya. 1997. Capacity of fading channels with channel side information. *IEEE Transactions on Information Theory* 43 (6): 1986–1992.
10. Jiang, L., L. Thiele, and V. Jungnickel. 2007. On the modeling of cross-polarized channel. European Wireless 2007.
11. Oestges, C., V. Erceg, and A. Paulraj. 2004. Propagation modeling of MIMO multipolarized fixed wireless channels. *IEEE Transactions in Vehicular Technology* 53: 644–654.
12. Haustein, T. et al. 2007. MIMO-OFDM for cellular deployment—Concepts, real-time implementation and measurements towards 3GPP-LTE. EUSIPCO 2007, Poznan, Poland.
13. Haustein, T., C. Zhou, A. Forck, H. Gabler, C. Helmolt, V. Jungnickel, and U. Kruger. 2004. Implementation of channel aware scheduling and bitloading for the multiuser SIMO MAC in a real-time MIMO demonstration test-bed at high data rate. *Proceedings of Vehicular Technology Conference (VTC2004-Fall)* 2:1043–1047.
14. Viswanath, P., D. Tse, and R. Laroia. 2002. Opportunistic beam forming using dumb antennas. *IEEE Transactions on Information Theory* 48 (6): 1277–1294.
15. Wirth, T., V. Jungnickel, A. Forck, S. Wahls, H. Gaebler, T. Haustein, J. Eichinger, D. Monge, E. Schulz, C. Juchems, et al. 2008. Real-time multiuser multiantenna downlink measurements. *Proceedings of IEEE Wireless Communications and Networking Conference (WCNC),* Las Vegas, NV, March 2008.
16. Schiffermüller, S., and V. Jungnickel. 2006. Practical channel interpolation for OFDMA. *IEEE GlobeCom 2006,* San Francisco, CA, November 2006.

Chapter 12

Measuring Performance of 3GPP LTE Terminals and Small Base Stations in Reverberation Chambers

M. Andersson, A. Wolfgang,
C. Orlenius, and J. Carlsson

Contents

12.1 Introduction

Multipath fading has for many years been a problem for mobile communication engineers, to be overcome by designing systems that have sufficient signal-to-noise ratio (SNR) or signal-to-interference ratio (SIR) margins. The fading is caused by multiple reflections in walls and ceilings (indoors) or cars and buildings (outdoors) that add constructively or destructively at the location of the terminal or base station antennas. The global system for mobile communications (GSM) and third generation (3G) both require fading margins of 20–30 dB to provide good voice quality at most locations in the cell [1]. The 3rd Generation Partnership Project (3GPP) long term evolution (LTE) and, to some extent, evolved HSPA (high-speed packet access) and WiMAX (worldwide interoperability for microwave access) will use a completely different approach to multipath fading. Instead of being something to be combated to provide robust voice communication everywhere in the cell, the fading conditions will be used to increase the bit rate for data communication for single users or increase the system throughput in order to increase the overall spectral efficiency [2].

A key technology to provide this flexibility at reasonable costs in 3GPP LTE systems is multiantenna solutions at the terminal or base station [3,4]. The multiantenna configuration can be used to increase the SNR and SIR using beam forming or different antenna diversity algorithms or utilized for higher throughput using partially uncorrelated transmission channels if multiple antennas are used at both base station and terminal—that is, MIMO (multiple input, multiple output) antennas. The use of more advanced antennas at the terminal and base station has been considered also for GSM and 3G for many years [5] but has not led to large deployments of terminals or base stations with advanced multiantenna solutions. One of the main reasons is the big difference in the required data rates between mobile broadband services such as 3GPP LTE compared to mobile telephony. In 3GPP LTE the bit rate in the down- and uplinks will be directly dependent on the available SNR/SIR received at the terminal from the base station and vice versa. In a GSM system, the SNR/SIR just has to be above a certain threshold value for the service to work. Even if there is a 20-dB extra link margin available—for example, due to smart antenna solutions—the bit rate will always be the same 9.6 kb/s in the up- and downlinks. Of course, the coverage will be better, but that does not affect the majority of users in urban areas and affects only a few users in rural areas.

The business case for advanced antennas to extend the coverage for a few users in rural areas has, in most cases, not been good enough for the extra cost of using advanced antenna solutions. The business case for higher throughput and better spectral efficiency for all users is much better. For the new mobile broadband systems—HSPA, WIMAX, and 3GPP LTE—extra SNR or SIR can be used in discrete steps to increase the bit rate through higher modulation rates (e.g., going from QPSK (quadrature phase shift keying) to 16 QAM (quadrature amplitude modulation) or 64 QAM) or reduced coding. Alternatively, the system capacity can be increased by adding more users with the same bit rate using narrower or fewer channels. Just as a few users in rural areas today discover that some mobile phones work and some do not work (often depending on how well the antennas have been implemented in the phones), all users of 3GPP LTE terminals will experience different up- and downlink speeds with different terminals at the same location in both urban and rural areas. The implementation of the antenna solutions will thus immediately affect the user perception, and good single- or multiantenna solutions will be a significant competitive advantage for 3GPP LTE terminal and base station antenna developers and manufacturers.

One of the main challenges facing the antenna engineer choosing an antenna configuration is how to reach a design with good performance and how to verify this performance in a reasonable time and at a reasonable cost. One of the differences of small antennas compared to large antennas is that they are much harder to model accurately with software. This is mainly because large antennas normally are used in an open environment without any neighboring objects to disturb the antenna function. Small antennas, on the other hand, are usually integrated in a chassis containing materials that absorb radiation or in different ways disturb the antenna function. Furthermore, in a mobile phone, there are a number of antennas that all affect each other.

For small antennas, in a single- or multiantenna configuration, the most important parameter is the antenna efficiency [6] (i.e., the parameter that directly influences how much of the transmitter power is radiated into space or how much of the radiation incident on the antenna reaches the receiver). By optimizing the antenna function directly to as high antenna efficiency as possible to influence such important parameters as coverage, battery lifetime, and bit rate in the up- and downlinks. For small antennas, this optimization is very hard to do accurately with software, but it is easy with measurements. Because most of the small antennas should have high efficiency over a number of frequency channels and sometimes over several frequency bands, there is a need for a large number of measurements during development and evaluation of wireless products.

12.1.1 Test Chambers for Terminals and Base Stations with Small Antennas

Despite their common usage, the traditional test instruments for measuring the antenna efficiency of small antennas or the total radiated power (TRP) and total isotropic sensitivity (TIS) of wireless devices with small antennas (i.e., anechoic chambers) have never been a suitable method for wireless devices to be used in a multipath environment because there are no reflections in an anechoic chamber. Anechoic chambers were originally developed to measure radar antennas during the Second World War. The method is very suitable for large antennas, which, in addition to radar antennas, can be antennas for microwave links, satellite antennas, etc. These antennas have in common that they are used in an environment with few or no reflections—so-called line-of-sight (LOS) conditions. Still, people who developed small antennas had no other choice than using anechoic chambers for characterizing them because no alternative measurement chambers were available. The anechoic chamber technology has continually been improved by using (e.g., near field probe measurements) multiprobe measurements to speed up measurement time. However, devices with small antennas are mainly used indoors or in urban environments where there are multiple paths, which are very different environments from those found in anechoic chambers.

A multipath environment is much easier to emulate in a reverberation chamber. The reverberation chamber is also a reference environment—a reflective reference environment with Rayleigh statistics that is much more suitable for devices with small antennas than the anechoic reference environment. The reverberation chamber has the advantages that it can be made much smaller and that the measurements can be performed much faster than in an anechoic chambers. Another very large advantage, especially important for 3GPP LTE, is the possibility to make direct measurements in the reflective (Rayleigh) reference environment of diversity gain and MIMO capacity for products with multiple antennas. The alternative "drive test" has several drawbacks:

- unreliable (one can never be sure to drive exactly the same route and through the same environment more than once);
- expensive (one needs to make simultaneous measurements for each of the multiple antennas in the terminal as well as for a reference antenna); and
- time consuming.

We will see in this chapter that, for the first time ever, the reverberation chamber makes it possible to do repeatable system throughput tests in a reflective environment as well as completely new tests of importance for 3GPP LTE technologies—for example, multiuser MIMO (MU-MIMO) and scheduling tests that are mainly only simulated in software today. System throughput tests could, as time passes, become more important for designers of 3GPP LTE terminals and small base stations than the traditional TRP and TIS tests done in anechoic or reverberation chambers because they show the actual performance in the downlink and uplink in a Rayleigh environment.

12.1.2 Brief Background on Reverberation Chambers

For about 30 years, reverberation chambers (or mode-stirred chambers) have been used to test how much electrical devices radiate (normally referred to as electromagnetic compatibility [EMC] measurements) to avoid interference with other electrical devices [7]. The reverberation chamber is usually a metal-box cavity, with different sizes in its three dimensions and with some type of mode-stirrer mechanism. When one excites one or several antennas in the chamber at a suitable frequency, a number of standing wave modes will be generated. By placing the device under test (DUT) in the cavity, one makes sure that all the radiation generated stays in the cavity. By changing the boundary conditions for the modes in the cavity using a movable metal plate (often in the form of a propeller), it is possible to ensure that the radiated power can be detected regardless of the direction in which it is sent. Traditional reverberation chambers for EMC usually have an accuracy of not better than 3 dB in standard deviation (STD). This is much too high an uncertainty to measure antenna efficiency, radiated power, or receiver sensitivity, although it is quite sufficient for EMC measurements.

At the end of the 1990s, Professor Per-Simon Kildal at the antenna group at Chalmers University of Technology, Gothenburg, Sweden, had an idea of how to improve the accuracy of reverberation chambers so that they could be used to measure antenna efficiency, radiated power, and receiver sensitivity of small antennas and wireless terminals with small antennas [8]. The origin of the idea was the comprehension that the traditional way of measuring antennas in anechoic chambers was not at all suited for small antennas or wireless terminals with small antennas (e.g., mobile phones) because they are normally used in environments with multiple paths (i.e., indoors or in urban areas). It also opened up the possibility to develop a very small test chamber that the antenna engineer could use at his desk (see Figure 12.1).

The vision of a very small measurement facility that would give future antenna engineers the possibility to design antennas at their desks led to the creation of the company Bluetest AB in 2000 [9]. Bluetest AB and Chalmers University of Technology worked very closely together for many years to develop and gradually improve the reverberation chamber technology. Today, Bluetest has three products

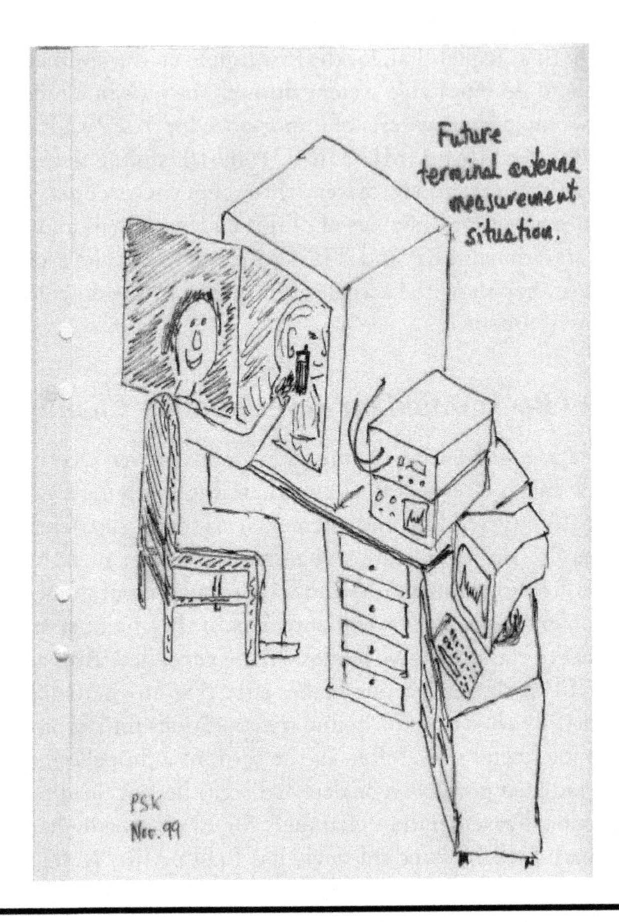

Figure 12.1 Per-Simon Kildal's vision in 1999.

in its portfolio: a small standard chamber suitable for antenna efficiency and total radiated power (TRP); a high-performance chamber with very high isolation (100 dB), which also is suitable for receiver sensitivity (TIS) and throughput measurements; and tailor-made mode stirrers to fit any size existing shielded room. Although some companies, institutes, and universities have developed reverberation chambers and software for measuring wireless devices with small antennas, by late-2008, Bluetest was the only company that commercially provided reverberation chambers and software for measuring antenna efficiency, TRP, TIS, and throughput.

Bluetest's customers can be found in Asia, Europe, and North and South America, and they are a mixture of mobile phone developers/manufacturers, mobile operators, companies that develop small antennas, test institutes, and universities. Eight of the world's largest mobile phone manufacturers, as well as two of the world's largest mobile phone operators are among the customers. With the increasing interest in this technology, it is likely that that competing companies will soon be supplying similar solutions. It is our belief that, in a few years, reverberation chambers will be the preferred test method for fast development and evaluation of wireless devices with small antennas.

12.1.3 Standardization of Measurements in Reverberation Chambers

One of the first organizations to use measurements in reverberation chambers for their quality standard of mobile phones TCO '01 was TCO Development [10]. They regularly publish TCP (telephone communication power) measurements of mobile phones made in Bluetest's reverberation chambers at their website (www.mobilelabelling.com) [11]. TCP is the average transmitted power over five communication channels within one GSM or 3G band. Mobile phones with high TCP and low SAR (specific absorption rate), are recommended by TCO Development.

The 3GPP has included measurements of UMTS (universal mobile telecommunications system) terminals in reverberation chambers in the 3GPP TR 25.914 V7.0.0. technical report [12] as an alternative to measurements in anechoic chambers. In this report, basic parameters of reverberation chambers for measurements of the radio performance of a 3G terminal are described. The calibration of reverberation chambers and the measurement method are described in some detail.

CTIA (formerly Cellular Telephone Industries Association, now CTIA—The Wireless Organization [13]) has, since late 2007, worked to standardize measurements of TRP and TIS in reverberation chambers. Traditionally, all CTIA standards for measurements of TRP and TIS of various wireless systems have been specified in anechoic chambers. Measurements of TRP and TIS of mobile terminals in anechoic chambers are described in reference 14. The new standard for TRP and TIS measurements in reverberation chambers is scheduled to be ready in 2009.

In mid-2008, no standards existed to measure MIMO capacity, diversity gain, or system throughput for wireless devices; this includes the effects of multiantenna solutions, which can be significant in a reflective environment. The reverberation chamber is a very good candidate for a suitable reflective reference environment, where repeatable tests of these parameters can be done rapidly and with good accuracy.

12.1.4 Chapter Outline

This chapter will use a hands-on approach and start by reviewing basic properties of reverberation chambers and giving an overview of ongoing research and benchmarking activities. It will then describe how to calibrate reverberation chambers and how they are used to measure antenna efficiency, TRP, and TIS—important parameters for all wireless devices with small antennas, including 3GPP LTE terminals and small base stations. It will also describe how, using a multiport network analyzer, diversity gain and MIMO capacity can be measured directly in the Rayleigh environment in about 1 minute, compared to hours or more using other technologies. The first repeatable system throughput measurements in a reflective reference environment (i.e., the reverberation chamber) will be described. The chapter will end by describing several new ways of measuring 3GPP LTE parameters and technologies mainly simulated in software today because real-life measurements (e.g., drives tests) are too complicated and time consuming. However, many of these parameters, such as MU-MIMO, opportunistic scheduling, and multibase station handover, should be easy to measure in the reverberation chamber.

12.2 Basic Properties of Reverberation Chambers

The reverberation chamber [7] has been used for EMC testing of radiated emissions and immunity for about three decades. It is basically a metal cavity that is sufficiently large to support many resonant modes, and it is provided with means to stir the modes so that a statistical field variation appears. It has been shown that the reverberation chamber represents a multipath environment similar to what we find in urban and indoor environments. Therefore, during recent years it has been developed as an accurate instrument for measuring desired radiation properties for small antennas as well as active mobile terminals designed for use in environments with multipath propagation.

When a signal is transmitted from a base station to a mobile terminal in a complex environment, the signal will take different wave paths because large, smooth objects between the antennas will cause reflections, edges on objects will cause diffraction, and small irregular objects will cause scattering of the waves originating from the transmitting antenna. The wave contributions via these paths will add at the receiver, and, because their complex amplitudes (i.e., amplitudes and phases) are independent, they may add up constructively or destructively, or anything else between these two extremes. The wave paths and the complex amplitudes will also change in time, due to the moving of the terminal or objects in the environment, and therefore the received signal will vary with time. This is referred to as fading.

The multipath environment at the receiver can be characterized by several independent plane waves. The independence means that amplitudes, phases, and polarizations, as well as the angles of arrival (AoA), are arbitrary to each other. If an LOS signal is absent and the number of incoming waves is large enough (typically a few hundreds), the in-phase and quadrature components of the received complex signal become normally distributed. This means that their associated magnitudes get a Rayleigh distribution, the power gets an exponential distribution, and the phases get a normal distribution over 2π. This is a direct result of the central limit theorem. In the left part of Figure 12.2 is shown an example of a Rayleigh signal as a function of time. We see that the signal level varies by more than 25 dB. Another way to illustrate the Rayleigh fading is to plot the cumulative (probability) density

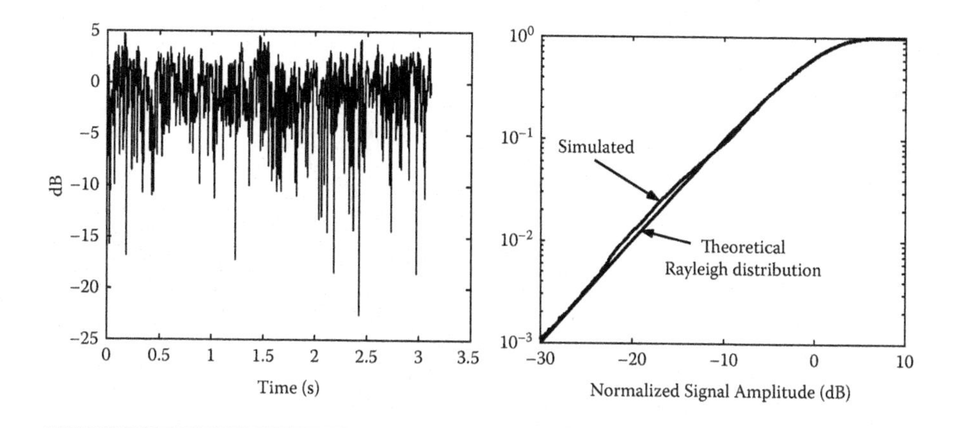

Figure 12.2 Example of a fading signal (left) and its CDF function (right). The signal levels are presented in decibels after being normalized to the time-averaged power.

function (CDF), as shown in the right part of Figure 12.2. The plot shows the cumulative probability of signal amplitudes in decibels, where the reference level is defined as the total received power divided by the number of samples (i.e., the average received power). As an example, we can see that the probability of having dips deeper than –20 dB is about 1%.

In a real environment, we might have a certain AoA distribution in both the elevation and azimuth planes. However, it is natural to assume that the mobile terminal can be arbitrarily oriented relative to directions in the horizontal plane, which means that the azimuth angle is uniformly distributed. We can also assume that the terminal has a preferred or most probable orientation relative to the vertical axis and, in addition, common environments (in particular, outdoor) have a larger probability of waves coming in from close-to-horizontal directions rather than close-to-vertical directions. Therefore, we might need an elevation distribution function to describe real multipath environments. Real environments normally also have a larger content of vertical polarization than horizontal because most base station antennas are vertically polarized. In propagation literature, this is characterized by cross-polar power discrimination (XPD) [1].

Because both the AoA distribution and XPD are different in different real environments, the performance of antennas and mobile terminals depends on where they are used or measured. This means that results from measurements in one environment cannot easily be transferred to another. Instead of measuring in real environments, it could be advantageous to do measurements in a reference environment with well-defined properties that gives repeatable results. One such reference environment is the isotropic environment, which has uniformly distributed polarization as well as a uniform AoA distribution over the whole sphere.

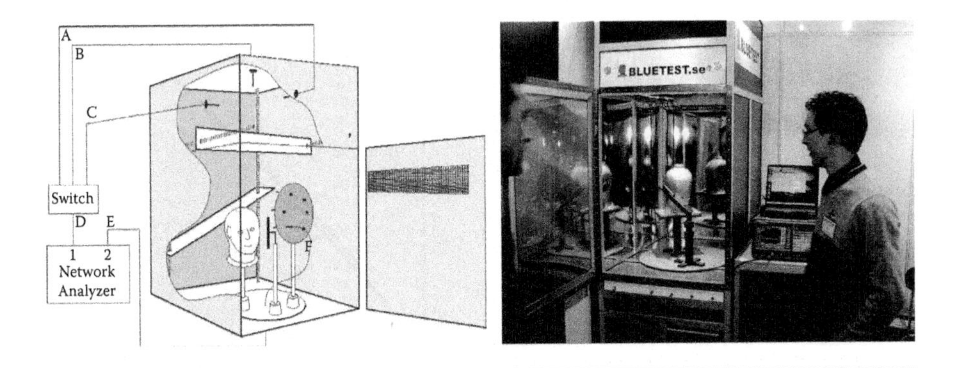

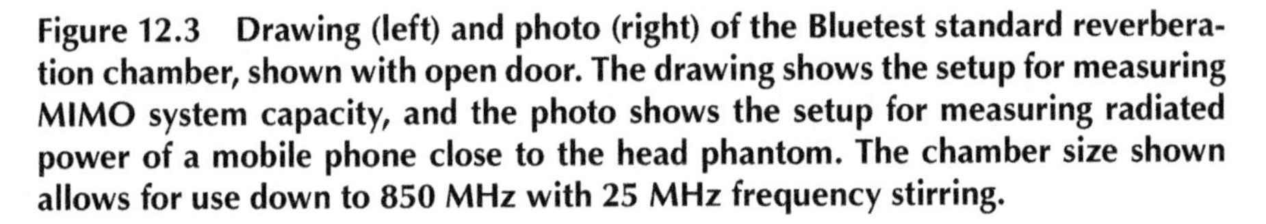

Figure 12.3 Drawing (left) and photo (right) of the Bluetest standard reverberation chamber, shown with open door. The drawing shows the setup for measuring MIMO system capacity, and the photo shows the setup for measuring radiated power of a mobile phone close to the head phantom. The chamber size shown allows for use down to 850 MHz with 25 MHz frequency stirring.

The reverberation chamber emulates such an isotropic environment. The isotropic reference environment has no counterpart in reality, but still it could be representative because any environment will appear isotropic if the mobile terminal is used with arbitrary orientation in the environment.

The basic measurement setup in a reverberation chamber that is used for calibration as well as for measuring passive antenna performance is shown in the left part of Figure 12.3.

The reverberation chamber is large in terms of the wavelength at the frequency of operation, which means that several modes are excited. It can be shown that each mode in a rectangular cavity can be expressed as eight plane waves [15] and that the directions of arrival of these plane waves are uniformly distributed over the unit sphere, if enough modes are excited. By stirring the modes, the set of incoming plane waves seen by an antenna in the chamber will change, as will how they combine to the received signal. The result is a signal that is fading. When the stirring is done properly, the resulting environment in the chamber will represent an isotropic multipath environment (i.e., the desired reference environment mentioned earlier). The key is the mode stirring, which can be done in different ways. The chamber shown in Figure 12.3 has the following stirring capabilities:

- Mechanical stirring is done by means of two metal plates that can be moved along a complete wall and along the ceiling by step motors. The larger the volume is that the stirrers cover, the better.
- In platform stirring [16], the antenna or mobile terminal under test is placed on a rotatable platform that moves the device under test to different positions in steps. This stirring method is very effective, in particular in small chambers.

- In polarization stirring [17], the transmissions between the device under test and each of three orthogonal wall-mounted antennas are measured successively. Thus, a good polarization balance is achieved.

An additional, very effective stirring method is referred to as frequency stirring. This corresponds to averaging the measured quantity over a frequency band and is done during the processing of measured results.

12.2.1 Basic Measurements—Reflection and Transmission Coefficients

Figure 12.4 shows how the reflection S_{11} and the transmission S_{12} between two antennas inside the chamber are measured. From physics it can be argued that S_{11} of the antenna under test must consist of two contributions: one the S_{11}^a from the antenna itself, as if it were located in free space, and the other S_{11}^c due to the chamber; thus:

$$S_{11} = S_{11}^a + S_{11}^c \tag{12.1}$$

The S_{11}^a contribution is deterministic, whereas S_{11}^c is random as a result of the mode stirring in the chamber. If the number of independent values of S_{11}^c is large enough, it must get a complex Gaussian distribution with a zero mean. Therefore, it is possible to determine S_{11}^a by complex averaging the measured S_{11} over all stirrer positions [18]:

$$S_{11}^a = \frac{1}{M} \sum_{\substack{\text{stirrer} \\ \text{pos}}} S_{11} = \overline{S_{11}} \tag{12.2}$$

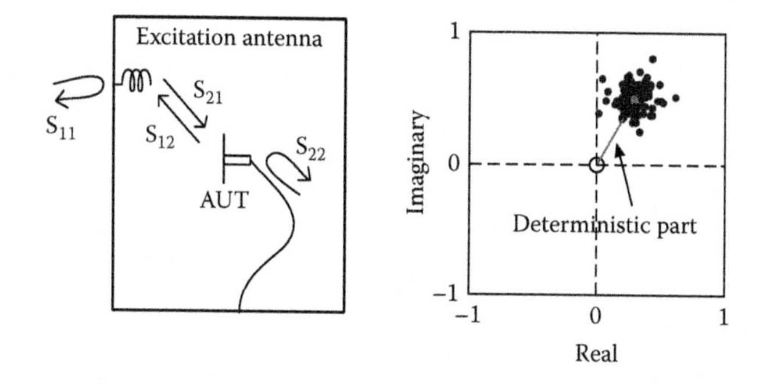

Figure 12.4 Example of measured *S*-parameters of two antennas in a reverberation chamber as a result of mode stirring and illustration of the deterministic and random contributions to them.

Using the same argument, the transmission coefficient S_{21} must also consist of two parts: one deterministic S_{11}^d the same as in free space, and another statistic part S_{11}^c due to the chamber; that is,

$$S_{21} = S_{21}^d + S_{21}^c \tag{12.3}$$

The deterministic part S_{11}^d in Equation (12.3) represents the direct coupling between the transmitting and receiving antennas in the chamber; because we would like the chamber to provide an isotropic reference environment with Rayleigh fading, it should be as low as possible. The direct coupling is given by the free space transmission formula, also known as Friis transmission formula (see, for example, Kildal [19]):

$$\left|S_{21}^d\right|^2 = \left(\frac{\lambda}{4\pi r}\right)^2 G_t G_r \tag{12.4}$$

where r is the distance between the antennas, λ is the wavelength, and G_t and G_r are the realized gains of the two antennas, respectively, in the direction of the opposite antenna. From Equation (12.4) it can be understood that the main lobes of the antennas should not point toward each other. If the direct coupling, S_{11}^d, is much smaller than the chamber contribution, S_{11}^c, to the total transmission, S_{21}, the chamber provides a Rayleigh fading environment. Effective methods to reduce the direct coupling include platform stirring, polarization stirring, and using fixed antennas that have nulls in the direction of the antenna under test (AUT).

The transmission between two antennas in free space is governed by the Friis transmission formula (12.4). The corresponding formula for transmission between two antennas in a reverberation chamber is described by Hill's transmission formula [20], which is valid under the assumptions that the direct coupling is negligible and that the chamber is large with many modes excited. We choose to refer to Hill's formula as the chamber transfer function and to present it in the following form:

$$G_{chamber} = \left|S_{21}^c\right|^2 = \frac{P_r}{P_t} = \frac{c^3 e_{rad1} e_{rad2}}{16\pi^2 V f^2 \Delta f} \tag{12.5}$$

where

f is the frequency;
c is the speed of light;
V is the chamber volume;
e_{rad1} and e_{rad2} are the total radiation efficiencies (i.e., including mismatch) of the two antennas; and
Δf is the average mode bandwidth.

The mode bandwidth Δf consists of four additive contributions due to losses in the walls, leakage through apertures and slots, antennas in the chamber, and absorbing objects in the chamber:

$$\Delta f = \sum_{\substack{all \\ walls}} \Delta f_{wall} + \sum_{\substack{all \\ slots}} \Delta f_{leakage} + \sum_{\substack{all \\ antennas}} \Delta f_{antenna} + \sum_{\substack{all\ lossy \\ objects}} \Delta f_{object} \tag{12.6}$$

with

$$\Delta f_{wall} = \frac{2A}{3V}\sqrt{\frac{c\rho f}{\pi\eta}}, \quad \Delta f_{leakage} = \frac{c\sigma_1}{4\pi V}, \quad \Delta f_{antenna} = \frac{c^3 e_{rad}}{16\pi^2 V f^2}, \quad \Delta f_{object} = \frac{c}{2\pi V}\sigma_a \tag{12.7}$$

where

V is the chamber volume;
η is the free space impedance;
A is the area of conducting surfaces (chamber walls, plate stirrers, etc.) with surface resistance ρ;
σ_l is the leakage cross-section of apertures and slots; and
σ_a is the absorption cross-section of absorbing objects.

The total mode bandwidth, Δf, for practical chambers can be very different, corresponding to Q-values between 30 and several thousands ($Q = f/\Delta f$). When measuring active terminals, it is important that Δf is larger than the modulation bandwidth; otherwise, measurement errors might appear.

12.2.2 Transmission Level and Calibration

In practice, a measurement of radiation efficiency, radiated power, or receiver sensitivity is based on first determining the chamber transfer function given by Equation (12.5) by using a reference antenna with known radiation efficiency e_{rad2}. Because we would like to have the same mode bandwidth, Δf, during the calibration as well as during the actual measurement of the AUT, it is important that the AUT is present in the chamber with its port terminated in a matched load during the calibration. The actual measurement will then be done by terminating the reference antenna in a matched load and measuring the transmission to the AUT. The ratio between the average transfer functions for these two cases will be equal to the ratio between the radiation efficiency of the reference antenna and the AUT. The radiation efficiency e_{rad1} of the fixed antenna does not need to be known because it will be the same for both measurements. See specific sections that follow for detailed calibration and efficiency measurement procedures.

It is also possible to calibrate the chamber without having the AUT in the chamber. For this case, the reference antenna should be removed when the AUT is measured. This is a simplified procedure with the advantage that we can use the same calibration when measuring several AUTs. In particular, this procedure is simpler when measuring many active terminals. It should be noted that this simplified method only works if the chamber is loaded so much that the reference antenna and AUT do not represent any significant contribution to the mode bandwidth Δf of the chamber. If this is not the case, the results of the measurements may be wrong because the chamber transfer function will not have a linear dependence on e_{rad} of the reference antenna and the AUT. This may in particular happen if the reference antenna or the AUT is made of materials that absorb radiation and thereby increase the mode bandwidth even when the antenna ports are open or short-circuited.

If we remove the free space mismatch efficiencies from Equation (12.5), we get an average chamber transfer function that varies more slowly with frequency. This can then be frequency stirred for better accuracy without losing resolution due to variations in the mismatch efficiency. The free space input reflection coefficients of the two antennas can be obtained by complex averaging of the S_{11} and S_{22}, as given by Equation (12.2), and the mismatch-corrected chamber transfer function is defined by

$$G_{chamber} = \frac{1}{N} \sum_{N} \frac{|S_{21}^{c}|^2}{\left(1 - |\overline{S_{11}}|^2\right)\left(1 - |\overline{S_{22}}|^2\right)} \tag{12.8}$$

where N is the number of stirrer positions and $|S_{21}^{c}|^2$ is given by Equation (12.5).

12.2.3 Accuracy and Number of Independent Samples

In order to perform accurate measurements in the reverberation chamber, we need the chamber transfer function (12.5) to be proportional to the radiation efficiency independently of which antenna or terminal we measure. This is possible only if the mode stirring creates enough independent samples. The S_{21} samples are, for sufficient independent samples, complex normally distributed. Then the relative accuracy by which we can estimate G_{chamber} has a standard deviation of [21]

$$\sigma = 1/\sqrt{N_{ind}} \tag{12.9}$$

where N_{ind} is the number of independent samples. This means that we need $N_{ind} = 100$ for an accuracy of ±10% (i.e., ± 0.5 dB). The number of independent samples is determined primarily by the mode density in the chamber (i.e., by the number of modes per megahertz). This is, to a good approximation, given by the classical formula

$$\partial N_{\text{mode}}/\partial f = Vf^2 8\pi/c^3 \tag{12.10}$$

The number of independent samples is proportional to the mode density, but the proportionality constant is not known because it depends on chamber loading, mode stirring methods, mechanical stirrer shapes, and the shape of the chamber. A preliminary experimental study has been performed [22], but more work is required to get unique and simple conclusions. However, we can preliminarily state the following relation between the mode density and the number of independent samples:

$$N_{ind} \leq 8[\partial N_{mode}/\partial f](\Delta f + B) \tag{12.11}$$

where B is the bandwidth of the frequency stirring. The factor 8 is due to platform stirring and is based on empirical experience as well as physical reasoning. It is clear from Equations (12.9) and (12.11) that the accuracy in a given chamber can be controlled by Δf and B. However, it should be noted that other effects are coming in when the loading is very large, so some care must be taken. Still, we have experienced a good Rayleigh distribution with many independent samples in a heavily loaded chamber with Q as low as 30 [23]. The problem with frequency stirring B is that the frequency resolution becomes worse. For antenna measurements, this means that we cannot resolve variations in the radiation efficiency that are faster than B. In practice, the resolution is somewhat better by using the mismatch-corrected chamber transfer function (12.8) before power averaging because mismatch efficiencies normally vary more quickly with frequency than efficiencies due to ohmic losses.

12.2.4 Research Activities on Reverberation Chambers

The reverberation chamber has been used for more than 20 years for EMC measurements like electromagnetic susceptibility and emissions, as described in the overview article by Bäckström, Lundén, and Kildal [7]. Its basic theories are well understood due to several papers by D. Hill at the National Institute of Standards and Technology (NIST) in Boulder, Colorado [20,25,26]. The main EMC application has been to generate high field strength for susceptibility testing. During the last 7 years, Kildal's antenna group at Chalmers University of Technology has developed the reverberation chamber into a more accurate instrument for measuring the characteristics of desired radiation of small antennas and active mobile terminals—in particular when these are intended for use in multipath environments with Rayleigh fading, such as for wireless/mobile communications in urban or indoor environments.

The reverberation chamber was known to create Rayleigh fading when the modes were stirred [24] by mechanical movement of plates or shaped wires (mechanical stirrers). Kildal's group showed that the modes represent plane waves with an omnidirectional distribution of directions of incidence [15]. Thus, the reverberation chamber corresponds to an isotropic multipath environment with uniform elevation distribution of the angles of arrival. Real environments may be more complex, with a certain polarization imbalance and an elevation distribution, but the uniform isotropic environment is an excellent, well-defined reference environment, making it possible to characterize the antennas and terminals in terms of classical antenna quantities such as radiation efficiency and radiated power. Therefore, the goal was to improve the isotropy. In particular, a polarization imbalance was detected and removed by using three orthogonal wall antennas [17].

The accuracy of the measurements increases with the number of modes that are excited (i.e., with the mode density). This generates a certain lowest frequency of operation for specified measurement accuracy. Kildal's group found that the accuracy could be considerably improved by making use of a new stirring method, referred to as platform stirring [16]. When this stirring method was used, the radiation efficiency of small antennas and radiated power of mobile terminals could be measured with an accuracy of 0.5-dB RMS (root mean square) [27] in a small reverberation chamber. This was carefully validated by comparison with measurements in anechoic chambers [28] and larger reverberation chambers [27].

The reverberation chamber generates a Rayleigh distribution. Therefore, the chamber is particularly well suited for characterizing antenna and terminal functions specific for such environments, as described in Kildal and Rosengren [29]. Actually, the radiation patterns play no role in such environments because many interfering waves contribute to a statistical wave channel. The characterizing parameter is, for single-port antennas, the classical radiation efficiency, including contributions from mismatch, ohmic losses in nonperfect materials of the antenna and terminal itself, and losses in nearby objects such as a user's hand or head. It is important to note that these quantities will be the same as when they were measured in a "free space" environment like an anechoic chamber. Even the antenna impedance measured in the reverberation chamber will appear the same as when it is measured in free space, after the processing of the complex S-parameters acquired over all stirrer positions [18]. Future terminals will, to higher degree, make use of antenna diversity to remove the problems of fading dips where the terminal may not work (i.e., outage), and the corresponding diversity gain (or, in other terms, reduction in outage probability) can directly be measured by reverberation chamber tests when evaluating the multiport antennas themselves [30,31] and when diversity is implemented in active mobile terminals [32].

The future mobile communication systems could make use of multiport antennas with even more advanced functions than antenna diversity for both uplink and downlink (i.e., MIMO antenna systems). MIMO antenna systems are characterized in terms of the maximum available capacity, which also can be measured in the reverberation chamber [29,33]. In order to characterize both diversity and MIMO systems correctly, complete equivalent circuits of antennas are needed on reception (see Section 2.5 in Kildal [19]).

Mobile terminals were initially characterized only on transmit, whereas in the last few years the receive function has gained more attention. This is characterized by the input signal level giving a certain specified bit error rate (BER) or frame error rate (FER) for code division multiple access (CDMA) systems. When averaged over directions in the radiation pattern, the limiting static signal level is referred to as the total isotropic sensitivity (TIS). An alternative in reverberation chambers is to measure BER/FER during continuous fading to determine what Kildal's group refers to as average fading sensitivity (AFS) [34]. The AFS can also be used to determine the TIS because there seems to be a fixed, system-specific relation between the two values. The AFS is a more realistic performance parameter than TIS and can be measured much faster.

The reverberation chamber needs further development, in particular in relation to controlling bandwidth and improving accuracy. The latter is needed in order to allow measurements of terminals for some wideband mobile communication systems. Therefore, in a research project we have developed a numerical model of the chamber to study accuracy in loaded chambers. The focus was not on detailed modeling of a specific real chamber but, rather, on defining a simplified numerical chamber that can be simulated much more efficiently and thereby is more convenient for studying fundamental properties of mode stirring and loading [35,36]. This work was supported by experimental work on chamber loading [23] and studies of Hill's formulas for how loading affects the Q and average mode bandwidth [37]. At the moment, no one else in the world is able to do such extensive computations on reverberation chambers.

The reverberation chamber group at FOI in Sweden has recently conducted important studies related to stirrer efficiencies and accuracy [38,39]. Over the past years, the reverberation chamber group at NIST has started to do studies of wireless measurements in reverberation chambers as well [40]. Other groups are also taking up research on wireless measurements in reverberation chambers, such as the group at the University of Naples [41]. One PhD student from the group in Naples has recently spent 3 months at Chalmers doing research together with Kildal's group. The interest in the reverberation chamber measurement technology has spread worldwide. Several leading world mobile phone and wireless systems companies are today using Bluetest reverberation chambers for measuring antennas, radiated power, or receiver sensitivity when developing wireless terminals.

The main interest of most university groups using reverberation chambers is that the chamber can be used to measure diversity gain and capacity of multiport antennas. We have coauthored one journal article [42] and several conference papers about such diversity antenna characterization.

CTIA—The Wireless Organization—an international nonprofit membership organization founded in 1984 that represents all sectors of wireless communications (cellular, personal communication services, and enhanced specialized mobile radio)—has started standardization of reverberation chambers for wireless measurements and has invited Bluetest AB to participate as a guest member. The main objective has been to develop initially good calibration procedures, and Bluetest has received an initial acceptance for the approach that it is using, which is the best that can be done at the moment. NIST is also in the CTIA Reverberation Chamber Subgroup and supports the Bluetest approach, which is natural because both research groups are based on the same fundamental papers by Hill [20,25,26].

12.2.5 Benchmarking of Reverberation with Anechoic Chambers

During the first years of development of the reverberation chamber into an accurate instrument for antenna and terminal measurements, it was found that the accuracy could be considerably improved by making use of new stirring methods, referred to as platform stirring [16] and polarization stirring [17]. Using these stirring methods, the radiation efficiency of small antennas and radiated power of mobile terminals could be measured with an accuracy of 0.5-dB RMS [27] in a small reverberation chamber. This was carefully validated by comparison with measurements in anechoic chambers [28] and larger reverberation [27]. More recently, several reverberation chambers of different sizes participated in a round-robin test within the framework of ACE (European Antenna Center of Excellence) [43]. The basic idea was to collect data from different test facilities around Europe for passive as well as active test devices and to compare the results. The focus was on the most popular frequency bands typically used in mobile communications systems. In total, 12 organizations participated in the measurement campaign and 5 used reverberation chambers. In order to guarantee comparable results, only one test kit was sent around (see Figure 12.5).

In order to define simple and repeatable test cases for the round robin, passive antennas consisting of half-wavelength dipoles for 900-, 1800-, and 2400-MHz bands and a slot antenna for the 5.2-GHz band were used. In order to account for losses in objects close to the terminal antenna (e.g., resembling the head of a mobile phone user), the test fixture consisted of a lossy cylinder and an adjustable antenna mount (Figure 12.5). The cylinder contains a lossy liquid and is sealed so that all participants measure under the same conditions. The distance between the antenna and the cylinder can be adjusted, so the radiation efficiency can be varied. The distance can accurately be read on a scale on the antenna mount so that repeatability is guaranteed. All antennas are measured for free space conditions as well as at different distances from the lossy cylinder using the fixture.

In order to measure diversity gain, the same fixture can be used, but in this case with two dipoles mounted. As active devices, a triple-band GSM phone, a WCDMA phone, and a small active device for the 1.785-GHz band were used. The phones are standard phones equipped with special software that makes them radiate at full power. The channel can be set by using the key pad, so the phones

Figure 12.5 Test kit used in the round-robin campaign, together with suitcase used for shipping.

are usable even without a base station emulator. It is therefore possible to measure them at test facilities that do not have such emulators. Both phones have an external antenna connector so that the output power delivered directly from the transmit amplifier can be measured before and after the radiated power measurements. This is done in order to ensure that changes in battery power do not affect measured results and to ensure that nothing has happened to the phones during the course of the round robin.

For both the phones and the small active device, the total radiated power is measured for free space conditions and when the phone is located in a talk position close to a head phantom. The same types of phones are used for the receiver sensitivity measurements. For this case, the phones are standard phones with the original software, and it is therefore necessary to use a base station emulator. These measurements are also done for free space conditions and for a talk position with the phone close to a head phantom.

The round robin was organized in such a way that the participants would not know which organizations had already performed their measurements, in order to guarantee that no one had access to results beforehand. The participants used different measurement methods, including three-dimensional radiation pattern integration in fully anechoic chambers, spherical near-field methods, a random positioner system, and reverberation chambers. The results from the measurement campaign show that the reverberation chamber gives accurate and repeatable results for passive antenna parameters as well as active terminal parameters such as total radiated power and total isotropic sensitivity. Results can be found in the journal article by

Carlsson [44], the conference articles [45–48], and the final ACE report, which can be found via the ACE website [43].

12.3 Calibration of the Reverberation Chamber

The main parameter to calibrate in a reverberation chamber is the average power transmission over a complete stirrer sequence. This will be determined by the amount of loss present in the chamber cavity (as described in the preceding sections), so it is of great importance to keep the same power-absorbing objects in the chamber during the calibration measurement as when the measurement of the actual test unit is performed.

For the calibration measurement, a reference antenna with known radiation efficiency must be used. Dipole antennas are convenient for this because of their predictable efficiency and low gain. For broadband measurements, antennas with higher bandwidth are preferred. An example of a broadband antenna suitable for use as a reference antenna in reverberation chamber measurements is the discone antenna. The reference antenna is preferably mounted on a low-loss dielectric stand to avoid reduction of the radiation efficiency of the reference antenna. The reference antenna must also be placed in the chamber in such a way that it is far enough from any walls, mode stirrers, chamber loading, or other object so that the environment for the reference antenna, taken over the complete stirring sequence, resembles a free space environment. For low-gain antennas, like dipoles, the free space condition is achieved by keeping the reference antenna a distance of half a wavelength from any reflecting object (such as metallic walls and stirrers) and 70% of a wavelength from any absorbing object (such as a head phantom).

The procedure presented here is based on using a vector network analyzer (VNA). The instrument is configured so that port 1 is connected to the fixed-measurement antenna and port 2 to the reference antenna (see example in Figure 12.6). The procedure can be performed in the following steps:

1. Place all objects that will be used in the test measurement inside the chamber.
2. Place the reference antenna in the chamber, keeping the distance from other objects as described before.
3. Calibrate the network analyzer with a full two-port calibration so that transmission between the antenna ports of the fixed-measurement antenna and the reference antenna can be measured accurately.
4. Connect the antennas and measure S_{11}, S_{22}, and either S_{21} or S_{12} for each of the mode-stirrer positions defined in the stirrer sequence.
5. Calculate the average power transfer function and antenna mismatches as described next.

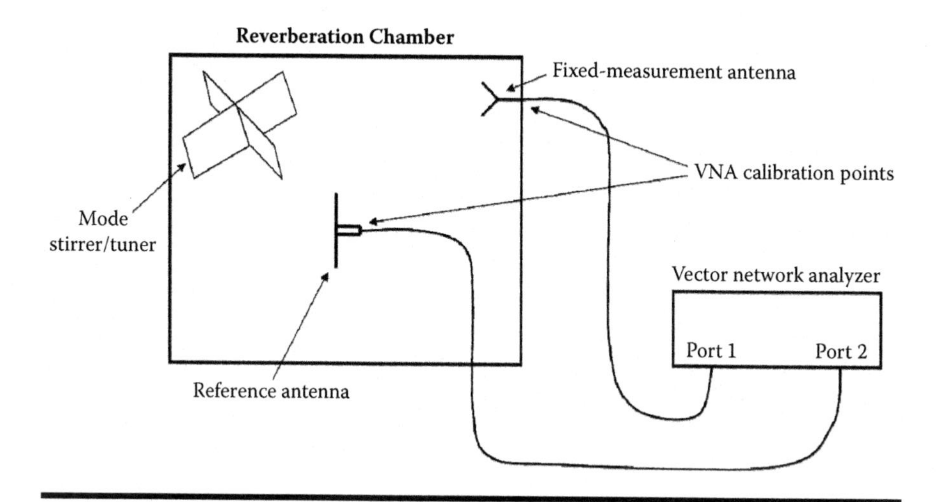

Figure 12.6 Schematic setup for calibration measurements in the reverberation chamber.

Calculate the mismatch correction parameter for the fixed-measurement antenna by taking a complex average of the measured S_{11} samples:

$$\overline{S_{11}} = \frac{1}{N}\sum_{n=1}^{N} S_{11}(n) \tag{12.12}$$

where N is the total number of measured samples, and n is the sample index. Equation (12.12) will give a reflection value equivalent to the free space reflection of the antenna because the complex averaging will cancel chamber reflections, as described in the basic measurements section earlier.

The corresponding power correction factor R_{fix} (i.e., the power lost due to reflections on the antenna port) can then be calculated as

$$R_{fix} = 1 - |\overline{S_{11}}|^2 \tag{12.13}$$

Equivalently, the free space mismatch and power correction coefficient for the reference antenna can be calculated by the following equations:

$$\overline{S_{22}} = \frac{1}{N}\sum_{n=1}^{N} S_{22}(n) \tag{12.14}$$

$$R_{ref} = 1 - |\overline{S_{22}}|^2 \tag{12.15}$$

The reference transfer function G_{ref} represents the average net power transmission in the chamber (i.e., corrected for mismatch on both transmit and receive antenna), as well as radiation efficiency of the reference antenna. Calculate the reference transfer function by the following equation:

$$G_{ref} = \frac{\frac{1}{N}\sum_{n=1}^{N} |S_{21}(n)|^2}{R_{fix} R_{ref}} \cdot \frac{1}{\eta_{ref}} \tag{12.16}$$

where η_{ref} is the radiation efficiency of the reference antenna, and N, n, R_{fix}, and R_{ref} are as defined before. Note that G_{ref} can be conveniently used to determine the quality factor or, equivalently, the average mode bandwidth of the cavity, as shown in Equation (12.5). The application of the calibration on measurement data is shown next for each specific measurement case.

12.3.1 Measurement of Antenna Efficiency

Radiation efficiency is an important parameter for all different types of antennas. For electrically small antennas used in various multipath scattering environments, it is, in fact, the dominating factor for the antenna performance. Classically, *radiation efficiency* is defined as the radiated power from an antenna structure divided by the power delivered to the antenna [19] (i.e., a measure of the loss in the antenna structure itself). In this text, the term *total radiation efficiency* is used to describe the efficiency of an antenna including its mismatch loss. Total radiation efficiency is thus defined as the radiated power from an antenna structure divided by the total available power at the antenna port.

A basic property of the reverberation chamber is that the average received power of an antenna, taken over a complete stirrer sequence, is proportional to the total radiation efficiency of the antenna. Radiation efficiency can therefore be measured by a relative measurement, where the average received power for the test antenna is compared to the average received power for a reference antenna with known radiation efficiency.

The radiation efficiency measurement procedure is similar to the calibration measurement procedure described earlier and can be summarized in the following steps:

1. Perform the calibration measurement procedure as described in the previous section, including the calculations of mismatch and reference transfer function.
2. Replace the reference antenna with the test antenna and repeat the measurement procedure (i.e., sample S_{11}, S_{21}, and S_{22}) for each stirrer position. If the fixed-measurement antenna is the same as in the calibration measurement, there is no need to remeasure S_{11}, and only S_{21} and S_{22} need to be samples.
3. Calculate the antenna mismatch and radiation efficiency for the test antenna as described in the following.

Equivalently to the calculations for the preceding reference transfer function, the test antenna mismatch, mismatch power correction, and average transfer function are calculated as shown in the following equations:

$$\overline{S_{22,test}} = \frac{1}{N}\sum_{n=1}^{N} S_{22,test}(n) \tag{12.17}$$

$$R_{test} = 1 - |\overline{S_{22,test}}|^2 \tag{12.18}$$

$$G_{test} = \frac{\frac{1}{N}\sum_{n=1}^{N} |S_{21}(n)|^2}{R_{fix} R_{test}} \tag{12.19}$$

The radiation efficiency is then directly given by

$$\eta = \frac{G_{test}}{G_{ref}} \tag{12.20}$$

and the total radiation efficiency is given by

$$\eta_{tot} = \eta \cdot (1 - |\overline{S_{22,test}}|^2) = \frac{G_{test}}{G_{ref}} \cdot R_{test} \tag{12.21}$$

12.3.2 Measurement of Total Radiated Power

For active test units, the TRP is determined by the power output from the amplifier and the radiation efficiency of the antenna. The TRP is therefore often used as a performance parameter, especially when a cabled connection to the antenna port is difficult. The following description is based on measurements of user equipment (UE) but can be modified to measurements of any kind of equipment by using the appropriate instrumentation.

The TRP measurement procedure is similar to the radiation efficiency procedure described earlier; the main difference is that the network analyzer is replaced by a base station simulator and power meter. The base station simulator is used to establish and maintain a connection to the UE and control its traffic channel and output power. The power meter is used to sample the transmitted power and could be a spectrum analyzer, base station simulator with integrated power meter, or a regular power meter—whichever is most suitable in the specific measurement case. Figure 12.7 shows the schematic setup for TRP measurements.

Handheld units are often tested in simulated-use position (e.g., with a head or hand phantom in the close vicinity of the test unit). The reverberation

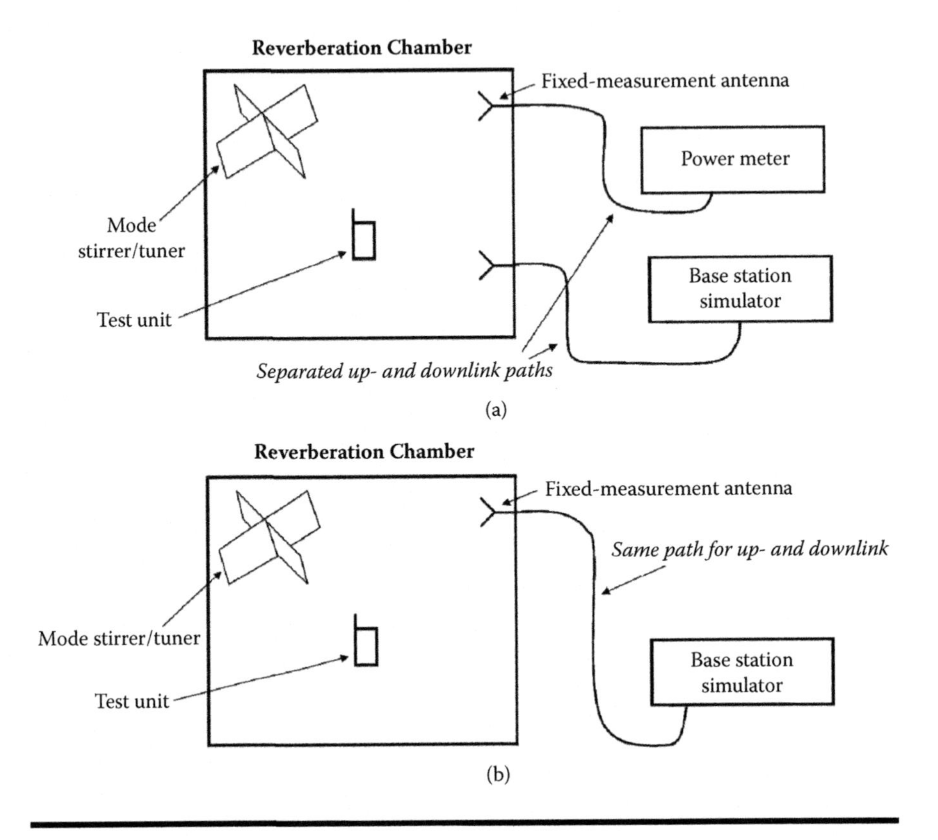

Figure 12.7 Schematic setups for TRP measurements. Alternative b, with the same up- and downlink paths, can also be used for sensitivity measurements.

chamber is very suitable for simulated-use testing due to the simple positioning of test units. In a reverberation chamber, the statistical field uniformity makes it possible to place the test object/antenna in any position and orientation in the chamber, as long as it is far enough from any other object not to give a reduction of the antenna's radiation efficiency. As described in the calibration procedure earlier, a distance of half of a wavelength from walls and other reflecting objects and 70% of a wavelength from power-absorbing objects is enough for low-gain antennas. This is also true for the combination of a test unit and a head phantom, for example; this means that as long as the combination of the two is placed far enough from other objects in the chamber, the effective radiation efficiency that will determine the TRP value is the efficiency of the test unit and the head phantom combined, in the specific relative position they are placed. Note that when head or hand phantoms are used, these objects must also be present in the chamber during the calibration, as described in the calibration measurement section previously.

The TRP measurement is performed in the following steps:

1. Perform the calibration measurement procedure as described in the previous section, including the calculations of mismatch and reference transfer function.
2. Measure the transmission loss of the cable connecting the power meter to the fixed-measurement antenna.
3. Page the test unit and put it to radiate with maximum output power on the traffic channel of interest.
4. Place the unit in the chamber. In the case of simulated-use testing, place the unit in the intended position relative to the tissue-simulating object (e.g., head or hand phantom).
5. Use the power meter to sample the transmitted power in each position of the mode stirrers.
6. Calculate the TRP value by taking an average of all power samples and applying the calibration as described next.

As in the radiation efficiency measurement, the TRP of the test unit is proportional to the average transferred power in the chamber. Through the calibration measurement, the average power transmission in the chamber is known, and therefore the TRP can be calculated with the following equation:

$$TRP = \frac{1}{N}\sum_{n=1}^{N} P(n) \Big/ G_{ref} R_{fix} T_{cable} \tag{12.22}$$

where

$P(n)$ is the power samples in stirrer position n;

N is the total number of stirrer positions;

T_{cable} is the transmission in the cable connecting the power meter to the fixed-measurement antenna; and

G_{ref} and R_{fix} are as defined in the calibration section before.

Note that the summation is done on decimal power values, even though the results are normally presented in decibels of measured power (dBm) units.

12.3.3 Measurement of Total Isotropic Sensitivity

Total isotropic sensitivity is a common way of characterizing an active unit at downlink frequencies. The TIS measurement procedure is standardized by the CTIA organization [14] and an equivalent parameter with measurement procedure;

however, under the name "total radiated sensitivity" (TRS), is standardized by the 3GPP organization. TIS can be measured in a reverberation chamber by the procedure described here. Instrumentation and setup are similar to the TRP measurement case described earlier and shown in Figure 12.7b.

Simplified, the TIS parameter is equal to the conducted sensitivity of the test unit's receiver degraded by the radiation efficiency of the antenna to which it is connected. In reality, the TIS parameter will also be affected by any interference guided to the receiver by the mechanical structure (including the antenna) of the test unit.

TIS is measured under static signal conditions (i.e., mode stirrers are kept still at each sample point) and the idea is to measure sensitivity due to noise impairment only. Due to the multiple reflections in the reverberation chamber, a delay spread of the incoming signals to the receiver will occur; this can be an added effect to the BER. In cases where the impairment due to signal replicas is unwanted, the actual effect of the delay spread in the present case can be tested prior to the TIS measurement. The test consists of putting the test unit in the chamber, setting up a loopback connection with the base station simulator, and then starting a BER measurement while the output powers of the test unit and base station simulator are both high enough to avoid bit errors due to low SNR. If the BER is zero in this setup, the effect of delay spread is negligible.

The TIS procedure is based on searching for the lowest base station simulator output power in each position of the mode stirrers that gives a BER better than the specified target BER level. The procedure can be done in the following steps:

1. Perform the calibration measurement procedure as described in the calibration measurement section, including the calculations of mismatch and reference transfer function.
2. Measure the transmission loss of the cable connecting the base station simulator to the fixed-measurement antenna.
3. Page the test unit, direct it to the traffic channel of interest, and put it in loopback mode to enable BER measurement.
4. Place the test unit in the chamber. In the case of simulated-use testing, place the unit in the intended position relative to the tissue-simulating object (e.g., head or hand phantom).
5. Set the base station simulator to a specific output power and perform a BER measurement.
6. Increase or decrease the base station output power as needed and repeat step 5 until the lowest output power is found that gives a BER better than the specified target BER.
7. Repeat steps 5 and 6 for each position of the mode stirrers.
8. Calculate the TIS value as described next.

The TIS parameter is calculated by the following equation:

$$TIS = \left(\frac{1}{N} \sum_{n=1}^{N} \frac{1}{P_{BSS}(n)} \right)^{-1} \Bigg/ G_{ref} R_{fix} T_{cable} \tag{12.23}$$

where

$P_{BSS}(n)$ is the output power from the base station simulator when this is adjusted to give the specified BER in the test unit for position n of the mode stirrers;
N is the total number of stirrer positions;
T_{cable} is the transmission in the cable connecting the base station simulator to the fixed-measurement antenna; and
G_{ref} and R_{fix} are as defined in the calibration section.

The advantage of the TIS parameter is that it is well defined and commonly used, and the relation to the radiation efficiency of the test unit's antenna is intuitive. However, TIS measurements are inherently time consuming due to the large number of BER measurements that have to be done during the measurement sequence. The fact that TIS is a static measurement and does not include any fading performance can also be considered a disadvantage because the actual TIS performance is not reflected in a real-use performance. This means that units that are optimized for TIS may not be the units that perform best in a real scenario.

12.3.4 Measurement of Average Fading Sensitivity

Average fading sensitivity is defined as the lowest average available power during a fading sequence for which the test units have an average BER better than a specified BER limit. As opposed to TIS measurements, AFS is measured under real-time fading conditions. This means that the mode stirrers are continuously moved during the measurement sequence, while the output power of the base station simulator is fixed. By doing this, the test unit will effectively experience a signal varying with Rayleigh distributed amplitudes.

Due to the multiple reflections in the reverberation chamber and the constantly moving stirrers, it is important to have control of the time domain properties of the channel created in the chamber, which may have an effect on the measured BER. This effect may be wanted or unwanted, depending on the specific test situation. For measurements where only noise impairment is considered, it is important to test that there are no irreducible bit errors present (e.g., by a similar procedure as mentioned in the TIS section). It is worth noting that, for AFS measurements where SNR is the limiting factor, TIS can be derived from the AFS value because there is a theoretical connection between the two values.

The instrumentation and setup for AFS measurements are the same as for the TIS measurements. The procedure can be performed in the following way:

1. Perform the calibration measurement procedure as described in the calibration measurement section, including the calculations of mismatch and reference transfer function.
2. Measure the transmission loss of the cable connecting the base station simulator to the fixed-measurement antenna.
3. Page the test unit, direct it to the traffic channel of interest, and put it in loopback mode to enable BER measurement.
4. Place the test unit in the chamber. In the case of simulated-use testing, place the unit in the intended position relative to the tissue-simulating object (e.g., head or hand phantom).
5. Set the base station simulator to a specific output power. The output power of the base station simulator will determine the average power available for the test unit during the measurement.
6. Put the mode stirrers in a constant movement mode and perform a BER measurement during the sequence. Adjust the number of bits tested so that the BER measurement spans the full mode-stirrer sequence.
7. Increase or decrease the base station output power as needed and repeat steps 5 and 6 until the lowest base station output power is found that gives a BER better than the specified target BER. This specific output power is denoted $P_{BS,lim}$.

The AFS value is then found by applying the calibration correction data as in the following equation:

$$AFS = \frac{P_{BS,lim}}{G_{ref} R_{fix} T_{cable}} \tag{12.24}$$

Alternatively, step 7 in the previous procedure can be replaced by performing steps 5 and 6 for a number of predefined output powers for the base station simulator. This will give a relation between available power and BER for the test unit; by interpolating the data points, the lowest average power needed for a BER better than the specified target limit can be determined.

Among the advantages of AFS measurements are the potential for very fast sensitivity measurements and that the fading property of the measurement gives the possibility to test in a wide range of real-time conditions. The possibility to test in actual fading conditions is very important for modern, multiantenna systems where, for example, the effect of diversity will only show up in such a test.

12.3.5 Measuring Antenna Diversity Gain

Diversity is a technique based on the use of more than one antenna that experiences different fading. By choosing or combining the signals of two different antennas, it is possibly to gain more than 10 dB of diversity gain in the worst fading dips appearing typically 1% of the time [49]. With additional antennas, even higher diversity gains are possible [50]. It is possible to measure diversity gain with drive tests (i.e., driving or walking with multiple antenna configurations through a fading environment). The problem when one wants to optimize the antenna configuration is that the fading in real environments is always changing and it is not possible to know if the results depend on changes in the environment or changes in the antenna configuration. It is also possible to measure diversity gain by measuring each antenna in the antenna configuration separately in an anechoic chamber. After the measurements, by using software it is possible to add any type of fading and then estimate the diversity gain [51]. This takes a relatively long time, on the order of 1–2 hours in total. A very efficient alternative is to use the repeatable Rayleigh distribution available in the reverberation chamber. The antenna configuration is positioned in the reverberation chamber, as shown in Figure 12.8.

If possible, a multiport network analyzer is used to measure amplitude and phase of the different antennas in the antenna configuration and the three fixed antennas in the reverberation chamber S_{1j}. For a diversity antenna with two branches, S_{12} and S_{13} are measured simultaneously. Both these antenna branches will show a certain probability to have fading below a certain level, usually referred to as the cumulative distribution function (CDF). By choosing the best of the measured S_{12} and S_{13} at every point in time, one gets a CDF that is called selection combining. It is, of course, possible to use the S_{12} and S_{13} to get the CDF of any diversity scheme (e.g.,

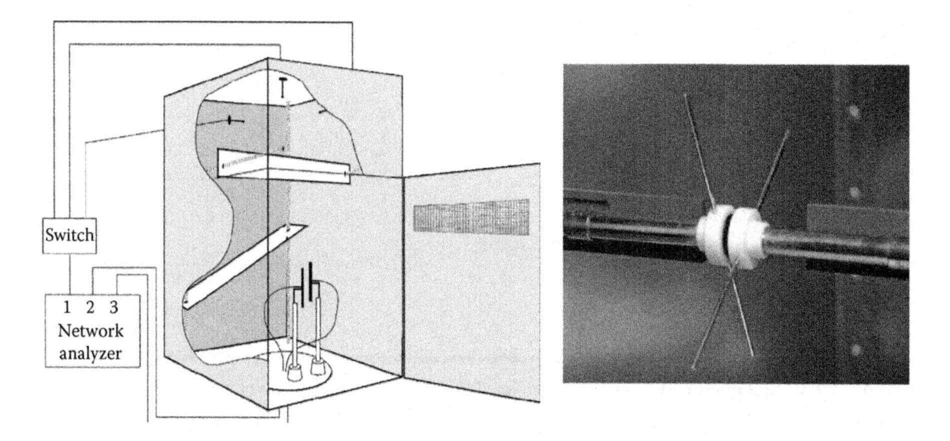

Figure 12.8 Left: drawing of the standard-sized Bluetest reverberation chamber, shown with open door. The drawing shows a setup for measuring diversity gain with a multiport network analyzer. Right: photo of 900-MHz dipoles used in the tests.

maximum ratio combining [MRC]) because both amplitude and phase have been measured. However, in this section we use selection combining.

By taking the difference between the CDF for either S_{12} or S_{13} and the CDF for selection combining, it is possible to estimate the "apparent" diversity gain (i.e., how much it is possible to gain in the deepest fading dips), normally at a probability level of 1%, by choosing the best antenna. The most relevant value, though, should be how much one gains in comparison to using an ideal antenna (i.e., to compare the CDF for a single antenna with 100% antenna efficiency with the CDF for selection combining). This is the "effective" diversity gain [30]. If one compares with the CDF for a real antenna (i.e., with losses), this is the "actual" diversity gain.

For antenna configurations with large mutual coupling, such as two dipoles moved very close to each other, antenna efficiency will become very low, so what may look like a very good diversity gain (i.e., "apparent" diversity gain) is in reality compared to using just a single antenna that is not at all very good. In Figure 12.9 below one can see that at 11-mm distance between two 900-MHz dipoles, the "effective" diversity gain is only 1.5 dB at the 1% probability level. In this example, more than 90% of the time the diversity configuration will lose signal strength compared to just using a single antenna. In the case of using dipoles, the "effective" and "actual" diversity gains are very similar because a dipole typically has an antenna efficiency of about 95%.

The measurements in Figure 12.9 take only 1 minute to perform in the Bluetest high-performance chamber using a multiport network analyzer. In addition to the

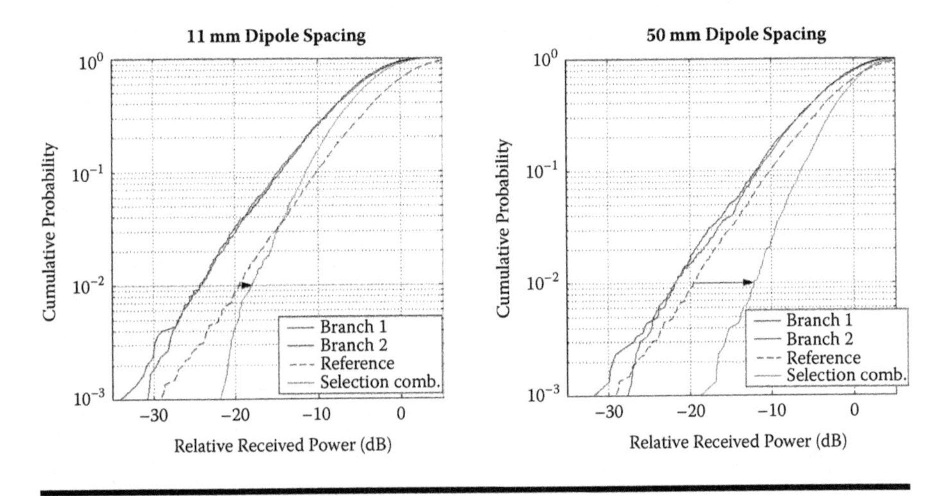

Figure 12.9 "Apparent" and "effective" diversity gain for two parallel dipoles at distances of 11 and 50 mm at 900 MHz. The dashed line corresponds to the fading of a single dipole normalized to 100% efficiency. The distance between the dashed line and the line for selection combining are the "effective" diversity gain, here shown with an arrow. The distance between one of the antenna branches and the line for selection combining is the "apparent" diversity gain.

Table 12.1 Apparent Diversity Gain, Radiation Efficiency, and Effective Diversity Gain at 1% CDF as a Function of Relative Polarization Angle and Distance between Two Dipoles

Relative Angle (deg.)	*Distance (mm)*	*Effective Diversity Gain (dB)*	*Apparent Diversity Gain (dB)*	*Radiation Efficiency (dB)*	*Correlation*
0	15	3	8.0	−4.2	0.63
30	15	6.5	10.8	−2.5	0.15
45	15	7.8	9.4	−1.3	0.09
60	15	9	9.8	−0.6	0.12
90	15	9.6	9.8	−0.3	0.1
0	30	5.8	10.1	−3.2	0.21
30	30	8	9.9	−1.4	0.02
45	30	8.1	8.8	−0.9	0.03
60	30	9.2	9.7	−0.4	0.07
90	30	9.9	10.9	−0.3	0.08

Note: At a relative angle of 0°, the dipoles are parallel, and at a relative angle of 90°, they are perpendicular.

"apparent" and "effective" diversity gain, one obtains the radiation efficiency of each antenna as well as the correlation of each. Table 12.1 shows these parameters for 10 different antenna configurations (i.e., with two dipoles at five different relative angles to each other and at a distance of 15 and 30 mm from each other, respectively).

In Table 12.1, we can clearly see that although the apparent diversity gain is relatively high for all configurations, the reduced antenna efficiency makes the effective diversity gain much smaller, especially at close distance or small relative angles. It is also interesting to note that although the correlation between the diversity antennas is high only when they are parallel, the effective diversity gain is still affected by mutual coupling at relative angles of 45° and less.

The traditional ways of measuring diversity gain using drive tests or anechoic chambers with software channel modeling are expensive and time consuming. The time spent to measure or estimate the diversity gain can easily be hours. The diversity measurements presented in this section have been made in 1 minute—one to two orders of magnitude faster. This kind of measurement speed will allow antenna designers of diversity antennas to work in completely new ways. They will be able to make small changes in the antenna positions or design and more or less immediately get feedback of whether or not there has been improvement. In a few years, it will be hard to imagine designing diversity antennas any other way. Companies with reverberation chambers almost always use this method only when designing or evaluating diversity gain.

12.3.6 Measuring MIMO Capacity

A necessary condition for MIMO antennas to provide higher throughput is that the antennas be relatively uncorrelated—the more so, the better. The communication must also take place in a fading environment, so that the different propagation paths are relatively uncorrelated. The more fading that occurs, the better. Furthermore, the most important parameter for good MIMO performance is that the antennas used have high antenna efficiency [29]. We saw in the preceding diversity case that the antenna efficiency is decreased when antennas are in close proximity. In the reverberation chamber, it is possible to measure what the propagation paths in a Rayleigh fading environment look like between the MIMO antennas and the three fixed wall antennas (i.e., the channel matrix **H**). If the matrix is known, it is possible to calculate the throughput capacity with Shannon's capacity formula:

$$C = \log 2(\det(\mathbf{I}M + (\mathrm{SNR}/3)\mathbf{H}^*)) \tag{12.25}$$

With a multiport network analyzer, such a measurement can be performed in only 1 minute in the Bluetest high-performance chamber. It is the same measurement as described in the section on diversity gain, but the postprocessing of the *S* parameters is different. If one does not have a multiport network analyzer or if the MIMO antenna consists of more antenna branches than one has ports, fast MIMO measurements can still be taken by measuring the MIMO channel columns of the **H** matrix one by one. Figure 12.10 shows an example of a MIMO antenna with six branches.

In this case, one starts by connecting one of the antenna ports of the test antenna to the network analyzer and the other ports and the reference antenna are

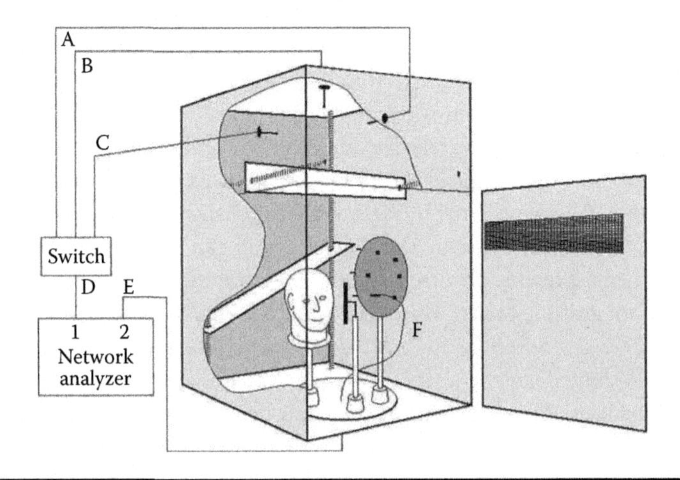

Figure 12.10 Setup for measuring a six-element MIMO antenna.

terminated in 50 Ω. The S parameters between the connected port in the MIMO configuration and each of the fixed wall-mounted antennas, used for polarization stirring, are measured for all positions of the platform and mechanical stirrers and all frequency points. The measurement procedure is then repeated for every antenna port, using exactly the same platform and stirrer positions. Thus, the field inside the chamber is exactly the same when measuring every port. The complex transmission coefficients S_{21} between the connected port and each of the fixed wall antennas, as well as the reflection coefficients S_{11} of each wall antenna and S_{22} of the test object ports, are stored for every stirrer position and frequency point.

Finally, we connect the reference antenna to the network analyzer and perform the same measurements. In a small chamber, it is advantageous to use frequency stirring (averaging) to improve accuracy. In such cases, we correct the complex samples of S_{21} with mismatch factors due to both S_{11} and S_{22} before frequency stirring, as explained in Kildal and Carlsson [52, Equation 1]. We also normalize the corrected S_{21} samples to a reference level corresponding to 100% radiation efficiency. This is obtained from the corrected S_{21} samples measured for the reference antenna and its known radiation efficiency. We refer to these corrected and normalized samples of S_{21} as normalized S_{21} values. The normalized S_{21} values can now be used to calculate the MIMO capacity as described previously.

12.3.7 Measuring System Throughput

Multiantenna solutions at the terminal or base stations will be key technologies in 3GPP LTE to ensure high throughput and low delays to users also at the cell edge. There are no accepted or well-known standards today to measure system throughput of multiantenna solutions. One of the reasons for this is that established standards use measurements in anechoic chambers, where the effect of multiantenna solutions is not directly measurable. It is clearly seen from drive tests that multiantenna solutions do provide significant improvements of throughput in an environment with rich scattering (i.e., indoors or in urban areas) compared to SISO (single input, single output) solutions.

The problem with drive tests, however, is that most of the time they give different results because the environment is under constant change. It is therefore normally not possible to compare the performance of two different multiantenna devices because they will show different throughput results in different environments and at different times. Thus, there is a need for a repeatable testing methodology to check the throughput of multiantenna devices in an environment with rich scattering. This environment should preferably have rich scattering that is repeatable as well as isolated from outside interference that may affect the measurements. The reverberation chamber fulfills these requirements. It is a shielded room (often with better than 100-dB isolation), with a repeatable fading environment following a Rayleigh distribution (see Figure 12.11), similar to the distribution in real indoor or urban environments. Objects in the reverberation chamber located more

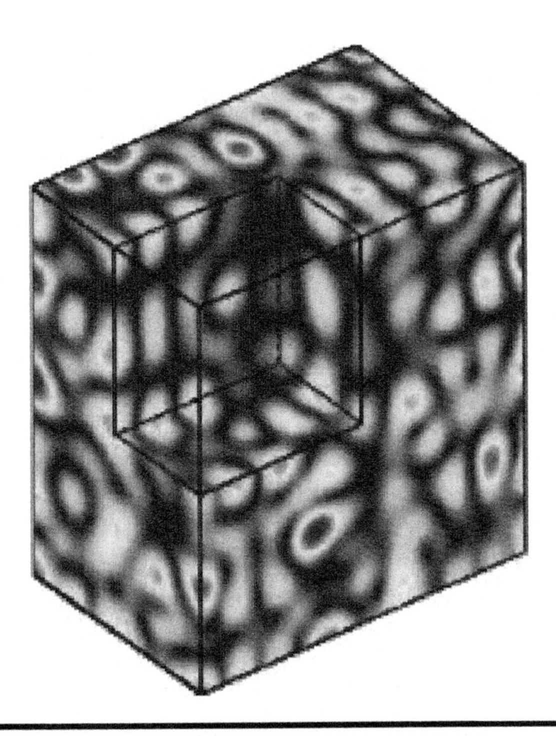

Figure 12.11 Simulation of the field in a reverberation chamber. It is Rayleigh distributed. This is similar to the distribution found in office environments and urban areas.

than 0.5 wavelength from each other will experience independent fading. By placing a multiantenna base station and multiantenna terminal inside the chamber a few wavelengths apart, it is possible to measure the throughput in the uplink and downlink in a repeatable way.

To test that this works in praxis, Bluetest, together with Sony Ericsson [53], performed a project [54], briefly described later, within the Swedish Chase (Chalmers Antenna Systems VINN Excellence Center) program [55]. A MIMO WLAN router and laptop were used to test throughput. WLAN devices were used because it was difficult to find prototype 3GPP LTE base stations and terminals in 2008. WLAN has some of the same properties (e.g., orthogonal frequency division multiplexing [OFDM], 20-MHz bandwidth, and support for multiantenna solutions [802.11n]) as 3GPP LTE, and the results should therefore be relevant for designers of 3GPP LTE equipment. The setup for the throughput measurements is shown in Figure 12.12. The WLAN router under test was a D-Link RangeBooster N 650 wireless router (model DIR-635) with operation restricted to the 2.4-GHz ISM band. The laptop used was equipped with a D-Link RangeBooster N USB adapter (model DWA-142), which is a two-antenna 802.11n draft-compliant WLAN client. The main software used for the throughput measurements was Iperf—open source

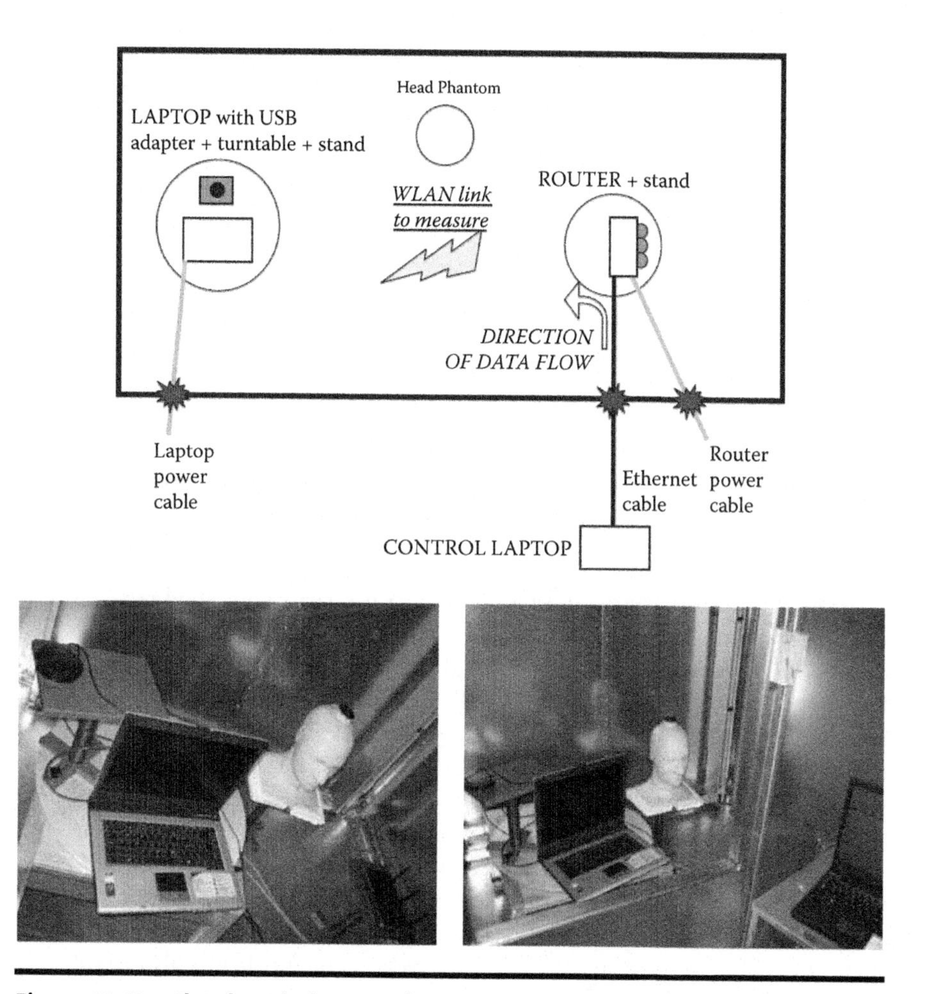

Figure 12.12 Sketch and pictures of measurement setup.

software developed by the National Laboratory for Applied Network Research of the United States [56] that can create TCP and UDP data streams and measure network throughput—allowing modification of various parameters for network testing.

The first measurements were done to verify the repeatability of the reverberation chamber. The measurements were performed over a 90-s continuous stirring sequence with sampling frequency of two samples per second. This was found to be long enough for a 95% confidence interval for a sample average smaller than 2 Mb/s. Figure 12.13 shows nine consecutive measurements of three channels with good repeatability. This is the first time, as far as we know, that repeatable throughput with this small spread in the results was performed in a rich scattering environment.

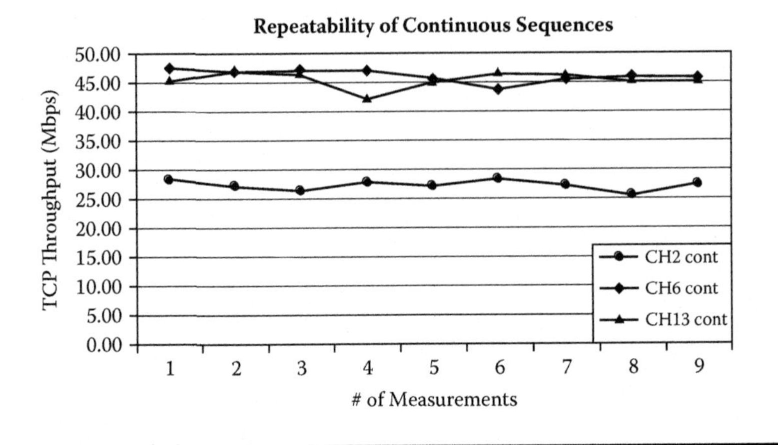

Figure 12.13 Nine consecutive measurements showing good repeatability.

Having shown that it is possible to do measurements with good repeatability, the next step was to compare a MIMO 802.11n configuration with a standard SISO 802.11g configuration. The throughput in the 802.11n MIMO 3 × 2 configurations was compared with the 802.11g mode of the router. For both operation modes, the modulation format was set to the most complex implemented—for 802.11n MCS 15 with 20-MHz bandwidth, 130 Mb/s rate medium access control (MAC) data, and for 802.11g, 54 Mb/s physical link rate. Three measurements were taken and averaged for each of the frequency channels.

The results are shown in Figure 12.14 and show how 802.11n with MIMO dramatically improves the throughput compared to 802.11g, which uses only SISO. This means an absolute increase of at least 25 Mb/s and represents on average nearly three times the throughput offered in the older standard. It was also possible to

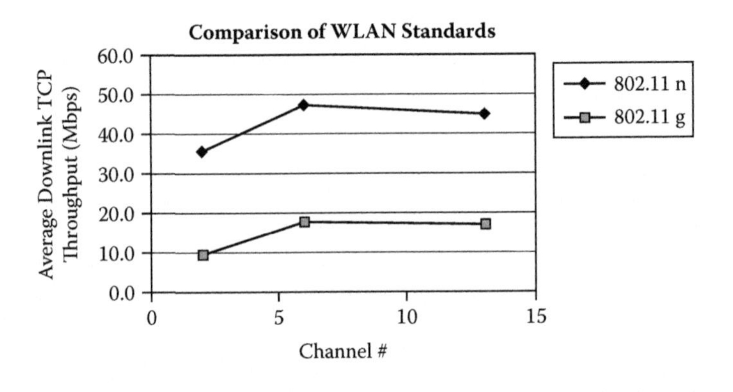

Figure 12.14 Comparison of WLAN standards.

see effects of removing or attenuating one or two antennas at the router. For more information about this, as well as how different types of antennas affected the throughput, see Olano et al. [57].

A new setup for active measurements of throughput of MIMO systems in the reverberation chamber has been described. The chosen communication standard was WLAN 802.11n, which was the first communication standard to commercialize devices implementing MIMO. WLAN has traditionally been a computer networking standard and, compared to cell phone standards, presents some special features regarding its operation and characterization. However, we believe that the results shown for WLAN are very encouraging and that the method used will be of great benefit to designers of 3GPP terminals as well as small indoor 3GPP LTE base stations. The measurements are fast and easy to set up, and can save much time and reduce uncertainties in comparison with current field tests. They offer more reliable assessment on the performance than measurements in other shielded chambers without stirring mechanisms and measurements in channel emulators.

12.4 New Tests for LTE-and-Beyond Systems in the Reverberation Chamber

In contrast to existing second- and third-generation systems, LTE-and-beyond systems are designed to exploit the spatial domain of the channel to a higher degree than what is done today. The spatial domain of the channel is exploited by using multiple antennas at the receiver and the transmitter. In LTE, multiple antennas are used for diversity, beam forming, spatial multiplexing, and interference suppression. The network decides the way in which the multiple antennas are used [58].

For equalization of the frequency-selective channel in the downlink, LTE-and-beyond systems are mostly based on OFDM-like transceiver structures. In addition to combating—and in effect exploiting—the frequency-selective nature of the channel, OFDM-like schemes allow for a flexible spectrum allocation by the base station to different mobile users.

The channel relevant for evaluating such a system, which exploits the frequency as well as the spatial domain of the channel, is the channel transfer function (CTF) $h(t, f, \theta)$. It is a function of time t, the frequency f, and the incident angle of the signal θ. It can be written as [59]

$$h(t,f,\theta)=C_R(f,\theta)h(t,f,\theta)C_T(f,\theta), \tag{12.26}$$

where C_R and C_T are the complex valued antenna responses of the receive and transmit antennas, respectively. It can be seen that not only the physical channel h itself but also the antenna response influence the link and therefore system-level performance. Taking into account the antenna at the mobile terminal becomes particularly important if the antenna operates in close vicinity to other objects, such as the hand of a mobile terminal user, which affects not only the antenna response but also the matching of the antenna. For beyond-LTE systems, including the antenna in transceiver performance evaluation may become even more important as the spectrum used by systems increases and potentially is scattered over multiple bands as identified for IMT-Advanced systems on the WRC-2007. In such systems, antennas become more suboptimal because only one antenna might be used for multiple frequency bands.

Having identified the need for system/link measurements including the antennas, different setups for such measurements are presented next. All these setups are based on measuring the throughput of wireless links as described earlier. The result is that if parameters, components or algorithms are optimized, the effect on throughput is directly visible. Alternatively, these throughput measurements might be replaced with BER measurements for a fixed throughput. The choice of measuring BER or throughput will ultimately depend on the problem investigated.

Note that the different possible setups described in this section have not been used for performing measurements yet. However, the throughput measurements described in the previous section open up a completely new set of measurement setups, which are particularly interesting in the context of LTE- and- beyond systems. A selection of such possible setups is described in this section.

The remainder of this section is organized as follows. In the first subsection, link-level measurements (i.e., measurements focusing on the communication between a single terminal and a base station) are discussed. These measurements are developed further in the subsequent section, where setups involving multiple mobile terminals are considered. The multiple terminal setup mimics a communication cell in a cellular system. The final step is then to include multiple base stations in the measurements, which is termed system measurements.

12.4.1 Link-Level Measurements

Although many traditional measurement environments, such as an anechoic chamber, are not capable of accounting for time, frequency, and space variations of the channel, the reverberation chamber is capable of providing such a test environment [8]. It is therefore suitable for measuring transceivers using multiple antenna techniques in combination with OFDM in a repeatable environment.

Measurement setup. If a base station equipped with small antennas operating in a rich scattering environment is considered, the measurement setup for a MIMO-OFDM link would look like that described in the earlier section on throughput measurements. However, in order to test systems at their limits (i.e., very low SNR), it might be necessary to introduce additional path loss between the base station and the terminal. This can be done by directly connecting attenuators at the base station and terminal antenna ports or, alternatively, by a setup using two connected

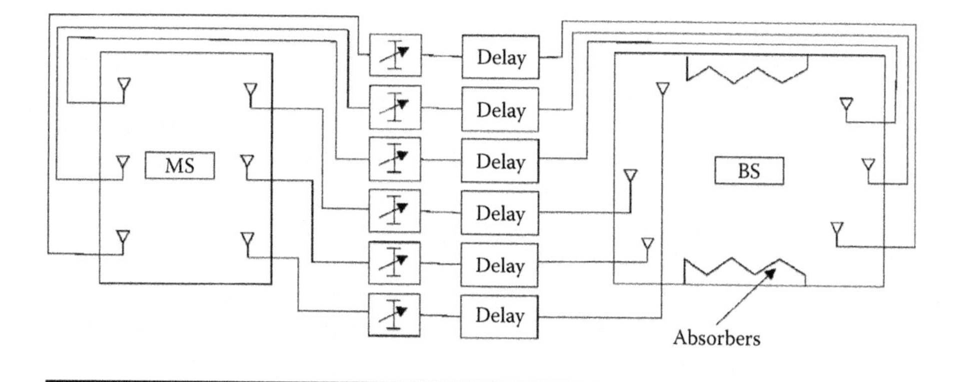

Figure 12.15 Two-chamber setup, where a mobile station (MS) is placed in the left chamber and the base station (BS) is placed in the right chamber. Both chambers are connected via transmission lines with different delays and attenuators to manipulate the frequency-selective behavior of the channel. The right chamber additionally uses absorbers to manipulate the angular spread of the channel. Note that the number of pickup antennas connecting the two chambers needs to be sufficiently high in order not to limit the rank of the MIMO channel.

reverberation chambers, as illustrated in Figure 12.15. In this setup, the mobile terminal is placed in the left chamber, which is equipped with a turntable and stirrers on the wall as described previously. The chamber on the right-hand side hosts the base station (antennas).

Both chambers are connected to each other via pickup antennas in each chamber, which are connected through a transmission line [5]. Rather than using only

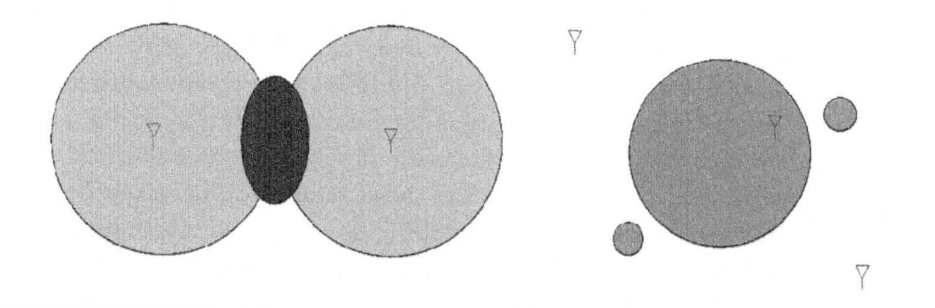

Figure 12.16 Two base station setups with handover region. Note that cells are based on receiver power rather than on geometry. In the handover area, the average power received by a mobile from two different base stations is approximately the same. This corresponds to the uniform average power distribution over the area in the reverberation chamber. Note that the handover area or cell edge in a communication system is not defined by geometry but rather by signal power levels.

transmission lines, one might in addition use attenuators and, potentially, elements introducing a propagation delay. If the delay elements impose a different propagation delay in different transmission lines, the channel becomes frequency selective, similar to the classical power delay profile known from channel modeling. By controlling the delays and attenuators in the cables, the frequency selectiveness of the channel can be easily controlled without modifying the chamber.

Depending on the loading of a chamber and whether it is equipped with some absorbing material, the different pickup antennas might even be used to control the angular spread of the signal. Note that the number of these pickup antennas used needs to be sufficiently large in order not to limit the rank of the channel between the two chambers.

In the described two-chamber setup, a complete transceiver, from digital algorithm to antennas, can be optimized and the performance of different setups can be evaluated using, for example, link throughput or BER as an optimization criteria. The measurements will be independent of the transceiver implementation. This allows us to characterize systems with, for example, integrated antenna and LNA designs, which are difficult to measure in current setups.

Of particular interest in link-level measurements in the LTE context are measurements with channel state information at the transmitter that can be used for link adaptation. In this case, the performance of power allocation on the OFDM or MIMO subchannels created through precoding, adaptive coding, and modulation can be evaluated. In contrast to other measurement environments, such as those created in drive tests, the performance of MIMO systems using spatial multiplexing can be investigated in a repeatable environment.

12.4.2 Single-Cell System-Level Measurements

Although the link-level tests discussed in the previous section allow for designing antennas and for optimizing transceiver designs, these measurements say little about the performance of the overall system. The link-level measurements, however, can be taken a step further towards system-level measurements by using multiple terminals in the chamber in conjunction with one base station connected to the chamber or placed inside the chamber. With such a setup, multiple access technologies such as OFDMA or MU-MIMO, including the associated scheduling, can be evaluated and the performance of different scheduling algorithms can be measured. Again, OFDMA and spatial division medium access (SDMA)/MU-MIMO rely on a space- and frequency-selective channel, respectively, which can be provided by the reverberation chamber.

Because scheduling algorithms generally also need a channel, which evolves over time, it can be desirable to move the stirrers in the chamber as well as the platform during the measurements. This channel evolution over time, however, needs to be sufficiently slow to allow SDMA/MU-MIMO to function. In a chamber of reasonable size, it might be possible to place numerous mobile terminals because they already experience independent fading, if the distance between them is larger than lambda-half.

Measurement setup. For the single-cell measurements, a base station is connected to the chamber (or, if it is small, potentially placed inside the chamber). In addition, mobile terminals are added, with spacing between them of at least half a wavelength to ensure independent fading. In the next step, communication is established between the mobile terminals and the base station and throughput measurements of all terminals in parallel are started. Then the platform and stirrers are moved slowly.

Depending on the objective of the measurements, a different number of terminals may be required. For the investigation of SDMA or MU-MIMO, it can be sufficient to use only two terminals and force them to operate on the same frequency–time resource block so that they need to be separated in space using MIMO techniques. If, however, scheduling algorithms are to be evaluated, the number of mobile terminals needs to be large in comparison to the number of available frequency–time slots because the quality of a scheduling algorithm shows best at a high system load. The results of these measurements are the individual throughputs of each terminal. They can be used not only to analyze sum-throughput in this artificial cell but also to investigate, for example, the fairness characteristics of the scheduling algorithm.

12.4.3 Multicell System-Level Measurements

Rather than connecting only a single base station to a reverberation chamber, it is also possible to connect more than one base station. This setup can be used to investigate the performance of the system when the terminal is moving in the handover area, as indicated in Figure 12.16.

If two base stations with identical output powers are connected to a reverberation chamber, the average received power from the two base stations at the terminal is also identical. The environment in the chamber therefore corresponds to exactly the handover region between two base stations. To change this, attenuators can be used to adjust the powers of the base stations connected to the chamber and, by that, "move" closer to one of the base stations. In the extreme, these attenuators might even be made time variant. Alternatively to using attenuators, it would also be possible to use a cylinder filled with lossy material and move it over the terminal in order to create fading on a slower timescale, which should trigger handover between two base stations.

With such a setup, it is possible to simulate handover scenarios from one base to another. Such a handover decision should not be made on the basis of short-term fading but, rather, of average power levels. The short-term fading in this case would be created by the chamber. In addition to the base station/system algorithms, this setup will allow for evaluating the performance of the interference rejection at the terminal. If connected to one base station, the other base station creates interference

Table 12.2 Different Potential Measurement Setups

	Link	*Single Cell*	*Multicell*
Multiantenna	Diversity, multiplexing (precoding), beam forming	MU-MIMO, SDMA, channel-dependent scheduling	Cooperative base stations, interference suppression at terminal, handover
OFDM(A)	Power allocation	Channel-dependent scheduling	

for the terminal, which it can combat using interference rejection based on multiple antenna schemes.

As an alternative to the classical handover scenario, the two base stations could also be used as cooperating base stations serving the terminals in the handover region jointly, as discussed in the context of IMT-Advanced.

In Table 12.2, different potential measurement setups discussed in the previous section, including the parameters/properties of the system that can be evaluated, are summarized.

12.5 Summary

In order to evaluate the performance of multiantenna wireless communication systems, classical measurement techniques known from antenna measurements, such as anechoic chambers, are no longer sufficient. The measurement setups for (multi) antenna measurements presented in the first part of this chapter show how reverberation chambers can be used as an alternative to current techniques. The accuracy of the reverberation chamber measurements has been analyzed and compared to classical antenna measurements where this is possible.

The methods developed for single-antenna characterization can be taken a step further to evaluate the performance of different multiantenna setups and algorithms, such as diversity combining. These diversity measurements require a fading environment and are therefore perfectly suited for reverberation chambers. If base station simulators are used for measurements, it is even possible to investigate the performance of different antenna setups, in terms not only of power gains but also of BER. Current state-of-the-art measurements in reverberation chambers even include throughput measurements of wireless links.

Although link measurements of wireless systems in reverberation chambers are nowadays well established and have even made their way into standardization, many research activities are ongoing that relate, for example, to improving the accuracy of the chamber. With the evolution of wireless standards, the attractiveness of reverberation chambers for link/system evaluation increases. Possible extensions of existing measurement techniques with a focus on LTE-and-beyond systems were presented at the end of this chapter. A lot of potential for further development lies particularly in manipulating the channel in the reverberation chamber to a larger extent and using the chamber for multiple terminal and multiple base station setups.

References

1. Vaughan, R., J. B. Andersen, and P. C. Clarricoats. 2003. Channels, propagation and antennas for mobile communications electromagnetic waves. Series 50, Institute of Electrical Engineers.
2. http://www.3gpp.org/Highlights/LTE/lte.htm
3. Andersson, M., C. Orlenius, and M. Franzén. 2007. Measuring the impact of multiple terminal antennas on the bit rate of mobile broadband systems using reverberation chambers. *International Workshop on Antenna Technology (iWAT07),* Cambridge, UK, pp. 368–371.
4. Andersson, M. et al. 2006. Antennas with fast beam steering for high spectral efficiency in broadband cellular systems. *Proceedings of the 9th European Conference on Wireless Technology,* Manchester, England, September 2006.
5. El Zooghby, A. 2005. *Smart antenna engineering.* Mobile communications series. Norwood, MA: Artech House Inc.
6. Wolfgang, J., C. Carlsson, C. Orlenius, and P.-S. Kildal. 2003. Improved procedure for measuring efficiency of small antennas in reverberation chambers. *IEEE AP-S International Symposium,* Columbus, OH, June 2003.
7. Bäckström, M., O. Lundén, and P.-S. Kildal, Reverberation chambers for EMC susceptibility and emission analyses. *Review of Radio Science* 1999–2002, (19): 429–452.
8. Kildal, P.-S. 2007. Overview of 6 years' R&D on characterizing wireless devices in Rayleigh fading using reverberation chambers. *International Workshop on Antenna Technology (iWAT07),* Cambridge, UK, 2007, pp. 162–165.
9. http://www.bluetest.se/
10. http://www.tcodevelopment.com/
11. http://www.mobilelabelling.com/
12. Annex E. 2006. Alternative measurement technologies: Reverberation chamber method, 3rd Generation Partnership Project: Technical specification group radio access network; measurements of radio performances for UMTS terminals in speech mode, 3GPP TR 25.914 V7.0.0 (2006–06).
13. http://www.ctia.org/
14. Test plan for mobile station over the air performance CTIA certification, rev. 2.2. November 2006.
15. Rosengren, K., and P.-S. Kildal. 2001. Study of distributions of modes and plane waves in reverberation chambers for characterization of antennas in multipath environment. *Microwave and Optical Technology Letters* 30 (20): 386–391.
16. Rosengren, K., P.-S. Kildal, C. Carlsson, and J. Carlsson. 2001. Characterization of antennas for mobile and wireless terminals in reverberation chambers: Improved accuracy by platform stirring. *Microwave and Optical Technology Letters* 30 (20): 391–397.

17. Kildal, P.-S., and C. Carlsson. 2002. Detection of a polarization imbalance in reverberation chambers and how to remove it by polarization stirring when measuring antenna efficiencies. *Microwave and Optical Technology Letters* 34 (2): 145–149.
18. Kildal, P.-S., C. Carlsson, and J. Yang. 2002. Measurement of free space impedances of small antennas in reverberation chambers. *Microwave and Optical Technology Letters* 32 (2): 112–115.
19. Kildal, P.-S. 2000. *Foundations of antennas—A unified approach.* Sweden: Studentlitteratur.
20. Hill, D. A., M. T. Ma, A. R. Ondrejka, B. F. Riddle, M. L. Crawford, and R. T. Johnk. 1994. Aperture excitation of electrically large, lossy cavities. *IEEE Transactions on Electromagnetic Compatibility* 36 (3): 169–178.
21. Ludwig, A. C. 1976. Mutual coupling, gain and directivity of an array of two identical antennas. *IEEE Transactions on Antennas Propagation* November: 837–841.
22. Marzari, E. 2004. Physical and statistical models for estimating the number of independent samples in the Chalmers reverberation chamber. Master's thesis, Chalmers University, Goteberg, Sweden.
23. Orlenius, C., M. Franzen, P.-S. Kildal, and U. Carlberg. 2006. Investigation of heavily loaded reverberation chamber for testing of wideband wireless units. *IEEE AP-S International Symposium,* Albuquerque, NM, July 2006.
24. Kostas, J. G., and B. Boverie. 1991. Statistical model for a mode-stirred chamber. *IEEE Transactions on Electromagnetic Compatibility* 33 (4): 366–370.
25. Hill, D. A. 1999. Linear dipole response in a reverberation chamber. *IEEE Transactions on Electromagnetic Compatibility* 41 (4): 365–368.
26. Hill, D. A. 1994. Electronic mode stirring for reverberation chambers. *IEEE Transactions on Electromagnetic Compatibility* 36 (4): 294–299.
27. Kildal, P.-S., and C. Carlsson (C. Orlenius). 2002. TCP of 20 mobile phones measured in reverberation chamber: Procedure, results, uncertainty and validation. Available from Bluetest AB, www.bluetest.se, Feb. 2002.
28. Serafimov, N., P.-S. Kildal, and T. Bolin. 2002. Comparison between radiation efficiencies of phone antennas and radiated power of mobile phones measured in anechoic chambers and reverberation chamber. *IEEE AP-S International Symposium,* San Antonio, TX, June.
29. Kildal, P.-S., and K. Rosengren. 2004. Correlation and capacity of MIMO systems and mutual coupling, radiation efficiency and diversity gain of their antennas: Simulations and measurements in reverberation chamber. *IEEE Communications Magazine* 42 (12): 102–112.
30. Kildal, P.-S., K. Rosengren, J. Byun, and J. Lee. 2002. Definition of effective diversity gain and how to measure it in a reverberation chamber. *Microwave and Optical Technology Letters* 34 (1): 56–59.
31. Kildal, P.-S., and K. Rosengren. 2003. Electromagnetic analysis of effective and apparent diversity gain of two parallel dipoles. *IEEE Antennas and Wireless Propagation Letters* 2 (1): 9–13.
32. Bourhis, R., C. Orlenius, G. Nilsson, S. Jinstrand, and P.-S. Kildal. 2004. Measurements of realized diversity gain of active DECT phones and base stations in a reverberation chamber. *IEEE AP-S International Symposium,* Monterey, CA, June 2004.

33. Rosengren, K., and P.-S. Kildal. 2005. Radiation efficiency, correlation, diversity gain, and capacity of a six monopole antenna array for a MIMO system: Theory, simulation and measurement in reverberation chamber. *Proceedings IEEE, Microwave Antennas Propagation* 152 (1): 7–16. See also Erratum published in August 2006.
34. Orlenius, C., P.-S. Kildal, and G. Poilasne. 2005. Measurements of total isotropic sensitivity and average fading sensitivity of CDMA phones in reverberation chamber. IEEE AP-S International Symposium, Washington, D.C., July 3–8.
35. Carlberg, U., P.-S. Kildal, and J. Carlsson. 2005. Study of antennas in reverberation chamber using method of moments with cavity Green's function calculated by Ewald summation. *IEEE Transactions on Electromagnetic Compatibility* 47 (4): 805–814.
36. Karlsson, K., J. Carlsson, and P.-S. Kildal. 2006. Reverberation chamber for antenna measurements: Modeling using method of moments, spectral domain techniques, and asymptote extraction. *IEEE Transactions on Antennas and Propagation* 54 (11), Part 1: 3106–3113.
37. Carlberg, U., P.-S. Kildal, A. Wolfgang, O. Sotoudeh, and C. Orlenius. 2004. Calculated and measured absorption cross-sections of lossy objects in reverberation chamber. *IEEE Transactions on Electromagnetic Compatibility* 46 (2): 146–154.
38. Wellander, N., O. Lundén, and M. Bäckström. 2006. Design parameters for efficient stirring of reverberation chambers. *Proceedings of IEEE EMC Society International Symposium* 2: 263–268.
39. Wellander, N., O. Lundén, and M. Bäckström. 2007. Experimental investigation and mathematical modeling of design parameters for efficient stirrers in mode-stirred reverberation chambers. *IEEE Transactions on Electromagnetic Compatibility* 49 (1): 94–103.
40. Holloway, C. L., D. A. Hill, J. M. Ladbury, P. F. Wilson, G. Koepke, and J. Coder. 2006. On the use of reverberation chambers to simulate a Rician radio environment for the testing of wireless devices. *IEEE Transactions on Antennas and Propagation* 54 (11), Part 1: 3167-3177.
41. Ferrara, G., M. Migliaccio, and A. Sorrentino, 2007. Characterization of GSM non-line-of-sight propagation channels generated in a reverberating chamber by using bit error rates. *IEEE Transactions on Electromagnetic Compatibility* 49 (3): 467-473.
42. Diallo, A., P. Le Thuc, C. Luxey, R. Staraj, G., Kossiavas, M. Franzén, and P.-S. Kildal. 2007. Diversity characterization of optimized two-antenna systems for UMTS handsets. *EURASIP Journal on Wireless Communication and Networking*, Hindawi Publishing Corporation, 2007.
43. ACE—Antenna Centre of Excellence: http://www.ist-ace.org
44. Carlsson, J. 2007. Benchmarking for measurement facilities for small antennas and active terminals within ACE—Results and experience from the first two years. *Frequenz, Journal of RF-Engineering and Telecommunications* 61 (3–4/2007, 59–62).
45. Carlsson, J. 2004. Benchmarking of small terminal antenna measurement facilities. *Proceedings of the JINA Conference*, Nice, France, 2004.
46. Carlsson, J., and P.-S. Kildal. 2005. Round robin test of active and passive small terminal antennas. *Proceedings of the LAPC 2005 Conference*, Loughborough, UK, April 4–6, 2005.
47. Carlsson, J. 2005. Benchmarking of facilities for small terminal antenna measurements. *Proceedings of the AP-S*, Washington, D.C., 2005.

48. Carlsson, J. 2005. Benchmarking of facilities for measuring small mobile terminals and their antennas: Results of a round robin test in ACE. *Proceedings of the ICEcom2005 Conference,* Dubrovnik, Croatia, 2005.
49. Andersson, M., C. Orlenius, and M. Franzen. 2007. Very fast measurements of effective polarization diversity gain in a reverberation chamber. *EuCAP 2007,* Edinburgh, UK.
50. Diallo, A., C. Luxey, P. Le Thuc, R. Staraj, G. Kossiavas, M. Franzén, and P.-S. Kildal. 2007. Evaluation of the performance of several four-antenna systems in a reverberation chamber. *International Workshop on Antenna Technology (iWAT07)*, Cambridge, UK, pp. 166–169.
51. Bolin, T., A. Derneryd, G. Kristensson, V. Plicanic, and Z. Ying. 2005. Two-antenna receive diversity performance in indoor environment. *Electronic Letters* 41, (22): 1205–1206.
52. Kildal, P.-S., and C. Carlsson. 2002. TCP of 20 mobile phones measured in reverberation chamber—Procedure, results, uncertainty and validation. Bluetest AB report, Feb 2002.
53. http://www.sonyericsson.com/
54. Olano, N. 2008. WLAN MIMO terminal test in reverberation chamber. Master's thesis at the Antenna Group, Department of Signals and Systems, Chalmers University of Technology, Göteborg, Sweden.
55. http://www.chalmers.se/s2/cha-en/chase
56. Iperf Version 2.0.2. Distributed applications support team of the National Laboratory for Applied Network Research of the United States of America: http://dast.nlanr.net/Projects/Iperf/#whatis
57. Olano, N., C. Orlenius, K. Ishimiya, and Z. Ying. 2008. WLAN MIMO throughput test in reverberation chamber. Accepted at IEEE AP-S International Symposium, San Diego, CA, 2008.
58. Dahlman, E., S. Parkvall, J. Sköld, and P. Beming. 2007. 3G *evolution—HSPA and LTE for mobile broadband.* New York: Academic Press.
59. 3GPP TR 25.996 V6.1.0. September 2003. Spatial channel model for multiple input multiple output (MIMO) simulations [online].

Links

1. The TCO developments homepage: http://www.tcodevelopment.com/
2. Link to Bluetest: http://www.bluetest.se/
3. Link to CTIA: http://www.ctia.org/
4. Link to Chase: http://www.chalmers.se/s2/cha-en/chase

Index

A

B

C

D

I

L

N

O

P

Q

R

S

T

U

V